CHURCHIL
CAMB
25 S

CHURCHILL COLLEGE LIBRARY
CAMBRIDGE
CANCELLED
FROM STOCK

M/E HDG: ADVANCES
CLASS: 595.701 COPY #
LOCATION: BRACKEN R.R. EXT.
BOOK# I0120242141

Advances in Insect Physiology

Volume 14

Advances in Insect Physiology

edited by

J. E. TREHERNE
M. J. BERRIDGE
and V. B. WIGGLESWORTH

Department of Zoology, The University
Cambridge, England

Volume 14

1979

ACADEMIC PRESS
LONDON NEW YORK SAN FRANCISCO
A Subsidiary of Harcourt Brace Jovanovich, Publishers

ACADEMIC PRESS INC. (LONDON) LTD
24/28 Oval Road
London NW1 2DX

United States Edition published by
ACADEMIC PRESS INC.
111 Fifth Avenue
New York, New York 10003

Copyright © 1979 by
ACADEMIC PRESS INC. (LONDON) LTD

All Rights Reserved

No part of this book may be reproduced in any form by photostat, microfilm, or any other means, without written permission from the publishers

British Library Cataloguing in Publication Data
Advances in insect physiology
Vol. 14
1. Insects – Physiology
I. Treherne, John Edwin
II. Berridge, Michael John
III. Wigglesworth, *Sir* Vincent
595.7′01 QL495 63–14039
ISBN 0–12–024214–1

PRINTED IN GREAT BRITAIN BY
BUTLER & TANNER LTD, FROME AND LONDON

Contributors

Franz Engelmann

Department of Biology, 405 Hilgard Avenue, University of California, Los Angeles, California 90024, California, USA

Arthur M. Jungreis

Department of Zoology, University of Tennessee, Knoxville, Tennessee 37916, USA

John Machin

Department of Zoology, 25 Hardbord Street, University of Toronto, Toronto M551A1, Ontario, Canada

K. Djie Njio

Pharmacological Laboratory, Polderweg 104, University of Amsterdam, Amsterdam 6, The Netherlands

John Palka

Department of Zoology, University of Washington, Seattle, Washington 98195, USA

Tom Piek

Pharmacological Laboratory, Polderweg 104, University of Amsterdam, Amsterdam 6, The Netherlands

Brian W. Staddon

Zoology Department, University College Cardiff, PO Box 78, Cardiff CFI 1XL, Wales, UK

Contents

Atmospheric Water Absorption in Arthropods

John Machin

Department of Zoology, University of Toronto, Canada

1 Introduction

"There are temptations, on aesthetic grounds, to give too much weight to broad unifying principles which deserve to be used only as clues for suggesting further enquiry . . ."

Professor Sir Andrew Huxley,
Presidential Address British Association
for the Advancement of Science, 1977

As Huxley pointed out the principle of uniformity in nature has been applied too enthusiastically in many branches of science. In the study of atmospheric absorption a cuticular theory for uptake of water vapour has been uniformly applied to a wide variety of insect and acarine species. At an early stage, water vapour absorption was treated as separate from other examples of water transport involving the liquid phase only. The proponents of a unique transport mechanism were persuaded by the much greater differences in water activity faced by animals absorbing water from the atmosphere and the fact that the humidity dependent weight adjustments observed in some of the smaller organisms resembled the behaviour of non-living hygroscopic materials. Since all of the known atmospheric absorbers were arthropods, it seemed most reasonable at the time to explain these observations in terms of a cuticular-epidermal pump in which the physical properties of solids played an important part. Indeed the popularity of this concept and its durability is indicated by the numerous reviews which accept some form of cuticular pump model (Beament, 1954; Edney, 1957; Beament, 1961, 1964; Knüle and Wharton, 1964; Lees, 1964; Locke, 1964; Beament, 1965; Edney, 1967a, b; Winston, 1967; Noble-Nesbitt, 1968; Winston, 1969; Ebeling, 1974; Locke, 1974; Neville, 1975).

This reviewer is persuaded by recent work with *Tenebrio molitor* larvae and *Arenivaga investigata* and by weaknesses in some of the evidence favouring the cuticular model that atmospheric absorption in all animals may fit instead within the framework of a water transport paradigm based on conventional cell structure and physiology. It is the purpose of this chapter to critically evaluate the original arguments in support of the cuticular absorption theory and to reinterpret the data in terms of conventional epithelial water transport. Such a reinterpretation has never been made and is now long overdue. It is hoped that the combination of both old and new data with current ideas might suggest to all workers in this field new and productive directions in which to proceed in the future.

2 The site of atmospheric absorption

2.1 THE EXTERNAL CUTICLE MODELS

Tenebrio molitor larvae were the first animals whose atmospheric water exchanges appeared to depart from equilibria dictated by the vapour pressure of the haemolymph (Buxton, 1930; Mellanby, 1932). The earliest explanations (Mellanby, 1932) suggested that water vapour could be absorbed by way of the tracheal system, seemed to be confirmed when Lees (1946) and Browning (1954), working with different acarines, showed that atmospheric absorption was prevented by blocking the spiracles. The observation that dehydrated ticks steadily gain weight in high humidity, a process which was inhibited by cuticular abrasion, together with the fact that cuticular water loss in gas or liquid form could apparently be increased when the animal was over-hydrated, convinced Lees (1946, 1947) of the existence of a sophisticated cuticular-epidermal water regulatory mechanism. The apparent connection between an aging tick's loss in absorptive capacity with the filling of the cuticular pore canals with wax also seemed to support this interpretation.

Beament was much impressed by Lees' evidence and greatly extended and elaborated the cuticular absorption theory (1954, 1961, 1964, 1965), basing his arguments on experiments with both insects and acarines and extending his theories uniformly to both groups. His reports of experiments (1964) in which he demonstrated that air in the tracheal system of cockroaches (*Periplaneta*) remained in passive equilibrium with the haemolymph (99% R.H.) convincingly put an end to the tracheal uptake theory. In further support of a cuticular site of atmospheric absorption Beament (1954) quoted Locke's (personally communicated) observations that mealworms stop taking up water vapour when the cuticle becomes separated from the epidermis during moulting. In addition Beament (1965) reported, with somewhat obscure and unsubstantiated authority, that blocking the mouth and anus of mealworms failed to prevent atmospheric uptake.

The Lees–Beament cuticular model of atmospheric absorption proposes that surface water activity is lower than haemolymph levels because of the structure and diameters of the pore canals. It was argued that the water repellent nature of the canal walls together with their small diameter combine to induce the contained water to form a concave meniscus of small radius of curvature at their free outside ends. The lowered vapour pressure at the surface of each meniscus (p_1) (Fig. 1a) varies with the radius of curvature (r) according to the following relationship, expressed graphically in Fig. 2:

$$\ln\frac{p_1}{p_0} = \frac{-2\sigma}{r}\cdot\frac{M}{RT\rho}$$

where p_0, the saturation vapour pressure, which exists at a flat free-water surface, 17.54 mmHg at 20 °C; σ, the surface tension of water, 75 dynes cm^{-1}; ρ, the density of water, 1 g cm^{-3}; M, the molecular weight of water, 18; R, the gas constant, 8.4×10^{-7} ergs mole^{-1} degree^{-1}; T, absolute temperature, 296 °K.

The fact that (p_1/p_0) is determined by the structure of the cuticle, explains why the ambient humidity from which absorption is possible is also a constant relative humidity at different temperatures despite widely varying saturation deficits. It was recognized that absorption into the cuticle was only part of the mechanism. The other half consisted of transferring the condensed water into the haemolymph, a process which would require osmotic forces

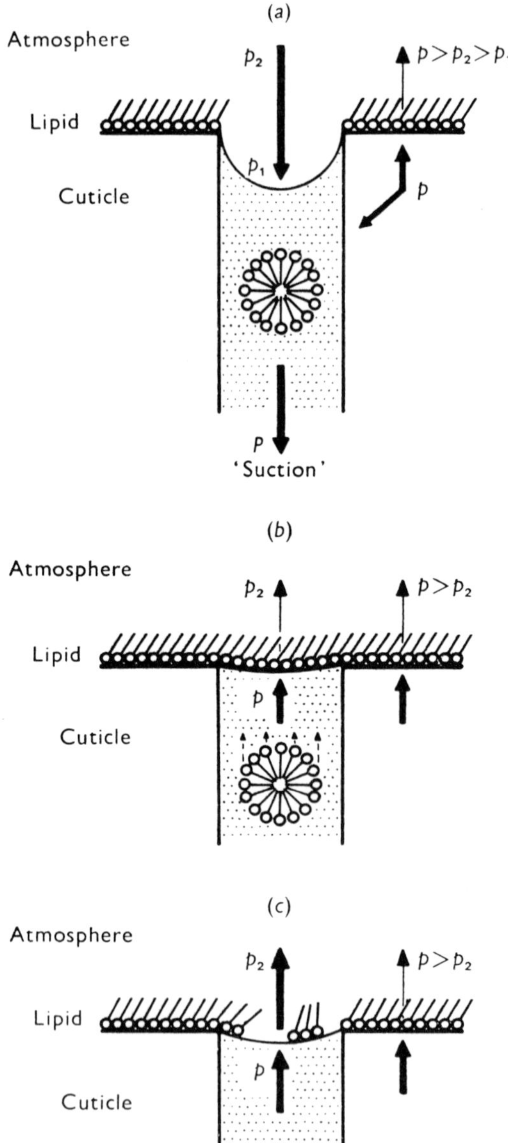

Fig. 1. Diagram of the cuticular absorption model showing the function of the pore canals. (*a*) High permeability phase during condensation on to the open curved surface of the meniscus. (*b*) Low permeability "recovery" phase with a complete protective layer of orientated lipid. This phase presumably coincides with absorption into the haemolymph. (*c*) Breakdown of the pump in the dead insect. (Taken from Noble-Nesbitt, 1969.)

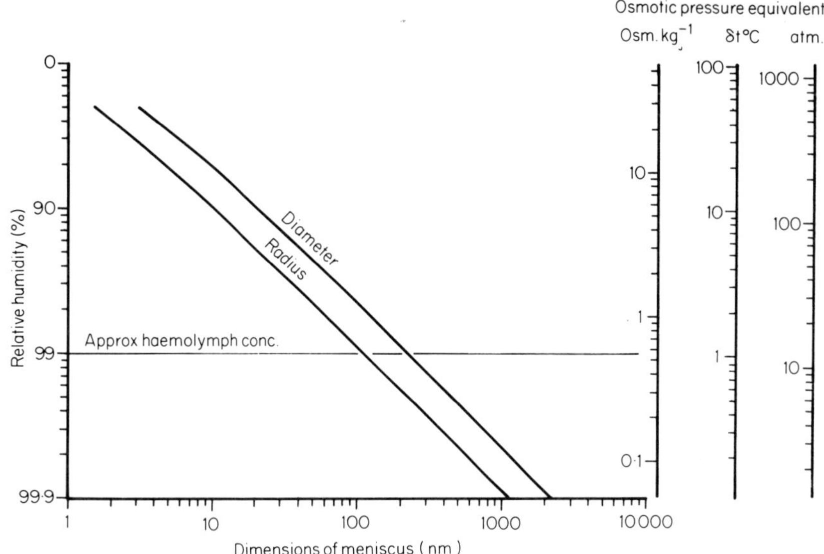

Fig. 2. Calculated relationship between curvature of a concave water surface and the relative humidity just above it at 20 °C (log scales). Equivalent water activities in osmotic units are also given.

equivalent to tens, even hundreds of atmospheres (Fig. 2). Beament exploited the well-established fact that insect cuticles and many other complex non-living membranes, show asymmetrical permeabilities favouring inward water movement. He proposed that inner cuticular layers underwent a cyclical change in water affinity, controlled and energetically dependent on the epidermal cells beneath. At the same time the impermeable, orientated layers of epicuticular lipids became disorientated or were otherwise displaced during suction and condensation (Fig. 1a) only to reform preventing outward water loss during the recovery phase of the cycle (Fig. 1b).

In support of a generally applicable absorption theory Beament (1964, 1965) cited experiments which showed that local areas of living cockroach cuticle (*Periplaneta americana*) rapidly absorbed liquid water drops leaving precipitated salts behind. In a related observation, Winston and Beament (1969) reported that the water content of freshly excised cuticles of *Periplaneta americana* and *Locusta migratoria* are below that required for passive equilibrium with haemolymph.

In the light of recent work which more convincingly establishes the rectum and other localized structures as being the site of atmospheric uptake, the evidence supporting external cuticular absorption does not appear so strong. Although Lees' evidence for cuticular-epidermal control of water loss is convincing, the next step in the argument, that various cuticular treatments or conditions actually prevent water vapour absorption cannot be considered to have been established. Abrasion of the cuticle is harsh treatment and Beament (1961) admits increased water loss could easily mask weight gain due to absorption. The observation that moulting insects are unable to absorb water vapour has been made a number of times (Edney, 1966; Noble-Nesbitt, 1970b; Machin, 1975) and seems a widespread phenomenon. However this is not due to cuticle separation, but rather that rectal concentration gradients necessary for absorption dissipate during the moult (Machin, 1975). Ryerse has observed (personal communication) that the Malpighian tubules of *Calpodes* also cease transporting during moulting.

Although Beament (1964, 1965) placed great emphasis on the temperature insensitive nature of the relative humidity threshold for absorption, it is significant that the alternative mechanism, lowering solvent activity by high solute concentrations, also shows the same characteristics. It is well known that the relative humidities above the saturated solutions of many salts and organic solutes are remarkably temperature insensitive (Winston and Bates, 1960). So too are the relative humidities above subsaturated solutions of, for example, NaCl (Fig. 3) and KCl. It is interesting to note that Beament (1961) was so convinced that absorption took place over the entire external surface and that simultaneous gains and losses were thereby impossible that he stated

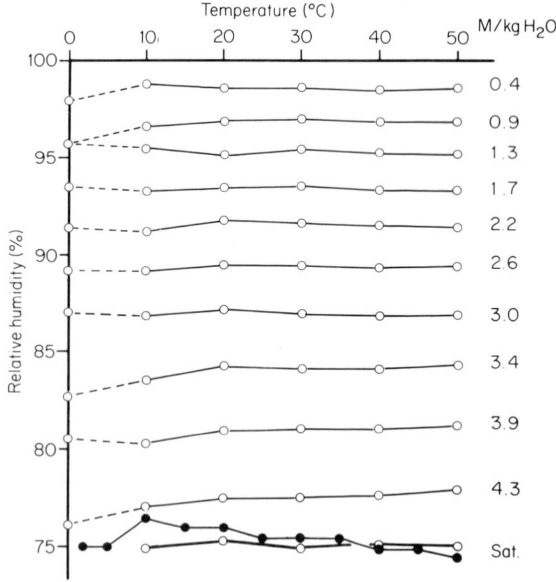

Fig. 4. Rates of weight loss (○) from *Periplaneta* with water drops placed on the cuticle in ambient chloride at different temperatures (calculated from data of Kracek, 1928). Saturated data (○) from Brönsted, 1928; (●) from Winston and Bates, 1960.

"... it seems obvious that it [absorption] will not be achieved by direct application of massive osmotic solutions, which are actively maintained, otherwise, the rate of evaporation from living and dead insects would be very different".

The results of ultrastructural studies of insect cuticle are equivocal. Although the necessary elements of a cuticular pump may be found in the epicuticle (Locke, 1964, 1974), direct proof of its operation which must be at the level of molecular interaction between water, lipids and cuticular proteins will, of course, never be obtained. The available data suggests that pore canals are generally greater in diameter than required by the Lees–Beament model. Noble-Nesbitt (1968, 1969) states that they are 80–100 Å (8–10 nm) in diameter in *Thermobia* whereas at 45% R.H. absorption threshold is consistent with a diameter of 3 nm (see Fig. 2). In *Tenebrio* an 88% R.H. threshold implies a 20 nm diameter where observed values are 90 to 100 nm (Gluud, 1968) or about 55 nm (Locke, 1974). In the cuticle of *Laelaps echidnina*, an animal with roughly the same threshold as *Tenebrio* (Wharton and Kanungo, 1962) the pore canals are about 30 to 50 nm in diameter (Wharton *et al.*, 1968). Beament (1965) explained the discrepancy by doubting whether numerical extrapolation could be made to tubes of very small diameter, suggesting that bound water molecules might reduce the effective diameter.

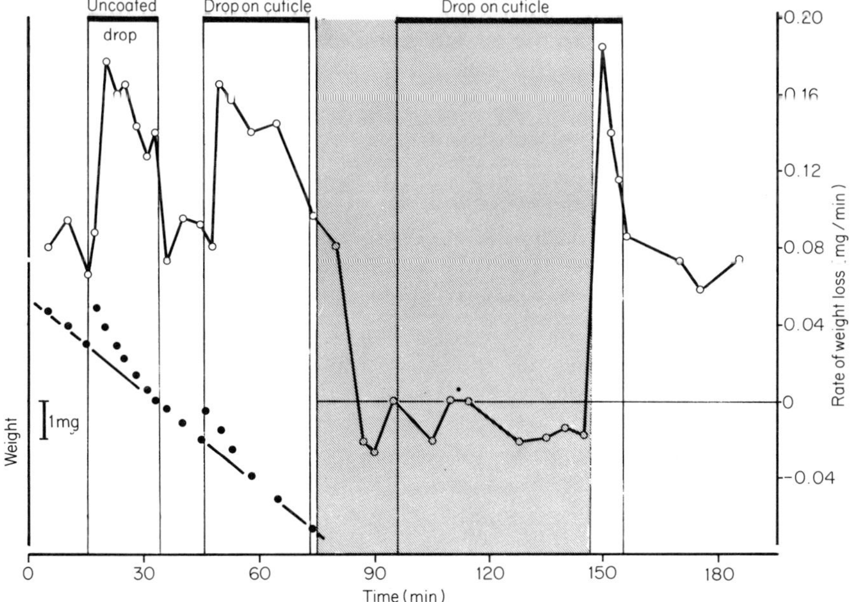

Fig. 4. Rates of weight loss (○) from *Periplaneta* with water drops placed on the cuticle in ambient humidity and in saturated air (shaded area). Negative values indicate gains. Weight measurements from which the rates were calculated, are partially indicated (●).

Noble-Nesbitt (1968) thought the answer might lie in the pore canal breaking up into finer tubules in the epicuticle.

Beament (1964, 1965) considered that his observations and experiments on liquid water drops placed on the cuticle of restrained *Periplaneta* supported the cuticle absorption theory. Even though his general observation that the rate of disappearance of the drops is very variable has been confirmed by the author, this does not necessarily mean that drops are short-lived because they pass into the cuticle. It can be seen in Fig. 4, which summarizes the author's experiments, that all drops placed on the balance with the cockroach, produce an increase in rate of weight loss, whether or not they come in contact with the cuticle. If water in the drops on the cuticle were substantially absorbed by the animal, no such increases in weight loss rates would have been observed. It was also observed that the rate of weight loss of drops on the cuticle never exceeded but sometimes approached the rate of loss of uncoated drops placed on wax beside the cockroach. The inescapable conclusion that the disappearance of the drop is caused by evaporation, presumably affected by variable lipid coating, is confirmed by the observation that a drop which stops evaporating in saturated air very rapidly disappears when the humidity is again lowered. Such marked humidity dependence would not be found if drops were actively transported across the cuticle. Beament's (1965) related observation that a tritium label in such drops passes across the cuticle and appears in the haemolymph cannot be considered proof of net water uptake.

Other aspects of Beament's (1961) theories of cuticle structure having a bearing on the uptake model, have also recently been questioned. The possible existence of valves to account for the low outward permeabilities during the non-absorptive phase of the pump would depend on a highly ordered but labile monolayer of lipids. That cuticular impermeability is principally due to the monolayer has recently been questioned by Lockey (1976) who doubts that cockroach cuticular lipid is strongly enough attracted to water to bring about orientation.

The strong similarity between the uptake kinetics of small animals capable of atmospheric absorption (see section 3.2) particularly in the minute flour mite *Acarus siro* (Knülle, 1962) and the hygroscopic behaviour of non-living materials (Knülle and Wharton, 1964) led to a slightly different cuticular absorption theory. This theory sought to explain how the equilibrium weight of an animal varied with the humidity in which it was kept and how rates of uptake at a given humidity were increased by previous hydration. Knülle (1962) and Wharton and Kanungo (1962) proposed a scheme in which the initial absorption of water vapour was due to the hygroscopic properties of the cuticle. These workers were aware that uptake in animals could not be explained by purely passive mechanisms, but did not explain why.

The arguments against exclusively physical exchanges of water vapour in small animals, without the use of energy, may be presented as follows. Animals could change their equilibrium weights with ambient humidity by means of a hygroscopic compartment separated from the rest of the body fluids. Unfortunately, the external cuticle alone is not sufficiently large in volume to account for the observed changes in weight. Then perhaps a second hygroscopic compartment, still isolated from the haemolymph exists in the animal. Experiments with tritiated water indicate that, although the body water is compartmentalized, each compartment readily exchanges with atmospheric water (Wharton and Devine, 1968; Knülle and Devine, 1972; Devine and Wharton, 1973; Arlian and Wharton, 1974; Arlian, 1975a, b; Ellingsen, 1975). The alternative possibility, that all fluid compartments in the body behave hygroscopically, also does not fit with the observations. The increase and decrease of equilibrium weights are not large enough to bring about the required changes in concentration of the haemolymph for completely passive movement of water. The loss in weight of dead animals in humidities where weight is regulated in live animals (Knülle, 1967) confirms that the haemolymph is not in passive equilibrium with ambient humidity, unless it is very high (about 99% R.H.).

As in the case of the earlier model, a hygroscopic theory of absorption from subsaturated atmospheres which is consistent with experimental observations, will only work if a substantial osmotic imbalance is actively maintained between cuticle and haemolymph. At the same time this imbalance must permit a unidirectional flow of water against the gradient. Once again this most essential aspect of any cuticular pump remains obscure.

There is some evidence that the cuticles of *Locusta migratoria* and the cockroaches *Periplaneta americana* and *Leucophaea maderae* (Winston, 1967, 1969; Winston and Beament, 1969; Winston and Hoffmeier, 1968), animals which are not known to take up water from the atmosphere, are not in equilibrium with the haemolymph. Rapidly excised cuticles from animals kept in a wide range of humidities all show gains in weight when placed in humidities equivalent to that of the haemolymph. The critical experiment (Winston and Beament, 1969) which seeks to establish the equilibrium humidity of the living cuticle shows that it is 98.5% R.H. (0.42 M NaCl) for *Periplaneta* at 25 °C and 98.2% R.H. (0.50 M NaCl) for *Locusta* compared with 99.4% for the haemolymph in both animals. Clearly this difference is slight compared with the imbalance required for absorption for much lower humidities. In this reviewer's opinion the disequilibrium between the cuticle and the haemolymph may be explained by the cuticle being partially dehydrated during life by the surrounding air, since the amount absorbed increases with humidity.

2.2 EVIDENCE IN FAVOUR OF LOCALIZED ABSORPTION SITES

Perhaps the principle argument against the general use of the external cuticle is the discovery that absorption occurs at localized sites. The finding by Noble-Nesbitt (1970a, b) that blocking the anus prevented water vapour absorption in *Thermobia domestica* (=*Lepismodes inquilinus*, see Noble-Nesbitt, 1970b) and *Tenebrio molitor* larvae represented an important change in direction. These results were later confirmed in *Tenebrio* by Dunbar and Winston (1975) and Machin (1975). Surprisingly, the identification of a rectal site of absorption was not seen at first as a threat to the cuticular absorption theory; the model was simply transferred to the rectal cuticle and epidermis (Noble-Nesbitt, 1970b). Those who have attempted to occlude a specific area of an arthropod's body with wax will be aware of the unsatisfactory nature of this technique. One can never be sure whether the failure of the application is due to minute cracking or an imperfect seal or whether its apparent effect is really due to the damage caused by applying the wax at too hot a temperature. Okasha (1971) questioned Noble-Nesbitt's results on the grounds that the application of wax in the anal region might inhibit sensory processes associated with atmospheric uptake. An alternative technique, evolved independently by Rudolph and Knülle (1974) and Noble-Nesbitt (1975) in which the wax simply provides an air-tight seal between a head and tail chamber, is much more satisfactory since the uptake of water vapour can be directly associated with exposure to high humidity. The disappearance of water from a drop of saturated KNO_3 (93% R.H. at 20 °C) observed only in the head chamber, together with various experiments in which wax was applied to the mouth parts (Rudolph and Knüle, 1974) conclusively demonstrated an anterior site of atmospheric uptake, which was close to the mouth in several ixodid ticks. On the other hand a posterior site of atmospheric absorption was confirmed in *Thermobia* when weight gains were found to occur only when the tail end of the animal was exposed to high humidity (Noble-Nesbitt, 1975). Machin (1976) and O'Donnell (1977a, 1978) have exploited an alternative technique for testing whether or not the rectum is capable of atmospheric absorption which does not involve restraining the experimental animal. The technique is based on Ramsay's (1964) method of determining the humidity in the rectal lumen of mealworms by measuring weight changes of freshly eliminated faecal pellets in known humidities. Machin found that faecal pellets produced by mealworms in high humidity before absorption began, gained weight above a threshold of about 90% R.H., demonstrating that conditions in rectum were compatible with atmospheric uptake. Pellets produced by mealworms during absorption did not subsequently change weight, demonstrating that luminal and ambient humidity differences were abolished by the inward diffusion of water vapour associated with absorption. On the contrary, O'Donnell (1977a, 1978) found that faecal pellets of *Arenivaga* lose weight in all humidities from which uptake is possible, indicating that absorption does not take place in the rectum. Using a technique similar to that of Rudolph and Knülle (1974) and Noble-Nesbitt (1975) he went on to demonstrate that water vapour was absorbed in the mouth region. The correspondence between weight gain and eversion of two bladder-like structures on either side of the mouth together with their surface temperatures measured by thermocouples, unequivocally identify these structures as the site of absorption in *Arenivaga*.

Clearly arthropods display a great variety of uptake mechanisms at different specifiic locations on the body. None of the examples just described involve the general external cuticle.

3 The kinetics of atmospheric absorption

The change from general surface absorption model to one involving a localized area is, at first sight, a minor one. Such a change was made by Noble-Nesbitt (1970b) who simply ascribed the Lees–Beament absorption model to the rectal cuticle without exploring its implications. The results of studies which identify limited sites of absorption are important because they require a rethinking about the forces involved in arthropod water exchange in the achievement of water balance. The rediscovery that the components of an animal's external surface can have different properties which define different simultaneous exchange processes, demands that we again view arthropod water relations in conventional terms of balancing separate gains and losses. Since water vapour absorption kinetics have previously been interpreted with cuticular absorption theories in mind, it is useful and hopefully revealing to reappraise the data knowing that several simultaneous exchange processes are possible in the same animal.

3.1 UPTAKE BY LARGER INSECTS

Because of their comparatively large size and low integumental permeability (see Table 1) *Tenebrio molitor* larvae, of the size range usually chosen for study, lose weight in dehydrating conditions only very slowly (Buxton, 1930; Mellanby, 1932; Machin, 1975). The mealworms' ability to oxidize stored fat during food deprivation to produce significant amounts of metabolic water (Johansson, 1920; Mellanby, 1932) permits the proportion of water to dry weight in these animals to remain almost constant during long periods without food. Their levels of hydration are rather consistent over a wide range of sub-absorption humidities (Buxton, 1930).

In favourably high humidities water vapour uptake is rapid enough to markedly increase weight (Dunbar and Winston, 1975) and cause measurable haemolymph dilution (Machin, 1975). Mealworms are capable of responding to existing levels of hydration and modulating the amount of atmospheric absorption accordingly. For example, animals taken from a culture where water is available from fruit and vegetables, will rarely absorb from the atmosphere unless they are first dehydrated. Animals from a dry meal culture will absorb water vapour almost immediately upon exposure to high humidity, the amount taken up increasing with animal size. It is important to note that dehydration increases the amount absorbed but not its rate of uptake (Machin, 1975). Water content and variability in osmotic pressure of the haemolymph (Buxton, 1930; Machin, 1975) together with the tendency for absorption following a period of dehydration to overshoot previous hydration levels (Dunbar and Winston, 1975) suggests the mealworm's capacity to regulate water content is rather poorly developed.

Continuous recording of the weight of intermoult mealworms indicate that the physiological parameters of the pump are remarkably consistent. Atmospheric uptake is primarily determined by the amount by which ambient humidity exceeds a threshold close to 88% R.H. Only slightly higher thresholds (90% R.H.) are observed at the onset and termination of prolonged uptake periods lasting many days. In animals of similar size, uptake rates, but not thresholds, apparently vary with the amount of faecal material in the rectum which interferes with absorption (Machin, 1976, 1978). Uptake rates in all animals increase linearly with ambient humidity above the threshold. For this reason there is no sign of pump saturation at high absorption rates, at least up to 98.7% R.H. (14.9% body wt day^{-1} at 20 °C). Since large variations in uptake rate in the same animal are only seen in studies using intermittent weighing techniques such variations must be experimental artefacts due to the adverse effects of handling the animal. It may be concluded that the modulation of atmospheric absorption in *Tenebrio* larvae is therefore brought about by an all or none switch, presumably the opening and closing of the anus, and not by changes in absorption rate. It follows that control of uptake appears to occur at a sensory level rather than at the level of the pump's physiology.

Using special experimental techniques water loss from mealworms can be seen to follow a profoundly different pattern from that normally observed. Weight loss in previously absorbing animals which are suddenly exposed to sub-absorption humidities with no further disturbance, occurs much more rapidly than is normally found of those humidities. The trend of such losses indicates that water is transpired from a fluid compartment of significantly lower vapour pressure (88% R.H.) than the haemolymph (Machin, 1976, 1978). These results refute the idea of a simple external exchange system, confirming instead that multiple surface exchange is possible.

Specimens of *Thermobia domestica* suitable for experimental work (30 mg) reach only half the weight of the smallest mealworm for which uptake data is available. Noble-Nesbitt's (1969) water loss data in sub-absorption humidities are consistent with the conventional passive cuticular model, behaving as a barrier between haemolymph and surrounding air. Transpirational permeabilities remain virtually unchanged from 1 to 43% R.H., about five times greater than in *Tenebrio* (Table 1). This, and the smaller size of *Thermobia*, contribute to comparatively rapid decline in weight observed in this animal following absorption or a period in sub-absorption humidities (Noble-Nesbitt, 1969; Okasha, 1971, 1972). In favourable humidities atmospheric absorption is able to make good previous losses and then control body weight and water content within much narrower limits than *Tenebrio*. Once again the equilibrium values which are established do not significantly vary with ambient humidity. Although daily weighings indicate that uptake rates decline as equilibrium is approached, continuous weight records indicate that this is again due to the decreasing length of intermittent uptake periods. As to be expected, Okasha (1971, 1972) found that water exchange in *Thermobia* was susceptible to a variety of experimental as well as natural variables such as moulting cycle. Although severe dehydration or starvation interfered with the animal's ability to maintain water balance by atmospheric absorption, he concluded that less extreme desiccation had no effect on the uptake mechanism. Once again it seems reasonable from Noble-Nesbitt's (1969) description of weight regulation in *Thermobia* to conclude that the necessary modulation of uptake is brought about by control of an on–off switch, presumably the anus, and not by physiological processes of the pump.

Beament (1961) regarded more rapid loss rates in dead compared to living specimens as important evidence that the special conditions in the cuticle necessary for atmospheric absorption were actively maintained (see Fig. 1c, section 2.1). Although many animals, including *Thermobia*, show this differential (Noble-Nesbitt, 1969) there are alternative explanations for it other than the breakdown of the cuticular pump after death. Noble-Nesbitt's explanation for *Thermobia* seems reasonable: that differing permeabilities in humidities close to the absorption threshold could be caused by losses masking a small uptake component when living, which disappear when the animal is killed (see section 3.3). Increased loss rates in dead *Thermobia* however persist in humidities which are too low for water vapour absorption. Noble-Nesbitt (1969) found this observation difficult to explain because evaporation from the tracheal system did not appear to be reduced in life by a spiracular closing mechanism. Surely the similarity in permeability between living and dead animals as seen in Noble-Nesbitt's data plotted in Fig. 5, followed by a progressive increase in permeability in the dead group indicates a similar but slower breakdown of the external impermeable barrier. This

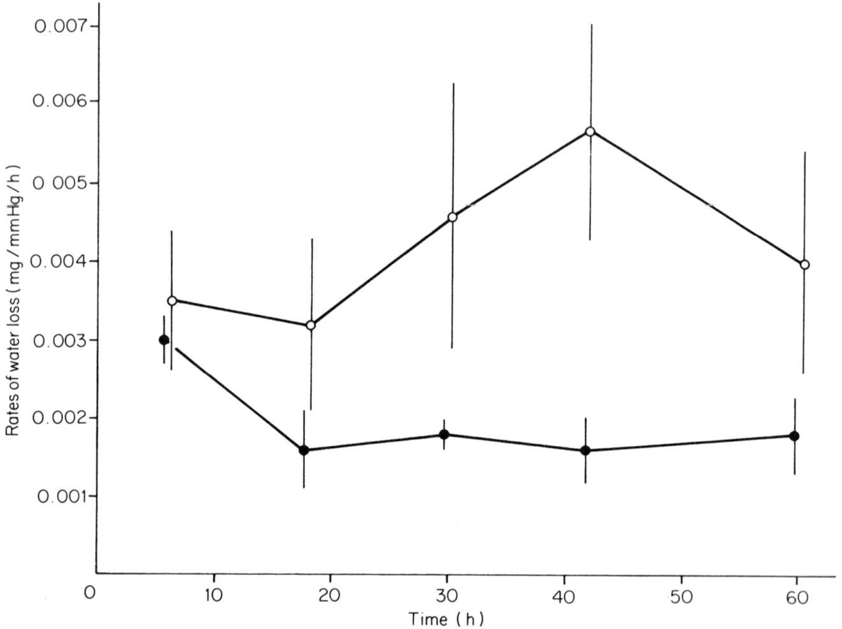

Fig. 5. Graph of changes of water loss permeabilities with time in living (○) and freshly killed (●) *Thermobia*. Error bars ± one standard deviation. (Data from Noble-Nesbitt, 1969.)

could take place without involving the tracheal system, for example by the exposure of previously protected permeable cuticular folds as the body dried and became distorted.

The way in which atmospheric absorption contributes to water balance in *Arenivaga investigata* is essentially the same as described for *Tenebrio* and *Thermobia*. Continuous uptake in laboratory specimens occurs after moderate dehydration in dry air. Uptake rate in the same individual increases linearly with ambient humidity above a threshold value of about 80% R.H. (Edney, 1966; O'Donnell, 1977a, b, 1978). Experiments in which transpirational losses are reduced by enclosing the body in wax or exposing it to high humidities show that some individuals gain weight in humidities as low as 71.5% R.H. (O'Donnell, 1977a, 1978) (see section 3.3). The uptake mechanism shows no saturation at high absorption rates (6.0% body wt day^{-1} in 98% R.H. at 25 °C) even in humidified air. During a prolonged period of absorption, uptake slowly declined to 60% of its initial value after 200 hours (O'Donnell, personal communication). However continuous weight records show that hydrated animals regulate their weights by intermittent absorption in constant favourable humidities.

3.2 UPTAKE BY SMALLER INSECTS AND "ACARINES"

The remaining animals capable of atmospheric absorption have body weights which are at least an order of magnitude smaller than *Thermobia*. The majority range from a few hundreds to a few tens of microgrammes. The smallest animal observed to take up water vapour is the flour mite *Acarus siro*, weighing about 5μg (Solomon, 1966). Small size and large surface to volume ratios mean that this group of animals must contend with more rapid changes in body water contents than in larger animals. In species which tend to have rather permeable integuments (*Ixodes ricinus* Lees, 1946, *Xenopsylla brasiliensis* pre-pupae Edney, 1948 and the larvae of *Dermacentor andersoni* and *Amblyomma cajennense* Knülle, 1965, and *Acarus siro* Knülle, 1962; Solomon, 1966) body weight shows great variability, even in humidities which are high enough for losses to be made good by absorption from the atmosphere. Other species such as *Echinolaelaps echidninus* (Wharton and Kanungo, 1962), *Xenopsylla cheopis* larvae (Knülle, 1967), *Dermacentor variabilis* (Knülle, 1966, and *Liposcelis rufus* (Knülle and Spadafora, 1969) are able to maintain fairly stable, so-called "equilibrium" weights. This facility is best developed in *Hyalomma dromedarii* and *Ornithodoros savignyi* (Hafez *et al.*, 1970) which appear to have very waterproof integuments. Nymphs and adults of these species show exceptionally stable body weights for 60 days or more, though the smaller larvae of *Hyalomma* show less stability.

Efforts by smaller animals to regulate their water contents by means of atmospheric absorption evidently yield different results than those already described for the larger insects. The "equilibrium" weights in most of the smaller species change significantly showing increased stable weights as the humidity increases. It is greatly regretted that workers with the smaller species have not chosen to use continuously recording balances. The lack of continuous weight records as the animal approaches "equilibrium" weight is particularly unfortunate since there is no other way of determining whether uptake reduction is brought about by modulating pump rate or duration. A further complexity in the interaction between water balance and ambient humidity is apparent in *Acarus siro*. Knülle (1962) showed that uptake rate at a given humidity was faster in animals which had previously been dehydrated. This phenomenon is not a general one however. Hair *et al.* (1975) working with much larger male and female ticks, *Amblyomma americanum*

(2.8 and 4.5 mg, respectively), *A. maculatum* (8.7 and 9.8. mg), *Dermacentor variabilis* (5.4 and 6.7 mg) found no increase in initial uptake rate following dehydration.

Efflux studies with tritiated water using the tick *Dermacentor variabilis* (Knülle and Devine, 1972) and the mites *Laelaps echidnina* (Devine and Wharton, 1973), and *Dermatophagoides farinae* (Arlian and Wharton, 1974) show that transpiration permeabilities change relatively little with ambient humidity. Net loss of water will therefore be expected to increase almost linearly with the difference in water activities between the haemolymph and surrounding air. Since uptake rates increase with ambient humidity, animals in fluctuating humidities will have great difficulty in balancing uptake with loss because the two exchanges are not automatically self-compensating. Weight stability in varying humidity can only come from elaborate regulatory mechanisms in which uptake is modulated to balance uptake weight losses. It follows that the smaller or more permeable species, where gains and losses bring about proportionally greater changes in body weight per unit time, will have more difficulty in regulating their water contents than the larger species. Put another way, in order to achieve the same level of regulation small animals must have more accurate and sensitive feedback control of uptake.

These difficulties in balancing gains with losses and their consequences for the weight of the animal are illustrated in Fig. 6. It can be seen that unless

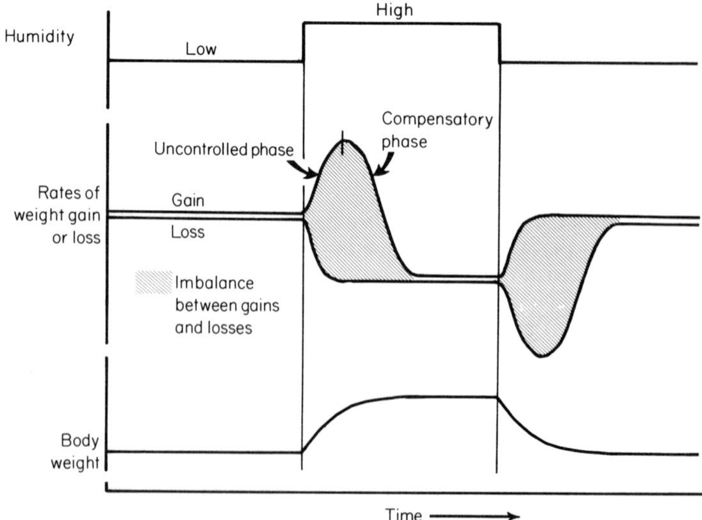

Fig. 6. Diagram illustrating hypothetical changes in water uptake and rates of loss following an alteration in humidity and the effect of exchange imbalance on body weight.

such a mechanism instantly matches a change in evaporation rate, the animal's weight, even though it eventually reaches a stable value, will increase as humidity is raised and decrease when humidity is lowered. The larger the ambient humidity change, the greater the initial imbalance between gains and losses and the greater the change in body weight. Since uptake is under biological control, it is also possible that dehydrated animals delay the modulation of uptake when humidity is raised in order to replenish lost water reserves. If this is the case, small animals like *Acarus siro* (Knülle, 1962) transferred from drier conditions to a higher humidity would gain weight faster than those from less severe conditions. It can be tentitatively concluded from this analysis that there need not be fundamental distinctions between the uptake kinetics of large and small species. What differences there are can be attributed to size: the smaller the animal, the more rapidly exchanges occur and the greater is the difficulty in maintaining a stable weight in fluctuating humidities.

As in the case of *Thermobia* the smaller atmospheric absorbers show increased rates of water loss after death (Lees, 1947; McEnroe, 1961; Knülle, 1967). Larvae of *Xenopsylla* species appear not to have spiracular closing mechanisms (Mellanby, 1934) but show progressive increases in water loss rate much the same way as *Thermobia*. On the other hand there is abundant evidence that *Acarines*, with the exception of *Acarus* which has no tracheal system, have well-developed spiracular closing mechanisms, as seen by the effects of high CO_2 and spiracle blocking experiments on water loss rates (Lees, 1946; Browning, 1954, Winston and Nelson, 1965; Ilcfnawy, 1970). It appears that the opening of the spiracles after death in these animals principally accounts for post-mortem increases in water loss. Mites killed with their spiracles open by hydrogen cyanide gas immediately show elevated rates of water loss. In spite of differences in the rate at which rapid post-mortem losses develop in *Acarines*, compared with *Thermobia* and *Xenopsylla* there seems no reason for believing the breakdown of a cuticular water pump is involved.

The kinetics of tritiated water in two mite species of the genus *Dermatophagoides* (Arlian and Wharton, 1974; Arlian, 1975b) have pointed to the existence of a second smaller water compartment whose exchange characteristics are different from those of the haemolymph. As with *Tenebrio*, lower water activities and more rapid rates of exchange strongly suggest the second compartment is associated with the absorbing mechanism, confirming once again the simultaneous existence of multiple exchange sites.

Fig. 7. Changes in water net uptake and loss rates (negative values) with relative humidity in a variety of animals. Humidity for zero net uptake is the "critical equilibrium" value. Uptake rates, corrected for significant losses are indicated by the dotted lines. Humidity for zero uptake is the "absorption threshold".

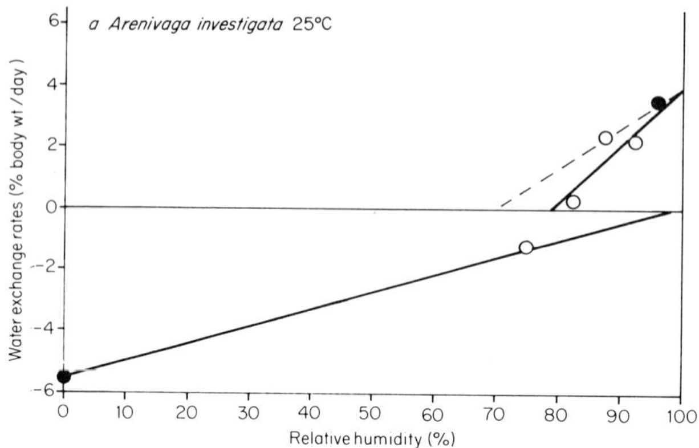

(*a*) *Arenivaga investigata* (data from O'Donnell, personal communication, ●); from Edney, 1966 (○).

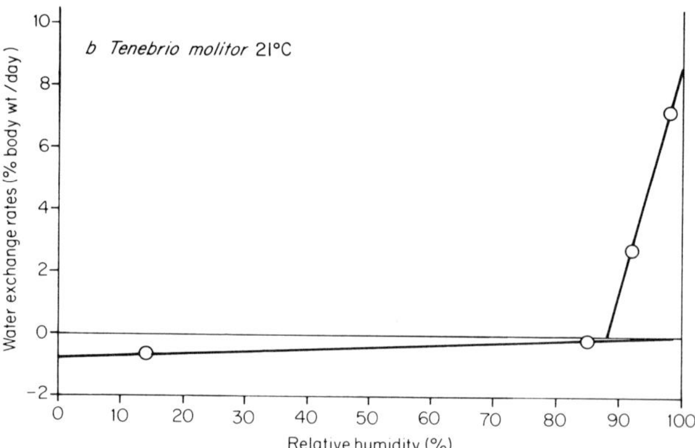

(*b*) *Tenebrio molitor* (data calculated from Machin, 1975).

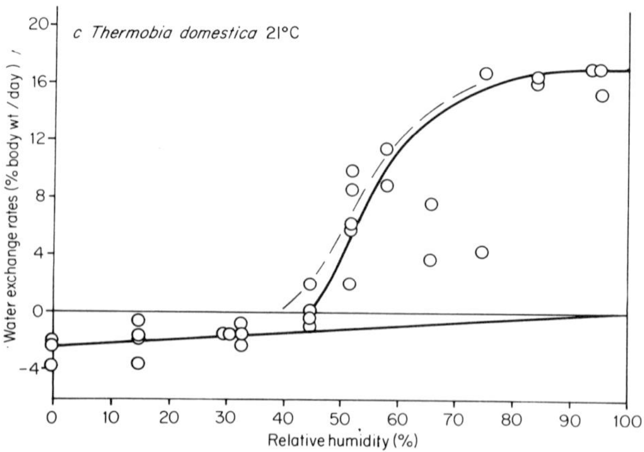

(*c*) *Thermobia domestica* (data from Beament *et al.*, 1964).

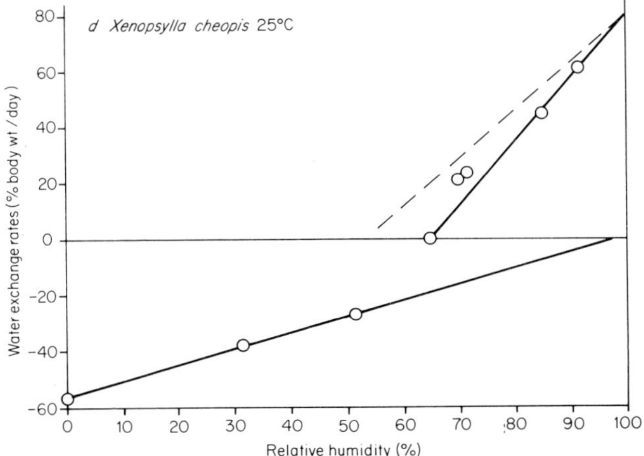

(*d*) *Xenopsylla cheopis* (data calculated from Knülle, 1967)

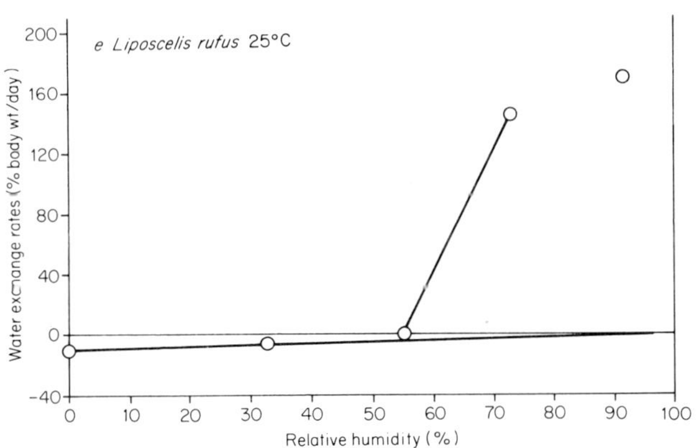

(*e*) *Liposcelis rufus* (data calculated from Knülle and Spadafora, 1969).

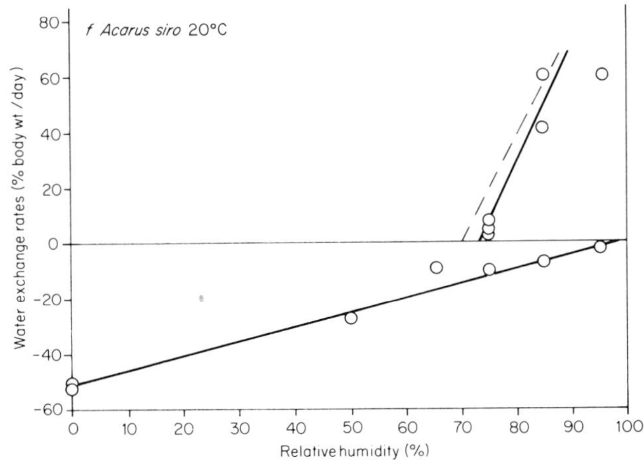

(*f*) *Acarus siro* (data calculated from Solomon, 1966).

3.3 ABSORPTION THRESHOLDS AND TEMPERATURE

It is apparent from the pattern of water vapour absorption seen in all animals that uptake ultimately depends on sufficiently high ambient water activities. Beament (1964) recognized the importance of determining the "humidity at which absorption ceased" treating it as a species specific parameter of the pumping mechanism. Many other terms defining the minimum humidity required for atmospheric absorption, and thereby the minimum values at which water balance without feeding can be maintained, have been subsequently introduced into the literature: "equilibrium relative humidity" (Wharton and Kanungo, 1962), "equilibrium humidity" (Wharton, 1963; Knülle and Wharton, 1964; Knülle, 1966), "Gleichgewichtsluftfeuchte" (Knülle, 1965), "critical equilibrium humidity, CEH" (Wharton and Devine, 1968; Noble-Nesbitt, 1969), "critical equilibrium activity, CEA" (Wharton and Devine, 1968; Wharton and Arlian, 1972; Arlian, 1975a, b). Although these concepts still have great ecological significance, the replacement of the cuticular absorption model with one in which gains and losses simultaneously occur, lessens their importance as unambiguous physiological parameters. This is because the minimum humidity necessary for water balance depends on both gains and losses whereas the minimum humidity for absorption depends on absorption alone. To emphasize this distinction it is proposed that the term "absorption site threshold humidity" or simply "pump threshold" be used to denote the true physiological threshold of the pump.

Of course "critical equilibrium humidity" and "pump threshold" are not always different, They can be identical if the rates of water loss from an animal are negligible in relation to rates of atmospheric absorption. Figure 7 shows six examples obtained from the literature, comparting separate gains and losses. This method of plotting the data contrasts with the single exchange approach used by Locke (1964, 1974), Beament *et al.* (1964), Wharton and Kanungo (1962), Edney (1966) and Hair *et al.* (1975). The new method is based on the previously discussed conclusion that a straight line drawn between zero at haemolymph equilibrium humidity (99%) and loss rates observed in low, preferably zero R.H.'s, reasonably describes water loss over the entire humidity range. The line describes losses only, because it depends on values from humidities which are too low for atmospheric uptake. The addition of corresponding loss rates to net uptake rates obtained from the literature will then give the true uptake rate values. Corrections made to CEHs in this way are also indicated in Fig. 7, where pump thresholds have previously been overestimated.

It can be seen in Table 1, which summarizes the water exchange parameters of animals capable of absorbing water from the atmosphere, that animals with the exception of *Thermobia*, having uptake/loss permeability ratios of less than ten, show lower pump thresholds than indicated by the CEH. *Thermobia* differs from the others by its exceptionally low CEH. Even though the loss permeability is one twenty-fifth the corresponding value for uptake, weight losses at 45% R.H. lead to significant errors in true uptake rates at this humidity (Fig. 7c). The new absorption thresholds for example of *Thermobia* (Noble-Nesbitt, 1969) and *Arenivaga* (O'Donnell, 1978) now fall more into line with the minimum thresholds suspected by these authors.

Despite its importance to the cuticular absorption theory, the evidence for a temperature independent relative humidity threshold for absorption has never been seriously evaluated. The idea was first based on the observation that humidities from which uptake at different temperatures was possible showed some consistency when expressed as relative humidity rather than saturation deficit (Buxton, 1930; Mellanby, 1932; Edney, 1948). These early studies did not actually determine threshold humidities at different temperatures. Unfortunately the collection and analysis of threshold values is rather unsatisfactory at present. Very little absorption data over a wide range of temperatures has been published. All of it is in the form of net uptake rates which makes conversion to absorption threshold unreliable without the primary data. Table 2, which summarizes threshold values obtained by extrapolating net uptake rates, confirms that much greater variation is found in saturation deficiencies than relative humidities.

According to purely physical principles, uptake rate by absorption systems which function by generating highly concentrated solutions, will be expected to increase with the difference in water activity between the pump and the air. Since the activities of solutions tend to be rather insensitive to temperature, uptake rate in different temperatures will be expected to follow a single trend when plotted against vapour pressure differences. Conversely regression lines describing uptake rates in different relative humidities will pass through a single threshold value but show increasing slopes with higher temperatures. However, physiological mechanisms are subject to biological limitations and there is evidence where uptake in a wide range of temperatures has been studied (5 to 45 °C; Sauer and Hair, 1971) that the rate falls off at 35 and 45 °C. In view of the well established dependence of uptake on metabolic activity (Browning, 1954; Kanungo, 1963, 1965; Arlian, 1975a) it seems likely that the inhibitory effect on metabolic process interferes with the animal's capacity to maintain the low water activities in the absorption system at high temperatures.

At present the use of temperatures is an under-exploited experimental tool for the study of absorption systems. Temperature has an important effect on metabolism, diffusion and the solubility of dissolved substances and increased use as an experimental variable might lead to the separation of basically different systems of atmospheric absorption.

TABLE 1 Summary of water exchange variables in a variety of insects and *Acarines*

Species and common names	Live wt (mg)	Temp. (°C)	Uptake perm. (% day^{-1} mmHg^{-1})	Loss perm. (% day^{-1} mmHg^{-1})	Uptake perm. / Loss perm.	C.E.H. (% R.H.)	Corrected pump threshold (% R.H.)	Authority
INSECTS								
Liposcelis rufus Booklouse	8×10^{-2}	25	34.1	0.44	78	55	55	Knüle and Spadafora (1969)
Xenopsylla cheopis larvae Oriental rat flea	1.8×10^{-1}	25	9.7	2.4	4	65	52	Knülle (1967)
Thermobia domestica Firebrat	30	21	3.8	0.15	25	45	40	Beament *et al.* (1964)
Tenebrio molitor larvae Mealworm	100	21	4.0	0.035	115	88	88	Machin (1975)
Arenivaga investigata Desert cockroach	310	25	0.8	0.24	3	79	70	Edney (1966); O'Donnell (pers. comm.)
"ACARINES"								
Acarus siro Flour mite	5×10^{-3}	20	24.5	3.0	8	72	70	Solomon (1966)
Ornithodoros savignyi nymphs	?	28	6.3	0.089	71	81	81	Hafez *et al.* (1970)

Table 1 (*contd.*)

Species and common names	Live wt (mg)	Temp. (°C)	Uptake perm. (% day^{-1} $mmHg^{-1}$)	Loss perm. (% day^{-1} $mmHg^{-1}$)	$\frac{\text{Uptake perm.}}{\text{Loss perm.}}$	C.E.H. (% R.H.)	Corrected pump threshold (% R.H.)	Authority
Amblyomma americanum, males	1.9	25	19.4	0.21	92	89	89	Sauer and Hair (1971)
Amblyomma americanum females Lone star tick	3.5	25	15.6	0.34	46	89	89	Sauer and Hair (1971)
Amblyomma americanum males	2.8	25	4.1	0.43	10	82	79	Hair *et al.* (1975)
Amblyomma americanum, females	4.5	25	3.2	0.43	7	81	74	Hair *et al.* (1975)
Dermacentor variabilis, males	5.4	25	4.2	0.21	20	83	83	Hair *et al.* (1975)
Dermacenter variabilis, females Rocky Mountain Wood tick	6.7	25	3.8	0.21	18	84	84	Hair *et al.* (1975)

TABLE 2 Critical equilibrium humidities (C.E.H.) at different temperatures, compared with corresponding saturation deficiencies (S.D.)

Species	Temp. (°C)	C.E.H.[a] (%)		S.D.[b] (mmHg)		Authority
Ixodes ricinus	9	94.0		0.52		Lees (1946)
	15	92.2		1.00		
	20.5	93.6		1.16		
	21	91.3		2.07		
Means plus Coeff. of Variation		92.8	1.35	1.19	54.6	
Amblyomma americanum males	15	87.0		1.66		Sauer and Hair (1971)
	25	88.5		2.73		
	35	89.0		4.64		
Means plus Coeff. of Variation		88.2	1.18	3.01	40.9	
Amblyomma americanum	15	85.0		1.92		Sauer and Hair (1971)
	25	90.0		2.38		
	35	88.0		5.06		
Means plus Coeff. of Variation		87.7	2.87	3.12	44.3	
Arenivaga investigata	10	83.1		0.64		Edney (1966)
	15	80.0		1.28		
	25	80.9		2.16		
	30	80.2		3.12		
Means plus Coeff. of Variation		81.1	1.75	1.80	59.9	
Thermobia domestica	23	44.6		13.78		Noble-Nesbitt (1969)
	28	46.9		17.89		
	33.5	46.9		24.48		
	37	47.3		29.51		
Means plus Coeff. of Variation		46.2	2.35	21.42	32.5	

[a] Determined by extrapolating uptake rates to zero

[b] Saturation deficiency is the difference between ambient and saturation vapour pressure at a given temperature

3.4 THE ALLOMETRY OF WATER EXCHANGE

Figure 8 shows the relationship between uptake rate and live weight. The weight data available ranges over a remarkable five orders of magnitude. To keep variability due to experimental conditions as small as possible values whenever possible were limited to those between 20 and 25 °C and at humidities greater than 93% to minimize errors due to simultaneous loss. It can

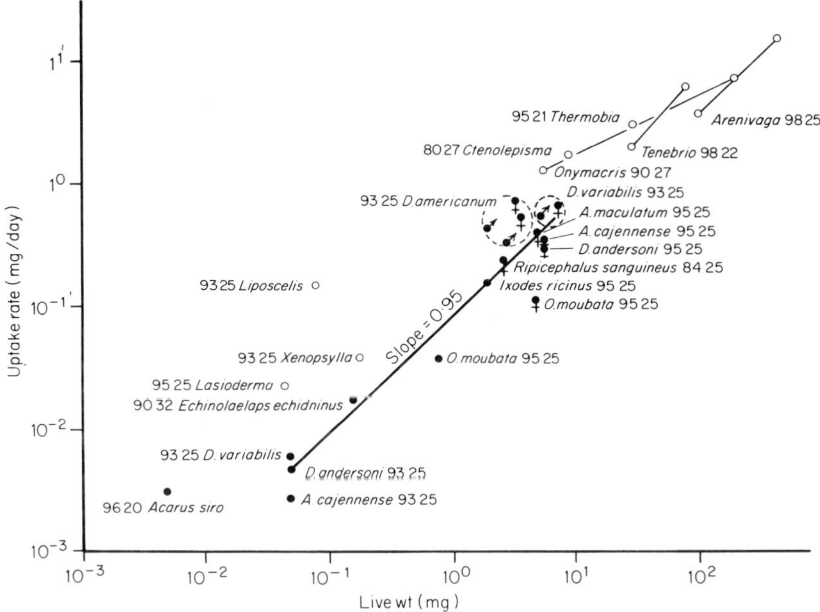

Fig. 8. Relationship between uptake rates in high humidity and live weights (log scales). Each point or family of points is identified by name, to genera in insects (○) and species in acarines (●). Values for adult acarines are separated by sex. (*A*, *Amblyomma*; *D*, *Dermacentor*; *O*, *Ornithodoros*.) The relative humidity (left) and temperature in °C (right) in which the values were determined, are also indicated. (Data calculated from Beament *et al.*, 1964; Browning, 1954; Coutchié, personal communication; Edney, 1971; Hair *et al.*, 1975; Knülle, 1966; Knülle and Spadafora, 1969; 1970; Lees, 1946; Machin, 1975; O'Donnell, personal communication; Solomon, 1966; Wharton and Kanungo, 1962.)

be seen that the regression line representing the majority of the mites and ticks which have similar body shapes, has a slope of 0.95 taking up approximately one tenth of the body weight per day. Although the fit to this relationship is not precise (index of fit, $r^2 = 0.90$) the general trend suggests that uptake mechanisms have evolved in proportion to animal bulk rather than to surface area. If the uptake mechanism were a phenomenon involving the cuticular surface a slope of 0.66 would be expected.

Animals at both weight extremes show an interesting divergence from the trend set by intermediate size animals. *Acarus*, *Liposcelis*, *Lasioderma* and *Xenopsylla* show a capacity for absorption which is greater than that predicted from their weights. This may reflect, particularly in *Liposcelis* which is also atypical in other respects, a novel and more efficient uptake mechanism. Alternatively, more efficient uptake mechanisms in animals such as *Acarus*, which can survive only a few hours in low humidities (Solomon, 1966) may be related to the possible need for rapid recovery from sub-lethal dehydration. The larger insects diverge from the trend in the opposite direction, showing that the ability of the uptake mechanism to increase with animal bulk is limited. A possible explanation for this, particularly with rectal absorption in the larger tenebrionid bettle larvae (*Onymacris marginipennis*) is that longer diffusion distances preclude the development of large absorptive mechanisms without loss of efficiency.

Different structural or physiological limitations are apparent in the way uptake rate varies with ambient humidity. Uptake in *Xenopsylla cheopis* (Knülle, 1967), *Tenebrio molitor* (Machin, 1976), *Arenivaga investigata* (O'Donnell, personal communication) (Fig. 7) and *Amblyomma americanum* (Hair *et al.*, 1975) increases linearly with ambient humidity above threshold at least up to 95, 98.7, 100 and 95% R.H., respectively. On the other hand the two thysanurans *Thermobia* (Beament *et al.*, 1964) and *Ctenolepisma* (Edney, 1971) show a falling off of the increase in absorption rate in high humidities. Whereas a linear increase in uptake must mean that the threshold for uptake remains constant in all humidities and absorption rates, nonlinearity in this relationship suggests that high flow rates adversely affect the animal's ability to maintain low vapour pressures required for absorption. Uptake kinetics suggesting saturation of the mechanism at high flow rates are also apparent in *Acarus* (Solomon, 1966) and *Liposcelis* (Knülle and Spadafora, 1969), small animals whose overall rates of absorption are very high (Fig. 7).

4 Mechanisms of atmospheric absorption

4.1 "TENEBRIO"

The rectum of *Tenebrio* larvae shows an unusual structural arrangement shared by a few other insects in which the ensheathed ends of the Malpighian tubules form a closed ring round the rectum. It has been established that this cryptonephridial complex is capable of generating extremely high osmotic pressures, principally by the transport of KCl from the haemolymph by special cells known as leptophragmata (Ramsay, 1964; Grimstone *et al.*,

1968). These osmotic pressures are used to extract liquid and gaseous water from the rectal lumen, eventually bringing the faeces to a humidity equivalent to about 90% R.H. (actually Ramsay, 1964, used an inaccurate value for the humidity given by saturated KCl and the substitution of a more reliable value from Winston and Bates, 1960, brings the mean faecal humidity closer to 88%). The obvious similarity between faecal levels of dehydration and the luminal humidities necessary for atmospheric uptake led Maddrell (1971) to speculate on the nature of atmospheric absorption by the rectum in *Tenebrio*. On the basis of Ramsay's data which indicate that water activities are higher in the perirectal compartment than in either the lumen or Malpighian tubules, he proposes a two-phase pumping mechanism. Water is actively transported by the rectal epithelium to the perirectal fluid before passing down osmotic gradients to the Malpighian tubules. The results of recent work

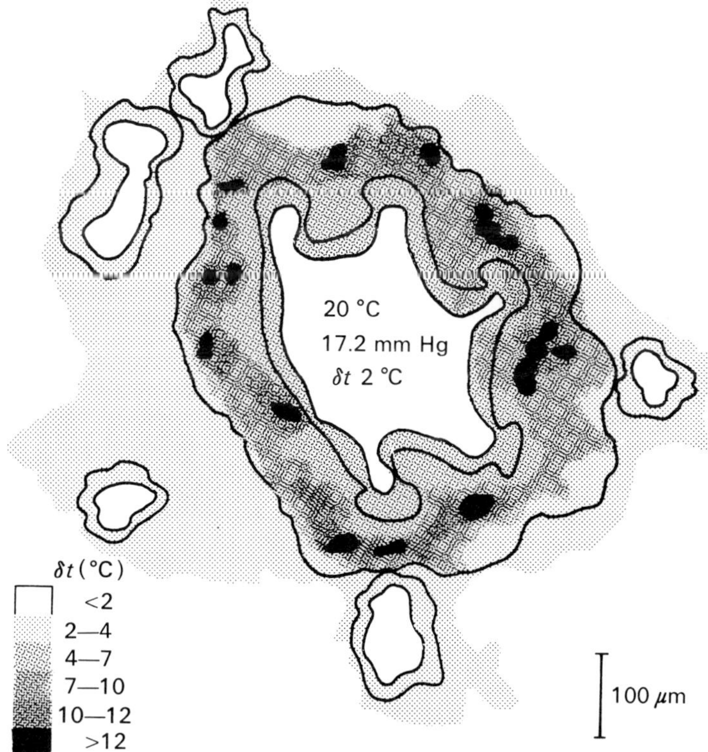

Fig. 9. Melting patterns in quick frozen, transverse section of the posterior cryptonephridial complex in an absorbing mealworm. Free Malpighian tubules, rectal lumen, epithelium, and the outer limit of the complex are outlined. Concentration equivalents of the humid air from which the animal was absorbing are indicated in the lumen. (From Machin, 1978.)

(Machin, 1978) contradicts this proposal and the data on which it is based. Melting patterns of frozen transverse sections of rectal complex indicate progessively decreasing water activities from lumen to Malpighian tubules (Fig. 9) suggesting that inward water flow takes place without the active involvement of the rectal epithelium. Ramsay's (1964) observations that osmotic pressures increase posteriorly, also confirmed by longitudinal and serial transverse frozen sections, suggests a technical reason for the discrepancy between methods. This results from problems incurred sampling fluids from compartments of small volume showing standing osmotic gradients. Because of mixing along the gradient, measurements will always yield values which fall short of the maximum. Since a greater proportion of the gradient is sampled in smaller compartments, values obtained from corresponding areas of adjacent compartments of markedly different volume will not necessarily be comparable. Although micropuncture yields more precise absolute values, the frozen section method is to be preferred here because comparisons from one region of the section to another are more reliable; this is particularly true in tissues where large osmotic pressure differences permit gross comparisons to be made.

If water is drawn from the rectal lumen by osmotic pressures actively generated by the Malpighian tubules, is it pertinent to ask what is the function of the protein-containing interstitial fluid? Maddrell (1971) states that large indiffusable molecules are ideally suited to transfer the effects of high osmotic pressure from one part of the complex to another. It follows also that protein solutions of high osmotic pressure which contain very little water, will function as rapidly responding coupling systems between rectal cuticle and Malpighian tubule because small changes in water content will greatly affect the solution's osmotic pressure. Preliminary measurements by the author (Fig. 10) showing that isolated drops of perirectal fluid undergo conventional, humidity dependent gains and losses with no hysteresis are consistent with this important but passive role of interstitial fluid. Volume measurements of drops, immediately after transfer from the mealworm, showed that the sample had an *in vivo* equilibrium humidity of 94% R.H., establishing that the fluid was capable of passively gaining water in humidities which exceed this value. Recent analyses (Machin, 1976) tend to support the idea that uptake is driven by a pumping system located some distance from the site of condensation. Pre-equilibrium weight adjustments following rapid humidity changes in absorbing mealworms indicate the existence of a sizable, passively exchanging fluid compartment which loosely couples the absorbing system with the rectal surface.

It is possible to propose a model which combines the transport and structural characteristics described by Ramsay (1964) and Grimstone *et al.* (1968) with those indicated by measurements on frozen sections (Fig. 11). As long

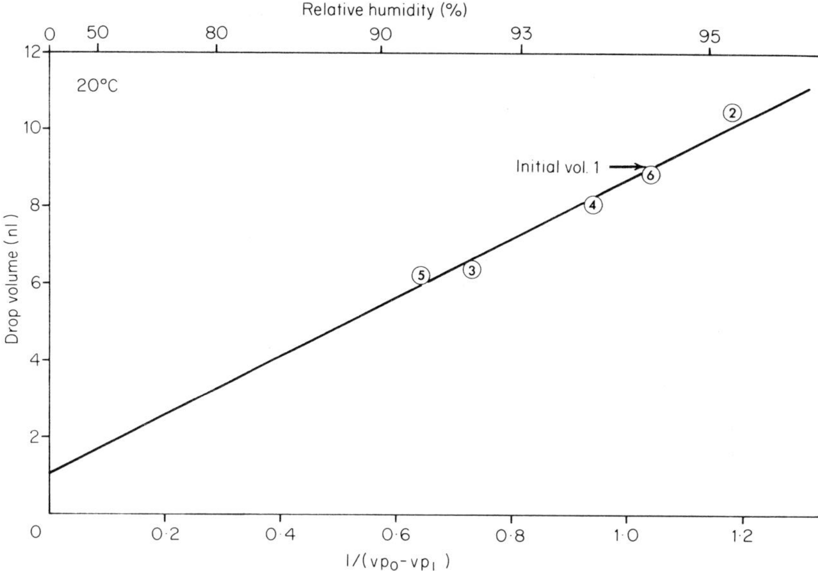

Fig. 10. Volume of a drop of perirectal fluid from an absorbing mealworm plotted against the reciprocal of vapour pressure lowering (solute concentration). The numbers refer to the order in which the points were determined in the humidity cell illustrated in Fig. 16. The initial volume of the drop is also indicated.

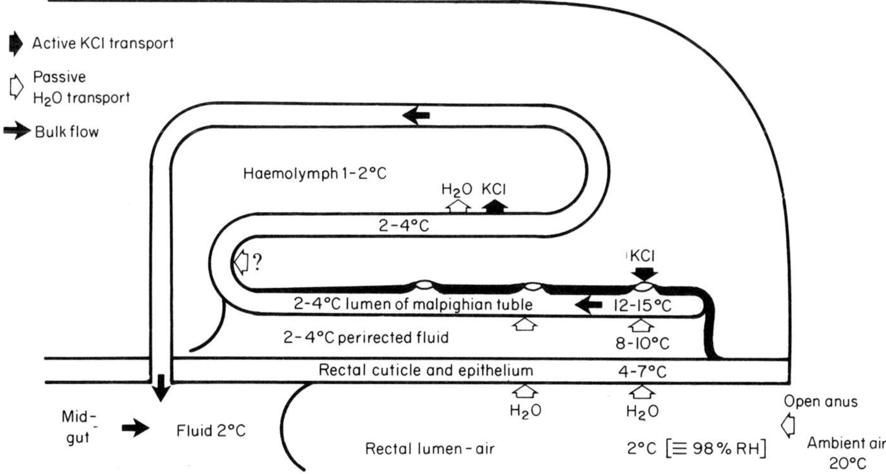

Fig. 11. Diagram of the osmotic pressures, water and solute flows in the cryptonephridial complex of an absorbing mealworm (osmotic pressures based on frozen sections, solute flows on Ramsay, 1964; Grimstone *et al.*, 1968).

as water activity of the surrounding air exceeds that of the posterior Malpighian tubules and the anus remains open, water vapour will condense on to the rectal epidermis and flow towards the tubules. Net water flow in this direction must be due to the fact that the peritubular membrane, which is constructed of up to 40 layers of flattened cells, and has low permeability to water in its posterior region (Grimstone *et al.*, 1968), is less permeable than the layers on the inner side of the tubules. As water enters the tubules, fluid is displaced anteriorly, gradually diluting it, presumably until gradients favouring uptake are abolished. Although there is no direct evidence of this, water and ions are presumably transferred from the diluted tubules fluid to the haemolymph by some form of coupled transport along the free loops of the Malpighian tubules. Patton and Craig (1939) were able to demonstrate that ^{24}Na passed from the rectal lumen to the fluid bathing the rectal complex, possibly by this route.

Comparisons between the rates at which water passes into (160 $\mu g\ h^{-1}\ mmHg^{-1}$) or out of (30 $\mu g\ h^{-1}\ mmHg^{-1}$) the rectal complex under equilibrium conditions show that there are further asymmetries in the layers between the Malpighian tubules and the rectal lumen (Machin, 1976). However it has not been possible to demonstrate that the rectification of flow occurs close to the surface where it would be of most benefit in reducing water loss in fluctuating humidities. Initial rates of non-equilibrium gain or loss following a given humidity change are similar and it seems therefore that superficial barriers such as the rectal cuticle are remarkably symmetrical in their permeability. This particular arrangement, in which superficial compartments are able to exchange with the atmosphere comparatively freely in either direction while deeper compartments are protected from loss may permit an independent identification of the active site of absorption. Comparisons between the total solvent volume of the passively exchanging superficial compartment and estimated water contents of the components of the rectal complex have so far eliminated the rectal cuticle as the site of active uptake. Progress towards a full understanding of the forces which govern water exchange in the other compartments is presently complicated by the discovery that rectal epithelial cells appear to regulate their volumes in a wide range of osmotic pressures (Machin, 1978).

4.2 "ARENIVAGA"

Water vapour is absorbed in *Arenivaga* by means of a novel system of highly specialized structures in the region of the mouth (O'Donnell, 1977a, 1978). The basic arrangement of these structures is illustrated diagrammatically in Fig. 12. A thin layer of hygroscopic fluid is apparently conveyed over the surface

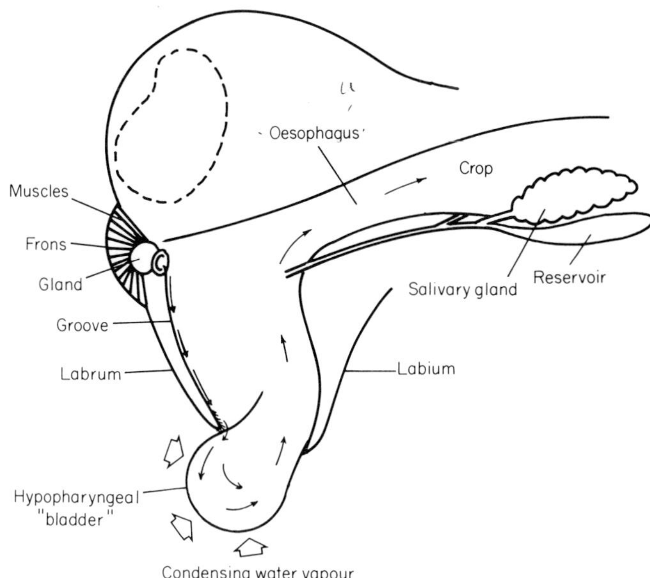

Fig. 12. Diagram showing the arrangements of structures involved in water vapour absorption in the head of *Arenivaga* (based on the work of O'Donnell).

of two eversible, integumental bladders associated with the hypopharynx (Fig. 13a). The bladder surfaces are coated with a dense mat of cuticular hairs (Fig. 13c) which hold and spread out the absorbing fluid by capillarity (O'Donnell, 1977b). The layer of fluid is so thin that it immediately dries out when the local humidity falls below absorption threshold. The animal restores the liquid layer on the surface by withdrawing both bladders and coating them with a fluid of low osmotic pressure, similar to that of the haemolymph, from a system of salivary glands and reservoirs (O'Donnell, 1977a).

The hygroscopic fluid is supplied to the bladders from two spherical glands located on the inside of the labrum (O'Donnell, 1977b). The fluid is conveyed from the glands by open grooves which pass to the lower margin of the labrum (see Fig. 13b). A group of flattened cuticular projections at the termination of each groove apparently engage with the hairs of the bladders, conducting fluid to them by capillarity. Lifting the labrum away from the bladders immediately causes them to become dry.

Very little is known at present about the structure and mode of operation of the two glands or of the forces which regulate the flow of fluid over the bladders. The glands are embedded in a massive muscular complex which can be seen to oscillate when the glands are secreting fluid. The rate of these

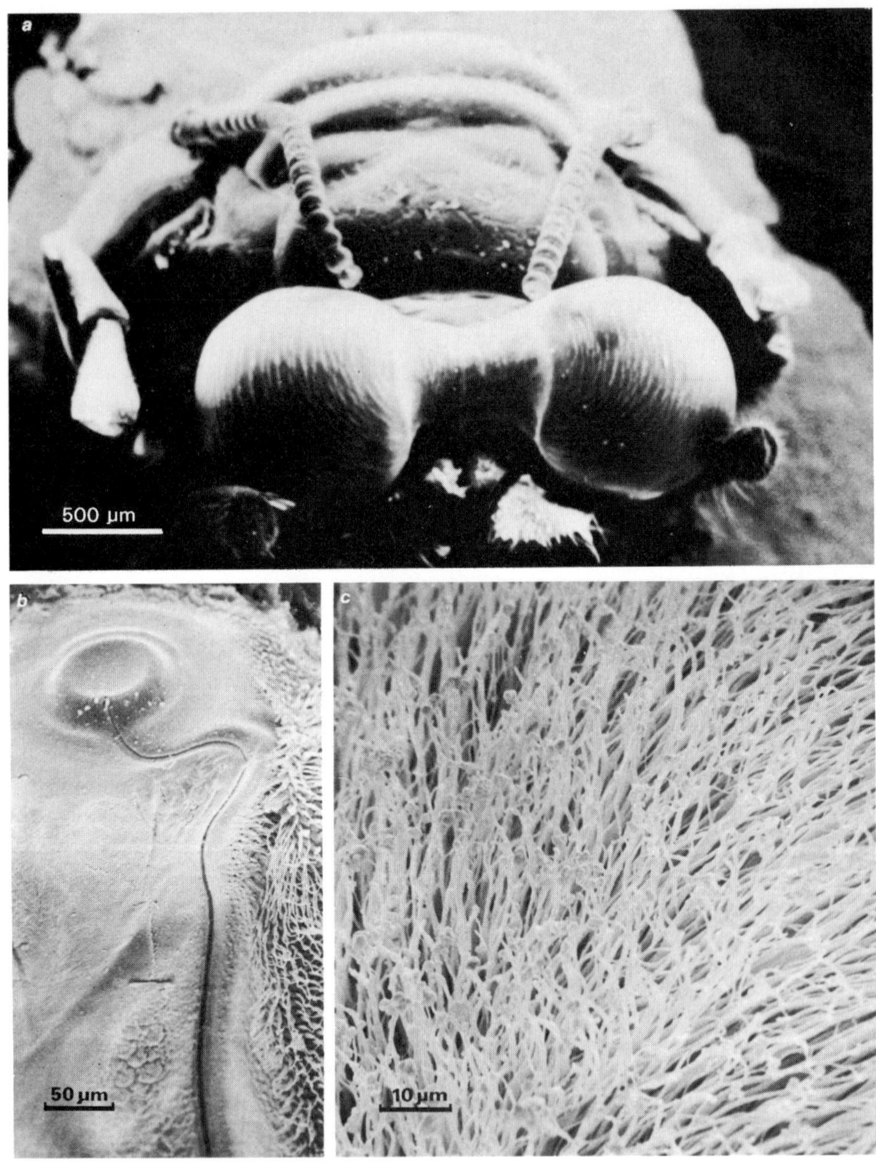

Fig. 13. Scanning electron micrographs of the structures involved in atmospheric absorption in *Arenivaga*. (*a*) The head seen from a low anterior view point, showing the paired hypopharyngeal "bladders". (*b*) Inside surface of the labrum showing groove for conveying fluid to the bladder and plate covering the gland. (*c*) Cuticular hairs covering the bladders. (Micrographs by M. J. O'Donnell.)

oscillations increases with ambient humidity and therefore with the condensation rate on the bladders. An application off distilled water to one of the bladders results in the temporary increase in the rate of oscillation of the muscles on that side. The accelerated flow rates over the bladders would presumably increase the efficiency of uptake by reducing the dilution caused by high rates of condensation (O'Donnell, personal communication).

The calculated pump threshold for *Arenivaga* is about 70% R.H., a value which agrees with a minimum of 71.5% suspected by O'Donnell (1978). Since the total concentration of inorganic solutes does not exceed about 1 mole (O'Donnell, personal communication), it must be assumed that low vapour pressure of the solution is largely due to organic solutes. Fluorescent dye placed on the bladders is transferred along the hypopharynx and oesophagus to the crop which suggests that condensed water is absorbed into the haemolymph through the gut (O'Donnell, 1977b). Once the hygroscopic fluid is diluted, water could be absorbed by coupled solute transport which would also make the solutes available for recycling.

4.3 "THERMOBIA"

Atmospheric water absorption in *Thermobia* combines high performance both in terms of rate (24.6% body wt day^{-1} at 85% R.H., 37 °C, Noble-Nesbitt, 1969) and low threshold, with deceptive structural simplicity of the rectum where it takes place. This consists of an irregularly shaped, cuticle lined terminal (anal) sac which, apart from its intrinsic musculature is composed of a single layer of epithelial cells (Noirot and Noirot-Timothée, 1971a, b). The sac communicates with the exterior by the anus which can be sealed with three valves (Noble-Nesbitt, 1973). Anteriorly, the sac leads through a sphincter into a more uniformly cylindrical rectum. A dimensional analysis of *Thermobia* hind gut indicates that diffusion can supply water vapour at sufficiently high rates to surfaces which are no more than 1 mm from the anus (Maddrell, 1971). However, Noble-Nesbitt (1973) suggests that regular pulsations of the rectum every 1 or 2 seconds might promote a tidal movement of air which would assist diffusion and perhaps make more of the rectal surface available for absorption.

The rectal site is the most likely site of condensation for other reasons. It is here that the epithelium shows the deepest and most regular folds in the apical plasma membrane, becoming intimately associated with large elongated mitochondria (Noirot and Noirot-Timothée, 1971b; Noble-Nesbitt, 1973). The swelling of extracellular fluid during fixation suggests that these structures are generating fluids of high osmotic pressure in desiccated specimens (Noble-Nesbitt, 1973). There is no indication in Noble-Nesbitt's (1974) weight traces of the non-equilibrium passive exchanges of the type seen in

Tenebrio (Machin, 1976), which confirms that the uptake mechanism in *Thermobia* is superficial and not much deeper than the site of condensation.

The problem of what mechanism is responsible for such a profound lowering of water activity (made even more remarkable by lowering the pump threshold of 45 to 40%, in this article) seems as intractable as ever. There are certainly no electrolytes that could be tolerated at concentrations as high as 30 moles, indeed inorganic compounds are not particlarly high in concentration in the rectal sac epithelium (Noble-Nesbitt, 1978). Maddrell (1971) has approached the problem from another direction, suggesting instead that the activity of luminal water is raised by compression. Unfortunately, the rate at which replacement and compression of the rectal contents would have to occur is very much faster than the frequency of anal opening and closing actually observed.

4.4 "ACARINES"

In spite of popularity of cuticular uptake at the time, the possibility that water vapour could be absorbed by hygroscopic saliva was anticipated by Wharton and Devine in 1968. At the present time five species of ixodid tick have been convincingly shown to absorb significant amounts of water vapour in the mouth region (Rudolph and Knülle, 1974; McMullen *et al.*, 1976). It is suggested by these workers that water vapour condenses onto a hygroscopic fluid, assumed to be saliva, which collects between the mouth parts and is subsequently taken into the body through the mouth. Rudolph and Knülle (1974) based this hypothesis on the observation that saliva, which was shown by chemical analysis to contain high concentrations of sodium and potassium, after being dried to a crystalline solid reabsorbed water at about 75% R.H. This value is well below the critical equilibrium humidities and absorption thresholds of most ixodid ticks (Table 1). Although all these observations are consistent with the above hypothesis, it is important to point out that they are far from conclusive evidence. Any drop of dried saliva, or haemolymph for that matter, if it contained predominantly NaCl, would exhibit the behaviour just described. Any secreted fluid will take up water in a higher humidity. Net gain of water from the atmosphere can only be established by demonstrating that the water activity of the secreted fluid corresponds to absorption threshold along the lines suggested earlier for perirectal fluid of *Tenebrio*.

The critical experiment which demonstrates that tick salivary glands are capable of generating fluid with very high osmotic pressures has not yet been performed. Previous studies of the mechanism of saliva secretion in ticks (Tatchel, 1969; Kaufman and Phillips, 1973a) have concentrated on the role of the salivary glands in water excretion following feeding. Saliva produced by the tick *Dermacentor andersoni* contains predominantly sodium and chloride with some potassium which together make up 80% of the osmotic pressure. The remaining anionic component is unknown. The secreted fluid is isosmotic to the haemolymph (Kaufman and Phillips, 1973b) normally about 350 mOsm kg^{-1} in engorged animals (Kaufman and Phillips, 1973a). Although McMullen *et al.* (1976) were able to demonstrate an increase in chloride concentration in *Amblyomma americanum* salivary gland after desiccation the increase was very small. Ultrastructural studies of the salivary glands of ticks (Kirkland, 1971; Meredith and Kaufman, 1973) provide no further suggestion that hyperosmotic fluids are secreted. Extensive basal folding in the gland cells are consistent with the Diamond and Bossert (1968) model of isosmotic fluid production. The possibility that other glands may produce the hygroscopic fluid has been suggested by Wharton (1976). He reports that on house dust mites, supracoxal as well as salivary glands open into the superficial podocephalic canal or groove which eventually delivers fluid into the mouth. Condensation could occur either as the fluid is conveyed down the canal or as it passes over the extensive surfaces of the pre-buccal cavity. The pumping action of the muscular pharynx in mites may be associated with the final absorption of fluid into the gut.

There are several other aspects of acarine water vapour absorption which invite further, more detailed study. Pump threshold of the flour mite *Acarus siro* (70%) is below the maximum humidity of even saturated NaCl (76% R.H. at 20 °C). The mechanics of water absorption in acarines is also poorly understood. It is not known for example, whether sophisticated continuous absorption mechanisms analagous to systems found in *Arenivaga* have been evolved or whether salivary secretion alternates with buccal uptake. This question could easily be resolved using modern recording electrobalances.

5 The basic elements of water absorption – speculations

It should be clear from the previous discussion that animals exploit many different, independently evolved water absorption systems, the mechanisms of which, as far as are understood, are consistent with currently accepted theories of solute coupled water transport. Even when the site of condensation is the same, important dissimilarities are apparent, for example, between rectal uptake in *Thermobia* and *Tenebrio* or between buccal uptake in *Arenivaga* and *Amblyomma.* Despite fundamental differences, the mechanisms of atmospheric absorption have in common, at least in the best known examples, three basic elements: hygroscopic fluid production, water vapour condensation and incorporation into the haemolymph. These elements represented diagrammatically in Fig. 14 are performed by different, spatially

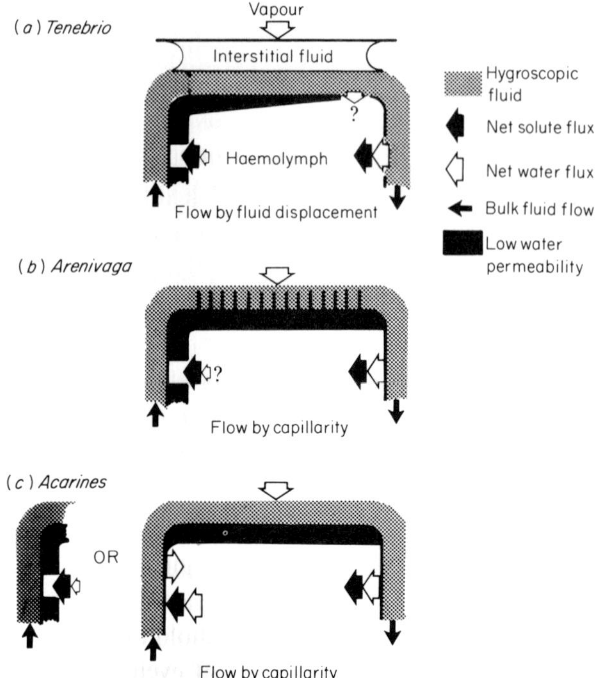

Fig. 14. Diagrammatic representation of water vapour absorption systems emphasizing the similarity between spatially separate components.

separate structures, which are functionally linked by the flow of hygroscopic fluid.

5.1 HYGROSCOPIC FLUID PRODUCTION

At the present state of our knowledge it seems likely that the majority of atmospheric absorption mechanisms have evolved by exploiting and refining already existing water transport systems such as the rectum or salivary gland. However, how such massively hyperosmotic solutions are produced and tolerated by living cells is poorly understood and when elucidated, will make a significant contribution to epithelial physiology. In *Tenebrio* and probably other tenebrionid beetle larvae (Coutchié and Crowe, 1975) high solute concentrations are apparently generated by the active transport of KCl by the leptophragmata. How these cells keep water transport to the minimum or how they resist depolarization in such high extracellular potassium levels is not known.

If other species also lower water activities according to the colligative properties of solutes, a number of mechanisms for generating suitably hyperosmotic solutions come to mind. An alternative way of producing a hyperosmotic solution would be to selectively reabsorb water from an isosmotic solution. Some organic molecules are attractive candidates for lowering solvent activities because they are soluble enough and could be tolerated by living cells in high concentration. Lowering of solvent activity could be achieved by either making use of the high osmotic coefficients and strongly non-ideal behaviour of macromolecules in high concentration or polymerizing them in more dilute solutions. The water barrier which isolates the hygroscopic fluid from body fluids of much lower osmotic pressure is a prominent and essential feature of many structures associated with atmospheric uptake. In *Tenebrio* this barrier is cellular while in animals having an external site of condensation it is cuticular.

5.2 WATER VAPOUR CONDENSATION

It is well known that a great many porous and fibrous solids hygroscopically absorb or desorb water vapour in response to changing humidity. Equilibrium water contents tend to vary non-linearly with relative humidity showing more rapid increases in the higher range (see *inter alia* Feldman and Sereda, 1965). Aqueous solutions also behave hygroscopically as they condense or evaporate water to bring about vapour pressure equilibrium with the prevailing humidity. Theoretically water content or solvent volume changes inversely with solute concentration or vapour pressure lowering. Figure 15 shows this relationship for KCl, at 20 °C, based on data in Lovelace *et al.* (1921). Although theoretical models of hygroscopic behaviour in solids depends on capillarity (Penman, 1955; Feldman and Sereda, 1965) and not Raoult's law, the non-linear relationship between water content and humidity in some solids, including insect cuticle also becomes nearly linear when plotted as the reciprocal of vapour pressure lowering (Machin, 1975, 1978). In view of the widespread occurrence of hygroscopic materials in animals, any study which seeks to establish a net gain of water vapour must go beyond merely demonstrating uptake of water in high humidities. Any hygroscopic material will gain water when transferred to a higher humidity, net gain can only be proved if secretory mechanisms actually produce material that has a lower vapour pressure than the prevailing ambient conditions. This is possible for example, using a special temperature regulated vapour cell illustrated in Fig. 16. The cell which is placed on a microscope stage, is made of heat conductive aluminium, which is temperature regulated by circulating water. An air stream, whose vapour pressure is regulated by heat exchangers described in Machin (1976) is temperature equilibrated to the aluminium before passing over the sample drop. The drop is placed on a silicone coated translucent base and viewed horizontally by means of the

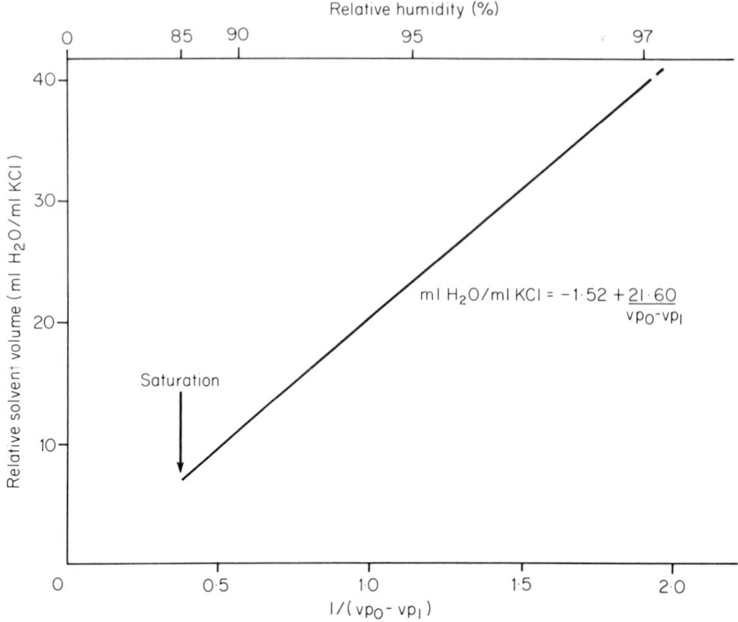

Fig. 15. Calculated solvent volume of a drop of potassium chloride solution in water (expressed relative to the volume of solute) plotted against the reciprocal of vapour pressure lowering (solute concentration). The empirical relationship (based on data in the "Handbook of Chemistry and Physics", 49th edition, 1968 and Lovelace *et al.*, 1921) included in the figure was used to calibrate the humidity cell illustrated in Fig. 16. The line's left-hand termination indicates the maximum solubility of KCl at 20 °C.

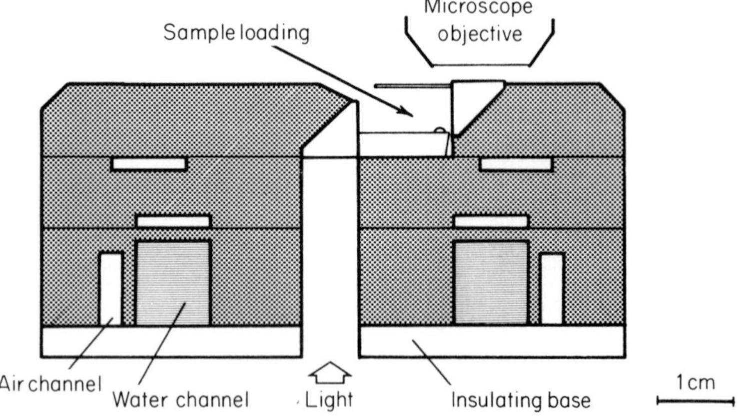

Fig. 16. Diagrammatic section of temperature regulated humidity cell used in studying the hygroscopic properties of fluids. Two 45° prisms permit the sample drop to be illuminated and viewed from the side. Humid air, after passing through the complex of channels, emerges out of section, into the sample well. The sample well is covered with a glass coverslip except for a slot where the samples are introduced from under oil. The shaded area represents aluminium.

45° prisms. Volume can be calculated by measuring its height and base diameter (Beament, 1958) with an ocular micrometer. The vapour pressure of the air can be calibrated using solutions of known volume/concentration into the haemolymph by conventional solute coupled transport. This volume on a regression line relating equilibrium volumes with known vapour pressures gives the drop's original concentration as well as describing its hygroscopic properties (Fig. 10).

At the present time one of the most serious deficiencies in our understanding of atmospheric absorption mechanisms is how a sufficiently low activity of water is achieved in the hygroscopic fluid. The problem is perhaps closest to being resolved in those animals such as *Tenebrio* and some of the ixodid ticks in which absorption threshold humidities exceed the saturation vapour pressures of salts such as KCl (85% R.H. at 20 °C) or NaCl (76% R.H. at 20 °C) (Winston and Bates, 1960) which have been shown to be highly concentrated in the absorbing fluids. The pump thresholds of *Acarus* (70%), *Liposcelis* (55%), *Xenopsylla* (52%) and *Thermobia* (40%) (Table 1) fall below these values, however. It seems doubtful that high levels of any other salts could be tolerated by the cells concentrating them (Maddrell, 1971). In *Arenivaga* (70%) the total inorganic content of the absorbing fluid amounts only to 1 mole (O'Donnell, personal communication). It seems likely that organic solutes are used to lower water activities in this and perhaps all of the animals listed above.

The majority of small molecular weight organic compounds are either too toxic or volatile (like alcohols) or too insoluble (like amino acids and sugars) to account for the vapour lowering required in some animals. The amino acid proline, even in the undissociated state is soluble enough (Hutchens and Kirby Hade, Jr., 1968) to condense water from humidities down to 78% R.H. at 26 °C. Another group of compounds, those hydrocarbons with multiple hydroxyl groups, are promising contenders because they are infinitely soluble in water at biological temperatures, and could potentially absorb for any humidities down to dry air. The humidities above different concentrations of two of such compounds, glycerol and polyethyl glycol, are compared in Fig. 17. Although solutions of the larger molecule eventually yield lower humidities per unit weight of solute, there may be disadvantages in employing large molecular weights. Figure 18 which indicates the different relative humidities obtainable with a number of representative solutions shows the advantages of keeping viscosity to a minimum by employing solutes of low molecular weight. Most of the atmospheric uptake systems described so far involve the flow of hygroscopic fluids where low viscosity would be an advantage. However, it is possible that the spacing of cuticular structures for example in *Arenivaga* and acarines would produce sufficient surface tensions to overcome the problem of high viscosity.

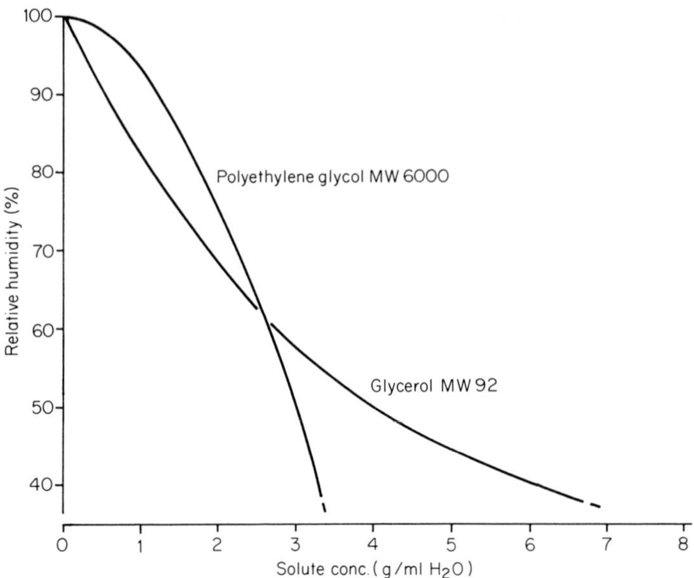

Fig. 17. Graph of equilibrium relative humidities obtained by different concentrations of glycerol and polyethylene glycol. Temperatures are 26.9 and 25°C, respectively. (Calculated from data of Braun and Braun, 1958 and Michel and Kaufmann, 1973.)

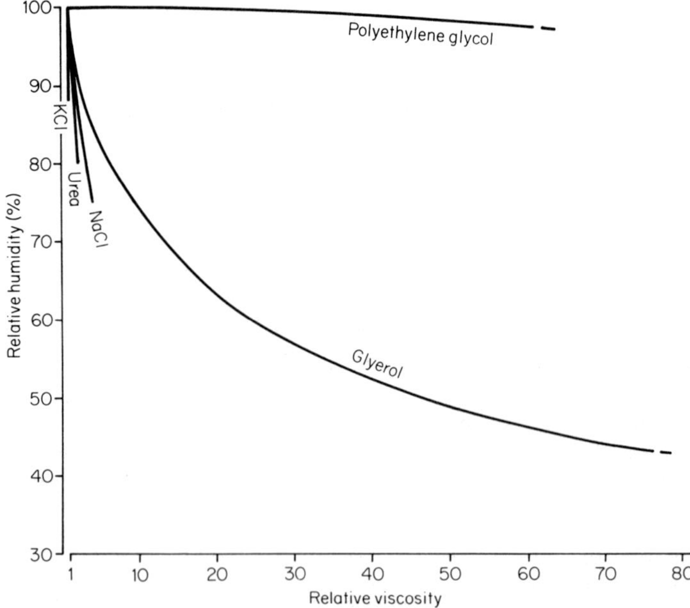

Fig. 18. Graph of equilibrium relative humidities of some representative solutions compared with their viscosities relative to that of water at the same temperature. Lines for KCl, NaCl and urea terminate at their maximum solubilities. (Calculated from data of Barr, 1928; Bates and Baxter, 1928.)

5.3 INCORPORATION INTO THE HAEMOLYMPH

The fact that the components of atmospheric absorption are separated from one another has important consequences. Spatial separation easily overcomes some of the difficulties with older absorption models whose temporally separated phases were necessitated by a system of fixed pore canals or hygroscopic solids.

The most important stumbling block to the cuticular absorption theory of how water is extracted by the epidermis from structures having very high water affinity is easily overcome. The gain of condensed water to the hygroscopic fluid and possible further additions from the gut or haemolymph, will dilute it and reduce osmotic gradients to the point that it can be absorbed into the haemolymph by conventional solute coupled transport. This mechanism does away with the need to have an elaborate valve system to operate unidirectional water flow, combining low water permeability in one direction with high permeability in the other. Water flow is rectified instead by the different fixed permeability characteristics of the fluid producing and absorption sites.

It is also possible that an absorption system with spatially separated components is more efficient because the continuous production and absorption of hygroscopic fluid promotes the continuous condensation of water vapour. Continuous uptake in a pumping system which involves alternate effective and recovery phases, can only be made continuous by having many unsynchronized functional units.

6 Summary

1 Several unrelated groups of insects and acarines absorb water vapour from humidities considerably below the water activities of the haemolymph.
2 Novel mechanisms involving the external cuticle were initially proposed to account for "uphill" water transport against such large forces.
3 Uptake was thought to be due to lowered water activities in the pore canals or to hygroscopic forces in cuticular solids. Water transport from cuticle to haemolymph, thought to be under energy dependent control of the epidermis, was poorly defined.
4 Recent experiments have seriously questioned cuticular transport models by demonstrating a rectal site of atmospheric uptake in *Thermobia* and *Tenebrio* and absorption in the mouth region in *Arenivaga* and *Acarines*.

5 In the absence of liquid water, animals regulate their water contents to some degree by means of atmospheric absorption. Water vapour absorption in larger species is modulated principally by the duration and not by the rate of uptake.

6 The water exchange kinetics of *Tenebrio* and certain *Acarines* show that water in these animals is distributed in two separate compartments. The second compartment which, by its low solvent activity, must be the pump, communicates to the exterior by more permeable barriers than the external cuticle.

7 Specific, localized absorption sites mean that water content is regulated in atmospheric absorbers by the balance of simultaneous gains and losses.

8 Smaller insects and acarines differ from larger species in having regulated water contents which change significantly with humidity. The smallest mites show uptake rates which appear to vary with the level of hydration of the animal.

9 It is proposed that there are no fundamental distinctions between the uptake mechanisms of large and small animals. Differences are ascribed to higher surface to volume ratios in smaller animals and the increased difficulty in balancing gains and losses in fluctuating conditions.

10 Faster losses by dead animals are associated simply with the breakdown of passive barriers in all species, with or without spiracular closing mechanisms.

11 Most species show linear uptake rates with ambient humidities. In contrast *Thermobia* shows some evidence of pump saturation at high rates of absorption.

12 Treatment of water vapour uptake and loss as separate processes has lead to the lowering of estimates of absorption threshold humidities in animals showing significant losses compared to uptake.

13 Rate of uptake in high humidity is related directly to body weight and not to surface area. Most animals absorb about 10% of their body weights per day. Uptake in small species tends to be more efficient than the average, and less efficient in the largest types.

14 *Tenebrio* appears to absorb water vapour through the cryptonephridial system in which high potassium chloride concentrations, generated by the Malpighian tubules, are osmotically coupled to the rectal lumen.

15 Structural and functional analysis of atmospheric uptake in *Thermobia* suggests that extremely low water activities are generated by the epithelium of the rectal sacs in this animal. The nature of activity lowering and how water is conveyed to the haemolymph remains unknown.

16 Water vapour is extracted from the air in *Arenivaga* by specialized hypopharyngeal evaginations of the integument. Hygroscopic fluid of un-

known composition from paired frontal glands flows over these condensing structures before absorption by the gut.

17 *Acarines* similarly extract water by external condensation before absorption into the mouth. The source of hygroscopic fluid and its mode of functioning has yet to be convincingly established.

18 All known atmospheric absorbing systems function with spatially separate sites of fluid secretion, condensation and absorption.

19 Organic molecules with multiple hydroxyl groups could be used to generate low water activities in animals without adequate electrolyte concentrations in their hygroscopic fluids or with exceptionally low absorption thresholds.

Acknowledgements

I gratefully acknowledge the financial support of the National Research Council of Canada and the many forms of practical and intellectual assistance from my associates, Pamela Coutchié, Michael O'Donnell, Andrew Forester, Lyanne Schlichter and Carol Shibuya.

References

Arlian, L. G. (1975a). Water exchange and effect of water vapour activity on metabolic rate in the dust mite, *Dermatophagoides*. *J. Insect Physiol.* **21**, 1439–1442.

Arlian, L. G. (1975b). Dehydration and survival of the European house dust mite, *Dermatophagoides pteronyssinus*. *J. Med. Ent.* **12**, 437–442.

Arlian, L. G. and Wharton, G. W. (1974). Kinetics of active and passive components of water exchange between the air and a mite, *Dermatophagoides farinae*. *J. Insect Physiol.* **20**, 1063–1077.

Barr, G. (1928). Viscosity of aqueous solutions of non-electrolytes. *In* "International Critical Tables" (Ed. E. W. Washburn), **5**, pp. 21–24. McGraw–Hill, New York.

Bates, S. J. and Baxter, W. P. (1928). Viscosity of aqueous solutions of strong electrolytes. *In* "International Critical Tables" (Ed. E. W. Washburn), **5**, pp. 12–19. McGraw-Hill, New York.

Beament, J. W. L. (1954). Water transport in insects. *Symp. Soc. exp. Biol.* **8**, 94–117.

Beament, J. W. L. (1958). The effect of temperature on the water-proofing mechanism of an insect. *J. exp. Biol.* **35**, 494–519.

Beament, J. W. L. (1961). The water relations of insect cuticle. *Biol. Rev.* **36**, 281–320.

Beament, J. W. L. (1964). The active transport and passive movement of water in insects. *In* "Advances in Insect Physiology" (Eds. J. W. L. Beament, J. E. Treherne, and V. B. Wigglesworth), **2**, pp. 67–129. Academic Press, New York and London.

Beament, J. W. L. (1965). The active transport of water: evidence, models and mechanisms. *Symp. Soc. exp. Biol.* **19**, 273–298.

Beament, J. W. L., Noble-Nesbitt, J. and Watson, J. A. L. (1964). The water-proofing mechanism of arthropods. III. Cuticular permeability in the firebrat, *Thermobia domestica* (Packard). *J. exp. Biol.* **41,** 323–330.

Braun, J. V. and Braun, J. D. (1958). A simplified method of preparing solutions of glycerol and water for humidity control. *Corrosion*, **14,** 117–118.

Brönsted, J. N. (1928). Solubility of salts and of strong acids and bases in water. *In* "International Critical Tables" (Ed. E. W. Washburn), **4,** pp. 216–249. McGraw-Hill, New York.

Browning, T. O. (1954). Water balance in the tick *Ornithodoros moubata* Murray, with particular reference to the influence of carbon dioxide on the uptake and loss of water. *J. exp. Biol.* **31,** 331–340.

Buxton, P. A. (1930). Evaporation from the meal-worm (*Tenebrio*: Coleoptera) and atmospheric humidity. *Proc. Roy. Soc. Lond.* **B. 106,** 560–577.

Coutchié, P. A. and Crowe, J. H. (1975). Absorption of water vapor from subsaturated air by tenebrionid beetle larvae. *Am. Zool.* **15,** 802.

Devine, T. L. and Wharton, G. W. (1973). Kinetics of water exchange between a mite *Laelaps echidnina* and the surrounding air. *J. Insect Physiol.* **19,** 243–254.

Diamond, J. M. and Bossert, W. H. (1968). Functional consequences of ultrastructural geometry in "backwards" fluid transporting epithelia. *J. Cell Biol.* **37,** 694–702.

Dunbar, B. S. and Winston, P. W. (1975). The site of active uptake of atmospheric water in larvae of *Tenebrio molitor*. *J. Insect Physiol.* **21,** 495–500.

Ebeling, W. (1974). Permeability of insect cuticle. *In* "The Physiology of Insecta" (Ed. M. Rockstein), **6,** pp. 271–343. Academic Press, New York and London.

Edney, E. B. (1948). Laboratory studies on the bionomics of the rat fleas, *Xenopsylla brasiliensis*, Baker, and *X. cheopis*, Roths. *Bull. Ent. Research*, **38,** 263–280.

Edney, E. B. (1957). "The Water Relations of Terrestrial Arthropods". Cambridge University Press, Cambridge.

Edney, E. B. (1966). Absorption of water vapour from unsaturated air by *Arenivaga* sp. (Polyphagidae, dictyoptera). *Comp. Biochem. Physiol.* **19,** 387–408.

Edney, E. B. (1967a). Water balance in desert arthropods. *Science, N.Y.* **156,** 1059–1066.

Edney, E. B. (1967b). The impact of the atmospheric environment on the integument of insects. *Proc. Int. Biometerol. Congr.* **4,** 71–81.

Edney, E. B. (1971). Some aspects of water balance in Tenebrionid beetles and a thysanuran from the Namid Desert of Southern Africa. *Physiol. Zoöl.* **44,** 61–76.

Ellingsen, I. (1975). Permeability to water in different adaptive phases of the same instar in the American House-dust mite. *Acarologia*, **17,** 734–744.

Feldman, R. F. and Sereda, P. J. (1965). Moisture content – its significance and interaction in a porous body. *In* "Humidity and Moisture. Measurement and Control in Science and Industry" (Ed. A. Wexler), **4,** pp. 233–243. Reinhold Publishing Corporation, New York.

Gluud, V. A. (1968). Zur Feinstruktur der Insektencuticula: ein Beitrag zur Frage des Eigengiftschutzes der Wanzencuticula. *Zool. Jb. Anat.* **85,** 191–227.

Grimstone, A. V., Mullinger, A. M. and Ramsay, J. A. (1968). Further studies on the rectal complex of the mealworm *Tenebrio molitor*, L. (Coleoptera, Tenebrionidae). *Proc. Roy. Soc. Lond.* **B. 253,** 343–382.

Hafez, M., El-Ziady, S. and Hefnawy, T. (1970). Biochemical and physiological studies of certain ticks (Ixodoidea). Uptake of water vapor by the different developmental stages of *Hyalomma (H.) dromedarii* Koch (Ixodidae) and *Ornithodoros (O.) savignyi* (Audouin) (Argasidae). *J. Parasit.* **56,** 354–361.

Hair, J. A., Sauer, J. R. and Durham, K. A. (1975). Water balance and humidity preference in three species of ticks. *J. Med. Ent.* **12,** 37–47.

Hefnawy, T. (1970). Biochemical and physiological studies of certain ticks (Ixodoidea). Water loss from spiracles of *Hyalomma (H.) dromedarii* Koch (Ixodidae) and *Ornithodoros (O.) savignyi* (Audouin) (Argasidae). *J. Parasit.* **56,** 362–366.

Hutchens, J. O. and Kirby Hade, Jr., E. P. (1968). Solubilities of amino acids in water at various temperatures. *In* "Handbook of Biochemistry" (Ed. H. A. Sober), pp. 390. The Chemical Rubber Company, Cleveland, Ohio.

Johansson, B. (1920). Der Gaswechsel bei *Tenebrio molitor* in seiner Abhängigkeit von der Nahrung. *Acta. Univ. Lund.* **16,** 1–36.

Kanungo, K. (1963). Effect of low oxygen tensions on the uptake of water by dehydrated females of the spiny rat-mite *Echinolaelaps echidninus Expl. Parasit.* **14,** 263–268.

Kanungo, K. (1965). Oxygen uptake in relation to water balance of a mite (*Echinolaelaps echidninus*) in unsaturated air. *J. Insect Physiol.* **11,** 557–568.

Kaufman, W. R. and Phillips, J. E. (1973a). Ion and water balance in the ixodid tick *Dermacentor andersoni.* I. Routes of ion and water excretion. *J. exp. Biol.* **58,** 523–536.

Kaufman, W. R. and Phillips, J. E. (1973b). Ion and water balance in the ixodid tick *Dermacentor andersoni.* III. Influence of monovalent ions and osmotic pressure of salivary secretion. *J. exp. Biol.* **58,** 549–546.

Kirkland, W. L. (1971). Ultrastructural changes in the nymphal salivary glands of the rabbit tick, *Haemaphysalis leporispalustris*, during feeding. *J. Insect Physiol.* **17,** 1933–1946.

Knülle, W. (1962). Die Abhängigkeit der Luftfeuchte-reaktionen der Mehlmilbe (*Acanis siro* L.) vom Wassergehalt des Körpers. *Z. Vergl. Physiol.* **45,** 233–246.

Knülle, W. (1965). Die sorption und transpiration des wasserdampfes bei der mehlmilbe (*Acarus siro* L.). *Z. vergl. Physiol.* **49,** 586–604.

Knülle, W. (1966). Equilibrium humidities and survival of some tick larvae. *J. Med. Ent.* **2,** 335–338.

Knülle, W. (1967). Physiological properties and biological implicaations of the water vapour sorption mechanism in larvae of the oriental rat flea, *Xenopsylla cheopis* (Roths.). *J. Insect Physiol.* **13,** 333–357.

Knülle, W. and Devine, T. L. (1972). Evidence for active and passive components of sorption of atmospheric water vapour by larvae of the tick *Dermacentor variabilis. J. Insect Physiol.* **18,** 1653–1664.

Knülle, W. and Spadafora, R. R. (1969). Water vapor sorption and humidity relationships in *Liposcelis* (Insecta: Psocoptera). *J. stored Prod. Res.* **5,** 49–55.

Knülle, W. and Spadafora, R. R. (1970). Occurrence of water vapour sorption from the atmosphere in larvae of some stored-produce beetles. *J. econ. Ent.* **4,** 1069–1070.

Knülle, W. and Wharton, G. W. (1964). Equilibrium humidities in arthropods and their ecological significance. *Proc. 1st int. Congr. Acarology. Acarologia,* **6,** 299–306.

Kracek, F. C. (1928). Pressure, temperature, concentration relations of two or more components and containing two or more phases. Two-component aqueous systems. *In* "International Critical Tables" (Ed. E. W. Washburn), **3,** pp. 351–385. McGraw-Hill, New York.

Lees, A. D. (1946). The water balance in *Ixodes ricinus* L. and certain other species of ticks. *Parasitology,* **37,** 1–20.

Lees, A. D. (1947). Transpiration and the structure of the epicuticle in ticks. *J. exp. Biol.* **23,** 379–410.

Lees, A. D. (1964). The effect of aging and locomotor activity on the water transport mechanism of ticks. *Proc. 1st int. Congr. Acarology. Acarologia,* **6,** 315–323.

Locke, M. (1964). The structure and formation of the integument in insects. *In* "Physiology of Insecta" (Ed. M. Rockstein), **3**, pp. 379–470. Academic Press, New York and London.

Locke, M. (1974). The structure and formation of the integument in insects. *In* "The Physiology of Insecta" (Ed. M. Rockstein), **6**, pp. 123–213. Academic Press, New York and London.

Lockey, K. H. (1976). Cuticular hydrocarbons of *Locusta*, *Schistocerca*, and *Periplaneta*, and their role in waterproofing. *Insect Biochem.* **6**, 457–472.

Lovelace, B. F., Frazer, J. C. W. and Sease, V. B. (1921). The lowering of the vapour pressure of water at 20 °C produced by dissolved potassium chloride. *J. Amer. Chem. Soc.* **43**, 102–110.

Machin, J. (1975). Water balance in *Tenebrio molitor*, L. larvae; the effect of atmospheric water absorption. *J. comp. Physiol.* **101**, 121–132.

Machin, J. (1976). Passive exchanges during water vapour absorption in mealworms (*Tenebrio molitor*): a new approach to studying the phenomenon. *J. exp. Biol.* **65**, 603–615.

Machin, J. (1978). Water vapour uptake by *Tenebrio*: a new approach to studying the phenomenon. *In* "Comparative Physiology – Water, Ions and Fluid Mechanics" (Eds. L. Bolis, K. Schmidt-Nielsen and S. H. P. Maddrell), pp. 67–77. Cambridge University Press, Cambridge.

Maddrell, S. H. P. (1971). The mechanisms of insect excretory systems. *In* "Advances in Insect Physiology" (Eds. J. W. L. Beament, J. E. Treherne and V. B. Wigglesworth), **8**, pp. 199–351. Academic Press, New York and London.

McEnroe, W. D. (1961). The control of water loss by the two-spotted spider mite (*Tetranychus telarius*). *Ann. ent. Soc. Am.* **54**, 883–887.

McMullen, H. L., Sauer, J. R. and Burton, R. L. (1976). Possible role in uptake of water vapour by ixodid tick salivary glands. *J. Insect Physiol.* **22**, 1281–1285.

Mellanby, K. (1932). The effect of atmospheric humidity on the metabolism of the fasting mealworm (*Tenebrio molitor* L., Coleoptera). *Proc. Roy. Soc. Lond.* **B. 111**, 376–390.

Mellanby, K. (1934). The site of loss of water from insects. *Proc. Roy. Soc. Lond.* **B. 116**, 139–149.

Meredith, J. and Kaufman, W. R. (1973). A proposed site of fluid secretion in the salivary gland of the ixodid tick *Dermacentor andersoni*. *Parasitology*, **67**, 205–217.

Michel, B. E. and Kaufmann, M. R. (1973). The osmotic potential of polyethylene glycol 6000. *Plant Physiol.* **51**, 914–916.

Neville, A. C. (1975). "Biology of the Arthropod Cuticle". Springer–Verlag, New York, Heidelberg and Berlin.

Noble-Nesbitt, J. (1968). Aspects of the structure, formation and function of some insect cuticles. *In* "Insects and Physiology" (Ed. J. W. L. Beament and J. E. Treherne), pp. 3–16. American Elsevier, New York.

Noble-Nesbitt, J. (1969). Water balance in the firebrat, *Thermobia domestica* (Packard). Exchanges of water with the atmosphere. *J. exp. Biol.* **50**, 745–769.

Noble-Nesbitt, J. (1970a). Water uptake from subsaturated atmospheres: its site in insects. *Nature, Lond.* **225**, 753–754.

Noble-Nesbitt, J. (1970b). Water balance in the firebrat, *Thermobia domestica* (Packard). The site of uptake of water from the atmosphere. *J. exp. Biol.* **52**, 193–200.

Noble-Nesbitt, J. (1973). Rectal uptake of water in insects. *In* "Comparative Physiology, Locomotion, Respiration, Transport and Blood" (Eds. L. Bolis, K. Schmidt-Nielsen and S. H. P. Maddrell), pp. 333–351. North-Holland Publishing Company, Amsterdam and London.

Noble-Nesbitt, J. (1975). Reversible arrest of uptake of water from subsaturated atmospheres by the firebrat, *Thermobia domestica* (Packard). *J. exp. Biol.* **62,** 657–669.
Noble-Nesbitt, J. (1978). Absorption of water vapour by *Thermobia domestica* and other insects. *In* "Comparative Physiology – Water, Ions and Fluid Mechanics" (Eds. L. Bolis, K. Schmidt-Nielsen and S. H. P. Maddrell), pp. 53–66. Cambridge University Press, Cambridge.
Noirot, C. and Noirot-Timothée, C. (1971a). Ultrastructure du proctodeum chez le Thysanoure *Lepismodes inquilinus* Newman (= *Thermobia domestica* Packard). I. La région antérieure (Iléon et rectum). *J. Ultrastruct. Res.* **37,** 119–137.
Noirot, C. and Noirot-Timothée, C. (1971b). Ultrastructure du proctodeum chez le Thysanoure *Lepismodes inquilinus* Newman (= *Thermobia domestica* Packard). II. Le sac anal. *J. Ultrastruct. Res.* **37,** 335–350.
O'Donnell, M. J. (1977a). Site of water vapour absorption in the desert cockroach, *Arenivaga investigata. Proc. natn. Acad. Sci. U.S.A.* **74,** 1757–1760.
O'Donnell, M. J. (1977b). Hypopharyngeal bladders and frontal glands: novel structures involved in water vapour absorption in the desert cockroach, *Arenivaga investigata. Am Zool.* **17,** 902.
O'Donnell, M. J. (1978). The site of water vapour absorption in *Arenivaga investigata. In* "Comparative Physiology – Water, Ions, and Fluid Mechanics" (Eds. L. Bolis, K. Schmidt-Nielsen and S. H. P. Maddrell), pp. 115–121. Cambridge University Press, Cambridge.
Okasha, A. Y. K. (1971). Water relations in an insect, *Thermobia domestica.* I. Water uptake from sub-saturated atmospheres as a means of volume regulation. *J. exp. Biol.* **55,** 435–448.
Okasha, A. Y. K. (1972). Water relations in an insect, *Thermobia domestica.* II. Relationships between water content, water uptake from subsaturated atmospheres and water loss. *J. exp. Biol.* **57,** 285–296.
Patton, R. L. and Craig, R. (1939). The rates of excretion of certain substances by the larvae of the mealworm, *Tenebrio molitor* L. *J. exp. Zool.* **81,** 437–457.
Penman, H. L. (1955). "Humidity". The Institute of Physics. Chapman and Hall Limited, London.
Ramsay, J. A. (1964). The rectal complex of the mealworm *Tenebrio molitor* L. (Coleoptera, Tenebrionidae). *Phil. Trans. Roy. Soc. Lond.* **B. 248,** 279–314.
Rudolph, D. and Knülle, W. (1974). Site and mechanism of water vapour uptake from the atmosphere in ixodid ticks. *Nature, Lond.* **249,** 84–85.
Sauer, J. R. and Hair, J. A. (1971). Water balance in the lone star tick (Acarina: Ixodidae): the effects of relative humidity and temperature on weight changes and total water content. *J. Med. Ent.* **8,** 479–485.
Solomon, M. E. (1966). Moisture gains, losses and equilibria of flour mites *Acarus siro* L., in comparison with larger arthropods. *Ent. exp. & appl.* **9,** 25–41.
Tatchell, R. J. (1969). The ionic regulatory role of the salivary secretions of the cattle tick, *Boophilus microplus. J. Insect Physiol.* **15,** 1421–1430.
Wharton, G. W. (1963). Equilibrium humidity. *Adv. Acarol.* **1,** 201–208.
Wharton, G. W. (1976). House dust mites. *J. Med. Ent.* **12,** 577–621.
Wharton, G. W. and Arlian, L. G. (1972). Utilization of water by terrestrial mites and insects. *In* "Insect and Mite Nutrition" (Ed. J. G. Rodriguez), pp. 153–165. North-Holland Publishing Company, Amsterdam and London.
Wharton, G. W. and Devine, T. L. (1968). Exchange of water between a mite, *Laelaps echidnina*, and the surrounding air. *J. Insect Physiol.* **14,** 1303–1318.
Wharton, G. W. and Kanungo, K. (1962). Some effects of temperature and relative

humidity on water-balance in females of the spiny rat mite, *Echinolaelaps echidninus* (Acarina: Laelaptidae). *Ann. ent. Soc. Am.* **55,** 483–492.

Wharton, G. W., Parrish, W. and Johnston, D. E. (1968). Observations on the fine structure of the cuticle of the spiny rat mite, *Laelaps echidnina* (Acari-Mesostigmata). *Acarologia,* **10,** 206–214.

Winston, P. W. (1967). Cuticular water pump in insects. *Nature, Lond.* **214,** 383–384.

Winston, P. W. (1969). The cuticular water pump and its physiological and ecological significance in the Acarina. *In* "Proceedings of the 2nd International Congress of Acarology" (Ed. G. O. Evans), pp. 461–468. Akadémiai Kiadó, Budapest.

Winston, P. W. and Bates, D. H. (1960). Saturated solutions for the control of humidity in biological research. *Ecology,* **41,** 232–237.

Winston, P. W. and Beament, J. W. L. (1969). An active reduction of water level in insect cuticle. *J. exp. Biol.* **50,** 541–546.

Winston, P. W. and Hoffmeier, P. (1968). The active control of cuticular water content in *Leucophaea maderae* L. *Am. Zool.* **8,** 391.

Winston, P. W. and Nelson, V. E. (1965). Regulation of transpiration in the clover mite *Bryobia praetiosa* Koch (Acarina: Tetranychidae). *J. exp. Biol.* **43,** 257–269.

NOTES ADDED IN PROOF

1 Coutchié and Crowe (*Physiol. Zool.* **52,** pp. 67 and pp. 88, 1979) have described water absorption by larvae of two tenebrionid beetles which show a much wider range than those of *Tenebrio* (6 to 600 mg) (Coutchié and Machin, *Am. Zool.* **18,** 627, 1978).

2 Results of further analysis of rectal melting point distribution suggest a single tubular pump driving both liquid and gaseous water absorption (Machin, *J. exp. Biol.* in press).

3 The work on which the personal communication of Ryerse (p. 6) is based has now been reported (Ryerse, *Proc. Miscrop. Soc. Can.* **4,** 48 (1977); *Nature Lond.* **271,** 745 (1978), *J. Insect Physiol.* **24,** 315 (1978)).

4 Wharton and Furumizo (*Acarologia* **19,** 112 (1977)) claim that the production of hygroscopic fluid by the supracoxal glands of mites is responsible for the absorption of water vapour. Since the water activity in freshly produced secretions was not determined, the demonstration of hygroscopic behaviour alone cannot provide definitive proof of net water uptake by supracoxal fluid.

5 Wharton and Richards (*Ann. Rev. Entomol.* **23,** 309 (1978)) have recently reviewed water vapour exchange kinetics of insects and *Acarines.* Their approach differs from the present review in emphasizing the "black box" kinetics of exchange rather than the site and mechanism of active water uptake.

Insect Vitellogenin: Identification, Biosynthesis, and Role in Vitellogenesis

Franz Engelmann

Department of Biology, University of California, Los Angeles, USA

1 Introduction

Vitellogenins are defined as the predominant yolk protein precursors which are produced extraovarially and taken up by the growing oocytes against a concentration gradient (Pan *et al.*, 1969). They are generally found only in females and consequently the term female specific proteins is also used for the same class of proteins. Neither of these terms is entirely satisfactory, because most blood proteins (female specific or not) contribute to the protein yolk and are, strictly speaking, vitellogenins, and vitellogenins are not always found only in females. Vitellogenins may be altered during uptake into the oocytes and have changed their physico-chemical characteristics. They may no longer be termed vitellogenins, yet most literature reports do not distinguish between the haemolymph and egg counterparts. Certainly, the haemolymph vitellogenin and the major yolk protein have identical electrophoretic mobility and are immunologically indistinguishable. When feasible and where the available data permit we will from now on make the distinction between the major yolk protein precursor and its counterpart in the oocyte or egg: vitellogenin and vitellin.

One of the most exciting aspects, and one which has attracted increasing attention during the last few years, is the mechanism of control of vitellogenin biosynthesis. Not only have we begun to understand how juvenile hormone(s) directs vitellogenin synthesis, but research in this area touches also on more general aspects of endocrinology and developmental biology, namely, how does a hormone control the *de novo* synthesis of a well-defined protein molecule produced by a fully differentiated tissue. The ramifications of interests, the level of sophistication of research, and the potential to yield gratifying results are reasons for a rapidly increasing wealth of publications. Furthermore, the close analogy to the control mechanisms known for avian and amphibian vitellogenins makes research on insect vitellogenin even more rewarding. In all of these cases hormones direct vitellogenin synthesis in extraovarian tissues. In insects the hormone is a sesquiterpenoid produced by the corpora allata and termed juvenile hormone (JH), and in vertebrates, ovarian steroid hormones stimulate vitellogenin synthesis in the liver of the female of the species. Interestingly, livers from the males of birds and amphibia also can be stimulated by oestrogen to produce vitellogenin. Fat bodies of the males of insects, on the other hand, are generally found not to produce vitellogenin; in the few cases in which low titers of vitellogenin are known to occur in males, it is unknown (with one exception) whether hormones direct its synthesis. In most insect species, male fat bodies do not synthesize vitellogenin even when exposed to high titres of juvenile hormone. A true sexual dimorphism of the adult fat bodies of insects (the target tissue for the hormone) does exist, whereas in vertebrates, sexual dimorphism is expressed at the hormonal level.

2 Identification of vitellogenin and vitellin

2.1 HISTORICAL PERSPECTIVE

It was first recognized by Telfer (1954) that the haemolymph of the female silkmoth *Hyalophora cecropia* contained a protein which is also found to be the predominant egg yolk protein; it is not present, or only in minor quantities, in haemolymph of the males or larvae of either sex. This protein was later appropriately termed vitellogenin (Pan *et al.*, 1969) to denote the relationship between precursor and final product in the eggs. For demonstrating identity of the haemolymph and egg proteins Telfer used the most reliable method available, namely, immunodiffusion. Subsequent to this pioneering work, vitellogenins and vitellins have been described in many different species using a variety of techniques, such as paper, starch gel, cellulose acetate, and polyacrylamide gel electrophoresis (PAGE). In Table 1 are listed, according to the chronological documentation, the known species, grouped by orders, for which vitellogenins have been described. It can be immediately seen that these species belong to only a few orders. There is, however, no reason to assume that vitellogenins do not occur in species of other orders as well.

TABLE 1 Identification of vitellogenin and vitellin

Species	Method used	Female specific	Identity of vitellogenin and vitellin verified	Authors
DICTYOPTERA:				
Leucophaea maderae	CI	+	+	Engelmann (1965); Engelmann and Penney (1966)
Periplaneta americana	C	±		Thomas and Nation (1966)
	PI	+	+	Bell (1969a, b)
Byrsotria fumigata	I	+		Barth and Bell (1970)
Eublaberus posticus	I	+		Bell and Barth (1971)
Blattella germanica	I	+	+	Tanaka (1973)

Table 1 (*contd*)

Species	Method used	Female specific	Identity of vitellogenin and vitellin verified	Authors
Nauphoeta cinerea	I	+	+	Bühlmann (1974)
ORTHOPTERA:				
Schistocerca gregaria	PpI	+	+	Hill (1962); Dufour *et al.* (1970)
Locusta migratoria	PI	+	+	Tobe and Loughton (1967); Bar-Zev *et al.* (1975)
Acheta domesticus	CP	+	+	Kunz and Petzelt (1970); Schneider (1973)
Schistocerca vaga	I	+	+	Engelmann *et al.* (1971)
Anacridium aegypticum	P	±	+	Girardie (1977)
HEMIPTERA:				
Rhodnius prolixus	SCPI	±	+	Coles (1964), 1965a); Baehr (1974)
Triatoma infestans	I	+		Perassi (1973)
Lygocoris pabulinus	P	+	+	Wightman (1973)
Oncopeltus fasciatus	PI	±	+	Kelly and Davenport (1976); Kelly and Telfer (1977)
Triatoma protracta	PI	+	+	Mundall and Engelmann (1977)
LEPIDOPTERA:				
Hyalophora cecropia	I	±	+	Telfer (1954)
Pieris brassicae	CI	+	+	Lamy (1967, 1970)
Antheraea polyphemus	P	+	+	Blumenfeld and Schneiderman (1968)
Celerio euphorbiae	P	+	+	Bielinska *et al.* (1971)
Danaus plexippus	I	+	+	Pan and Wyatt (1971)

Table 1 (*contd*)

Species	Method used	Female specific	Identity of vitellogenin and vitellin verified	Authors
Bombyx mori	PI	±	+	Doira and Kawaguchi (1972); Ono *et al.* (1975)
Thaumetopoea pityocampa	CP	+	+	Lamy (1973)
Manduca sexta	P	+	+	Nijhout and Riddiford (1974)
Nymphalis antiopa	P	+	+	Herman and Bennett (1975)
Philosamia cynthia	P	+	+	Chino *et al.* (1976)
COLEOPTERA:				
Tenebrio molitor	SI	±	+	Laverdure (1968, 1972)
Leptinotarsa decemlineata	I	+	+	De Loof (1969)
Dermestes frischi	P	+	+	Küthe (1972)
HYMENOPTERA:				
Apis mellifica	PCI	+		Lensky and Alumot (1969); Engels (1972)
Bombus terrestris	P	+	+	Röseler (1974, 1977)
Melipona quadrifasciata	C	+		Engels and Engels (1977)
Scaptotrigona postica	C	+		Engels and Engels (1977)
DIPTERA:				
Drosophila melanogaster	I	+	+	Fox and Yoon (1958); Gavin and Williamson (1976a)
D. virilis	I	+	+	Gingeras *et al.* (1973)
D. hydei	I	+	+	Gingeras *et al.* (1973)
D. silvestris	I	+	+	Gelti-Douka *et al.* (1974)
Musca domestica	PI	+	+	Bodnaryk and Morrison (1966, 1968)

Species	Method used	Female specific	Identity of vitellogenin and vitellin verified	Authors
Culex tarsalis	I	+		Chao (1969)
Sarcophaga bullata	I	+	+	Wilkens (1969)
Aedes aegypti	PI	+	+	Hagedorn and Judson (1972)
Phormia regina	A	+	+	Mjeni and Morrison (1973, 1976)
Culex pipiens	PI	+		Schumann (1973)

C, Cellulose acetate electrophoresis; P, Polyacrylamide gel electrophoresis; S, Starch gel electrophoresis; Pp, Paper electrophoresis; I, Immunodiffusion techniques; A, Agarose electrophoresis; +, Positive identification; ±, Vitellogenin present in females and males or nymphs (usually in low titres)

From Table 1 it is also apparent that after the first report by Telfer a rather long time interval elapsed before vitellogenins were demonstrated by immunological techniques for other insect species. Coles (1964, 1965a, b) observed in *Rhodnius prolixus* the immunological identity of female specific proteins of the haemolymph with that of the egg proteins; unfortunately no documentation was provided in these publications. Then, using immunoelectrophoresis, it was shown conclusively for the cockroach *Leucophaea maderae* (Fig. 1) that only the vitellogenic females contain a unique antigen in the haemolymph; the same antigen is also contained in the eggs (Engelmann and Penney, 1966).

2.2 TECHNIQUES USED FOR IDENTIFICATION

In many of the early observations, paper, starch gel, or cellulose acetate electrophoresis techniques have been used for identification of vitellogenin and vitellin. Generally, these techniques give rather poor resolution of proteins in a given sample and are therefore inadequate for a critical analysis. Nevertheless these methods provided important clues as to the occurrence of female specific proteins in several species of insects (Hill, 1962; Coles, 1964, 1965a; Engelmann, 1965). Cellulose acetate electrophoresis is still used and provides an easy and quick means for identification of vitellogenins, for example, in *Apis* spec. and several stingless bees (Engels and Engels, 1977).

From among the available techniques, polyacrylamide gel electrophoresis

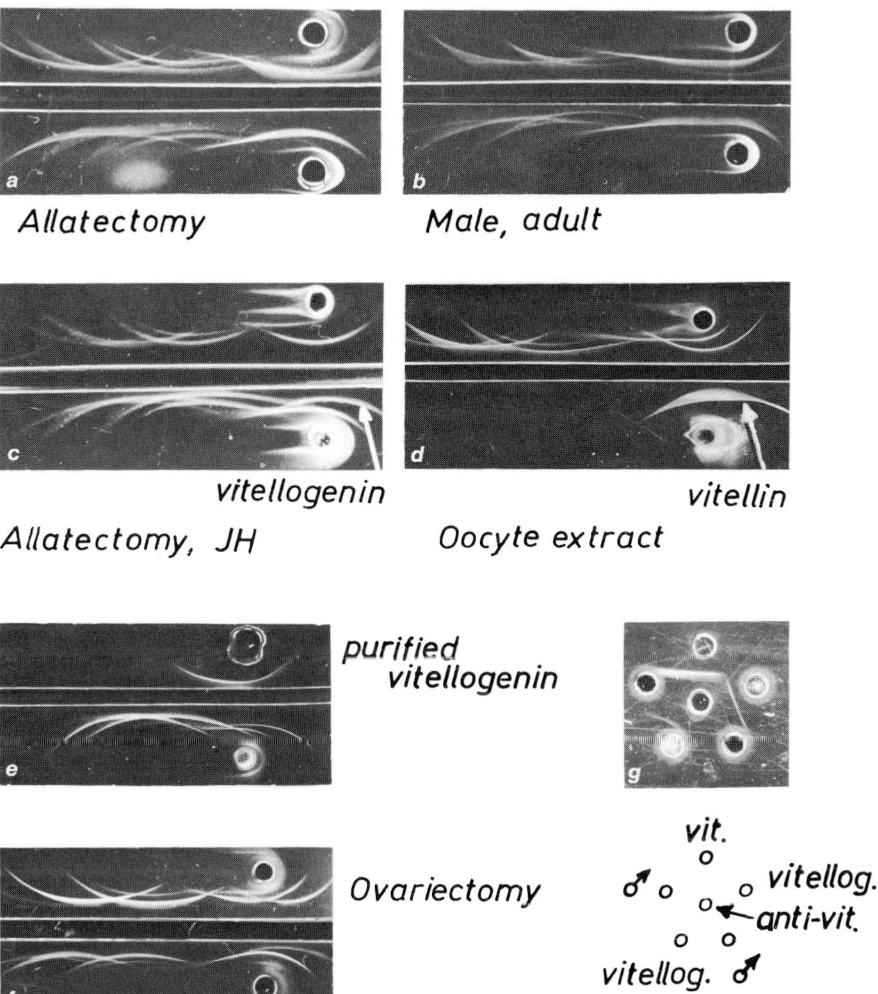

Fig. 1. Analysis of vitellogenin and vitellin of *Leucophaea* by immunoelectrophoresis and Ouchterlony immunodiffusion. The upper traces in *a–d* and the lower traces in *e* and *f* represent samples of haemolymph from normal vitellogenic females. In *g* the vitellin and vitellogenin precipitation lines are continuous, indicating immunological identity.

(PAGE) appears to be the preferred method for obvious reasons. The high resolution for many proteins of the samples has the potential for yielding simultaneously the maximum information on several protein species of any given sample, be it haemolymph or tissue homogenates. Indeed, fluctuating titers of vitellogenin and other haemolymph proteins were estimated throughout a reproductive cycle of *Leucophaea* (Scheurer, 1969) using densitometry measurements of stained bands. However, one has to keep in mind that PAGE in natural gels separates proteins according to their charges and size, and therefore, the sum of these properties may not provide an absolute measure for identity of given proteins. This is particularly apparent when a dozen or more soluble proteins are contained in a sample. Under these conditions PAGE may not separate two proteins whose sum of charge and size is nearly identical, and consequently one may fail to identify the true vitellogenin. Also, non-specific proteins may mask the female specific vitellogenin. Keeping these considerations in mind, one may need to take a second look at reports which fail to identify sex specificity for vitellogenins in certain species (even though titre differences are apparent) (e.g. Tobe and Loughton, 1967; Laverdure, 1968; Baehr, 1974; Girardie, 1977). Furthermore, since the synthesis of vitellogenin in most species is JH dependent, the presence of small amounts of this protein after allatectomy or decapitation leads one to suspect that either surgical techniques or protein determinations may have been inadequate (Laverdure, 1970; Baehr, 1974). These words of caution only imply a certain reservation about the incomplete technical documentation but do not necessarily suggest that the reported data are invalid. Refined procedures sometimes clarified the case, as for example for *Oncopeltus* in which the use of a low percentage polyacrylamide gel allowed a better resolution and consequent identification of the vitellogenin (Kelly and Davenport, 1976). In any event, if immunodiffusion techniques are not used in conjunction with PAGE or cellulose acetate electrophoresis, the data obtained are highly suggestive, at best, but inconclusive.

In PAGE the same migration distance of certain proteins from different samples electrophoresed under identical conditions may denote identity. Indeed, this is how identity of vitellogenin and vitellin is often documented. In most cases the conclusions drawn are valid, yet one cannot be entirely certain, particularly when the gels contained many proteins. Furthermore, since vitellogenins are very large molecules, and since haemolymph and tissue samples may contain proteolytic enzymes, degradation of the proteins may occur relatively easily during preparation of the samples. Degradation can also occur under alkaline electrophoretic conditions, and consequently one may identify a degradation product rather than the true vitellogenin. Whether artifacts have been produced during preparation of the sample can only be recognized when comparisons are made with purified vitellogenin, since artifacts and native proteins may blend and become indistinguishable from one another. For example, an artifactual breakdown product was identified in PAGE for the vitellogenin and vitellin of *Leucophaea*. In this case the purified proteins most often yielded two bands on the gels, only one of them being the native molecule (Engelmann *et al.*, 1976). It had been clarified by molecular size estimation and immunodiffusion of the eluted bands that one of these bands is an artifact which does, however, crossreact with antivitellogenin (Fig. 2). Similarly, two major vitellin bands were produced

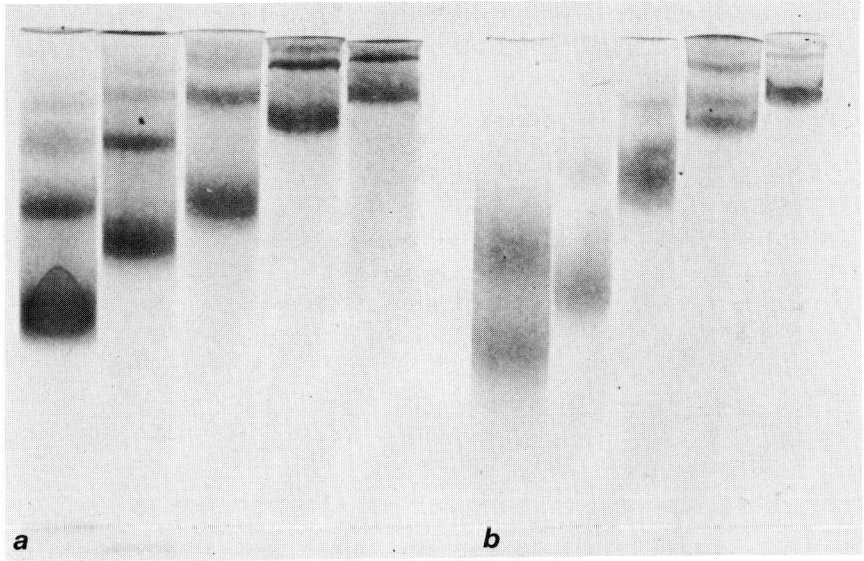

Fig. 2. Polyacrylamide gel electrophoresis of purified vitellin (*a*) and vitellogenin (*b*) on natural gels at concentrations ranging from 4 to 8%. Notice the variable degree of degradation in the various samples. The fastest moving band represents a polypeptide of about 2.8×10^5 daltons and is half the size of the native vitellogenin (*cf.* Engelmann *et al.*, 1976).

from *Triatoma* yolk proteins, and one of these bands had no counterpart in the haemolymph (Mundall and Engelmann, 1977); this second protein may be an artifact. While these considerations apply for *Leucophaea* and possibly for *Triatoma*, the identification of two vitellogenins in *Periplaneta americana* (Bell, 1969a, 1970) or *Byrsotria fumigata* (Barth and Bell, 1970) cannot necessarily be discounted as artifacts. In *Periplaneta* at least, PAGE and immunoelectrophoresis allowed a clear distinction of two vitellogenins (Fig. 3); in immunodiffusion the two antigens formed distinct spurs, showing non-identity (Bell, personal communication; Engelmann unpublished).

From the foregoing discussion it is clear that the most reliable method available for identification of the species' vitellogenins and vitellins is immunodiffusion, as already demonstrated in 1954 by Telfer for *Hyalophora*. In all cases known, vitellogenin and vitellin react identically against the anti-vitellogenin, thus demonstrating their identity as well as giving clues as to the origin of vitellogenin. Using this most stringent criterion, Table 1 shows that the definite documentation for vitellogenin is given only for six species of Dictyoptera, three Orthoptera, three Hemiptera, four Lepidoptera, two Coleoptera, and four Diptera.

While immunodiffusion techniques reliably document the identity of vitellogenin and vitellin of the same species, it also has to be shown that the anti-

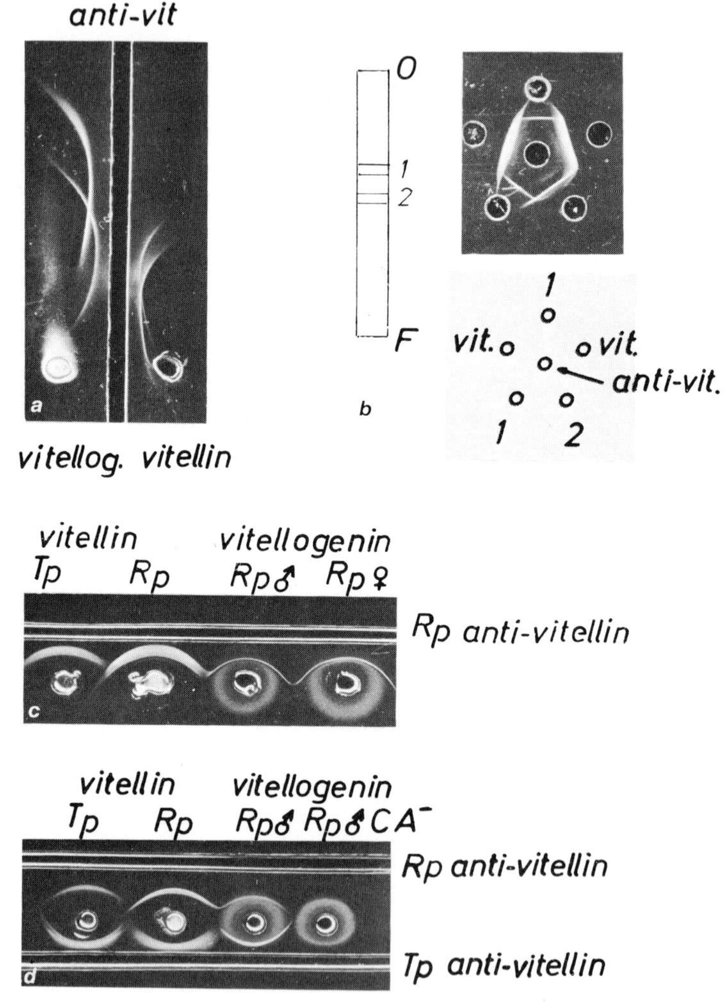

Fig. 3. Immunological identification of 2 vitellogenins and vitellins in *Periplaneta*. Both of the vitellogenins crossreact with anti-vitellin (*a* and *b*). In *b* is shown the non-identity of the 2 vitellogenins which have been eluted from polyacrylamide gels. *c* and *d* demonstrate the partial identity of vitellins from *Rhodnius* and *Triatoma* as well as the occurrence of vitellogenin in male haemolymph of *Rhodnius*. Following allatectomy of the male, vitellogenin could no longer be demonstrated in the haemolymph. Tp, *Triatoma protracta*; Rp, *Rhodnius prolixus*.

vitellogenin is species specific. For example, Telfer (1954) reported that *Antheraea* vitellogenin partially crossreacted with *Hyalophora* anti-vitellogenin, thus illustrating partial identity of the antigens. Partial identity of the vitellogenins from *Rhodnius* and *Triatoma* has also been demonstrated (Mundall, 1976). In this case anti-vitellogenins of either *Rhodnius* or *Triatoma* precipitated vitellogenin or vitellin of both species. In immunodiffusion assays one spur developed in either case, clearly showing only partial identity of the two vitellogenins (Fig. 3). For either of these two reports relatively closely related species were compared and, therefore, one can suspect that the respective vitellogenins are similar in their antigenicity.

2.3 VITELLOGENIN AND VITELLIN TITRES

Immunodiffusion techniques (Oudin's method on rates of antigen migration in agar) have been used initially to quantitate the changes in relative amounts of vitellogenin in the haemolymph of *Hyalophora* (Telfer and Williams, 1953; Telfer, 1954). In this species, vitellogenin (antigen 7) begins to appear at the larval/pupal moult, stays at a high titre throughout pupal diapause, but then decreases rapidly during adult development and still further after the imaginal moult (Fig. 4). The onset of decrease in titre coincides with beginning yolk deposition in the growing oocytes. Interestingly, vitellogenin is found in the haemolymph several months before vitellogenesis is initiated,

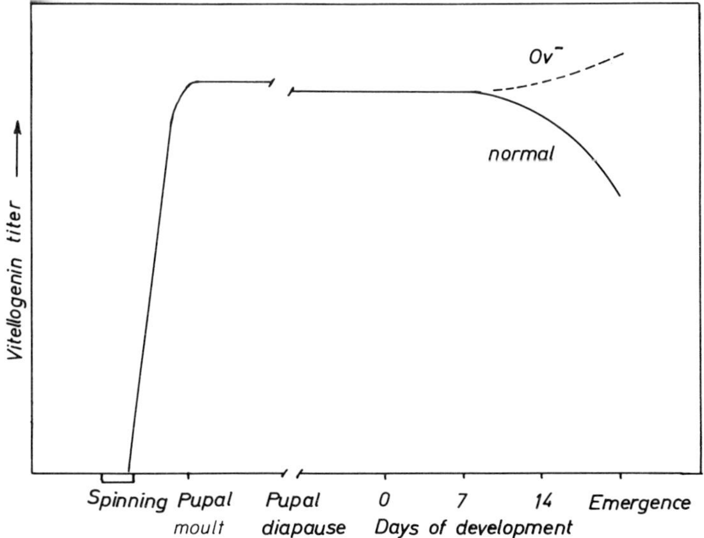

Fig. 4. Ontogeny of haemolymph vitellogenin titres in the silkmoth *Hyalophora* (modified from Telfer, 1954).

a fact which argues for a control mechanism of vitellogenin uptake that is operative during adult development and not before. Nothing is known concerning this control mechanism.

Similar immunodiffusion techniques were used for the studies on the relative and absolute vitellogenin titres in haemolymph of *Periplaneta* during a reproductive cycle (Bell, 1969b). While the total protein concentration in the haemolymph did not change dramatically, the concentration of the two vitellogenins combined measured about 0.2% during the height of vitellogenesis and about 0.65% during ootheca formation. Clearly vitellogenins make up only a small fraction of the total haemolymph proteins at any time. Analogous findings have been reported for the cockroach *Leucophaea* (Engelmann, 1978). In this latter case rocket immunoelectrophoresis has been used, and it was found that the haemolymph contained on the average 0.29% vitellogenin whereas total protein concentration was about 4.5% during the height of egg growth. According to Gavin and Williamson (1976a) haemolymph from females of *D. melanogaster* during vitellogenesis contains about two percent vitellogenin as determined by rocket immunoelectrophoresis, whereas Kambysellis (1977) reported only a 0.29% vitellogenin titre 24 hours after eclosion. Various sterile mutants of *Drosophila* contained similar or even higher amounts of vitellogenin in the haemolymph than normal wild type females (Kambysellis and Craddock, 1976). Large variations of vitellogenin titres may be found depending on the time of the animals' reproductive cycle in which estimations are made, and it may also depend on the animals' nutritional status. Discrepancies in the precise values for the same species determined by different laboratories may be easily explained in this fashion. The rather low vitellogenin titre during the most active phase of vitellogenesis – in spite of the high rate of vitellogenin synthesis – is obviously the result of rapid drainage of this molecule into the growing oocytes. When this drainage was prevented following ovariectomy a dramatic increase in the relative proportions of vitellogenin to total proteins did occur in *Hyalophora* (Telfer, 1954), *Periplaneta* (Bell, 1969b), and *Leucophaea* (Engelmann, 1978); further discussion concerning this aspect will be found below (section 4.5).

In contrast to the observations of a rather low vitellogenin titre in the haemolymph of the two cockroach species and *Drosophila* mentioned above is the honeybee *Apis mellifica* which is found to have an extremely high vitellogenin titre in the blood (Engels, 1972, 1974). Young egg-laying queens were found to have a vitellogenin content of 70% and non-laying workers up to 40–50% of the haemolymph proteins (Rutz and Lüscher, 1974; Engels, 1974). In stingless bees similar magnitudes of vitellogenin titres have been reported (Engels and Engels, 1977). In these latter cases titre determinations have been made by densitometry measurements of stained bands after cellulose acetate or polyacrylamide gel electrophoresis, techniques which appear to provide reproducible results. The functional significance of such high vitellogenin titres in the haemolymph of several bee species is obscure, but it may denote an extremely high rate of synthesis in view of the enormous reproductive potential of the honeybee queen.

The predominant yolk protein species is vitellin, as can readily be shown by the appropriate methods. For example, 88% of the yolk proteins of *Periplaneta* is vitellin (Bell 1969b), and for *Blattella* it is 93% (Oie *et al*., 1975); the same magnitude is found in *Leucophaea* (unpublished data). In several species of *Drosophila* the vitellin makes up between 50 and 80% of the yolk proteins (Gelti-Douka *et al*., 1974), and for *Aedes* 75% of the extractable proteins is reportedly vitellin (Hagedorn and Judson, 1972). These figures illustrate clearly that vitellogenin accumulates in the growing oocytes and must have been taken up against a concentration gradient. The mechanism by which this is achieved is not fully understood, but the known facts will be discussed below (section 7).

2.4 RATES OF VITELLOGENIN SYNTHESIS

Rates of synthesis of vitellogenin in normal and JH treated animals were determined for *Leucophaea* (Engelmann, 1971a) by immunoprecipitation assays. For this species dose response curves were established for JH treatments by this method (Fig. 5). Similar immunoprecipitation assays were used for rate studies in *Danaus plexippus* (Pan and Wyatt, 1976), *Locusta* (Chen *et al*., 1976), and *Triatoma protracta* (Mundall and Engelmann, 1977). It could be shown that immunoprecipitation of labelled vitellogenin is the most reliable approach for gathering information on rates of vitellogenin synthesis. It is obvious that the determinations of vitellogenin titres and rates of synthesis could be used for a maximum yield of information under normal and any experimental conditions. Unfortunately only limited use of these techniques in combination has been made to date for any species.

In summary we can conclude that the method of choice for identification and quantification of vitellogenin and vitellin is immunodiffusion or immunoprecipitation. Certainly, other methods such as PAGE are very useful tools for many aspects of studies on the yolk protein precursor and the final product. However, artifacts may be produced in an uncontrollable fashion during preparation of the samples and this may preclude their use for quantitative studies. Exhaustive controls have to be made in any particular case before application of these techniques become reliable tools.

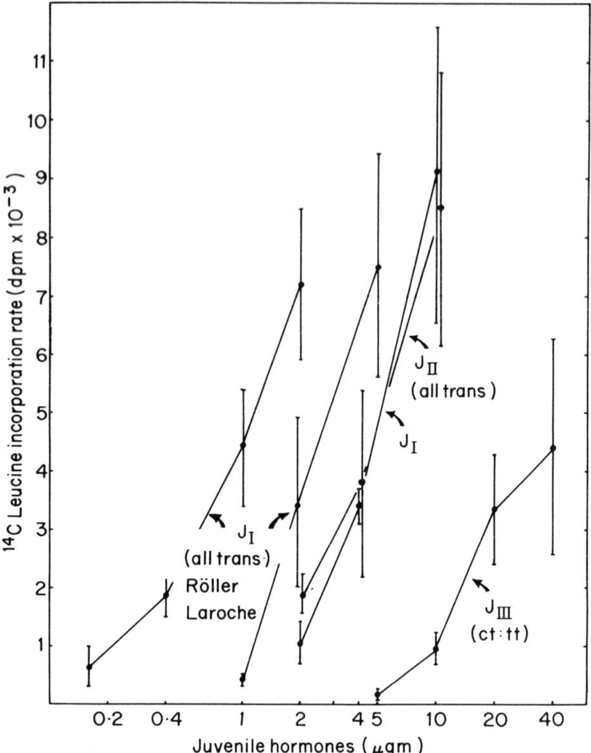

Fig. 5. Induction of vitellogenin in allatectomized females of *Leucophaea* following topical application of various JHs. Induction is measured as the rate of incorporation of ^{14}C-leucine into anti-vitellogenin precipitable vitellogenin.

3 Characterization of vitellogenin and vitellin

As pointed out above, vitellogenin and vitellin are immunologically identical in all species investigated. This is seen in Ouchterlony immunodiffusion assays or by immunoelectrophoresis. Identity is demonstrated by the absence of any spurs and complete fusion of the precipitation arcs (Figs 1, 3). To date purification of vitellogenin and vitellin has, however, been accomplished in only a few species and, consequently, comparisons can only be of tentative nature. Nevertheless it is useful to assemble the known facts (Table 2).

3.1 SOLUBILITY

For *Leucophaea* it has been known that vitellin is virtually insoluble in low ionic strength media (Dejmal and Brookes, 1968, 1972), whereas the haemolymph precursor, vitellogenin, is quite soluble (Engelmann *et al.*, 1976). This property of vitellin has been used to extract it from the eggs in rather pure form in species such as *Leptinotarsa* (De Loof and de Wilde, 1970a), *Acheta* (Schneider, 1973), *Blattella* (Oie *et al.*, 1975), *Nauphoeta cinerea* (Bühlmann, 1976), and *Locusta* (Cohen *et al.*, 1976). This procedure for vitellin isolation failed, however, in *Apis* (Engels, personal communication) and *Triatoma* (Mundall and Engelmann, 1977). At present it is unclear whether low solubility of vitellin in low ionic strength media can be found in many more species of different orders. Certainly, the fact that vitellogenin is quite soluble in low ionic strength solutions makes this procedure for isolation of vitellogenin unsuitable (Engelmann *et al.*, 1976). At present, nothing is known concerning the basis for the change in solubility properties. For *Leucophaea* one may speculate that it is caused by the modification of the subunit composition or it may happen in conjunction with the aggregation of the native vitellogenin molecules within the oocytes (Dejmal and Brookes, 1968, 1972). An aggregation of three 14S vitellogenin molecules to one 28S unit occurs shortly after uptake into the oocytes. The dynamics of aggregation were analysed by Brookes (1969) when he followed the flow of radiolabel over time from the 14S to the 28S unit employing a sucrose density gradient. It was found that the 28S unit is rather unstable and can be disaggregated by simple one time freezing or alkaline treatment (Koeppe and Ofengand, 1976a; Engelmann, unpublished). Under these conditions the molecules are irreversibly disaggregated to the 14S units. These units are immunologically indistinguishable from the native vitellogenin. Nothing is reported for other species on aggregations of vitellogenin molecules within the oocytes. As a matter of fact, no report has mentioned vitellogenin units larger than 16S for any species. Since vitellins of several species are known to be insoluble in low ionic strength media just as in *Leucophaea*, it is unlikely that aggregation of the native vitellogenin molecules is the basis for this property.

3.2 LIPID AND CARBOHYDRATE MOIETIES

With regard to the lipid content of vitellogenin, one observes a wide range between 6.9 and 15.7% (Table 2). Interestingly, the two cockroaches studied in detail exhibit the extremes in this respect, whereas the three Lepidoptera and *Locusta* have just about the same amount of lipid, 10%. Carbohydrate moieties range from 1 to 14%, and it is seen that vitellogenins of the two Lepidoptera and *Drosophila* contain the lowest amounts. Whether any significance can be attached to these findings on the relative amounts of carbohydrate moieties remains to be seen. The possibility exists that carbohydrates play a role in recognition of the vitellogenin molecules by the oocyte membranes. As a matter of fact, the vitellogenin of *Locusta* could be isolated from the haemolymph by affinity chromatography using concanavalin A which binds to the carbohydrate moiety of the molecule (Cohen *et al.*, 1976).

3.3 MOLECULAR WEIGHTS AND SUBUNIT COMPOSITION

Even though the overall physico-chemical characteristics of the known vitellogenins appear to be similar, comparisons cannot be made because only a few vitellogenin species have been adequately characterized. Furthermore, we do not have even the complete set of data for both vitellogenin and vitellin of a single insect species. It is seen, with two exceptions (Table 2), that the molecular weights of the native molecules of insect vitellogenin range between 5.0×10^5 and 6.5×10^5 daltons. These are very large molecules for which one may expect a complex assembly of subunits. In two *Drosophila* species vitellogenin and vitellin consist of one small subunit (Gavin and Williamson, 1976a; Kambysellis, 1977). It is not known how many of these make up the native vitellogenin. Two subunits, one large and one small, are found in the three Lepidopteran, one Dictyopteran, and one Dipteran species (Kunkel and Pan, 1976; Mundall and Law, 1977, Chino *et al.*, 1977; Atlas *et al.*, 1978). In none of these species is it known how many subunits build the vitellogenin since, as pointed out by Kunkel and Pan (1976), simple stoichiometry cannot be applied on the basis of densitometry of the bands seen on polyacrylamide gels. Vitellogenins of *Leucophaea* (Koeppe and Ofengand, 1976a) and *Locusta* (Chen *et al.*, 1976) appear to be made up of four and five subunits respectively. These subunits may have been derived from a large primary translation product by proteolytic cleavage during secretion into the haemolymph. This hypothesis is supported for *Locusta* by the observation that in a pulse chase experiment the radiolabel could be chased within 30 to 60 minutes into five "subunits". This observation points to the interesting phenomenon that the primary translation product may be processed before it reaches the haemolymph and oocytes. It also indicates that it may be easily degraded during isolation procedures, and it is therefore difficult to identify the actual translation product.

While it appears that the native translation product is being proteolytically processed upon secretion into the haemolymph, there may be even further modifications made during the pinocytotic uptake by the oocytes. This is suggested for *Leucophaea* in which vitellogenin is reportedly composed of four subunits whereas vitellin contains only three clearly identifiable subunits. The question may be raised whether we have here the first indication of a transformation of the native vitellogenin molecule into vitellin. Koeppe and Ofengand (1976a) postulated a proteolytic conversion of one of the subunits within the oocytes; this subunit then attains a molecular weight similar to one already present in vitellogenin. The two subunits – the original and the derived one – are no longer resolved by SDS polyacrylamide gel electrophoresis. An analysis of the amino acid composition or that of the peptide map of the various subunits may provide evidence for support of this hypothesis. Nothing is known for any other species about the modification of vitellogenins during uptake by the growing oocytes. The finding for *Leucophaea* argues most convincingly for the desirable distinction between vitellogenin and vitellin. Certainly the usage of these terms will find more and more support as additional results become available.

3.4 AMINO ACID COMPOSITION

So far we have seen that all known vitellogenins are macromolecules of glycolipoprotein properties and that in some species the derived molecules, the vitellins, are insoluble in low strength media. A survey of the amino acid composition of both vitellogenin and vitellin shows further common features (Table 3). They all have relatively high contents of aspartic acid, serine, glutamic acid, and leucine. This is, however, no distinctive feature which would set aside vitellogenins from other haemolymph proteins of insects or other large proteins in general (Kunkel and Pan, 1976). For two *Drosophila* species a very high glycine content was reported (Kambysellis, 1977); it was, however, noted that this may be an artifact due to the use of a glycine buffer during isolation of the vitellogenins. For *Leucophaea* it was shown that the protein contains a small amount of phosphoserine (Engelmann and Friedel, 1974). The amount is so small that the appearance of the small absorbancy peak in amino acid analysis may have been overlooked or discarded as an artifact were it not for the observation that practically all ^{32}P label was recovered from these fractions. No data are available for other species which would indicate that vitellogenins in general may be phosphoproteins, except for a casual reference to *Drosophila* (Kambysellis, 1977). In this latter case no data have been made available. Without radiolabelling, the small amounts of phosphorus would not be detectable; histochemical techniques are probably not reliable, because the lipid moiety of vitellogenin contains phospholipids. The proteins have to be extensively delipidized before determinations of the phosphorus content can be made.

Vitellogenins are taken up preferentially by the growing oocytes against a concentration gradient. Other haemolymph proteins are found in the oocytes in about the same proportion as in the haemolymph. The question arises then as to what properties of the vitellogenin molecule allow its preferential incorporation into the oocytes. It does not seem to be the unusual amino acid composition, the size of the molecule, or its relative abundance in the haemolymph. As a matter of fact, for the few species for which we have conclusive data, vitellogenin occurs in the haemolymph at a far lower

TABLE 2 Characteristics of vitellogenin and vitellin

Species	Method of purification	Native molecule MW × 10^5	Subunits MW × 10^5	Lipids %	Carbohydrates %	Authors
DICTYOPTERA:						
Leucophaea maderae						
Vitellogenin	QAE, PAGE	5.25				Engelmann *et al.* (1976)
Vitellin	QAE, PAGE	5.35				Engelmann *et al.* (1976)
Vitellin	H_2O ppt	5.59		6.9	8.3	Dejmal and Brookes (1972)
Vitellin	H_2O ppt PAGE		1.18 0.87 0.96 0.57			Koeppe and Ofengand (1976a)
Blattella germanica						
Vitellogenin	DEAE, sucrose gradient	6.59	1.00 0.52	15.7	4.5	Kunkel and Pan (1976)
ORTHOPTERA:						
Locusta migratoria						
Vitellin	DEAE	5.50 ± 0.40		9.6	14.0	Chen *et al.* (1976)
Vitellin	DEAE, PAGE	5.30 ± 0.30	1.30 1.20 1.00 0.65 0.55		11.0	Gellissen *et al.* (1976)
Vitellin	PAGE	5.70 ± 0.10				McGregor and Loughton (1974)
HEMIPTERA:						
Triatoma protracta						
Vitellin	PAGE	4.37				Mundall and Engelmann (1977)
Rhodnius prolixus	PAGE					
Vitellin	PAGE	4.60				Mundall (1976)

LEPIDOPTERA:							
Philosamia cynthia							
Vitellogenin	$(NH_4)_2SO_4$ DEAE, PAGE	5.00	2.50	0.55	10.0	2.5	Chino *et al.* (1969, 1976, 1977)
Hyalophora cecropia							
Vitellogenin	DEAE, sucrose gradient, PAGE	5.16	1.20	0.43		1.0	Kunkel and Pan (1976)
Vitellin	DEAE	5.09			9.4		Pan and Wallace (1974)
Manduca sexta							
Vitellogenin	DEAE, sucrose gradient, PAGE	2.60	1.80	0.50	12.0	3.0	Mundall and Law (1977)
DIPTERA:							
D. melanogaster							
Vitellin	DEAE, PAGE		0.49				Gavin and Williamson (1976a)
Vitellin	$(NH_4)_2SO_4$, PAGE		0.46				Kambysellis (1977)
D. virilis							
Vitellin	$(NH_4)_2SO_4$, PAGE		0.49				Kambysellis (1977)
Culex pipiens							
Vitellin	Biogel, PAGE	3.80	1.60	0.82			Atlas *et al.* (1978)

TABLE 3 Amino acid compositions of vitellogenin and vitellin (expressed as mole per cent)

		Leucophaea		*Blattella*	*Locusta*		*Hyalophora*	*Philosamia*		*Leptinotarsa*	*D. melanogaster*	*D. virilis*
	Vitellogenin[1]	Vitellin[1]	Vitellin[2]	Vitellogenin[3]	Vitellin[4]	Vitellin[5]	Vitellogenin[3]	Vitellogenin[6]	Vitellin[6]	Vitellogenin[7]	Vitellin[8]	Vitellin[8]
P Ser	0.26	0.39	–	–	–	–	–	–	–	–	–	–
Asp	14.72	14.67	15.41	12.18	11.13	10.4	9.68	10.2	10.2	9.97	10.1	7.8
Thr	5.63	5.51	5.09	5.97	4.82	4.9	5.15	5.2	5.2	5.36	5.6	4.5
Ser	8.13	7.37	6.92	9.55	7.57	8.8	9.57	7.8	7.7	9.70	11.8	13.0
Glu	11.27	10.56	11.63	10.97	12.71	12.7	14.32	15.7	15.6	12.81	11.7	11.5
Pro	4.74	4.51	4.83	4.47	6.03	7.9	4.42	5.4	5.4	5.19	3.8	2.5
Gly	3.39	3.19	3.46	3.07	5.10	5.2	4.39	4.6	4.7	5.17	18.1	30.2
Ala	5.69	5.18	5.55	4.97	8.46	7.8	7.50	7.3	7.2	5.50	8.2	7.2
Val	6.46	6.77	7.12	7.75	7.81	6.5	5.59	6.0	6.0	6.84	4.7	3.6
Met	1.79	1.99	2.02	2.82	1.70	1.3	1.96	2.1	2.1	2.52	1.1	0.5
Ile	4.54	4.85	5.22	4.74	5.10	3.9	4.45	4.8	4.3	5.68	2.9	2.0
Leu	8.96	9.43	9.27	8.69	10.24	9.4	6.02	6.4	6.5	7.05	5.8	4.7
Tyr	4.35	4.85	3.85	4.89	5.06	6.5	5.16	4.8	4.9	3.78	3.0	1.9
Phe	4.35	4.25	4.24	4.49	2.95	2.7	3.46	3.2	3.2	4.72	2.6	1.6
Lys	6.14	6.44	5.62	7.06	5.71	5.5	7.55	8.9	8.9	7.75	4.0	2.9
His	3.14	3.39	3.20	4.14	1.58	2.0	3.14	3.3	3.5	2.90	2.1	0.4
Arg	6.40	6.64	6.53	4.27	4.05	4.0	7.94	4.3	4.2	5.01	3.8	2.1
Cys/2	–	–	–	0.86	1.01	0.6	–	–	–	–	0.6	3.6
Try	–	–	–	0.79	2.19	–	0.62	–	–	–	–	–

[1] Engelmann and Friedel (1974)
[2] Dejmal and Brookes (1972)
[3] Kunkel and Pan (1976)
[4] Chen *et al.* (1976)
[5] Gellissen *et al.* (1976)
[6] Chino *et al.* (1976)
[7] De Loof and De Wilde (1970a)
[8] Kambysellis (1977)

concentration than other proteins. The few additional possibilities that can be discussed in this regard will be dealt with below (section 7).

4 Control of vitellogenin biosynthesis

Following the discovery of female specific proteins which make up the bulk of the egg proteins in *Hyalophora* (Telfer, 1954) it took more than ten years before the research began to focus on control mechanisms of synthesis of these macromolecules. It is apparent that in species in which juvenile hormone controls vitellogenesis that this hormone may also control vitellogenin synthesis. Interestingly, in *Hyalophora* in which the vitellogenin (antigen 7) was discovered, no endocrine or other means of control of either vitellogenin synthesis or vitellogenesis is known to date (Williams, 1952; Pan, 1977). The short lived *Hyalophora* moth emerges with a nearly complete complement of fully grown eggs; they have fully matured during adult development in the latter part of the pupal lifetime, at a time when the juvenile hormone titre is rather low. The same is presumably found in other species of moth, such as *Bombyx mori*. In still other species, like *Manduca sexta*, hormones may affect other phases of vitellogenesis but not synthesis of vitellogenin (Nijhout and Riddiford, 1974). In this chapter, then, we will deal primarily with the known control mechanisms that are amenable to experimental manipulation.

4.1 JUVENILE HORMONE

Probably the first indication of endocrine control of specific protein synthesis was implied in the finding that females of *Drosophila melanogaster* contained two unique antigens which were never seen in males (Fox and Yoon, 1958). This was shown by simple Ouchterlony immunodiffusion of extracts from whole males and females. No conclusions had been drawn from these findings and the implications are only seen in retrospect. A better foundation for implicating juvenile hormone in specific protein synthesis was given by Hill (1962) when he showed that haemolymph of vitellogenic females of *Schistocerca gregaria* contained a protein which was characterized by its slow electrophoretic mobility. This protein was never seen in males and non-vitellogenic females. Unfortunately Hill used starch gel electrophoresis which has an inherent poor resolution for many proteins. Subsequent to this, Coles (1964, 1965b) speculated that the corpora allata are involved in the production of yolk proteins in *Rhodnius*, because after decapitation these proteins were virtually absent from the haemolymph. Unfortunately, no adequate documentation for this speculation was given by Coles. As is readily apparent, these early reports could only be suggestive for endocrine control of synthesis of the specific yolk protein since, first of all, the applied electrophoretic methodology was inadequate and not specific enough for identification of specific proteins, and secondly, no controls, like specifically removing the corpora allata and their reimplantation, had been made at that time.

For *Leucophaea* it was then shown by immunoelectrophoresis that following allatectomy one antigen was absent from the haemolymph of females, the same antigen which is never present in haemolymph of males or normal non-vitellogenic females (Engelmann and Penney, 1966) (Fig. 1). This so-called female specific protein (vitellogenin) was always present in vitellogenic females and is the prominent yolk protein of the eggs. Reimplantation of active corpora allata or application of the recently identified juvenile hormone to allatectomized females was followed by the reappearance of the vitellogenin (Engelmann, 1969), as documented by immunoelectrophoresis. Similarly, the dependence of the female specific protein on active corpora allata was reported nearly simultaneously or shortly thereafter for several additional species from different insect orders (Table 4). It should be noted that in some of these reports the stringent requirements for reliable identification of the vitellogenin and vitellin were not met and, consequently, verification of the endocrine control of vitellogenin synthesis must be considered tentative. Also, in some cases the necessary surgical manipulations of the animals were rather crude or incomplete, such as decapitation, and resulting conclusions may therefore be later subject to modification. Be this as it may, for every instance of known JH controlled vitellogenesis, JH control of vitellogenin synthesis appears to be the major link in the sequence of events leading to fully grown eggs.

Probably the most convincing set of data on JH control of vitellogenin synthesis comes from the demonstration of a graded response of allatectomized *Leucophaea* females to increasing doses of applied JH (Engelmann, 1971a) (Fig. 5). Rate of synthesis of vitellogenin increased exponentially with increasing doses of any one of the three known JHs applied to allatectomized females of *Leucophaea*. Essentially similar findings were reported for *Danaus* (Pan and Wyatt, 1976) and *Nauphoeta* (Bühlmann, 1976). In *Locusta*, fat body tissues of JH treated animals were incubated *in vitro* in the presence of radiolabel, and then assayed for vitellogenin synthesis during a standard incubation time (Chen *et al.*, 1976). The same type of dose response was obtained as for *in vivo* synthesis in the other species mentioned. In all of these species the animals responded to the hormone treatment with *de novo* synthesis of one clearly identifiable protein.

In absolute terms, various JHs stimulated vitellogenin synthesis to different degrees, as illustrated, for example, for *Leucophaea* (Engelmann, 1971a). In this case JH_{III}, which presumably is the animals' own species of

TABLE 4 Juvenile hormone controlled vitellogenin synthesis

Species	Surgical procedures and treatment	Method of identification	Authors
DICTYOPTERA:			
Leucophaea maderae	CA$^-$, JH	Immunol.	Engelmann and Penney (1966) Engelmann (1969); Brookes (1969)
Periplaneta americana	CA$^-$, CA$^+$	Immunol.	Bell (1969a)
Byrsotria fumigata	Decap. JH	Immunol.	Bell and Barth (1970)
Nauphoeta cinerea	Decap. JH	Immunol.	Bühlmann (1976)
ORTHOPTERA:			
Schistocerca vaga	CA$^-$, JH	Immunol.	Engelmann *et al.* (1971)
Locusta migratoria	CA$^-$, JH	Immunol.	Chen *et al.* (1976)
HEMIPTERA:			
Rhodnius prolixus	CA$^-$, JH	PAGE	Baehr (1974)
Triatoma protracta	CA$^-$, JH	Immunol.	Mundall and Engelmann (1977)
Oncopeltus fasciatus	JHA	Immunol.	Kelly and Telfer (1977)
LEPIDOPTERA:			
Danaus plexippus	CA$^-$, JH	Immunol.	Pan and Wyatt (1971, 1976)
COLEOPTERA:			
Leptinotarsa decemlineata	CA$^-$, JH	Immunol.	De Loof (1969); De Loof and De Wilde (1970b)
HYMENOPTERA:			
Lasioglossum zephyrum	JH	Immunol.	Bell (1973)
Apis mellifica, worker	CA$^-$, JH	Immunol.	Imboden *et al.* (1976)
Bombus terrestris, worker	JH	PAGE	Röseler (1974, 1977)
DIPTERA:			
Sarcophaga bullata	CA$^-$, JH	Immunol.	Wilkens (1969)
Phormia regina	CA$^-$, JHA	Agarose electroph.	Mjeni and Morrison (1976)

CA$^-$, Allatectomy; JH, Juvenile hormone applied; JHA, Juvenile hormone analogue applied

hormone, was least effective in a bioassay which measured the rate of vitellogenin synthesis on the 5th day following topical application of the hormone; JH_I and JH_{II} were equipotent but were more effective than JH_{III}. Obviously, the mode of application and type of assay affect the conclusions drawn on the relative potencies of the various hormones. This is seen in the report on another cockroach species, namely, *Nauphoeta* (Lüscher and Lanzrein, 1976). Here, total amounts of vitellogenin produced within a standard time interval following JH injection (not topical application) were estimated by Ouchterlony immunodiffusion. Under these assay conditions it appeared that JH_{III} was the most potent of the hormones tested. The discrepancy on the relative potency of the hormones in the two cockroach species may be only apparent and simply the reflection of different techniques and criteria used. In either case we know nothing on rates of inactivation of the various hormones (these may be different after the different modes of application), and we know nothing on the rates of removal of the vitellogenin from the circulation through uptake by the oocytes (which may affect the total amount of vitellogenin in the haemolymph at any time). Furthermore, any one of the hormones may have different rates of release into the circulation, since in the one case the hormone has to penetrate the cuticle, and in the other, it is released at the surface of injected carrier oil droplets. These considerations clearly make it futile to attempt to judge the true relative potency of the three JHs with respect to induction of vitellogenin.

From this discussion it follows that more reliable data would be obtained by an *in vitro* culture of fat body tissues in the presence of JH. This is a difficult task, since fat bodies of cockroaches and some other insect species contain symbiotic bacteria, and therefore, long-term culture of healthy tissues is not easy to accomplish. In fact, no successful induction of vitellogenin by JH under *in vitro* organ culture conditions has been reported as yet. Long-term cultures will be necessary, since it has been shown that the first marginal sign of *in vivo* induction can be recognized only after about 24 h in *Leucophaea*, and it takes at least 48 h before a significant amount of vitellogenin is produced (Engelmann, 1971a).

With regard to the control of vitellogenin synthesis, some social bees pose an interesting case. Haemolymph of both queens and workers of *Apis* contains vitellogenin, yet in a queenright colony only the queen produces eggs (Engels, 1972; Rutz and Lüscher, 1973, 1974). One may assume that vitellogenin uptake by the oocytes in the workers is controlled by some yet unknown mechanism. In allatectomized workers vitellogenin could no longer be detected immunologically in the haemolymph (Rutz *et al.*, 1976). Application of JH_3 in quantities of 1–10 μg to allatectomized workers caused a reappearance of vitellogenin and JH treatment of normal workers increased the vitellogenin titre (Rutz *et al.*, 1976; Imboden *et al.*, 1976). Obviously, JH controls vitellogenin synthesis in the worker honeybee, which is not surprising since it was shown earlier (Lukoschus, 1956) that workers have large corpora allata, denoting JH production. In contrast to this observation are reports that allatectomized queens of *Apis* still contained vitellogenin at a high titre (even though somewhat lower than unoperated controls) and did even lay eggs when treated with CO_2 (Engels and Ramamurty, 1976). Repeated application of JH caused a restoration of vitellogenin titres to those observed in normal queens (Ramamurty and Engels, 1977). It furthermore was shown that allatectomized queens incorporated injected radiolabel into vitellogenin and that JH treatment only enhanced the rate of incorporation. These results led Engels and Ramamurty to speculate that JH affects vitellogenin synthesis somewhat, but it may not be the sole factor which controls *de novo* synthesis of this macromolecule. The possibility was mentioned that vitellogenin induction in *Apis* may be independent of JH. These authors failed, however, to consider that queens may obtain JH from the workers either directly or indirectly by food transfer. Transfer of JH from workers to queens could be determined by the use of labelled JH.

Workers of a more primitive social Hymenopteran, *Bombus terrestris*, produced vitellogenin only when queenless (Röseler, 1974, 1977). This was also manifested in egg production by queenless workers. Typically, titres of yolk proteins, identified by polyacrylamide gel electrophoresis, increased dramatically after removal of the queen. Interestingly, workers of queenright colonies synthesized vitellogenin after treatment with 1–4 μg JH, but no eggs were produced. Only when 8 μg or more of JH per animal was given were eggs made. Röseler (1977) also showed that queenright workers have small corpora allata and that the haemolymph contains a low titre of JH, whereas queenless workers attain JH titres five times that of queenright animals. The conclusion is obvious: in this species the queen normally causes suppression of JH production in the worker, and furthermore that JH is essential for vitellogenin induction.

The finding of JH controlled *de novo* synthesis of vitellogenin in insects of several orders opened up an area of research which may have broader implication for endocrinology than heretofore envisioned. The relative ease by which a specific protein can be identified and quantified, as well as the production of large quantities makes the search for the finer details of JH control of protein synthesis manageable.

For *Leucophaea* it was shown that application of actinomycin D did block the JH induced vitellogenin synthesis (Engelmann, 1971a), just as was later reported for *Aedes* (Fallon *et al.*, 1974). This finding was taken to mean that JH may control vitellogenin synthesis via transcription of specific mRNA. The effect of actinomycin D is, however, not specific enough to allow an unequivocal conclusion. Application of α-amanitin, a drug which is known to inhibit RNA polymerase activity, provided then more conclusive data. Within four hours after application of the drug (20 μg per 2 g animal) vitellogenin synthesis was inhibited by 50% in *Leucophaea*, and within 12 hours virtually no vitellogenin was synthesized on ergastoplasmic membranes anymore (Fig. 6) (Engelmann, 1976). The dose of α-amantin applied was tolerated by these animals, since they recovered within a few days and exhibited normal behaviour such as response to courting males. This observation argues very strongly in favour of JH control of transcriptional events in vitellogenin mRNA production. Very likely the same applies for other insect species in which we know that JH controls vitellogenin synthesis. Indeed α-amanitin injected in *Drosophila* within 12 hours after eclosion reduced vitellogenin synthesis significantly and curtailed egg production (Gavin and Williamson, 1976a).

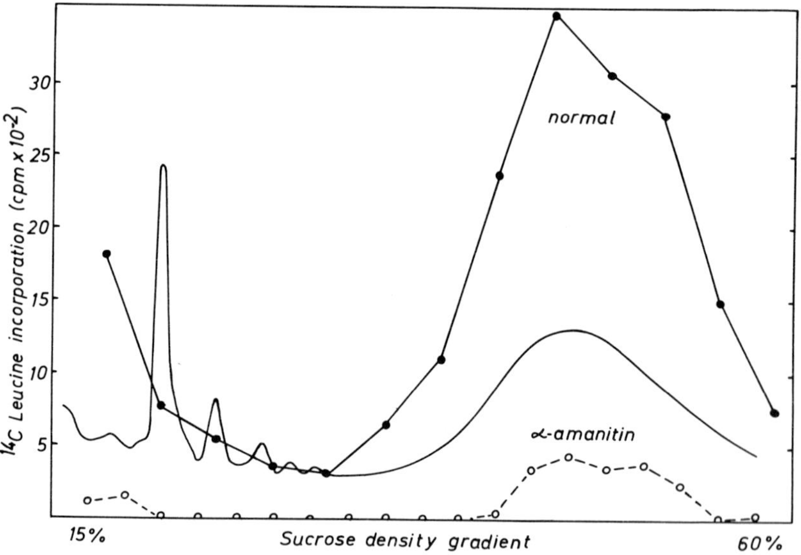

Fig. 6. Inhibition of vitellogenin synthesis on *Leucophaea* ergastoplasmic membranes after 12 h of α-amanitin treatment (20 μg per animal). The prominent absorbancy peak in the lower portion of the sucrose gradient represents the microsomes (Engelmann, 1976).

While the haemolymph of most insects contains rather low titres of vitellogenin during vitellogenesis (section 2.3), applied radiolabel enters mostly the vitellogenin molecule then. During the most active phase of vitellogenin production or after stimulation with high doses of JH, 80 to 90% of ^{14}C-leucine entered into the antibody precipitable vitellogenin in *Leucophaea* (Engelmann, 1971, 1978). Similarly, 80% of the protein label was identified in vitellogenin of *Locusta* (Baker-Grunwald and Applebaum, 1977), and 85% in that of *Apis* (Engels, 1974). These few reports on the intensity of vitellogenin synthesis, which undoubtedly could be expanded for other species, illustrate clearly that vitellogenin production must be one of the most active synthetic processes in the fat bodies of the adult female during times of vitellogenesis.

Synthesis of the complete native vitellogenin of the haemolymph may not be entirely under the control of JH in some species. This is, e.g. discussed for *Oncopeltus* in which an anti-vitellogenin precipitable vitellogenin was found during diapause when no eggs are produced (Kelly and Telfer, 1977). A second vitellogenin, precipitable by the same anti-vitellogenin, was produced after application of a JHA. This vitellogenin showed partial identity with the first one. The question then arises whether JH directs only the synthesis of one of the vitellogenins or perhaps only modifies in some way the first, which is only then taken up by the oocytes. One could perhaps argue that low JH titres during diapause cause synthesis of an incomplete vitellogenin which is not recognized by the oocytes. Also, oocyte uptake may require higher JH titres than vitellogenin synthesis. None of these questions has been answered as yet.

From all of these results the pertinent and most exciting question emerges: How does the JH actually affect transcription of the vitellogenin mRNA? The elucidation of hormone action at the molecular level in protein biosynthesis is certainly a multifaceted and difficult task in insect endocrinology, yet it is one of the most important aspects of research today. In this search the identification of the vitellogenin polysomes of the female fat bodies is one of the key elements for an understanding of the machinery of hormone induced protein synthesis. Judging from the size of the native vitellogenin and of that of the known subunits or primary translation products, one can expect to find a class of very large polysomes. Such large polysomes have been found to contain about 40 ribosomes in *Leucophaea* (Engelmann, 1977). From fat bodies of vitellogenic females a unique polysome profile was obtained with a prominent absorbancy peak in the lower portion of a sucrose density gradient (Fig. 7). This uniqueness is presumably the expression of the fact that in vitellogenic females 80% or more of protein synthesis in the fat bodies is identifiable as vitellogenin. Labelled nascent vitellogenin polypeptides could be precipitated with these large polysomes by anti-vitellogenin. Likewise, large amounts of adenosine labelled RNA were associated specifically with these polysomes. Since both nascent vitellogenin and adenosine labelled RNA could be precipitated by anti-vitellogenin in the same fractions, it is plausible that these fractions contained indeed the true vitellogenin polysomes. An artifactual aggregation of ribosomes or small polysomes is unlikely, because under identical procedural conditions male fat body tissues or those of non-vitellogenic females did not yield such large polysomes. The difficulties in obtaining the vitellogenin polysomes stem from the biochemical

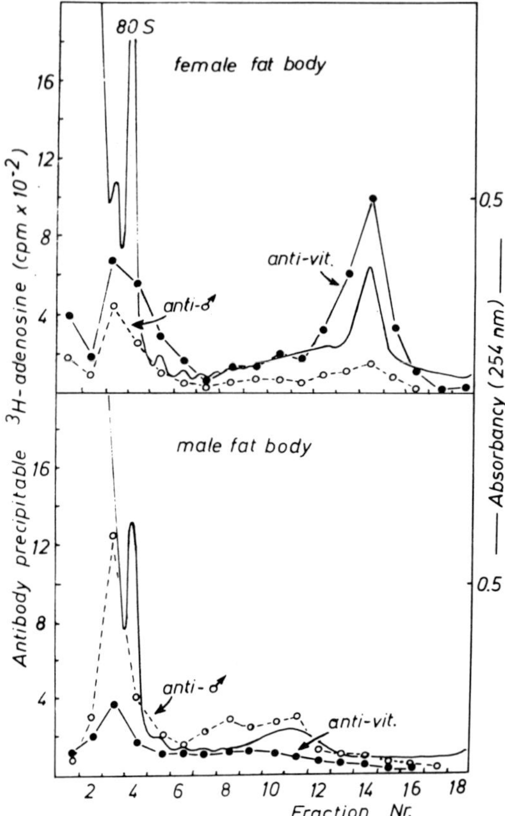

Fig. 7. Polysome profiles of fat bodies from vitellogenic females as well as males of *Leucophaea*. As shown, anti-vitellogenin precipitated adenosine labelled RNA primarily in the area of the large polysomes from female fat bodies. It could also be shown that nascent vitellogenin is present in the same fractions (Engelmann, 1977).

complexity of the fat body tissues. The abundance of ribonucleases in these tissues is probably one of the major reasons for the difficulties encountered in the past. Methods which were successfully used for vitellogenin polysome isolation in birds and amphibians were only marginally useful here (Engelmann, 1977). With this demonstration of insect vitellogenin polysomes it will now be possible to obtain the messenger in relatively pure form, since we can assume it must be contained nearly exclusively in these polysomes.

4.2 ECDYSONE

In 1971 Spielman and co-workers reported that injection of 10 μg of β-ecdysone into sugar fed *Aedes aegypti* was followed by vitellogenesis. This effect of ecdysone appeared to contrast the findings for several other insect species in which ecdysone administration inhibited vitellogenesis. The latter was clearly documented for two species of cockroaches (Engelmann, 1959, 1971b; Whitehead, 1974), beetles and houseflies (Robbins *et al.*, 1968), or a Lepidopteran (Herman and Baker, 1976). Furthermore, it was shown for *Leucophaea* (Engelmann, 1971b) that JH induced vitellogenin synthesis was curtailed when ecdysone was given simultaneously with JH. This latter observation appeared to show that ecdysone interfered with the primary action of JH. The finer details are, however, not known in this instance.

Aedes, the only species known in which ecdysone appeared to stimulate vitellogenin synthesis (Hagedorn *et al.*, 1973; Fallon *et al.*, 1974; Hagedorn, 1974), became an interesting case indeed, as shown by the rapidly increasing number of publications dealing with discoveries of ecdysone in adult insects. However, to date only the presence of ecdysone has been documented in most cases and attempts to demonstrate its specific effect on vitellogenin synthesis have been unsuccessful. In this discussion then we can limit ourselves to the one case, namely, *Aedes*. In this species newly synthesized vitellogenin from *in vitro* tissue cultures, after addition of ecdysone, was precipitated by its antibody, and the antigen-antibody precipitate then collected on Millipore filters for radioassay (Fallon *et al.*, 1974). In a dose response assay *in vitro* it was seen that ecdysone at a concentration of about 10^{-7} M caused the production of vitellogenin at half the maximal rate possible in this system (Hagedorn *et al.*, 1975). Unfed non-vitellogenic females contain about 80 to 100 pg of ecdysone, which corresponds to about a 10^{-7} M concentration; however, at this time no vitellogenin is produced by these animals. Approximately 20 hours after a blood meal the level of ecdysone in the whole animals rose to about 275 pg and 10 hours later had decreased again to the original level of unfed females. Interestingly, the fat body response in production of vitellogenin *in vitro* reached a maximum at 24 to 30 hours after a normal blood meal. Ovaries of normal females contained ecdysone, and it was shown that cultured ovaries from blood fed females released a certain amount of α-ecdysone into the medium, whereas those from unfed animals released only small quantities (Hagedorn *et al.*, 1975). In this context it is noteworthy that even ovariectomized females as well as males contained ecdysone in amounts of about 90 pg per animal (Schlaeger *et al.*, 1974). This is the level of ecdysone found in unfed females. The source of ecdysone or the tissue containing this steroid is not known in these cases.

In another insect species, namely *Locusta*, it was demonstrated that the ovaries indeed synthesize ecdysone. Synthesis took place only towards the termination of vitellogenesis, i.e. when vitellogenin production had nearly ceased (Lagueux *et al.*, 1977). For this species it was shown that ecdysone is presumably never released into the circulation and thus does not affect vitellogenin synthesis.

Interestingly, following emergence, fat bodies of *Aedes* appeared to respond to ecdysone only after an initial exposure to JH, since tissues taken from one to two day old allatectomized females could not be induced (Flanagan and Hagedorn, 1977). Normally, females feed for the first time several days after the imaginal molt. They begin to produce vitellogenin about 4 hours thereafter and peak production is observed 30 hours after the blood meal. Following ecdysone injection peak production of vitellogenin is seen a few hours earlier (Hagedorn, 1974).

From the available data on *Aedes* Hagedorn constructed a model in which he postulated that ecdysone (the ovarian hormone of earlier terminology) is produced and released into the circulation by the ovaries after a blood meal. The circulating steroid presumably stimulates the fat bodies to synthesize vitellogenin (Hagedorn, 1974). Since actinomycin D inhibited induction of vitellogenin synthesis (Fallon *et al.*, 1974) it was further postulated that ecdysone is primarily involved in transcriptional events, just as had been shown for JH in another insect species (Engelmann, 1971a). Upon close scrutiny of the data it is apparent that this model still lacks experimental evidence for several of its components. First of all, it remains to be seen whether the ovaries indeed release ecdysone *in vitro*, a fact which is crucial for the interpretation of the data. In addition, since ovariectomized females and males have about one-third the amount of ecdysone as fed normal females, it is clear that the ovaries are not the sole source of ecdysone. Surely this amount may be inadequate for stimulation of vitellogenin synthesis in these cases, and it may indeed be the 3-fold rise in ecdysone titre which is crucial for the observed effects. The question is then, are the ovaries the source of the necessary increased levels of ecdysone that can induce vitellogenin? With regard to the primary locus of action of ecdysone the reports of Fong and Fuchs (1976a, b) are of interest. They showed that α-amantin did not inhibit vitellogenin synthesis in *Aedes*, a finding which makes it likely that ecdysone influences post-transcriptional events rather than transcription of vitellogenin mRNA. The effect of ecdysone in *Aedes* is thus not comparable to that of JH in other species.

In addition to the above mentioned incomplete documentation for the model of how ecdysone may stimulate vitellogenin synthesis in *Aedes*, this work has come under severe criticism on technical grounds. Vitellogenin was always collected on Millipore filters after antibody precipitation. Since Millipore filters are known to adsorb proteins rather non-specifically, it is possible that non-specific proteins or even labelled amino acids have been adsorbed to the filter and then mistakenly identified as vitellogenin (Borovsky and van Handel, 1977). (See notes added in proofs, p. 418.)

How can one bring these findings on *Aedes* (assuming that vitellogenin was indeed synthesized under ecdysone control) into line with the facts that in several other species ecdysone inhibited rather than stimulated vitellogenesis and vitellogenin synthesis? Furthermore, in all species in which there is known hormonal control of vitellogenin synthesis it is the JH which controls this phase of vitellogenesis. It is noteworthy that ecdysone stimulated both female specific and general protein synthesis in fat body cultures of *Bombyx*, but egg specific proteins were not synthesized (Ono *et al.*, 1975). Similar findings were reported for *Danaus* (Herman and Baker, 1976) and *Locusta* (Hoffmann, personal communication). On the basis of these latter findings as well as the recent report that in *Aedes* JH is essential for "priming" the fat bodies (Flanagan and Hagedorn, 1977) before a blood meal or ecdysone response was demonstrable, one can formulate an alternative hypothesis: *Aedes* may be a unique case among insects in which both JH and ecdysone are involved in vitellogenesis. Vitellogenin messenger is possibly transcribed under the control of JH during the first few days of imaginal life of the female. This may be the "priming" of the fat bodies. Then, either a blood meal, providing the essential nutrient material for protein synthesis, or the artifactual application of relatively large doses of ecdysone will allow translation of the available vitellogenin messenger. This hypothesis finds support in the observation that α-amanitin did not curtail vitellogenin synthesis stimulated by injection of ecdysone, illustrating that vitellogenin messenger must have been available before the stimulus for vitellogenin synthesis was provided. For an *in vivo* effect rather large doses of ecdysone were required in *Aedes* (5 μg per female weighing 2 to 3 mg), but ecdysone at a concentration of 10^{-7} M was reported to stimulate vitellogenin synthesis in fat body tissue cultures. In contrast to this, as little as 40 μg of ecdysone inhibited the JH inducible vitellogenin synthesis in *Leucophaea* females weighing about 2 g (Engelmann, 1971b). Possibly, *Aedes* has more powerful ecdysone inactivating enzymes than *Leucophaea*, and therefore, very large doses were required for *in vivo* induction.

With the available information we do not have to invoke different mechanisms for induction of vitellogenin synthesis in *Aedes* compared to the remaining insect species in which endocrine control of vitellogenesis is proven. JH is presumably the essential stimulus for transcription of the vitellogenin messenger. Post-transcriptional regulation may be different in different species. In *Aedes*, ecdysone may be one of the agents which can turn on this translational machinery. Yet it has to be shown unequivocally that this is an event which takes place under normal conditions *in vivo*.

4.3 SECRETION OF VITELLOGENIN BY THE FAT BODIES

For some time before actual evidence was available, it was speculated that the fat bodies of the female are the site of vitellogenin synthesis. It was thought that vitellogenin is exported into the haemolymph from where it is taken up by the growing oocytes. The first reliable evidence for the correctness of this hypothesis was provided for the cockroaches *Periplaneta* (Pan *et al.*, 1969) and *Leucophaea* (Brookes, 1969; Engelmann, 1969) as well as for the moth *Hyalophora* (Pan *et al.*, 1969) and mosquito *Aedes* (Hagedorn and Judson, 1972). In these cases, newly synthesized vitellogenin was identified by immunoprecipitation or immunodiffusion of the proteins released from the fat bodies, or vitellogenin was characterized by sucrose density gradient centrifugation. Immunofluorescence techniques were also employed in the cockroach *Blattella* (Tanaka and Ishizaki, 1974) for visual documentation of the site of vitellogenin synthesis. The same applies for additional species.

The question then follows as to the mode of synthesis of the native vitellogenin molecules by the fat bodies of the vitellogenic female. Are there any cytological features which can be identified and which are associated with synthesis of vitellogenin? It was reported that fat body cells particularly of vitellogenic females of *Leptinotarsa* (De Loof and Lagasse, 1970) are rich in rough surfaced endoplasmic reticulum. Similar findings were documented for *Calliphora* (Thomsen and Thomsen, 1974), *Nauphoeta* (Wüest, 1975, 1976), and *Locusta* (Chen *et al.*, 1976; Lauverjat, 1977). These observations could be interpreted to mean that the rER is involved in vitellogenin synthesis. However, no direct proof for this hypothesis was provided by these electron microscopic studies; they were merely correlations possibly denoting a functional significance.

The majority of ergastoplasmic membranes of the fat bodies from vitellogenic females of *Leucophaea* are also of the rough type. Microsomes prepared from such fat bodies were practically all studded with numerous ribosomes and hardly any smooth membranes could be seen (Fig. 8) (Engelmann, 1974; Engelmann and Barajas, 1975). This observation presumably corresponds to the fact that during the most active phase of egg growth more than 80% of the newly synthesized fat body proteins is vitellogenin. Microsomes of *Leucophaea* contained the newly synthesized vitellogenin as shown by the fact that after dissolution of the membranes, vitellogenin could be precipitated with anti-vitellogenin (Engelmann, 1974; Engelmann and Barajas, 1975). Removal of all ribosomes and polysomes from the microsomes alone by simultaneous treatment of the preparation with puromycin (1 mM) and high KCl (500 mM) concentrations did not release the vitellogenin in any appreciable amounts. Puromycin is thought to terminate polypeptide

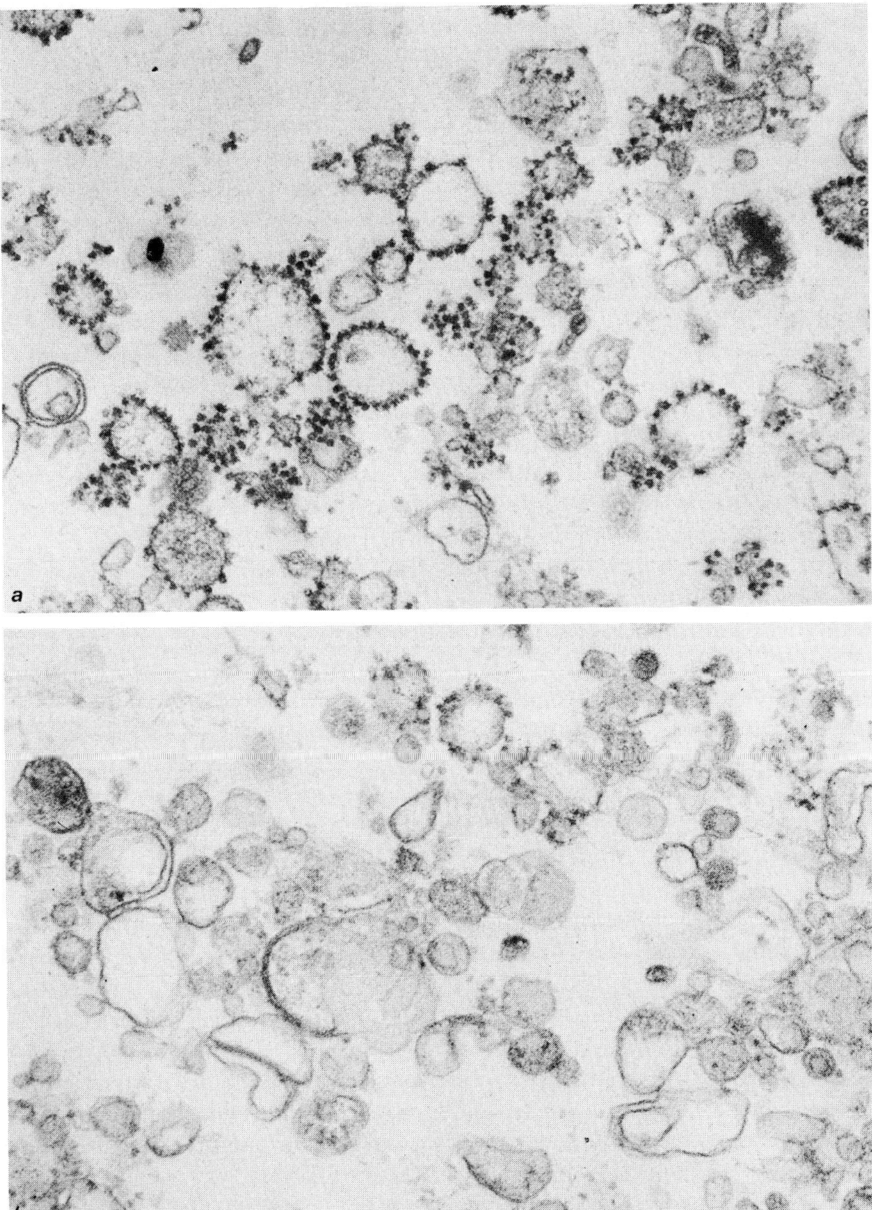

Fig. 8. Microsomes from fat bodies of vitellogenic (*a*) and allatectomized (*b*) females of *Leucophaea*. Notice the abundance of ribosomes and polysomes on membranes from vitellogenic females.

chain elongation and thus free the ribosomes which had been anchored by the nascent polypeptides to the membranes. In fat body cells of vitellogenic *Locusta* (Chen *et al.*, 1976), newly synthesized vitellogenin molecules were actually seen within the cisternae of the rER by electron microscopy.

These results then formed the basis for widening to insects the model for the mode of exportable protein secretion (Engelmann, 1974; Engelmann and Barajas, 1975), a model that had been formulated in definite terms for several vertebrate systems by Adelman *et al.* (1973) and Tata (1973). According to the model, active ribosomes (polysomes) are bound to ergastoplasmic membranes by two mechanisms, one that is sensitive to high concentrations of KCl and a second one which is provided by the nascent polypeptide chain while it is being vectorially secreted into the cisternae. There is at present no experimental evidence that would contradict the applicability of this model to the mode of secretion of insect vitellogenin. It will now be extremely interesting to further investigate the mode of assembly of the native molecule as well as that of the addition of lipid and carbohydrate moieties which are presumably added to the protein within the cisternae.

During research on vitellogenin synthesis in the fat bodies of *Locusta* an interesting finding emerged. Following a 10 minute radiopulse practically all anti-vitellogenin precipitable label of the fat body homogenates was contained in one polypeptide of 2.6×10^5 daltons (Chen *et al.*, 1976). It was furthermore seen that after a pulse chase from 30 minutes to three hours the label gradually moved into 5 subunits ranging from 1.4×10^5 to 0.52×10^5 daltons. These observations argue for proteolytic processing of the primary translation product within the fat bodies or during secretion into the haemolymph. If the primary translation product is indeed a 2.6×10^5 dalton peptide the native molecule of 5.5×10^5 daltons found in the haemolymph is presumably a dimer. The primary translation product of the vitellogenin messenger of other species may be of similar size as that seen for *Locusta*. This is supported by the finding that in *Leucophaea* the vitellogenin polysomes are rather large and composed of about 40 ribosomes (Engelmann, 1977). On the basis of the size of polysomes coding for known proteins one can extrapolate and arrive at a translation product for *Leucophaea* vitellogenin which may indeed be of comparable size to that shown for *Locusta*. It is however clear that we are manipulating the available data and extrapolating from them. Experimental evidence, on the other hand, can be provided to support the conclusion (Engelmann, unpublished).

For *Leucophaea* it is not known whether the primary translation product is processed before it reaches the haemolymph, as it appears to be the case in *Locusta*, or later within the haemolymph itself. Three, or possibly four, subunits of vitellin have been identified by SDS polyacrylamide gel electrophoresis (Koeppe and Ofengand, 1976a) and a simple stoichiometry has been applied to explain the size of the native molecule of about 5.6×10^5 daltons. From cultures of fat bodies of this species two antibody precipitable products of about 2.6 and 1.8×10^5 daltons were obtained, and it was speculated that these are the primary translation products which are processed and transformed into a total of four identifiable bands on SDS polyacrylamide gels (Koeppe and Ofengand, 1976b). Regardless of whether these findings will find confirmation or not, we certainly begin to find the basis for research on the fine details of vitellogenin synthesis and secretion in the fat bodies of insect species.

4.4 "IN VITRO" VITELLOGENIN SYNTHESIS

For any system in which one attempts to document endocrine controlled synthesis of a specific gene product it is essential that the identifiable product can be synthesized in a complete cell free incubation medium after the addition of the specific mRNA. This postulate naturally applies for the JH induced vitellogenin synthesis as well.

RNA extracted from fat bodies of vitellogenic females of *Locusta* or those which had been treated with a JH analog was added to a wheat germ system, and it was found that increased amounts of labelled proteins were produced that could be precipitated by anti-vitellogenin (Chen *et al.*, 1976). In several of the trials, antibody precipitable label was in the order of 10% of the trichloroacetic acid precipitable label. RNA extracted from males or allatectomized females did not produce anti-vitellogenin precipitable radioactive compounds. This finding could be taken to mean that vitellogenin had been produced during *in vitro* incubation. It should be noted, however, that after addition of *Drosophila* RNA or even rabbit globin mRNA locust anti-vitellogenin precipitated considerable amounts of radioactivity. In the absence of any further identification of the translation product by physico-chemical criteria one may therefore question whether the procedure truthfully identified locust vitellogenin.

Koeppe and Ofengand (1976b) reported for *Leucophaea* preliminary observations on vitellogenin synthesis, or that of its precursor, in a wheat germ translation system. The product could not be identified by either immunoprecipitation or PAGE. Apparently translation was incomplete. Since it was found for *Leucophaea* (Engelmann, 1974) and *Locusta* (Chen *et al.*, 1976) that vitellogenin is made *in vivo* on ergastoplasmic membranes, one may postulate that this is one of the essential components for the *in vitro* synthesis as well. Whether it is an absolutely necessary component could be debated. Support for the idea of the importance of membranes in translation of the vitellogenin mRNA is found in the observation that the use of microsomal membranes greatly enhanced the vitellogenin synthesis, at least the product that was precipitable by anti-vitellogenin (Engelmann, 1972, 1976). The addition of degranulated ergastoplasmic membranes, derived from either males or females, to the otherwise complete system significantly stimulated vitellogenin synthesis (Engelmann, 1976). Membranes from female fat body cells appeared to be more effective than those from male fat body cells. The translation product was precipitable by anti-vitellogenin but no further identification has been made. As in the cases cited above, bulk RNA was used in the *in vitro* protein synthesis. The identified vitellogenin messenger was not available for any of the reported *in vitro* studies.

At this time we do not have an unequivocal demonstration of *in vitro* synthesis of the complete vitellogenin for any insect species, since antibody precipitation of products made under the conditions described are highly suggestive evidence only and not conclusive.

4.5 EFFECT OF OVARIECTOMY

Vitellogenin, the major yolk protein precursor, is rapidly removed from the haemolymph by the growing oocytes during periods of vitellogenesis. As shown for *Periplaneta* (Bell, 1969b) and *Leucophaea* (Engelmann, 1978), vitellogenin concentrations in the haemolymph were of the order of 0.2 to 0.65%, which corresponds to between 4 and 10% of the haemolymph proteins. On the other hand, it is also reported that vitellogenin content of *Apis* haemolymph can reach up to 70% of the proteins. Levels of vitellogenin in circulation reflect the balance of rates of synthesis and rates of uptake by the oocytes, processes that may be controlled independently by hormones such as JH.

From these considerations the question naturally follows: Do the ovaries directly or indirectly influence vitellogenin production? This question, which has been asked many times before, gained renewed significance when it was shown that ovariectomy in *Aedes* was followed by a reduced rate of vitellogenin synthesis (Hagedorn and Fallon, 1973). Generally it had been shown for many species that vitellogenin titres in the haemolymph had increased after ovariectomy. This was, for example, documented for *Hyalophora* in which the haemolymph of ovariectomized adults had a five to ten times higher vitellogenin concentration than normal animals (Telfer, 1954). Similarly, vitellogenin concentrations had increased twofold in *Periplaneta* (Thomas and Nation, 1966; Bell, 1969a), *Leptinotarsa* (De Loof and De Wilde, 1970), and *Locusta* (Goltzené, 1977), and fivefold in *Nauphoeta* (Wilhelm and Lüscher, 1974). In *Leucophaea*, vitellogenin titres increased 20 to 40 times the normal concentration during a long-term observation period (Engelmann, 1978). In addition to these quantitative reports it was noticed that vitellogenin titres were high after ovariectomy in *Rhodnius* (Coles, 1964,

1965b), *Sarcophaga* (Wilkens, 1969), *Byrsotria* (Barth and Bell, 1970), *Pieris* (Lamy, 1970), *Bombyx* (Doira and Kawaguchi, 1972), *Triatoma* (Mundall and Engelmann, 1977), and others. All of these findings can be interpreted to mean that vitellogenin is produced unabated after ovariectomy and then accumulated in the haemolymph, because drainage into the oocytes no longer occurred. Since at the same time the corpora allata appear to remain active (Scharrer and von Harnack, 1961) and liberate JH (Tobe and Stay, 1977) one may assume that production of vitellogenin continues. This assumption was supported by the observation for *Nauphoeta* in which the extensive rER of fat body cells remained highly developed after ovariectomy (Wüest, 1975).

In none of the cases mentioned above were any data provided that would indeed give direct evidence for continued vitellogenin production. The conclusions drawn were based only on vitellogenin titre determinations in the haemolymph. A firm foundation and, in general, a confirmation for the accepted implication was then given for the cockroach *Leucophaea* (Engelmann, 1978). In this species the amount of vitellogenin of the haemolymph increased enormously in animals ovariectomized shortly after emergence, and reached a concentration of 5 to 8% within four weeks; this is 20 to 40 times higher than ever observed in normal vitellogenic females. At the same time that one recognizes this accumulation of vitellogenin it is also noted that the release of the newly synthesized protein had decreased to 50% of the original value within four weeks and further to 12% within the next five months. One could interpret this to denote a decreased synthesis of vitellogenin and/or release from the fat bodies. While incorporation of radiolabel into haemolymph vitellogenin was indeed reduced, no significant reduction of radiolabel was noticed in microsomal vitellogenin of the fat bodies, even after long-term ovariectomy (Engelmann, 1978). This was thought to indicate that synthesis was not appreciably reduced under these conditions, but release into the haemolymph was restrained. Apparently the accumulation of vitellogenin to about 65 to 75% of the total protein in the haemolymph caused a reduction of release. Vitellogenin consequently accumulated in enormous quantities within the fat bodies.

From these studies it is shown that monitoring either haemolymph vitellogenin titres or rates of release from the fat bodies alone could lead to misinterpretations. For *Leucophaea* at least, if only titres are monitored one could conclude a continued production whereas monitoring the labelling of haemolymph vitellogenin would lead to the conclusion that synthesis had been reduced to a low rate. In reality, no significant reduction of vitellogenin synthesis did occur as long as JH production was not restrained. The general conclusion therefore is that the ovaries do not exert any control over vitellogenin production, and any noticeable deviation from the normal vitellogenic female is based on secondary causes.

4.6 GENETIC CONTROL

The ultimate goal in studies of synthesis of specific gene products naturally is the identification of the gene and the ability to manipulate the coded information at will. Probably the first report which implied that the information on female specific proteins is contained on or controlled by genes on the X chromosome, but that the Y chromosome does not influence its expression was given by Fox and Yoon (1958) for *Drosophila*. According to this report a xx/y female produced two antigens which were also found in xx females, but not in males (x/y). The product appeared to be produced exclusively by the female. It is also shown that gynandromorphs of the cockroach *Byrsotria* with mixed abdomen produced vitellogenin (Barth and Bell, 1971). The same was demonstrated for gynandromorphs of the honeybee *Apis* (Engels *et al*., 1975). In this latter case, as long as the abdomen was substantially female, vitellogenin could be identified. This implied that the fat bodies must have the female genetic constitution for production of vitellogenin. Whether this is indeed the essential prerequisite for all insect species is questionable, since males of several species are said to contain at least small quantities of vitellogenin; this is shown, for example, for *Hyalophora* (Telfer, 1954), *Tenebrio* (Laverdure, 1968, 1972), *Rhodnius* (Mundall, 1976), or *Oncopeltus* (Kelly and Telfer, 1977).

Ultimately, the species which will lend itself for a genetic analysis of vitellogenin production will be *Drosophila*. A number of female sterile mutants of this species are known (King and Mohler, 1975; Kambysellis and Craddock, 1976). Sterility, i.e. no production of fully grown eggs, could be the result of several genetic lesions, such as lack of JH production, inability to produce vitellogenin, or various defects in the ovaries themselves. Several of these sterile mutants have been checked for the presence of vitellogenin, and it was observed that all had vitellogenin titres in the range, or even above that, of normal females (Kambysellis and Craddock, 1976; Kambysellis, 1977). Vitellogenin was, however, not incorporated, because the ovaries were either not fully differentiated or incapable of incorporating yolk for unknown reasons. Reportedly, in the sterile mutant ap^4 no eggs are matured (Postlethwait and Weiser, 1973; Gavin and Williamson, 1976b); yet following the application of the JH analog ZR 515, fully grown eggs were produced. Upon close scrutiny of these mutants it was shown that ap^4/ap^4 females contained vitellogenin (Gavin and Williamson, 1976b). These results could possibly mean that the JH titre is too low in these mutants to stimulate uptake of the vitellogenin into the oocytes and thus not allow complete vitellogenesis. The application of the JHA provides perhaps this added stimulus.

It would be rewarding to find mutants whose only genetic lesion is the lack of vitellogenin synthesis. For a genetic analysis of vitellogenin synthesis and its endocrine control this type of mutant would be essential.

5 Vitellogenesis in the male milieu

The very fact that only the female of the species contains a protein which is also the predominant yolk protein may allow us to speculate that ovaries transplanted into males may not incorporate yolk. However, as early as 1907 Meisenheimer found for the Lepidopteran *Porthetria* (= *Ocneria*) *dispar* that larval ovaries transplanted into male larvae produced fully grown and chorionated eggs after metamorphosis. This was confirmed by Kopeč (1912) for the same species. In these cases it is of interest to note that undifferentiated ovaries had been transplanted and only after differentiation of ovaries and genitalia during metamorphosis was full growth recognized. This occurred whether the hosts had been castrated or not. Telfer (1954) reported for *Hyalophora* that fully matured eggs were made by ovaries transplanted into males, except that these eggs were somewhat smaller than those produced by females. Likewise, eggs produced by *Antheraea polyphemus* males, following implantation of ovaries, weighed on the average only 2.5 mg, which is considerably less than those made by females (Telfer and Rutberg, 1960). It is possible then that the eggs did not attain the full weight in the male milieu because males are largely lacking the vitellogenin. This seemed a logical interpretation, since after transfusion with female blood the weight of the eggs now measured 3.6 mg, well within the range of normal, female produced eggs. No analysis of the nature of the egg proteins had been made then, just as no information is available on egg proteins produced by males of *Galleria* (Lender and Duveau, 1960) and *Bombyx* (Doira and Kawaguchi, 1972).

It is known that the haemolymph of males of certain Lepidoptera, such as *Hyalophora* (Telfer, 1954) or *Bombyx* (Doira and Kawaguchi, 1972), contained at least a low titre of vitellogenin, which may allow complete vitellogenesis to occur within the male milieu even though the process was slow and small-sized eggs were produced. However, upon analysis of the yolk proteins it became apparent that male produced eggs of *Pieris brassicae* (Lamy and Karlinsky, 1974) and those of *Bombyx* (Lamy and Julien-Laferriere, 1974) did not contain vitellogenin in any measurable amount, as shown by immunological techniques. This observation implied that vitellogenin may not be the necessary prerequisite for production of mature eggs in certain species.

Until recently all the species in which it was shown that eggs could grow to full size in the male milieu were members of the order Lepidoptera. It was then reported for *Drosophila* that implanted ovaries will mature eggs within the male of this species (Kambysellis, 1977). For *Drosophila* it was demonstrated that vitellogenin is being synthesized by the male and incorporated into the growing oocytes. Synthesis of vitellogenin in the male reportedly occurred only after implantation of ovaries (Kambysellis, 1977), a finding which may point to an interesting role of the ovaries in this species. The question then has to be answered in this connection whether the presence of the ovaries is also required in the female for vitellogenin production. It should be recalled that ovariectomy in other species, including some Diptera, did not curtail or inhibit vitellogenin synthesis (section 4.5), a finding which is at variance with that reported for *Aedes* (Hagedorn and Fallon, 1973).

In contrast to these reports for some Lepidoptera and *Drosophila*, ovaries transplanted into males of other species did not incorporate any yolk. In most of these cases it was also shown that the ovarian implants stayed healthy for long periods of time. This is seen in *Locusta migratoria manilensis* (Quo, 1959), *L. m. migratoriodes* (Vogel, 1968), *Nauphoeta* (Wilhelm and Lüscher, 1970), *Eublaberus posticus* (Bell and Barth, 1971), *Periplaneta* (Bell, 1972a), or *Tenebrio* (Laverdure, 1967, 1968). In any one of these cases one could argue that the complete absence of vitellogenin in the male was the reason for the failure to mature eggs. For *Leucophaea*, at least, this question was experimentally approached (Engelmann and Ladduwahetty, 1974; Engelmann, unpublished). In this species, repeated injections of vitellogenin in large quantities did not result in deposition of any yolk in the competent implanted ovaries. As a matter of fact, vitellogenin was immunologically no longer detectable in the male system on the 3rd day following the injection; vitellogenin was eliminated by the male. Implanted adult female fat bodies together with competent ovaries did not produce vitellogenin any longer within the male environment. The implanted female fat bodies became necrotic within a few days, whereas male controls or tissues of female nymphs did not. This latter result is certainly not the consequence of a low JH titre in the male, since the same was observed after additional implantation of active corpora allata or treatment with JH. The underlying cause for the elimination of the female components in the male milieu is obscure. In *Nauphoeta* males the additional supply of JH likewise did not induce any yolk deposition in the ovarian implants (Wilhelm and Lüscher, 1970). From the available information, which is restricted to a few species, one still could conclude that it is the unavailability of vitellogenin in the male that limits the incorporation of yolk into the implanted oocytes. This conclusion is perhaps only applicable to species outside the order of Lepidoptera.

In this context, a set of observations on *Rhodnius* gains significance. First of all, it could be shown immunologically that males of this species contain vitellogenin in high concentrations (Mundall, 1976) and that vitellogenin synthesis by the male is JH dependent. However, even in this species, implanted competent ovaries did not incorporate any yolk (Engelmann, unpublished). This was also seen for ovaries that had been implanted into male nymphs and which had metamorphosed with the host. These implants had completely healed in, had differentiated, and even showed rythmic pulsating movements as in normal animals. No trace of yolk was detectable in the implanted ovaries after several months and after several blood meals had been given to the adult host stimulating vitellogenin production; the haemolymph contained large quantities of vitellogenin. These data suggest that the male milieu lacks another component which, in this species, is essential for uptake of the vitellogenin by the growing oocytes. It is worth mentioning here that the males of the closely related blood sucking bug *Triatoma* do not have the vitellogenin of the female (Mundall and Engelmann, 1977).

In summary then, these observations on vitellogenesis in the male show us several interesting features. Males of the orders Dictyoptera, Orthoptera, Hemiptera, and Coleoptera apparently will not mature eggs within implanted ovaries regardless of the availability of vitellogenin and the high titres of JH. On the other hand, species of the order Lepidoptera can produce fully grown eggs within the male milieu even in the absence of vitellogenin which normally is the precursor for the predominant yolk protein in the female. Furthermore, for *Drosophila* and *Rhodnius* the conclusion must be that the males contain the genetic information for vitellogenin and the same is possibly true for several members of the Lepidoptera. From all of these reports it is also clear that males of most species studied cannot be used to serve as an "*in vivo* substrate" for a study of vitellogenin synthesis and vitellogenesis because transplanted oocytes will not incorporate vitellogenin.

6 Contribution of non-specific proteins to protein yolk

While it is obvious for the species studied that one or, at most, two female specific proteins make up the bulk (80 to 90%) of the protein yolk, it is also seen that other proteins contribute to the yolk. The question is whether these latter proteins are essential for making the egg. Very little attention has been given to these proteins, presumably because the issue is not as clear-cut as in the case of vitellogenin. Generally it is assumed that the so-called "non-specific" proteins are likewise derived from extraovarian sources. However, it is equally possible that proteins derived from the follicular epithelium contribute to the protein pool. Certainly the chorionic proteins are made by the epithelium, but how much of the epithelial proteins actually enter the oocytes is difficult to assess.

For several species of blood sucking insects it was recognized that foreign proteins, as well as the animal's own proteins, may enter the oocytes. Eggs of *Rhodnius*, *Triatoma infestans*, *T. brasiliensis*, *Cimex lectularius*, and *Pediculus humanus* appear reddish, suggesting that host haemoglobin was taken in (Wigglesworth, 1943). In these cases no quantitative estimation was made of how much haemoglobin contributed to the yolk. In *Hyalophora* two major proteins are taken up by the growing oocytes, antigen 7 and a carotenoid (Telfer, 1960). Both of these proteins are produced by the animal itself. Antigen 7 (vitellogenin) is taken up against a concentration gradient, and it is shown that the oocytes have a vitellin concentration about 28 times that of the haemolymph. On the other hand, the carotenoid is present in approximately the same proportions in haemolymph and oocytes. Similar observations have been made in other species, i.e. non-specific haemolymph proteins are taken up by the oocytes but they are not concentrated within them. This is, for example, shown by polyacrylamide gel electrophoresis or immunodiffusion techniques for *Periplaneta* (Nielsen and Mills, 1968), *Leucophaea* (Dejmal and Brookes, 1968; Scheurer, 1969), *Acheta* (Schneider, 1973), *Locusta* (McGregor and Loughton, 1974), or *Drosophila* (Kambysellis, 1977). For *Chironomus thummi* it is reported that residual amounts of larval haemoglobin, the animal's own product, are incorporated into the growing oocytes and thus removed from the circulation (Travis and Schin, 1976). Adults no longer produce haemoglobin.

When it was shown for several blood sucking insects that a foreign protein, i.e. host haemoglobin, was taken up by the oocytes it was highly suggestive for the mode of protein sequestration by the growing eggs. Oocytes of *Hyalophora* even incorporated injected proteins such as bovine gamma globulin, bovine serum albumin, ovalbumin, or lobster oocyte proteins, species of proteins the animal normally never encounters (Telfer, 1960). The amounts of the foreign proteins taken up were relatively small, yet the fact illustrated very well the basic principle that oocytes incorporate extraovarian proteins. In this context it is also interesting that the male may supply proteins during copulation which later end up in the oocytes. The only species for which this has been conclusively shown immunologically is the grasshopper *Melanoplus sanguinipes* (Friedel and Gillott, 1977). The amounts of protein transferred are presumably rather small and insignificant in terms of supply of basic nutrient material for the future embryo.

In all of this discussion the question is whether these proteins, which do not belong to the class "vitellogenin", represent a vital component for the eggs. It would be difficult to test this point, and therefore, it probably will remain a moot question. It is likely that the only essential protein is the vitellin, because it represents the bulk of the proteins and thus serves as the major source of amino acids during embryogenesis. Other haemolymph proteins are merely incorporated coincidentally together with the vitellogenin.

7 Uptake of haemolymph proteins by the oocytes

After it had been recognized that the majority of yolk proteins are derived from extraovarian sources and that, in particular, one specific protein is the major yolk protein precursor, it was, of course, interesting to gain more information on the details that lead to fully grown eggs.

7.1 THE MODE OF VITELLOGENIN ENTRY

One of the most direct demonstrations for the entry of vitellogenin into the oocytes was given for *Hyalophora* by treating frozen sections of ovarioles with fluorescin labelled anti-vitellogenin (Telfer, 1961; Telfer and Melius, 1963). It was seen that follicular epithelial cells separate and leave "gaps" through which the vitellogenin appeared to enter. The same fluorescence was also found associated with the yolk spheres within the oocytes. Vitellogenin does not pass through the epithelial cells themselves. This presumably allows a very fast passage of the yolk protein precursor from the haemolymph into the oocytes. Intercellular spaces between follicular epithelial cells were also seen during vitellogenesis in species such as *Bombus terrestris* (Hopkins and King, 1966), *Anagasta kühniella* (Cruickshank, 1971, 1972), *Leptinotarsa* (De Loof *et al.*, 1972), or *Rhodnius* (Davey and Huebner, 1974). For the latter species the degree of opening up of spaces between epithelial cells, termed patency, was taken as a criterion for the "readiness" of the follicle to incorporate vitellogenin. Undoubtedly the same can be documented for other species.

In conjunction with the formation of protein yolk, the oolemma is thrown into many folds and pits that contain the vitellogenin. These pits then pinch off and form membrane coated vesicles that coalesce with one another to form large yolk spheres. It was first illustrated for *Aedes* (Roth and Porter, 1962, 1964), *Lygaeus Kalmii* (Kessel and Beams, 1963), and *Periplaneta* (Anderson, 1964). The same pattern of micropinocytosis was found in additional species belonging to different orders, such as *Hyalophora* (Stay, 1965) or *Leptinotarsa* (De Loof *et al.*, 1972). In all of these species, demonstration of micropinocytosis was taken as evidence for uptake of extraovarian proteins by the oocytes. This visual documentation confirmed biochemical data that show the immunological identity of the haemolymph yolk protein precursor vitellogenin with vitellin, the major yolk protein (section 2.2).

Electron micrographs of the follicular epithelium in certain species reveal a well-developed rough-surfaced endoplasmic reticulum indicating an intense protein synthesis. Generally this is observed towards the termination of egg growth, i.e. during formation of the vitelline membranes and chorion. It is agreed that proteins used for these membranes are derived from the follicular epithelium. This applies, for example, in *Bombus* (Hopkins and King, 1966), *Leptinotarsa* (De Loof *et al.*, 1972), *Locusta* (Goltzené, 1977), as well as in species of other orders. While the necessary machinery for protein synthesis is best developed during the final stages of vitellogenesis, it was also shown by radiolabelling that, for example, in *Calliphora* and *Musca* RNA production is very intense during earlier stages of vitellogenesis, denoting a corresponding protein synthesis (Bier, 1963). Later it was shown for *Hyalophora* that the follicular epithelium produces a protein that is secreted into the intercellular spaces of the follicular epithelium during sequestration of vitellogenin (Telfer and Anderson, 1968; Anderson and Telfer, 1969). In this species, following a histidine pulse much of the label was found in autoradiographs over the cells themselves and in the spaces in between these cells. However, most of the ^{3}H-glucosamine label was seen autoradiographically over the gaps suggesting that the follicle protein is a glycoprotein. Pulse chase experiments showed that the label later entered the oocytes together with the vitellogenin and thus contributed to the yolk protein pool (Anderson and Telfer, 1970). This protein, distinct from vitellogenin, comprises about 5 to 10% of the extractable egg proteins (Telfer *et al.*, 1976). The term "paravitellogenin" was proposed. Isolated follicular epithelial cells of *Hyalophora* produced *in vitro* a protein that was similar to the one identified from *in vivo* observations (Bast and Telfer, 1976). It may be the "paravitellogenin". It had an apparent molecular weight of 55 000 daltons on SDS polyacrylamide gels.

Follicle cell proteins appear also to be released into the intercellular spaces in *Anagasta* (Cruickshank, 1971, 1972) and in *Periplaneta* (Bell and Sams, 1974). In all of these species a functional significance is suggested for the follicular protein. For *Hyalophora* it was shown that vitellogenin in the intercellular spaces was concentrated twofold compared to the haemolymph (Anderson and Telfer, 1970). It was also shown that extracted follicular protein had a higher affinity to female protein than to male proteins (Anderson, 1970). This suggested then that as part of the sequestration process vitellogenin binds to the "paravitellogenin" within the intercellular gaps, thus facilitating a following further concentration in the oocytes (Anderson and Telfer, 1970; Telfer and Smith, 1970). The precise mode of entry of the yolk precursor and the complex mechanisms that govern the movement of particular proteins against a concentration gradient are unclear. A few current findings and ideas may, nevertheless, be worth mentioning here. In *Hyalophora*, uptake of vitellogenin appeared to be dependent on the presence of Ca^{++} and Mg^{++} (Anderson and Telfer, 1970). The same was shown to be the case for *Locusta* (Wajc *et al.*, 1977). In the latter species it was also documented that injected serum albumin caused a reduction of vitellogenin uptake, presumably because it competed with the binding sites for vitellogenin. From this the conclusion could be drawn that binding of vitellogenin may be electrostatic in nature. Another potentially interesting property of vitellogenin was found when it was shown that *Locusta* vitellogenin could be isolated by affinity chromatography using concanavalin A (Cohen *et al.*, 1976). The question has been asked whether the carbohydrate moiety of vitellogenin plays a role in recognition by the oocyte membranes and thus facilitates the uptake. Whatever the exact mechanisms of vitellogenin uptake may be it is clear that in most species vitellogenin is the protein that triggers the uptake of protein by the oocytes. In *Culex*, as in other species, the phenomenon appears to be a receptor mediated energy requiring process (Roth *et al.*, 1976).

7.2 SPECIFICITY OF VITELLOGENIN UPTAKE

As shown many times for many insect species (Table 1), the eggs contain one or two predominant yolk proteins which originate in extraovarian tissues. These yolk proteins are taken up against a concentration gradient. The question then is, what is it that makes vitellogenin so specific that the follicles incorporate this species of protein in preference to others? In *Periplaneta* (Bell, 1969b) and *Leucophaea* (Engelmann, 1978) non-specific haemolymph proteins occur outside the ovary in much higher concentrations than vitellogenin, yet they are not preferred. Concentration of a specific protein therefore cannot be the reason for its preferential incorporation into the oocytes. The preferential uptake of vitellogenins does not, however, have to be restricted to the species own vitellogenin, as shown already by Telfer (1954) for *Antheraea* ovaries which incorporated *Hyalophora* vitellogenin in preference to other *Antheraea* proteins available. It is, of course, possible that this is a particular situation since *Antheraea* and *Hyalophora* may be considered somewhat related.

Ovaries from *D. melanogaster* or *D. simulans* transplanted into *D. annanassae* deposited yolk just as did those from *D. simulans* transplanted into *D. melanogaster* (Monod and Poulson, 1937). Ovaries from *D. annanassae* failed to mature eggs in the *D. melanogaster* milieu. Fully grown eggs were obtained in transplanted ovaries from *D. montana* in *D. virilis* hosts (Kambysellis, 1968). This type of research was expanded to 22 *Drosophila* species from which the ovaries were transplanted into either *D. grimshawi* or *D. hawaiiensis* (Kambysellis, 1970). In all but one of the combinations, fully grown eggs were obtained. For any one of these cases we probably can correctly assume that the transplanted ovaries incorporated the vitellogenin of the host; naturally, this is not known with absolute certainty. In a similar study, ovaries of 18 species of cockroaches were transplanted between species of various phylogenetic relationships (Bell, 1972b). It was generally found that yolk was deposited in ovarian transplants in cases where the ovaries had originated from members of the same subfamily. It was shown by immunodiffusion assays for some of these cases that the yolk indeed contained the host's vitellogenin. Implants between species of different subfamilies most often did not incorporate yolk and in some instances the implants had degenerated.

Vitellogenins of species of different orders presumably are not recognized by the foreign ovary and therefore are not incorporated at a higher rate than the non-specific species proteins. This was shown in a recent study in which competent ovaries had been exchanged between *Hyalophora* and *Blattella* (Kunkel and Pan, 1970, 1976). In an even more elegant *in vitro* experiment, ovaries of *Hyalophora* were incubated in media containing radiolabelled vitellogenins of both species. In this case essentially only *Hyalophora* vitellogenin was taken up by the *Hyalophora* ovary (Fig. 9).

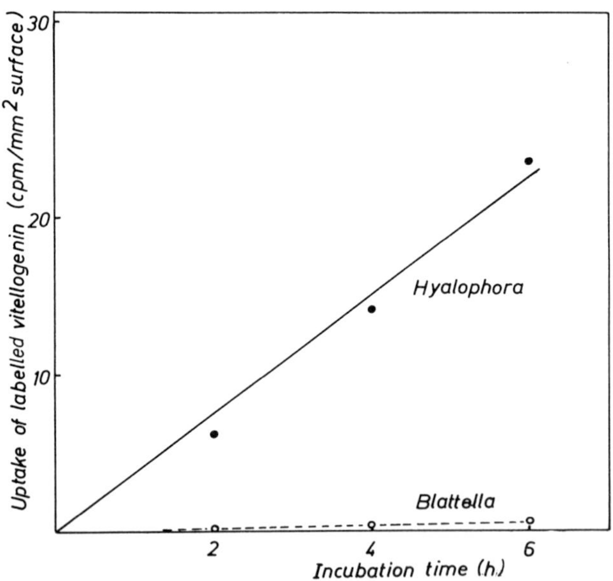

Fig. 9. Demonstration of the selective uptake of the species specific vitellogenin by *Hyalophora* ovaries *in vitro*. *Blattella* vitellogenin was not taken up by these ovaries. (From Kunkel and Pan, 1976.)

In contrast to all of these reports is the brief comment by Roth and Levitt (1976) pertaining to *Culex pipiens* in which chicken yolk protein competed successfully with mosquito vitellogenin for uptake by the oocytes on an equimolar basis. This finding may be recognized as an unusual exception, but it may also serve as a basis for the search for common molecular characteristics of both of these yolk proteins.

The sum of the available information certainly argues for the notion that the specific vitellogenin species must be recognized by the ovary. Presumably a receptor located on the oolemma must recognize the species specific or closely related vitellogenin. It is also possible that the follicle protein of a given species is involved in the recognition of the "own" vitellogenin. Any one of these possibilities is testable and data will come forth in due time that will shed more light on the details of vitellogenin recognition by the ovarian follicle. It is unclear what may actually be the recognition signal from the species' vitellogenin. All known vitellogenins are rather large molecules (Table 2), they all have carbohydrate moieties, and the amino acid compositions are not too dissimilar in species of the cockroach and flies (Table 3). Recognition by the species ovaries may be possible by subtle differences in the primary structure of the vitellogenins. In the light of this discussion we cannot ignore the fact that certain Lepidoptera produce fully grown eggs that do not contain vitellogenin or only traces of it. What may the mechanism of protein uptake be in these cases?

7.3 CONTROL MECHANISMS FOR VITELLOGENIN UPTAKE

Ovaries differentiate and attain the adult characteristics during metamorphosis. Just as other adult features of the insect such as cuticular patterns, wings, and genitalia, and ovarioles differentiate during a period in which ecdysone is acting in the absence of appreciable titres of JH. Differentiation of the ovarioles is essential for acquiring the potential of vitellogenin uptake. This was exemplified for *Periplaneta* in which it was shown that ovaries from last instar nymphs transplanted into adults did not incorporate the available vitellogenin (Bell and Sams, 1975).

It is generally found that prior to or simultaneously with the onset of vitellogenin uptake, follicular epithelial cells move apart opening up gaps between them. These intercellular spaces are presumably the channels through which the yolk precursor is taken up (Telfer, 1960; Telfer and Melius, 1963). In *Rhodnius* this so-called "patency", which appears to be under the control of JH, could only develop if the animals had a prior period of exposure to a low titre of JH which presumably causes further differentiation (Abu-Hakima and Davey, 1975); a blood meal then triggers an enhanced corpus allatum activity and patency of the ovarian follicle. Patency appears to be under the direct control of JH and is independent of the availability of vitellogenin for uptake by the oocytes, as shown for *Rhodnius* (Davey and Huebner, 1974; Abu-Hakima and Davey, 1977). Synthesis of proteins by the follicular epithelium apparently does not accompany the development of patency in this species.

Uptake of vitellogenin by the growing oocytes appears to be under control of the JH, as evidenced for *Periplaneta* when it was shown that injected vitellogenin was not incorporated into the oocytes unless JH was made available as well (Bell, 1969a). Presumably JH directs synthesis of the follicle protein (Bell and Sams, 1974). This protein may be essential for vitellogenin uptake (section 7.1). Circumstantial evidence for JH control of vitellogenin uptake was provided for *Oncopeltus* in which it was shown that short day, non-vitellogenic females had the vitellogenin, yet no oocyte incorporation was observed until the animals were subjected to long day conditions (Kelly and Davenport, 1976). Long day is known to cause activation of the corpora allata. An analogous situation was reported for the diapausing *D. grisea*, a cave-dwelling species, which did not produce eggs under short day light exposure, but appeared to contain vitellogenin nonetheless (Kambysellis and Heed, 1974); 14 hours of light induced vitellogenesis. No further details were made available for these species and it remains speculative how vitellogenin uptake by the oocytes is controlled. It could be that vitellogenin uptake requires high JH titres, levels of hormone which are not produced by the diapausing animals. Low JH titres may be adequate for vitellogenin synthesis.

From the rather scanty number of reports on control of vitellogenin uptake we cannot be confident that we know how uptake is effected. It is nevertheless tempting to propose that JH may control the synthesis of the follicle protein in those species for which we know that vitellogenesis is JH dependent. Unequivocal evidence for this is still lacking. If, however, it will be proven to be correct we will begin to understand the role of JH in vitellogenesis as an agent that has multiple sites of action, synergistically leading to an accelerated egg production.

Incorporation of vitellogenin can also be genetically controlled as was shown for the mutant sm in *Bombyx* (Kawaguchi and Doira, 1973). In this mutant vitellogenin is available in high titres, yet hardly any incorporation occurs. The genetic lesion is not fully understood. Similarly, several mutants of *Drosophila* are known to contain vitellogenin but do not produce eggs (Kambysellis and Craddock, 1976; Kambysellis, 1977). Certain of these mutants appear to have histologically normal and competent ovarioles. In these cases lack of vitellogenin incorporation may be caused by an inadequate JH titre, as shown by the fact that ap^4/ap^4 sterile mutants did produce eggs after treatment with the JH analogue ZR-515 (Postlethwait and Weiser, 1973), just as did the mutant 12–222 (Kambysellis and Craddock, 1976). It was speculated that some of the mutants may lack the "receptor" for vitellogenin in the ovaries (Kambysellis, 1977) or may not produce the follicle protein. No details are known concerning the actual basis for lack of egg production in the majority of the many sterile mutants of *Drosophila*. It is quite clear, however, that *Drosophila* is the species of choice for a study on genetic control of vitellogenin uptake just as it is for a study on control of vitellogenin synthesis.

8 A model for endocrine controlled vitellogenin synthesis and vitellogenesis

Since the original report by Wigglesworth (1936), who unequivocally documented endocrine control of reproduction in *Rhodnius*, the research in this area has, in recent years, expanded to molecular aspects of hormone action. While the overall control of egg maturation in insects by JH is well established for the majority of insect species, our knowledge of the precise details is still fragmentary (cf. Engelmann, 1970; Wyatt, 1972). In species for which JH control of vitellogenesis is established it appears that JH controls vitellogenin synthesis as well, and it may be thought that this is the single most important event leading to the production of fully grown eggs. It is true that synthesis of vitellogenin and its control can be most conveniently monitored by the use of immunodiffusion and radiolabelling techniques. Vitellogenin is also the most actively produced protein during vitellogenesis and consequently unequivocal data can be readily obtained. It is more difficult to establish the experimental procedures for a study of the so-called non-specific proteins which may also be under control of JH, but in this case hormones may only enhance the synthesis rather than initiate a *de novo* synthesis.

Having identified endocrine control of vitellogenin synthesis does not, however, imply that we know for certain how the hormone is involved. It is reasonable to assume that JH controls transcription of the vitellogenin messenger RNA. At the same time we realize that we have not even identified this molecule for any species of insect, let alone defined what we mean by control of transcription. This latter aspect undoubtedly will be the focus of research for the next few years. While we assume that JH controls mRNA transcription it may also affect concomittantly the processing of ribosomal RNA species. An increased proliferation of ergastoplasmic membranes during synthesis of vitellogenin is, in addition, controlled by JH in *Leucophaea* (dela-Cioppa and Engelmann, unpublished). This finding corroborated the observed increase in rER in the fat body cells of vitellogenic females of several species of insects. The fine details of the mechanism by which JH influences membrane proliferation are unclear at this time. Certainly, more membranes allow binding of more vitellogenin polysomes and consequently translation of more vitellogenin molecules. Possibly secretion of vitellogenin into the circulation is also controlled by JH. Circumstantial evidence for this is available for *Leucophaea* (Engelmann and Ladduwahetty, 1974). And last but not least, uptake of vitellogenin by the oocytes is very

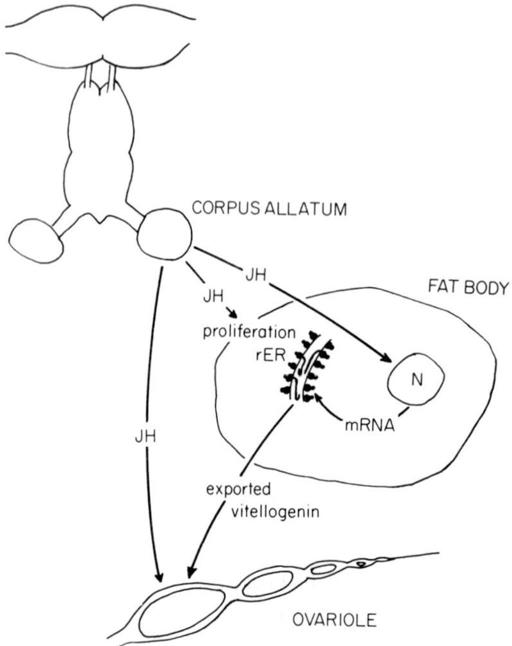

Fig. 10. Diagramatic representation of the actions of JH in conjunction with vitellogenin synthesis and vitellogenesis.

likely a process controlled by JH. In this case, the synthesis of a specific protein within the follicular epithelium is presumably controlled by JH. This may be part of the actual mode of hormone controlled vitellogenin uptake.

Taking all these events together, we see the multiplicity of steps in which JH may be the key hormone (Fig. 10). Nothing is known concretely on the involvement of additional hormones (presumably neurosecretory hormones) in vitellogenin synthesis and vitellogenesis. All of the steps, in which we identify JH as the important hormone, are important for vitellogenesis and synergistically enhance the rate of egg production. Some of these events, such as transcription of the vitellogenin mRNA or that of the follicle protein, are all or none processes controlled by the hormone. Others, such as enhanced production of ergastoplasmic membranes, only allow an increased rate of vitellogenin synthesis but may not be absolutely essential.

Acknowledgement

I thank Ms Judith Bradshaw for her help in the preparation of this review and the many useful suggestions made during many discussions.

References

Abu-Hakima, R. and Davey, K. D. (1975). Two actions of juvenile hormone on the follicle cells of *Rhodnius prolixus* Stål. *Can. J. Zool.* **53,** 1187–1188.

Abu-Hakima, R. and Davey, K. D. (1977). Effects of hormones and inhibitors of macromolecular synthesis on the follicle cells of *Rhodnius. J. Insect Physiol.* **23,** 913–917.

Adelman, M. R., Sabatini, D. D. and Blobel, G. (1973). Ribosome-membrane interaction. Nondestructive disassembly of rat liver rough microsomes into ribosomal and membranous components. *J. Cell Biol.* **56,** 206–229.

Anderson, E. (1964). Oocyte differentiation and vitellogenesis in the roach *Periplaneta americana. J. Cell Biol.* **20,** 131–155.

Anderson, L. M. (1970). Protein synthesis and incorporation by isolated *cecropia* oocytes. *Am. Zool.* **10,** 526.

Anderson, L. M. and Telfer, W. H. (1969). A follicle cell contribution to the yolk spheres of moth oocytes. *Tissue & Cell* **1,** 633–644.

Anderson, L. M. and Telfer, W. H. (1970). Extracellular concentrating of proteins in the *cecropia* moth follicle. *J. cell. Physiol.* **76,** 37–54.

Atlas, S. J., Roth, T. F. and Falcone, A. J. (1978). Purification and partial characterization of *Culex pipiens fatigans* yolk protein. *Insect Biochem.* **8,** 111–115.

Baehr, J. C. (1974). Contribution à l'étude des variations naturelles et expérimentales de la protéinémie chez les femelles de *Rhodnius prolixus* (Stål) Insecte Hemiptère Reduviidae. *Gen. comp. Endocr.* **22,** 146–153.

Baker-Grunwald, T. and Applebaum, S. W. (1977). A quantitative description of vitellogenesis in *Locusta migratoria migratorioides. J. Insect Physiol.* **23,** 259–263.

Barth, Jr., R. H. and Bell, W. J. (1970). Physiology of the reproductive cycle in the cockroach *Byrsotria fumigata* (Guerin). *Biol. Bull. mar. biol. Lab., Woods Hole,* **139,** 117 160.

Barth, R. H. and Bell, W. J. (1971). Reproductive physiology and behavior of *Byrsotria fumigata* gynandromorphs (Orthoptera (Dictyoptera): Blaberidae). *Ann. ent. Soc. Am.* **64,** 874–879.

Bar-Zev, A., Wajc, E., Cohen, E., Sapir, L., Applebaum, S. W. and Emmerich, H. (1975). Vitellogenin accumulation in the fat body and haemolymph of *Locusta migratoria* in relation to egg maturation. *J. Insect Physiol.* **21,** 1257–1263.

Bast, R. E. and Telfer, W. H. (1976). Follicle cell protein synthesis and its contribution to the yolk of the *cecropia* moth oocyte. *Devl. Biol.* **52,** 83–97.

Bell, W. J. (1969a). Dual role of juvenile hormone in the control of yolk formation in *Periplaneta americana. J. Insect Physiol.* **15,** 1279–1290.

Bell, W. J. (1969b). Continuous and rhythmic reproductive cycle observed in *Periplaneta americana* (L.). *Biol. Bull. mar. biol. Lab., Woods Hole,* **137,** 239–249.

Bell, W. J. (1970). Demonstration and characterization of two vitellogenic blood proteins in *Periplaneta americana:* an immunological analysis. *J. Insect Physiol.* **16,** 291–299.

Bell, W. J. (1972a). Transplantation of ovaries into male *Periplaneta americana:* effects on vitellogenesis and vitellogenin secretion. *J. Insect Physiol.* **18,** 851–855.

Bell, W. J. (1972b). Yolk formation by transplanted cockroach oocytes. *J. exp. Zool.* **181,** 41–48.

Bell, W. J. (1973). Factors controlling initiation of vitellogenesis in a primitively social bee, *Lasioglossum zephyrum* (Hymenoptera: Halictidae). *Insectes Sociaux,* **20,** 253–260.

Bell, W. J. and Barth, Jr., R. H. (1970). Quantitative effects of juvenile hormone on reproduction in the cockroach *Byrsotria fumigata*. *J. Insect Physiol.* **16,** 2303–2312.

Bell, W. J. and Barth, Jr., R. H. (1971). Initiation of yolk deposition by juvenile hormone. *Nature, Lond.* **230,** 220–222.

Bell, W. J. and Sams, G. R. (1974). Factors promoting vitellogenic competence and yolk deposition in the cockroach ovary: the post-ecdysis female. *J. Insect Physiol.* **20,** 2475–2485.

Bell, W. J. and Sams, G. R. (1975). Factors promoting vitellogenic competence and yolk deposition in the cockroach ovary: Larval-adult transition. *J. Insect Physiol.* **21,** 173–180.

Bielinska, M. and Piechowska, M. J. (1971). The protein in haemolymph and homogenates of eggs of the hawk-moth *Celerio euphorbiae* L. (Lepidoptera). *Bull. Acad. Pol. Sci.* **19,** 441–444.

Bier, K. (1963) Autoradiographische Untersuchungen über die Leistungen des Follikelepithels and der Nährzellen bei der Dotterbildung und Eiweissynthese im Fliegenovar. *Arch. Entw. Mech. Org.* **154,** 552–575.

Blumenfeld, M. and Schneiderman, H. A. (1968). Effect of juvenile hormone on the synthesis and accumulation of a sex-limited blood protein in the *Polyphemus* silkmoth. *Biol. Bull. mar. biol. Lab., Woods Hole,* **135,** 466–475.

Bodnaryk, R. P. and Morrison, P. E. (1966). The relationship between nutrition, haemolymph proteins, and ovarian development in *Musca domestica* L. *J. Insect Pysiol.* **12,** 963–976.

Bodnaryk, R. P. and Morrison, P. E. (1968). Immunochemical analysis of the origin of a sex-specific accumulated blood protein in female houseflies. *J. Insect Physiol.* **14,** 1141–1146.

Borovsky, D. and van Handel, E. (1977). A specific radioimmunassay for vitellogenesis in mosquitoes. *J. Insect Physiol.* **23,** 655–658.

Brookes, V. J. (1969). The induction of yolk protein synthesis in the fat body of an insect, *Leucophaea maderae*, by an analog of the juvenile hormone. *Devl. Biol.* **20,** 459–471.

Brookes, V. J. and Dejmal, R. K. (1968). Yolk protein: structural changes during vitellogenesis in the cockroach *Leucophaea maderae. Science, N.Y.* **160,** 999–1001.

Bühlmann, G. (1974). Vitellogenin in adulten Weibchen der Schabe *Nauphoeta cinerea.* Immunologische Untersuchungen über Herkunft und Einbau. *Rev. Suisse Zool.* **81,** 642–647.

Bühlmann, G. (1976). Haemolymph vitellogenin, juvenile hormone, and oocyte growth in the adult cockroach *Nauphoeta cinerea* during first pre-oviposition period. *J. Insect Physiol.* **22,** 1101–1110.

Chao, J. (1969). Agar gel immundiffusion and immunoelectrophoretic studies with the extracts of male and female adult *Culex tarsalis. Mosquito News,* **29,** 432–435.

Chen, T. T., Couble, P., De Lucca, F. L. and Wyatt, G. R. (1976). Juvenile hormone control of vitellogenin synthesis in *Locusta migratoria. In* "The Juvenile Hormones" (Ed. L. I. Gilbert), pp. 505–529. Plenum Press, New York and London.

Chino, H., Murakami, S. and Harashima, K. (1969). Diglyceride-carrying lipoproteins in insect hemolymph. Isolation, purification and properties. *Biochim. biophys. Acta,* **176,** 1–26.

Chino, H., Yamagata, M. and Sato, S. (1977). Further characterization of Lepidopteran vitellogenin from haemolymph and mature eggs. *Insect Biochem.* **7,** 125–131.

Chino, H., Yamagata, M. and Takahashi, K. (1976). Isolation and characterization of insect vitellogenin. Its identity with hemolymph lipoprotein II. *Biochim. biophys. Acta,* **441,** 349–353.

Cohen, E., Emmerich, H. and Applebaum, S. S. (1976). Vitellogenic glycoproteins in the migratory locust *Locusta migratoria*. *Coll. internat. C.N.R.S.* **251,** 465–473.
Coles, G. C. (1964). Some effects of decapitation on metabolism in *Rhodnius prolixus* Stål. *Nature, Lond.* **203,** 323.
Coles, G. C. (1965a). Haemolymph proteins and yolk formation in *Rhodnius prolixus* Stål. *J. exp. Biol.* **43,** 425–431.
Coles, G. C. (1965b). Studies on the hormonal control of metabolism in *Rhodnius prolixus* Stål. – I. The adult. *J. Insect Physiol.* **11,** 1325–1330.
Cruickshank, W. J. (1971). Follicle cell protein synthesis in moth oocytes. *J. Insect Physiol.* **17,** 217–232.
Cruickshank, W. J. (1972). Ultrastructural modifications in the follicle cells and egg membranes during development of flour moth oocytes. *J. Insect Physiol.* **18,** 485–498.
Davey, K. G. and Huebner, E. (1974). The response of the follicle cells of *Rhodnius prolixus* to juvenile hormone and antigonadotropin *in vitro*. *Can. J. Zool.* **52,** 1407–1412.
Dejmal, R. K. and Brookes, V. J. (1968). Solubility and electrophoretic properties of ovarian protein of the cockroach, *Leucophaea maderae*. *J. Insect Physiol.* **14,** 371–381.
Dejmal, R. K. and Brookes, V. J. (1972). Insect lipovitellin. Chemical and physical characteristics of yolk protein from the ovaries of *Leucophaea maderae*. *J. biol. Chem.* **247,** 869–874.
Doira, H. and Kawaguchi, Y. (1972). Changes in haemolymph and egg protein by the castration and implantation of the ovary in *Bombyx mori*. *J. Fac. Agr. Kyushu Univ.* **17,** 119–127.
Dufour, D., Taskar, S. P. and Perron, J. M. (1970). Ontogenesis of a female specific protein from the locust, *Schistocerca gregaria*. *J. Insect Physiol.* **16,** 1369–1377.
Engelmann, F. (1959). Über die Wirkung implantierter Prothoraxdrüsen im adulten Weibchen von *Leucophaea maderae* (Blattaria). *Z. vergl. Physiol.* **41,** 456–470.
Engelmann, F. (1965). Endocrine activation of protein metabolism during egg maturation in an insect. *Am. Zool.* **5,** 673.
Engelmann, F. (1969). Female specific protein: biosynthesis controlled by corpus allatum in *Leucophaea maderae*. *Science, N.Y.* **165,** 407–409.
Engelmann, F. (1970). "The Physiology of Insect Reproduction". Pergamon Press, Oxford and New York.
Engelmann, F. (1971a). Juvenile hormone-controlled synthesis of female-specific protein in the cockroach *Leucophaea maderae*. *Archs Biochem. Biophys.* **145,** 439–447.
Engelmann, F. (1971b). Endocrine control of insect reproduction, a possible basis for insect control. *Acta Phytopathol. Acad. Sci. Hung.* **6,** 211–217.
Engelmann, F. (1972) Juvenile hormone-induced RNA and specific protein synthesis in an adult insect. *Gen. comp. Endocr.* (Suppl.) **3,** 168–173.
Engelmann, F. (1974). Juvenile hormone induction of the insect yolk protein precursor. *Am. Zool.* **14,** 1195–1206.
Engelmann, F. (1976). Induction of insect vitellogenin *in vivo* and *in vitro*. In "The Juvenile Hormones" (Ed. L. I. Gilbert), pp. 470–485. Plenum Press, New York and London.
Engelmann, F. (1977). Undegraded vitellogenin polysomes from female insect fat bodies. *Biochem. biophys. Res. Commun.* **78,** 641–647.
Engelmann, F. (1978). Synthesis of vitellogenin after long-term ovariectomy in a cockroach. *Insect Biochem.* **8,** 149–154.

Engelmann, F. and Barajas, L. (1975). Ribosome-membrane association in fat body tissue from reproductively active females of *Leucophaea maderae*. *Exp. Cell Res.* **92,** 102–110.

Engelmann, F. and Friedel, T. (1974). Insect yolk protein precursor: A juvenile hormone induced phosphoprotein. *Life Sci.* **14,** 587–594.

Engelmann, F., Friedel, T. and Ladduwahetty, M. (1976). The native vitellogenin of the cockroach *Leucophaea maderea*. *Insect Biochem.* **6,** 211–220.

Engelmann, F., Hill, L. and Wilkens, J. L. (1971). Juvenile hormone control of female specific protein synthesis in *Leucophaea maderae*, *Schistocerca vaga*, and *Sarcophaga bullata*. *J. Insect Physiol.* **17,** 2179–2191.

Engelmann, F. and Ladduwahetty, M. (1974). The female specific protein, an extra-ovarian protein essential for egg maturation in insects. *Zool. Jahrb. Physiol.* **78,** 289–300.

Engelmann, F. and Penney, D. (1966). Studies on the endocrine control of metabolism in *Leucophaea maderae* (Blattaria). I. The hemolymph proteins during egg maturation. *Gen. comp. Endocr.*. **7,** 314–325.

Engels, W. (1972). Quantitative Untersuchungen zum Dotterprotein-Haushalt der Honigbiene (*Apis mellifica*). *Arch. EntwMech. Org.* **171,** 55–86.

Engels, W. (1974). Occurrence and significance of vitellogenins in female castes of social Hymenoptera. *Am. Zool.* **14,** 1229–1237.

Engels, W. and Engels, E. (1977). Vitellogenin und Fertilität bei stachellosen Bienen. *Insectes Sociaux*, **24,** 71–94.

Engels, W., Fahrenhorst, H., Drescher, W. and Möbius, K. (1975). Weibchen-spezifische Haemolymph Proteine by Gynandern von *Apis mellifica*. *Arch. EntwMech. Org.* **178,** 79–88.

Engels, W. and Ramamurty, P. S. (1976). Initiation of oogenesis in allatectomized virgin honey bee queens by carbon dioxide treatment. *J. Insect Physiol.* **22,** 1427–1432.

Fallon, A. M., Hagedorn, H. H., Wyatt, G. R. and Laufer, H. (1974). Activation of vitellogenin synthesis in the mosquito *Aedes aegypti* by ecdysone. *J. Insect Physiol.* **20,** 1815–1823.

Flanagan, T. R. and Hagedorn, H. H. (1977). Vitellogenin synthesis in the mosquito: the role of juvenile hormone in the development of responsiveness to ecdysone. *Physiol. Entomol.* **2,** 173–178.

Fong, W. F. and Fuchs, M. S. (1976a). The long term effect of α-amanitin on RNA synthesis in adult female *Aedes aegypti*. *Insect Biochem.* **6,** 123–130.

Fong, W. F. and Fuchs, M. S. (1976b). The differential effect of RNA synthesis inhibitors on ecdysterone induced ovarian development in mosquitoes. *J. Insect Physiol.* **22,** 1493–1499.

Fox, A. S. and Yoon, S. B. (1958). Antigenetic differences between males and females in *Drosophila* not attributable to the Y chromosome. *Transpl. Bull.* **5,** 52–55.

Friedel, T. and Gillott, C. (1977). Contribution of male-produced proteins to vitellogenesis in *Melanoplus sanguinipes*. *J. Insect Physiol.* **23,** 145–151.

Gavin, J. A. and Williamson, J. H. (1976a). Synthesis and deposition of yolk protein in adult *Drosophila melanogaster*. *J. Insect Physiol.* **22,** 1457–1464.

Gavin, J. A. and Williamson, J. H. (1976b). Juvenile hormone-induced vitellogenesis in apterous[4], a non-vitellogenic mutant in *Drosophila melanogaster*. *J. Insect Physiol.* **22,** 1737–1742.

Gellissen, G., Wajc, E., Cohen, E., Emmerich, H., Applebaum, S. W. and Flossdorf, J. (1976). Purification and properties of oocyte vitellin from the migratory locust. *J. comp. Physiol.* **B 108,** 287–301.

Gelti-Douka, H., Gingeras, T. R. and Kambysellis, M. P. (1974). Yolk proteins in *Drosophila*: Identification and site of synthesis. *J. exp. Zool.* **187,** 167–172.
Gingeras, T. R., Gelti-Douka, H. and Kambysellis, M. P. (1973). Yolk proteins in *Drosophila. Drosophila Inform. Serv.* **50,** 58.
Girardie, J. (1977). Contrôle neuroendocrine des protéines sanguines vitellogenes d'*Anacridium aegyptium* sain et parasité. *J. Insect Physiol.* **23,** 569–577.
Goltzené, F. (1977). Contribution a l'étude de l'ovogenèse chez *Locusta migratoria* L. (Orthoptère)). Thèse, Université Louis Pasteur, Strasbourg.
Hagedorn, H. H. (1974). The control of vitellogenesis in the mosquito, *Aedes aegypti. Am. Zool.* **14,** 1207–1217.
Hagedorn, H. H. and Fallon, A. M. (1973). Ovarian control of vitellogenin synthesis by the fat body in *Aedes aegypti. Nature, Lond.* **244,** 103–105.
Hagedorn, H. H. and Judson, C. J. (1972). Purification and site of synthesis of *Aedes aegypti* yolk protein. *J. exp. Zool.* **182,** 367–378.
Hagedorn, H. H., Fallon, A. M. and H. Laufer, H. (1973). Vitellogenin synthesis by the fat body of the mosquito *Aedes aegypti*: Evidence for transcriptional control. *Devl. Biol.* **31,** 285–294.
Hagedorn, H. H., O'Connor, J. D., Fuchs, M. S., Sage, B., Schlaeger, D. A. and Bohm, M. K. (1975). The ovary as a source of α-ecdysone in an adult mosquito. *Proc. natn. Acad. Sci. U.S.A.* **72,** 3255–3259.
Herman, W. S. and Baker, J. F. (1976). Ecdysterone antagonism, mimicry, and synergism of juvenile hormone action on the monarch butterfly reproductive tract. *J. Insect Physiol.* **22,** 643–648.
Herman, W. S. and Bennett, D. C. (1975). Regulation of oogenesis, female specific protein production and male and female reproductive gland development by juvenile hormone in the butterfly, *Nymphalis antiopa. J. Insect Physiol.* **99,** 331–338.
Hill, L. (1962). Neurosecretory control of haemolymph protein concentration during ovarian development in the desert locust. *J. Insect Physiol.* **8,** 609–619.
Hopkins, C. R. and King, P. E. (1966). An electron-microscopical and histochemical study of the oocyte periphery in *Bombus terrestris* during vitellogenesis. *J. Cell Sci.* **1,** 201–216.
Imboden, H., Wille, H., Gerig, L. and Lüscher, M. (1976). Die Vitellogeninsynthese bei der Bienen-Arbeiterin (*Apis mellifera*) und ihre Abhängigkeit von Juvenilhormone (JH). *Rev. Suisse Zool.* **83,** 928–933.
Kambysellis, M. P. (1968). Interspecific transplantation as a tool for indicating phylogenetic relationship. *Proc. natn. Acad. Sci. U.S.A.* **59,** 1166–1172.
Kambysellis, M. P. (1970). Compatibility in insect tissue transplantations. I. Ovarian transplantations and hybrid formation between *Drosophila* species endemic to Hawaii. *J. exp. Zool.* **175,** 169–180.
Kambysellis, M. P. (1977). Genetic and hormonal regulation of vitellogenesis in *Drosophila. Am. Zool.* **17,** 535–549.
Kambysellis, M. P. and Craddock, E. M. (1976). Genetic analysis of vitellogenesis in *Drosophila. Genetics*, **83,** 38.
Kambysellis, M. P. and Heed, W. B. (1974). Juvenile hormone induces ovarian development in diapausing cave-dwelling *Drosophila* species. *J. Insect Physiol.* **20,** 1779–1786.
Kawaguchi, Y. and Doira, H. (1973). Gene-controlled incorporation of haemolymph protein into the ovaries of *Bombyx mori. J. Insect Physiol.* **19,** 2083–2096.
Kelly, T. J. and Davenport, R. (1976). Juvenile hormone induced ovarian uptake of a female-specific blood protein in *Oncopeltus fasciatus. J. Insect Physiol.* **22,** 1381–1393.

Kelly, T. J. and Telfer, W. H. (1977). Antigenic and electrophoretic variants of vitellogenin in *Oncopeltus* blood and their control by juvenile hormone. *Devl. Biol.* **61,** 58–61.

Kessel, R. G. and Beams, H. W. (1963). Micropinocytosis and yolk formation in oocytes of the small milkweed bug. *Exp. Cell Res.* **30,** 440–443.

King, R. C. and Mohler, J. D. (1975). The genetic analysis of oogenesis in *Drosophila melanogaster*. *In* "Handbook of Genetics" (Ed. R. C. King), **3,** pp. 757–791. Plenum Press, New York and London.

Koeppe, J. and Ofengand, J. (1976a). Juvenile hormone-induced biosynthesis of vitellogenin in *Leucophaea maderae*. *Archs Biochem. Biophys.* **173,** 100–113.

Koeppe, J. and Ofengand, J. (1976b). Juvenile hormone induced biosynthesis of vitellogenin in *Leucophaea maderae* from large precursor polypeptides. *In* "The Juvenile Hormones" (Ed. L. I. Gilbert), pp. 486–504. Plenum Press, New York and London.

Kopeč, S. (1912). Untersuchengen über Kastration and Transplantation bei Schmetterlingen. *Arch. EntwMech. Org.* **33,** 1–116.

Kunkel, J. G. and Pan, M. L. (1970). Insect vitellogenins: Comparative properties and specific uptake. *Am. Zool.* **10,** 513.

Kunkel, J. G. and Pan, M. L. (1976). Selectivity of yolk protein uptake: Comparison of vitellogenins of two insects. *J. Insect Physiol.* **22,** 809–818.

Kunz, W. and Petzelt, C. (1970). Synthese und Einlagerung der Dotterproteine bei *Gryllus domesticus*. *J. Insect Physiol.* **16,** 941–947.

Küthe, H. W. (1972). Veränderungen im Proteinmuster während Oogenese und Frühentwicklung von *Dermestes frischi* (Coleoptera). *J. Insect Physiol.* **18,** 2411–2423.

Lagueux, M., Hirn, M. and Hoffmann, J. (1977). Ecdysone during ovarian development in *Locusta migratoria*. *J. Insect Physiol.* **23,** 109–119.

Lamy, M. (1967). Mise en évidence, par électrophorèse sur acétate de cellulose, d'une protéine vitellogène dans l'hémolymphe de l'imago femelle de la piéride de chou (*Pieris brassicae* L.). *C. r. hebd. Seanc. Acad. Sci., Paris,* **265,** 990–993.

Lamy, M. (1970). Les protéines impliquées dans la vitellogenèse des Lépidoptères. *Ann. Endocr.* **31,** 485–503.

Lamy, M. (1973). Les protéines de l'hémolymphe et du vitellus de l'œuf de la processionnaire de chêne, *Thaumetopoea pityocampa,* Schiff. (Lépidoptère, Notodontidae). *C. r. hebd. Seanc. Acad. Sci., Paris,* **277,** 2557–2560.

Lamy, M. and Julien-Laferriere, N. (1974). Vitellogenèse protéique en milieu mâle chez le ver à soie *Bombyx mori* L. (Lépidoptères). *C. r. hebd. Acad. Sci., Paris,* **279,** 343–346.

Lamy, M. and Karlinsky, A. (1974). Vitellogenèse protéique au milieu mâle chez la piéride du chou, *Pieris brassicae* L. (Lépidoptère). *C. r. hebd. Acad. Sci., Paris,* **278,** 91–94.

Lauverjat, S. (1977). L'évolution post-imaginal du tissue adipeux femelle de *Locusta migratoria* et son contrôle endocrine. *Gen. comp. Endocr.* **33,** 13–34.

Laverdure, A. M. (1967). Mode d'action des corpora allata au cours de la vitellogenèse chez *Tenebrio molitor* (Coléoptère). *C. r. hebd. Acad. Sci., Paris,* **265,** 145–146.

Laverdure, A. M. (1968). Étude, par électrophorèse en gel d'amidon, de l'évolution des protéines au cours du dévelopment post-embryonnaire de *Tenebrio molitor* (Coléoptère). *C. r. hebd. Acad. Sci., Paris,* **267,** 1996–1998.

Laverdure, A. M. (1970). Étude électrophorétique des variations du protéinogramme au cours du développement post-embryonnaire de *Tenebrio molitor* (Coléoptère). Étude comparée du cas de *Galleria mellonella* (Lépidoptère). *Ann. Endocr.* **31,** 504–515.

Laverdure, A. M. (1972). L'évolution de l'ovaire chez la femelle adulte de *Tenebrio molitor* – La vitellogenèse. *J. Insect Physiol.* **18,** 1369–1385.

Lender, T. and Duveau, J. (1960). Le développement de la gonade larvaire de *Galleria mellonella* (Linné) après transplantation ou ligature de la larve. *C. r. hebd. Acad. Sci., Paris,* **250,** 3511–3513.

Lensky, Y. and Alumot, E. (1969). Proteins in the spermathecae and haemolymph of the queen bee (*Apis mellifica* L. *var. ligustica* Spin.). *Comp. Biochem. Physiol.* **30,** 569–575.

De Loof, A. (1969). Causale mechanismen bij de vitellogenese van de coloradokever, *Leptinotarsa decemlineata* Say. Thesis, Rijksfaculteit Gent.

De Loof, A. and Lagasse, A. (1970). Juvenile hormone and the ultrastructural properties of the fat body of the adult Colorado beetle, *Leptinotarsa decemlineata* Say. *Z. Zellforsch. mikrosk. Anat.* **106,** 439–450.

De Loof, A., Lagasse, A. and Bohyn, W. (1972). Proteid yolk formation in the Colorado beetle with special reference to the mechanism of the selective uptake of haemolymph proteins. *Proc. konigkl. Ned. Akad. Wet. Ser.* **C 75,** 125–143.

De Loof, A. and De Wilde, J. (1970a). The relationship between haemolymph proteins and vitellogenesis in the Colorado beetle, *Leptinotarsa decemlineata. J. Insect Physiol.* **16,** 157–169.

De Loof, A. and De Wilde, J. (1970b). Hormonal control of synthesis of vitellogenic female protein in the Colorado beetle, *Leptinotarsa decemlineata. J. Insect Physiol.* **16,** 1455–1466.

Lukoschus, F. (1956). Untersuchungen zur Entwicklung der Kastenmerkmale bei der Honigbiene (*Apis mellifica* L.). *Z. Morphol. Ökol. Tiere,* **45,** 157–197.

Lüscher, M. and Lanzrein, B. (1976). Differential effects of the three juvenile hormones (JH I-II-III) in the cockroach *Nauphoeta cinerea. Coll. internat. C.N.R.S.* **251,** 435–440.

McGregor, D. A. and Loughton, B. G. (1974). Yolk-protein degradation during embryogenesis of the African migratory locust. *Can. J. Zool.* **52,** 907–917.

Meisenheimer, J. (1907). Ergebnisse einiger Versuchsreihen über Exstirpation und Transplantation der Geschlechtsdrüsen bei Schmetterlingen. *Zool. Anz.* **32,** 393–400.

Mjeni, A. M. and Morrison, P. E. (1973). Changes in Haemolymph proteins in the normal and allatectomized blowfly *Phormia regina* Meig. during the first reproductive cycle. *Can. J. Zool.* **51,** 1069–1079.

Mjeni, A. M. and Morrison, P. E. (1976). Juvenile hormone analogue and egg development in the blowfly, *Phormia regina* (Meig.). *Gen. comp. Endocr.* **28,** 17–23.

Monod, J. and Poulson, D. F. (1937). Specific reactions of the ovary to inter-specific transplantation among members of the *melanogaster* group of *Drosophila. Genetics,* **22,** 257–263.

Mundall, E. (1976). The control of vitellogenesis and oviposition in the blood-sucking bug, *Triatoma protracta* (Hemiptera: Reduviidae). Thesis, University of California, Los Angeles.

Mundall, E. and Engelmann, F. (1977). Endocrine control of vitellogenin synthesis and vitellogenesis in *Triatoma protracta. J. Insect Physiol.* **23,** 825–836.

Mundall, E. and Law, J. H. (1977). An investigation into the composition and structure of *Manduca sexta* vitellogenin. *Am. Zool.* **17,** 863.

Nielsen, D. J. and Mills, R. R. (1968). Changes in electrophoretic properties of haemolymph and terminal oocyte proteins during vitellogenesis in the American cockroach. *J. Insect Physiol.* **14,** 163–170.

Nijhout, M. M. and Riddiford, L. M. (1974). The control of egg maturation by juvenile

hormone in the tobacco hornworm moth, *Manduca sexta*. *Biol. Bull. mar. biol. Lab., Woods Hole*, **146,** 377–392.
Oie, M., Takahashi, S. Y. and Ishizaki, H. (1975). Vitellogenin in the eggs of the cockroach, *Blattella germanica*: purification and characterization. *Devl. Growth Diff.* **17,** 237–246.
Ono, S. E., Nagayama, H. and Shimura, K. (1975). The occurrence and synthesis of female-and egg-specific proteins in the silkworm, *Bombyx mori*. *Insect Biochem.* **5,** 313–329.
Pan, M. L. (1977). Juvenile hormone and vitellogenin synthesis in the *cecropia* silkworm. *Biol. Bull. mar. biol. Lab., Woods Hole*, **153,** 336–345.
Pan, M. L., Bell, W. J. and Telfer, W. H. (1969). Vitellogenic blood protein synthesis by insect fat body. *Science, N.Y.* **165,** 393–394.
Pan, M. L. and Wallace, R. A. (1974). *Cecropia* vitellogenin: Isolation and characterization. *Am. Zool.* **14,** 1239–1242.
Pan, M. L. and Wyatt, G. R. (1971). Juvenile hormone induces vitellogenin synthesis in the monarch butterfly. *Science, N.Y.* **174,** 503–505.
Pan, M. L. and Wyatt, G. R. (1976). Control of vitellogenin synthesis in the monarch butterfly by juvenile hormone. *Devl. Biol.* **54,** 127–134.
Perassi, R. (1973). Female specific proteins in *Triatoma infestans* haemolymph. *J. Insect Physiol.* **19,** 663–671.
Postlethwait, J. H. and Weiser, K. (1973). Vitellogenesis induced by juvenile hormone in the female sterile mutant apterous-four in *Drosophila melanogaster*. *Nature, Lond.* **244,** 284–285.
Quo, Fu (1959). Reciprocal transplantation of gonads in the adult oriental migratory locust, *Locusta migratoria manilensis* Meyen. *Sci. Rec.* **3,** 567–572.
Ramamurty, P. S. and Engels, W. (1977). Allatektomie- und Juvenilhormon-Wirkungen auf Synthese und Einlagerung von Vitellogenin bei der Bienenkönigin (*Apis mellifica*). *Zool. Jahrb. Physiol.* **81,** 165–176.
Robbins, W. E., Kaplanis, J. N., Thompson, M. J., Shortino, T. J., Cohen, C. F. and Joyner, S. C. (1968). Ecdysones and analogs: effects on development and reproduction of insects. *Science, N.Y.* **161,** 1158–1159.
Röseler, P. F. (1974). Vergleichende Untersuchungen zur Oogenese bei weiselrichtigen and weisellosen Arbeiterinnen der Hummelart *Bombus terrestris* (L.). *Insectes Sociaux*, **21,** 249–274.
Röseler, P. F. (1977). Juvenile hormone control of oogenesis in bumblebee workers, *Bombus terrestris*. *J. Insect Physiol.* **23,** 985–992.
Roth, T. F., Cutting, J. A. and Atlas, S. B. (1976), Protein transport: a selective membrane mechanism. *J. Supramol. Structure*, **4,** 527–545.
Roth, T. F. and Levitt, R. (1976). Oocyte membrane receptors for yolk protein sequestration. *XVth internat. Congr. Ent. Washington, Poster Session.*
Roth, T. F. and Porter, K. R. (1962). Specialized sites on the cell surface for protein uptake. *In* "Electron Microscopy" (Ed. S. S. Breese), **2,** Academic Press, New York and London.
Roth, T. F. and Porter, K. R. (1964). Yolk protein uptake in the oocyte of the mosquito *Aedes aegypti* L. *J. Cell Biol.* **20,** 313–332.
Rutz, W. and Lüscher, M. (1973). The occurrence of vitellogenin in worker bees of *Apis mellifica* and the possibility of its transmission to the queen. *Proc. VII Congr. IUSSI, Lond.* 340–344.
Rutz, W. and Lüscher, M. (1974). The occurrence of vitellogenin in workers and queens of *Apis mellifica* and the possibility of its transmission to the queen. *J. Insect Physiol.* **20,** 897–909.

Rutz, W., Gerig, L., Wille, H. and Lüscher, M. (1976). The function of juvenile hormone in adult worker honeybees, *Apis mellifera*. *J. Insect Physiol.* **22,** 1485–1491.

Scharrer, B. and von Harnack, M. (1961). Histophysiological studies on the corpus allatum of *Leucophaea maderae*. III. The effect of castration. *Biol. Bull. mar. biol. Lab., Woods Hole*, **121,** 193–208.

Scheurer, R. (1969). Haemolymph proteins and yolk formation in the cockroach, *Leucophaea maderae*. *J. Insect Physiol.* **15,** 1673–1682.

Schlaeger, D. A., Fuchs, M. S. and Kang, S. H. (1974). Ecdysone-mediated stimulation of DOPA decarboxylase activity and its relationship to ovarian development in *Aedes aegypti*. *J. Cell. Biol.* **61,** 454–465.

Schneider, R. (1973). Protein-Analysen an einzelnen Eiern und Eiteilen während der Embryonalentwicklung von *Acheta domesticus* L. Thesis, University of Marburg, Germany.

Spielman, A., Gwadz, R. W. and Anderson, W. A. (1971). Ecdysone-initiated ovarian development in mosquitoes. *J. Insect Physiol.* **17,** 1807–1814.

Schumann, W. (1973). Immunogenetic and electrophoretic studies with extracts of different adult *Culex pipiens*. *J. Insect Physiol.* **19,** 1387–1396.

Stay, B. (1965). Protein uptake in the oocytes of the *cecropia* moth. *J. Cell Biol.* **26,** 49–62.

Tanaka, A. (1973). General accounts on the oocyte growth and the identification of vitellogenin by means of immunospecificity in the cockroach, *Blattella germanica* (L.). *Devl. Growth Diff.* **15,** 153–168.

Tanaka, A. and Ishizaki, H. (1974). Immunochemical detection of vitellogenin in the ovary and fat body during a reproductive cycle of the cockroach, *Blattella germanica*. *Devl. Growth Diff.* **16,** 247–255.

Tata, J. R. (1973). Ribosome-membrane interaction and protein synthesis. *Karolinska Symp. Res. Meth. reproduct. Endocr.* **6,** 192–224.

Telfer, W. H. (1954). Immunological studies of insect metamorphosis. II. The role of a sex-limited blood protein in egg formation by the *cecropia* silkworm. *J. gen. Physiol.* **37,** 539–558.

Telfer, W. H. (1960). The selective accumulation of blood proteins by the oocytes of Saturniid moths. *Biol. Bull. mar. biol. Lab., Woods Hole*, **118,** 338–351.

Telfer, W. H. (1961). The route of entry and localization of blood proteins in the oocytes of Saturniid moths. *J. biophys. biochem. Cytol.* **9,** 747–759.

Telfer, W. H. and Anderson, L. M. (1968). Functional transformation accompanying the initiation of a terminal growth phase in the *cecropia* moth oocyte. *Devl. Biol.* **17,** 512–535.

Telfer, W. H. and Melius, M. E. Jr. (1963). The mechanism of blood protein uptake by insect oocytes. *Am. Zool.* **3,** 185–191.

Telfer, W. H., Rubenstein, E. and Klein, M. T. (1976). Intrafollicular protein secretion during vitellogenesis in *Hyalophora cecropia*. *XVth internat. Congr. Ent. Washington, Poster Session.*

Telfer, W. H. and Rutberg, L. D. (1960). The effects of blood protein depletion on the growth of the oocytes in the *cecropia* moth. *Biol. Bull. mar. biol. Lab., Woods Hole*, **118,** 352–366.

Telfer, W. H. and Smith, D. S. (1970). Aspects of egg formation. *Symp. Roy. Ent. Soc. Lond.* **5,** 117–134.

Telfer, W. H. and Williams, C. M. (1953). Immunological studies of insect metamorphosis. I. Qualitative and quantitative description of the blood antigens of the *cecropia* silkworm. *J. gen. Physiol.* **36,** 389–413.

Thomas, K. K. and Nation, J. L. (1966). Control of a sex-limited haemolymph protein by corpora allata during ovarian development in *Periplaneta americana* (L.). *Biol. Bull. mar. biol. Lab., Woods Hole*, **130,** 254–264.

Thomsen, E. and Thomsen, M. (1974). Fine structure of the fat body of the female of *Calliphora erythrocephala* during the first egg-maturation cycle. *Cell Tiss. Res.* **152,** 193–217.

Tobe, S. S. and Loughton, B. G. (1967). The development of blood proteins in the African migratory locust. *Can. J. Zool.* **45,** 975–984.

Tobe, S. S. and Stay, B. (1977). Corpus allatum activity *in vitro* during the reproductive cycle of the viviparous cockroach, *Diploptera punctata* (Eschscholtz). *Gen. comp. Endocr.* **31,** 138–147.

Travis, J. L. and Schin, K. (1976). Evidence for haemoglobin uptake by oocytes of *Chironomus thummi. J. Insect Physiol.* **22,** 1601–1608.

Vogel, A. (1968). Résultats de transplantation d'ovaires d'imagos à *Locusta migratoria* (L.). *C.r. hebd. Seanc. Acad. Sci., Paris*, **267,** 1043–1046.

Wajc, E., Baker-Grunwald, T. and Applebaum, S. W. (1977). Binding and uptake of trypan blue by developing oocytes of *Locusta migratoria migratorioides. J. Embryol. exp. Morphol.* **37,** 1–11.

Whitehead, D. L. (1974). The inhibition of gonadotropin by ecdysone and tanning hormone in the colleterial gland of cockroaches. *Gen. comp. Endocr.* **22,** 412–413.

Wigglesworth, V. B. (1936). The function of the corpus allatum in the growth and reproduction of *Rhodnius prolixus* (Hemiptera). *Q. J. microsc. Sci.* **79,** 91–121.

Wigglesworth, V. B. (1943). The fate of haemoglobin in *Rhodnius prolixus* (Hemiptera) and other blood-sucking arthropods. *Proc. Roy. Soc. Lond.* **B131,** 313–339.

Wightman, J. A. (1973). Ovariole microstructure and vitellogenesis in *Lygocoris pabulinus* (L.). and other mirids (Hemiptera: Miridae). *J. Ent.* **A48,** 103–115.

Wilhelm, R. and Lüscher, M. (1970). Über die Reifung transplantierter Oocyten unter verschiedenen Bedingungen bei der Schabe *Nauphoeta cinerea. Rev. Suisse Zool.* **77,** 621–624.

Wilhelm, R. and Lüscher, M. (1974). On the relative importance of juvenile hormone and vitellogenin for oocyte growth in the cockroach *Nauphoeta* cinerea. *J. Insect Physiol.* **20,** 1887–1894.

Wilkens, J. L. (1969). The endocrine control of protein metabolism as related to reproduction in the fleshfly *Sarcophaga bullata. J. Insect Physiol.* **15,** 1015–1024.

Williams, C. M. (1952). Physiology of insect diapause. IV. The brain and prothoracic glands as an endocrine system in the *cecropia* silkworm. *Biol. Bull. mar. biol. Lab., Woods Hole*, **103,** 120–138.

Wüest, J. (1975). Les effets de la castration sur le cycle du tissue adipeux chez la blatte *Nauphoeta cinerea* femelle. *Mitt. schweiz. Ent. Ges.* **48,** 443–449.

Wüest, J. (1976). The reproductive cycle of the cockroach *Nauphoeta cinerea*: the fat body metabolism and its hormonal control. *Gen. comp. Endocr.* **29,** 298–299.

Wyatt, G. R. (1972). Insect hormones. *In* "Biochemical Actions of Hormones" (Ed. G. Litwack), **2,** pp. 385–490. Academic Press, New York and London.

NOTES ADDED IN PROOF (*See p. 418*)

Physiology of Moulting in Insects

Arthur M. Jungreis

Department of Zoology, University of Tennessee, Knoxville, USA

1 Introduction

The periodic requirement for moulting in arthropods is so well established that in recent years little attention has been given to details of this process in insects. In fact, the process of moulting has been perceived as "resolved" by many invertebrate physiologists. Of consequence, scant attention has been given toward the construction of a broadly conceived systematic approach to moulting as a coordinated process coupling (a) changes in the titres of hormones and target tissue responsiveness, with (b) alterations in the enzymatic capacity of the epidermis to synthesize or degrade cuticular components, and (c) with the physiology of digestion and resorption of the ecdysed cuticle. Unfortunately, lack of information precludes presenting such a story at this time. What I have attempted to do is briefly summarize the status of the various pieces of the moulting picture, which will lead, hopefully, to a better understanding of the coordinated processes of moulting.*

2 Hormone titres and the decision to moult

Hormonal control of inset development has been amply documented in the reviews by Wigglesworth (1970), Novak and Sláma (1971), Wyatt (1972), Doane (1973), Gilbert and King (1973), Sláma *et al.* (1974), Goldsworthy and Mordue (1974), Willis (1974), Morgan and Poole (1977), and Riddiford and Truman (1978). Even to summarize this vast literature as it affects moulting is beyond the scope of this chapter. However, the temporal relationship between hormone synthesis and release as it effects the commitment by the epidermis to synthesize larval, pupal or adult cuticle with their stage associated enzyme and protein specific complements is of considerable interest. Hormone levels associated with and responsible for the characteristic features of cuticle following larval–larval, larval–pupal or pupal–adult moults have recently been quantitated. Unfortunately, the complexity of the regulatory

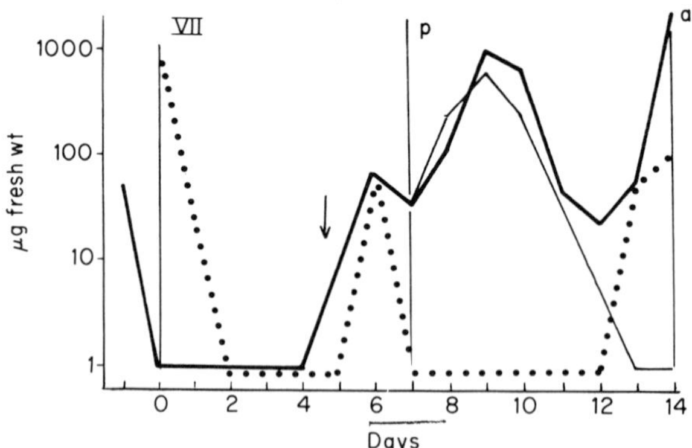

Fig. 1. Changes in ecdysone and juvenile hormone titres during development of the wax moth *Galleria mellonella*. VII, seventh instar larval and pharate pupal stages (days 0–7); p, pupal and pharate adult stages (days 7–14); a, adult stage (day 14). Arrow shows the time of cocoon spinning; (. . . .) juvenile hormone; (——) ecdysone levels in males and females (days 1–7) or in females only (days 7–14); (——) ecdysone levels in males (days 7–14). (Redrawn after Hsiao and Hsiao, 1977.)

* Characteristics of the various larval and pharate pupal stages in development for *Hyalophora cecropia* and *Manduca sexta* are presented in the Appendices.

mechanisms uncovered far exceed that expected from the simplistic models requiring only two-hormones (juvenile hormone and ecdysone).

The total body content of both juvenile hormone (de Wilde *et al.*, 1968) and ecdysone (Staal, 1967) have been determined simultaneously using bioassays in the wax moth *Galleria mellonella* during the larval–pupal and pupal–adult stages in development (Hsiao and Hsiao, 1977; Fig. 1). These data may or may not reflect rates or quantities of hormone released and provide little information regarding target tissue sensitivity or commitment. In the tobacco hornworm *Manduca sexta*, the haemolymph titres of these hormones have been determined independently using bioassays or radio-immunoassays during larval–larval, larval–pupal and pupal–adult stages in development (Truman *et al.*, 1973; Bollenbacker *et al.*, 1975; Fain and Riddiford, 1975; Nijhout and Williams, 1974b; Riddiford and Truman, 1978) (Figs. 2 and 3). In *Galleria*, the titre of JH is high during the early portion of the last larval

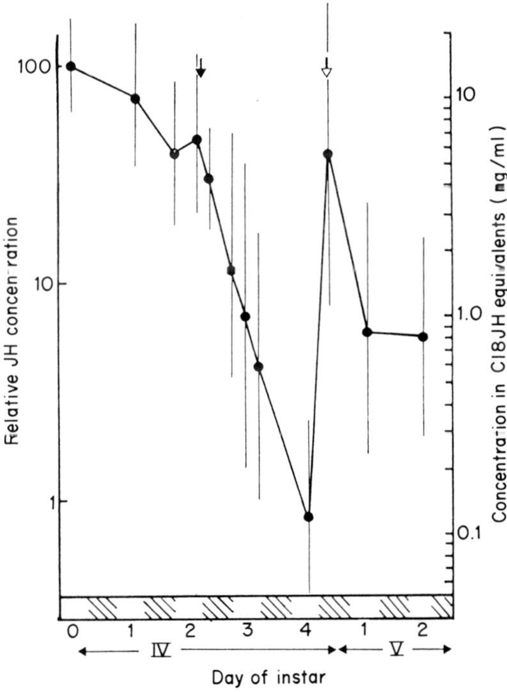

Fig. 2. Juvenile hormone titre in *Manduca sexta* larvae after ecdysis to the fourth instar through the first third of the fifth instar. The juvenile hormone concentration at each stage was calculated from the mean scores of initial dilutions of extracts. Each point is based on 3 to 5 extractions and about 25 assay larvae; bars indicate ± one standard deviation; (→) mean time of ecdysiotropin (=PTTH=prothoracicotropin) for Gate II larvae. (⇾) mean time of ecdysis of Gate II larvae to the fifth instar; cross-hatch represents darkness and represents the beginning of each day. (From Fain and Riddiford, 1975.)

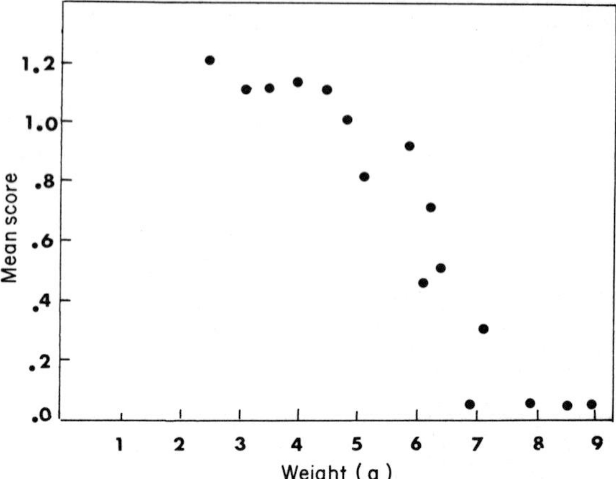

Fig. 3. The titre of juvenile hormone in haemolymph of *Manduca sexta* at various weights during the fifth larval instar. Each point is the average of 5 assays. (From Nijhout and Williams, 1974.)

instar, in agreement with observations by Nijhout and Williams (1974b) and Fain and Riddiford (1975) in *M. sexta*, and Varjas *et al.* (1976) in *Pieris brassicae* and *Barathra brassicae*. However, unlike the responses observed in *M. sexta* during the transition between the fourth to fifth larval instars, and the fifth larval instar to pupal stage in development, a burst of JH hormone synthesis (accumulation?) is noted preliminary to the moult in *G. mellonella*, *P. brassicae* and *B. brassicae* (Figs 2 and 3 contrasted with Fig. 1). In all of these species as well as in *Calliophora erythrocephala* (Young and Young, 1976) the point in development characterized by this late burst of JH release also corresponds to that period when ecdysone is being synthesized and released at its maximal rate (Figs 4 and 5). The increase in total body JH observed in *G. mellonella*, *P. brassicae* and *B. brassicae* has not been correlated with specific morphological events, and it is doubtful whether the accumulated hormone is actually being released at this time. In *H. cecropia* an increase in JH release during the larval–pupal transformation would have no effect on the epidermal commitment to synthesize pupal cuticle (Riddiford, 1972). On the other hand, the viscera would be affected since they are not fully committed to form pupal tissues and would be "incomplete" and retain morphological features in common with both larvae and pupae at the time of moulting (Riddiford, 1972, 1975). The situation in *M. sexta* would be similar, except that the epidermal commitment to produce pupal cuticle would also be disrupted or reversed (i.e. producing larval cuticle) if JH were given prior to "heart exposure" (=pink stripe) (Fain and Riddiford, 1975; Nijhout and Williams, 1974b; Nijhout, 1976).

The decision to initiate a larval–larval in contrast to a larval–pupal moult can be correlated with body weight or head capsule size (Nijhout, 1975;

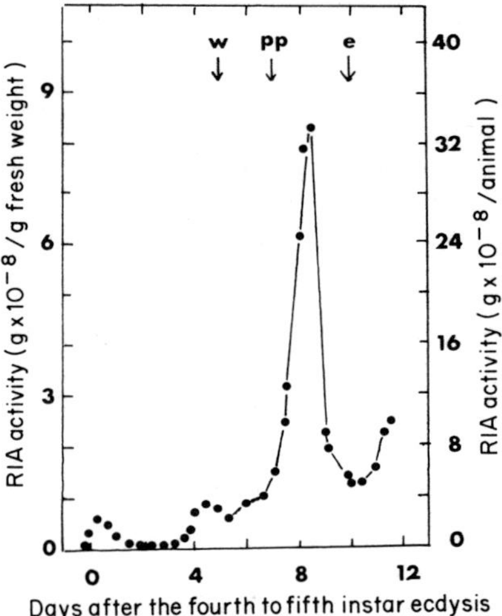

Fig. 4. The titre of ecdysone in haemolymph of *Manduca sexta* during the fifth larval instar (days 0–5), pharate pupal (days 5–10), pupal (day 10) and pharate adult (days 11–12) stages in development. Radioimmunoassay (RIA) activity is expressed per gram fresh weight or per animal. w, initiation of wandering stage; pp, initiation of pharate pupal development; e, day of the larval–pupal ecdysis. Each point represents an average of four animals, each determined independently in duplicate at two different concentrations. (Redrawn and modified after Bollenbacher *et al.*, 1975.)

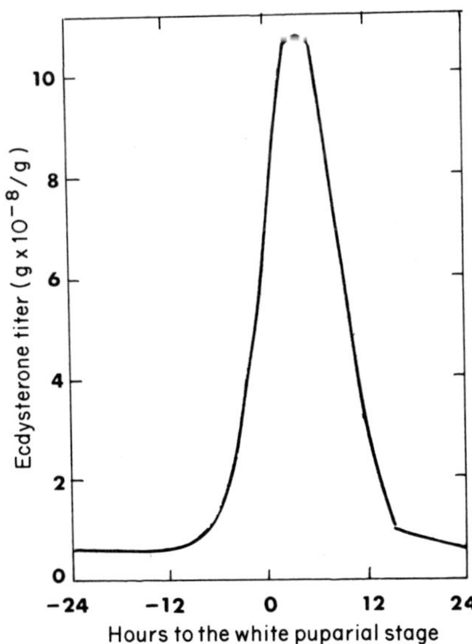

Fig. 5. The titre of ecdysone in the blowfly, *Calliphora erythrocephala*, measured for 24 h before to 24 h after the white puparial stage. (Redrawn and modified after Young and Young, 1976.)

Nijhout and Williams, 1974a). However, this decision to pupate is really related to the haemolymph titre of JH, since the minimal body weight (5 grams) or head capsule size required for a larval–pupal in contrast with a larval–larval moult is also the point in development when the titre of JH drops markedly (Nijhout and Williams, 1974b; Fig. 3). Why the titre of JH drops at this time (i.e. a cessation of corpora allata activity) is unknown.

Following upon the heels of specific photoperiod controlled releases of prothoraciotropic hormone (PTTH) (Gibbs and Riddiford, 1977), two peaks of ecdysone release are observed during the last larval instar in *M. sexta* (Fig. 4; Truman and Riddiford, 1974; Nijhout and Williams, 1975a), but not *C. erythrocephala* (Shaaya and Karlson, 1965; Young and Young, 1976). In Lepidoptera, only one burst of ecdysone occurs between larval–larvel moults, while in Diptera, only one burst of release is noted between larval–pupal moults. Since the release of ecdysone in Diptera occurs at a time corresponding to the 2nd release of ecdysone in Lepidoptera, it would be interesting to know whether early infusion of ecdysone in Diptera would cause ecdysis of the larval cuticle.

Each of the two periods of PTTH release cause increases in ecdysone synthesis by the prothoracic glands (Bollenbacker *et al.*, 1975; Fig. 6). However,

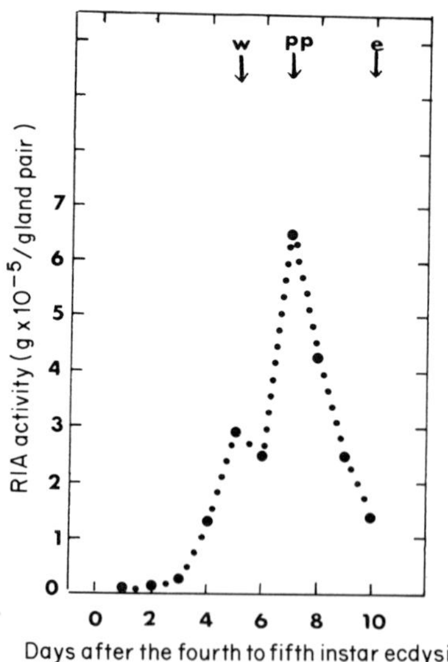

Fig. 6. Secretion of α-ecdysone by ecdysial glands of fifth larval instar *Manduca sexta* measured *in vitro*. Each point represents an average of four pairs of glands, each determined independently in duplicate at two different concentrations. Feeding fifth instar (days 0–5); w, initiation of wandering preliminary to pharate pupal development (day 5); pp, pharate pupal development (days 5–10); e, day of the larval pupal ecdysis (day 10). (Redrawn and modified after Bollenbacher *et al.*, 1975.)

whereas the initial release of PTTH in *M. sexta* is not accompanied by a decrease in the rate of ecdysone degradation (contrast the rate of ecdysone synthesis by the prothoracic glands in Fig. 6 with the haemolymph titre in Fig. 4), that synthesized following the second period of PTTH release as documented in *C. erythrocephala* is associated with a decrease in the rate of ecdysone degradation (Fig. 7; Young and Young, 1976).

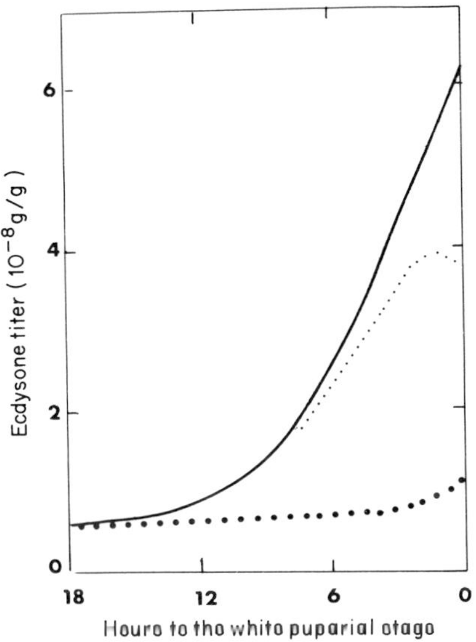

Fig. 7. The titre of ecdysone in the blowfly, *Calliphora erythrocephala*, expected under a variety of conditions of synthesis and catabolism during the initiation of puparium formation. (——) observed values; (. . . .) expected values calculated from the observed rate of ecdysone synthesis and a constant fractional rate of catabolism; (. . . .) expected values calculated from a constant rate of ecdysone synthesis and the observed fractional catabolic rate. (From Young and Young, 1976.)

Infusion of ecdysone into middle or later feeding fifth instar larvae of *H. cecropia* and *M. sexta* prior to the period of gut evacuation results in the formation of pupae with larval–pupal intermediate type viscera and/or cuticles (Fain and Riddiford, 1977; Nijhout and Williams, 1974b; Nijhout, 1976; Mitsui and Riddiford, 1976, 1978; Riddiford, 1972, 1975) while infusion of ecdysone into early fifth instar larvae results in additional larval instars (Nijhout, 1976). Thus, one striking difference between these two species lies in the fixity of epidermal commitment at the time of gut evacuation, *H. cecropia* epidermis being refractory to JH, while that of *M. sexta* retaining some responsiveness (see Riddiford, 1976; Fain and Riddiford, 1977). In terms of cuticular protein synthesis, prior to gut evacuation, the decision of the larval epidermis to produce pupal cuticular proteins is irrevocable in *H. cecropia* and variable in *M. sexta*.

3 Composition, synthesis and degradation of the integumentary epithelium

A description of hormonal changes occurring preliminary to and contiguous with moulting and ecdysis has been presented in the preceding section. However, without a knowledge of the cuticle's composition, one can neither evaluate the roles of these hormones in the synthesis of the respective stage specific integumentary epithelia nor can one glean the sequence of events that must precede digestion and reabsorption of the ecdysed cuticle. In light of the large numbers of reviews that have dealt with selected aspects of these topics (Beament and Treherne, 1967; Ebeling, 1974; Hackman, 1964, 1971, 1974; Hepburn, 1976; Laing, 1933; Locke, 1964, 1974, 1976; Neville, 1975; Noble-Nesbitt, 1967; Richards, 1951; Smith, 1968; Verson, 1902; Wachter, 1930; Weis-Fogh, 1970; Wigglesworth, 1933, 1959, 1972, 1976; Zacharuk, 1976), only a general summary of the composition of cuticle and the sequence of synthetic and degradative events will be presented in this paper. Unfortunately, studies on hormonal involvement in the moult cycle have not been correlated with specific changes in cuticle composition or the sequence of events that precede ecdysis. Consequently, changes in the cuticle during moulting can not be correlated with structural changes. Therefore, cuticle synthesis and degradation will be described as though it occurred independent of changes in the respective hormone titres.

The inter-moult cuticle is composed of three or four basic layers (see Fig. 8) (from outside towards epidermals cells): 1. cuticulin including superficial, cement and/or lipid layers, the outer epicuticle and the membrane epicuticulin layers), 2. inner, internal or dense epicuticle, 3. fibrous or procuticle (including the outer-most exocuticle, intermediate meso-cuticle and innermost endocuticle – see Shatz, 1952), and 4. subcuticle (Schmidt, 1956; Zacharuk, 1972). The cuticulin layer is resistant to enzymatic attack during cuticle digestion and is normally ecdysed with the exuvium at the beginning of each instar, stadium or stage in development. The internal epicuticle is the pigmented layer within the cuticle. The fibrous cuticle represents the bulk of the cuticle, consisting as it does of exo-, meso- and endo-cuticle. (The "well-characterized" distinctions between layers of the fibrous cuticle are based only on observations in the light microscope, such distinctions being absent in the electron microscope (Delechambre, 1971; Zacharuk 1972).) The sub-cuticle is a histochemically distinct layer between the pro-cuticle and the epidermis serving to bind cuticle to epidermis during the inter-moult period. However, its existence as a separate structural component independent of the fibrous cuticle has been questioned by Locke (1964) who proposed that the sub-cuticle is composed of unordered fibres within the newly secreted endocuticle. In many insects, pore canals are also found in the cuticle. These pore canals

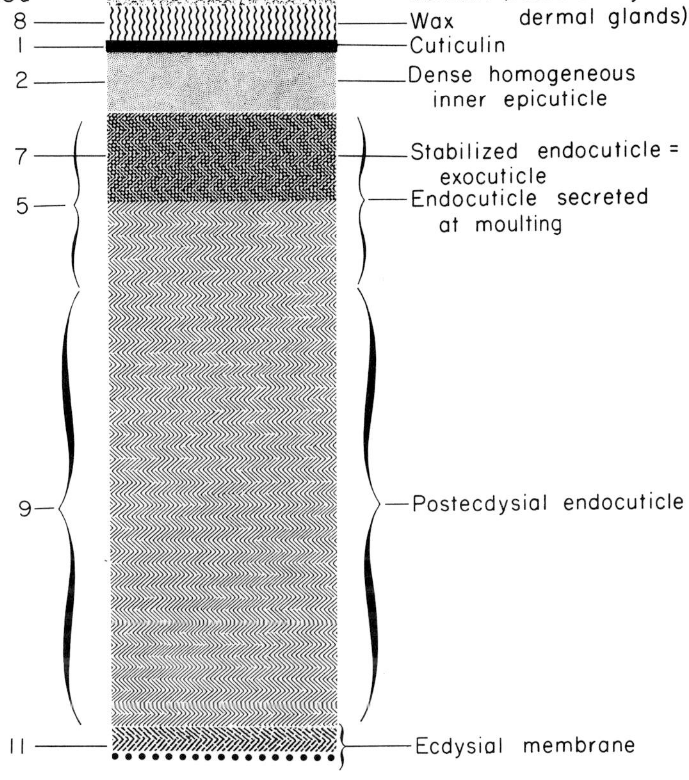

Fig. 8. Components in the cuticle (from Locke, 1974).

function as conduits through which flow proteinaceous materials for incorporation into the fibrous cuticle, inner epicuticle and cement layer of the cuticulin. The epidermal layer (Fig. 9) is composed of six types of cells; simple columnar epithelial (involved in cuticle synthesis, repair and degradation), epithelial tendon (cells in which tonofibrillae extend between the underlying musclature and the overlying cuticle to anchor the cuticle – see Lai-Fook, 1967; Zacharuk, 1972), oenocytes (synthesize lipids in the cuticulin layer – Wigglesworth, 1976), trichogen cells (seta forming cells), tormogen cells (socket forming cells) and dermal glands (synthesize the cement layer of the cuticulin; Locke, 1974; Wigglesworth, 1976).

At the onset of each moult cycle (i.e. those events associated with separation of the "old" from "new" cuticles and leading up to the shedding or ecdysis of the old cuticle), a characteristic sequence of events is initiated which results both in the formation of a new cuticle and in the digestion and ecdysis of the old cuticle (Figs. 10–18). The partial freeing of the "old" cuticle from

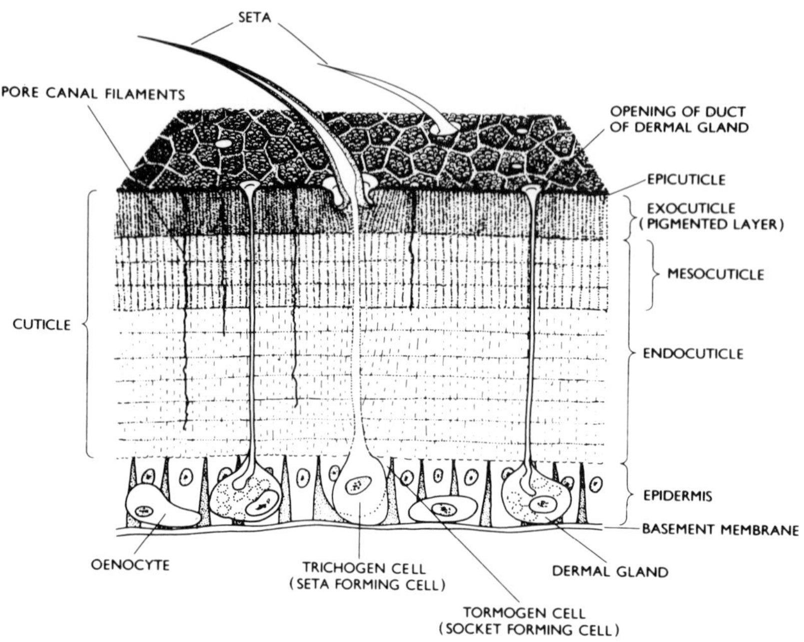

Fig. 9. Epidermal cells and their locations in a generalized insect integument. One type of specialized epidermal cell not shown in this figure are epithelial tendon cells (Lai-Fook), 1976). These cells serve to anchor the cuticle to the epidermis (from Hackman, 1971).

the underlying epidermis (the tonofibrillae of the epithelial tendon cells remaining intact until right before ecdysis) has been given the term apolysis by Jenkins and Hinton (1966). The larval–pupal apolysis appears to follow an increase in epidermal cell mitoses, which in turn had been preceded by increases in epidermal cell volume (Locke, 1970; Barbier, 1971; Zacharuk, 1972, Kunkel, 1975; Wigglesworth, personal communication).

A characteristic sequence of events takes place before, during and immediately after apolysis. Following upon the heels of apolysis proper (i.e. cell replication) is the formation of a moulting or exuvial space (Verson, 1902; Wachter, 1930; Wigglesworth, 1933; Laing, 1935) (Figs 12 and 13). The exuvial space is a region bounded by the newly synthesized ecdysial membrane and the endocuticular portion of the "old" fibrous cuticle. Considerable controversy exists regarding the origin of the exuvial space, although cytologists appear to agree that it is formed following secretion of ecdysial droplets (resulting from the fusion of secretory vesicles from the Golgi apparatus into the ecdysial membrane – Passonneau and Williams, 1953; Malek, 1958; Locke and Krishnan, 1973) and the cuticulin layer (Locke, 1969a) of the cuticle (Fig. 14). How the initial separation between fibrous

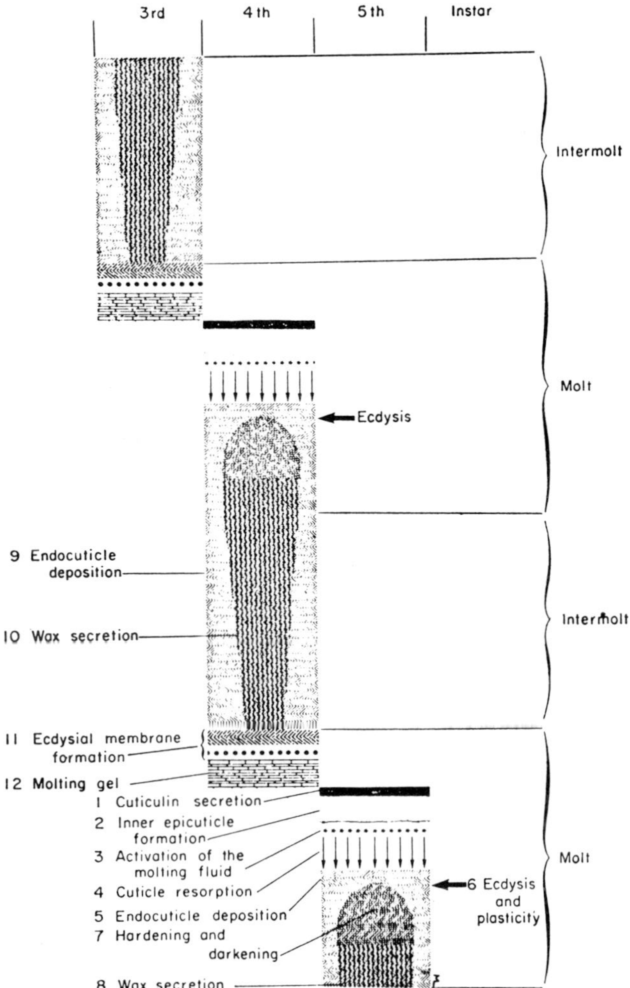

Fig. 10. Order in which cuticle components are deposited (from Locke, 1974).

endocuticle and the new ecdysial membrane in initiated remains a source of conjecture (Noble-Nesbitt, 1963b; Filshie, 1970; Zacharuk, 1976), although agreement has been reached regarding the presence of a "moulting gel" within this space (Passonneau and Williams, 1953; Noble-Nesbitt, 1963b; Taylor and Richards, 1965; Barbier, 1967; Barra, 1969; Filshie, 1970; Mitchell *et al.*, 1971; Zacharuk, 1972; Locke and Krishnan, 1973). Following synthesis of the cuticulin layer, formation of the inner and fibrous cuticles is initiated (Figs. 8 and 15).

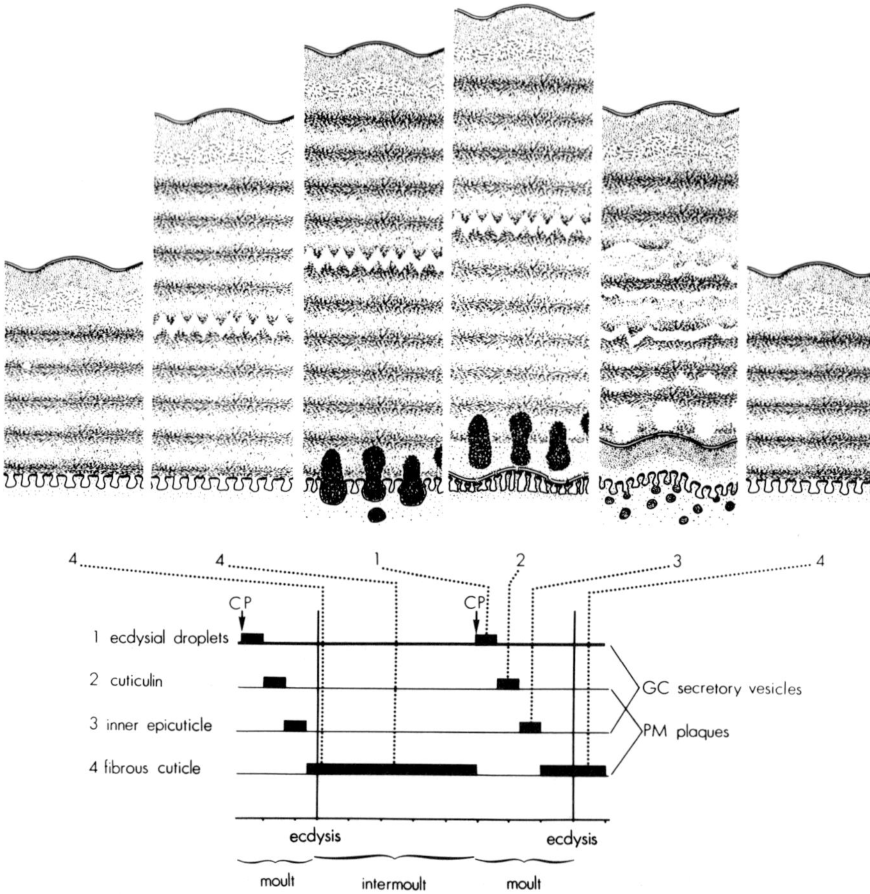

Fig. 11. The main sequence of events in cuticle deposition in *Calpodes ethlius* during the larval-pupal transformation. CP, critical period for the operation of the prothoracic (=ecdysial) glands, after which the moult sequence proceeds independently of their presence. (From Locke, 1976.)

Concurrent with the formation of the fibrous cuticle in pharate pupal *H. cecropia* is a change in cell shape (Fig. 16). The epidermal cells change from cuboidal to short and then tall columnar over a 4 day period with the shape of newly ecdysed pupal integumentary epithelial cells only having just returned to the pre-apolysis cuboidal shape (Hikida and Jungreis, 1978, 1979). This change in cell shape results in a decrease in lateral membrane interactions between epidermal cells, and a tremendous increase in cell surface exposed to the extra-cellular space.

Changes in cell shape preliminary to the larval–pupal ecdysis are accom-

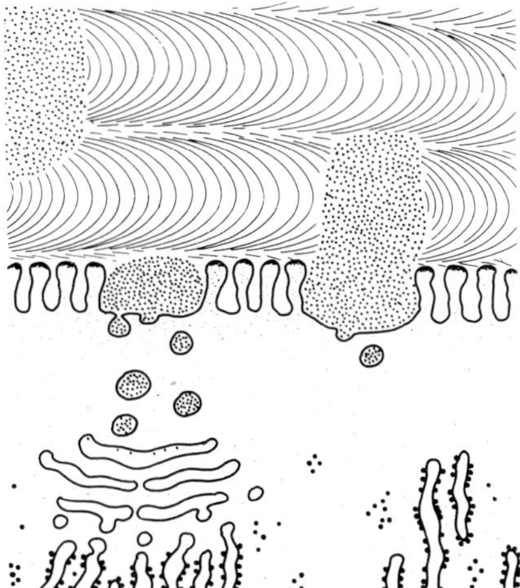

Fig. 12. Summary of the formation in *Calpodes ethlius* during the larval–pupal transformation of ecdysial droplets by the fusion of secretory vesicles from the Golgi complex. (From Locke and Krishnan, 1973.)

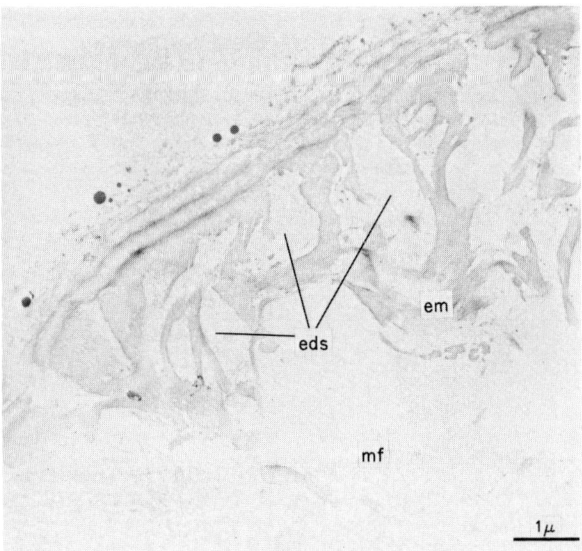

Fig. 13. The ecdysial membrane (em) showing clear spaces (eds) in the membrane which were occupied by the ecdysial droplets before their dissolution. After complete dissolution, a fenestrated ecdysial membrane resists digestion by moulting fluid (mf). (From Locke and Krishnan, 1973.)

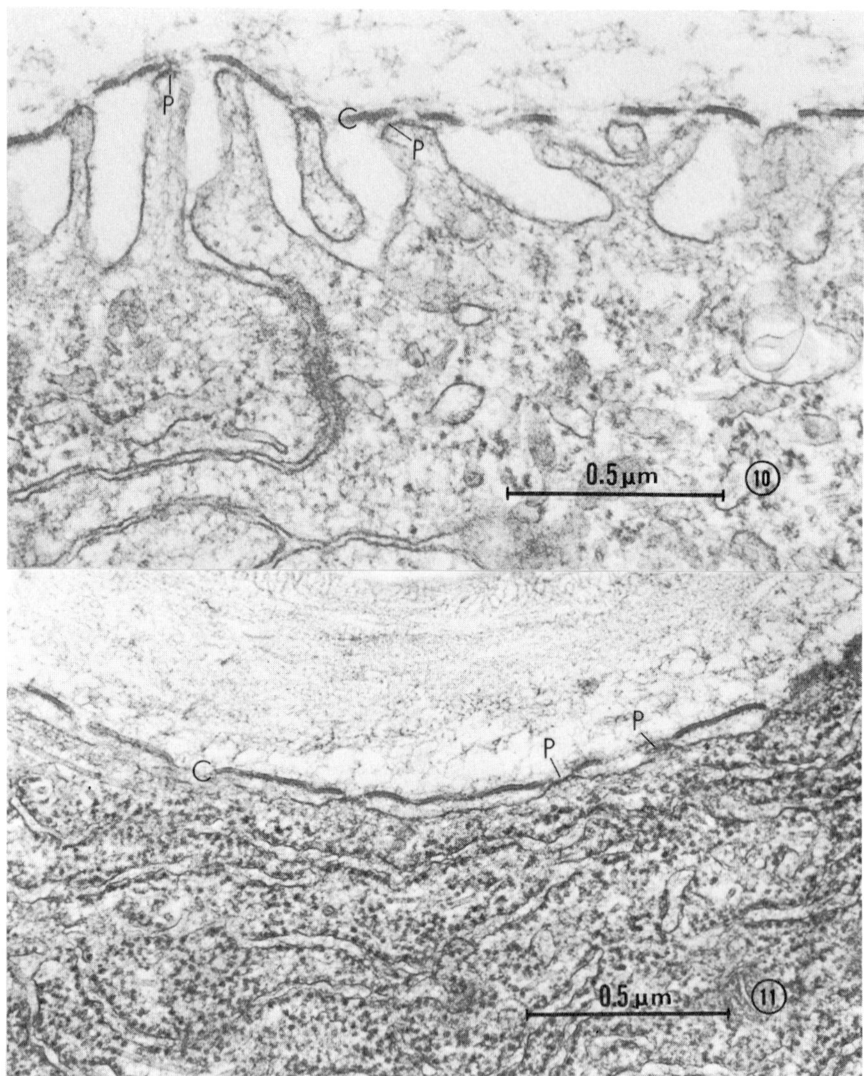

Fig. 14. The formation in *Calpodes ethlius* of the cuticulin layer (C) at the surface of the plasma membrane plaques (P). The cuticulin layer overlying the integument forms at plasma membrane plaques at the tips of the microvilli. (From Locke, 1976.)

panied in *H. cecropia* and *M. sexta* by the active transport of a potassium bicarbonate salt solution and cuticle chitinase into the exuvial space (Jungreis, 1973, 1974, 1978b; Jungreis and Harvey, 1975; Bade, 1978; Bade, personal communication). This potassium bicarbonate solution serves to activate pro-enzymes and indirectly to provide substrates for pre-activated enzymes present in the moulting gel in much the same way that cocoonase is normally activated by a potassium bicarbonate secretion (Kafatos, 1968; Lensky *et al.*, 1970; Filshie, 1970; Katzenellenbogen and Kafatos, 1970, 1971a, b, c; Mitchell *et al.*, 1971; Locke and Krishnan, 1973; Jungreis and Harvey 1975; Bade, 1978). Hydrolysis of the larval cuticle, for example, then proceeds with alacrity. Resorption of the cuticular hydrolysate takes place via pinocytotic activity through the apical surface of the epidermal cells (Hikida and Jungreis, 1978) (Fig. 18). The cuticular hydrolysate is digested in multivesicular bodies and lysosomes and released into the cytosol (Fig. 17). These degradative products are then sequestered by the rough endoplasmic reticulum (which become dilated) and synthesized into material having a flocculent appearance in the electron microscope. During the resorption of the moulting

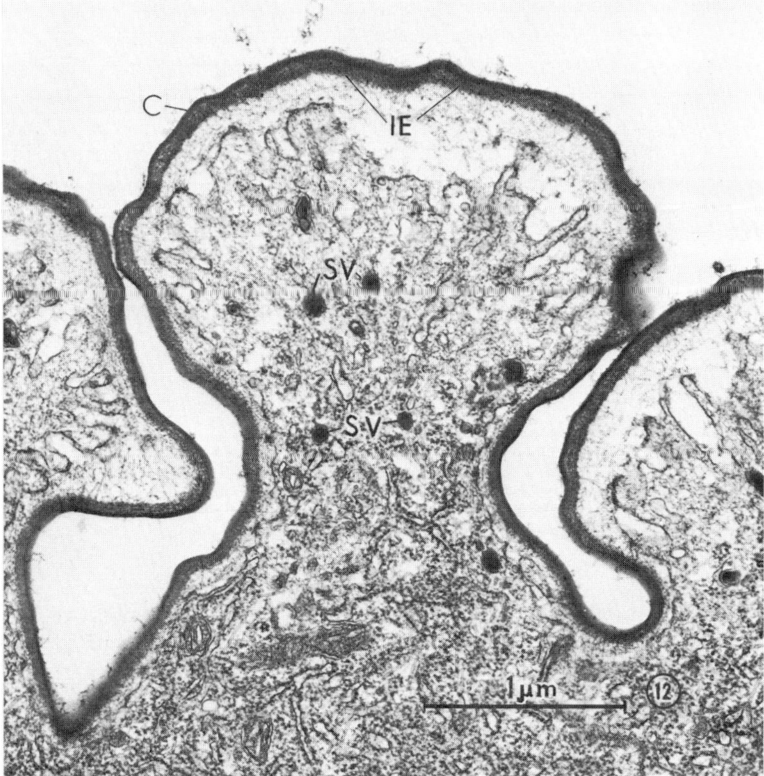

Fig. 15. The formation in *Calpodes ethlius* of the inner epicuticle by the discharge of secretory granules from the Golgi complex. The apical plasma membrane withdraws from the completed cuticulin surface (C), and secretory vesicles (SV) discharge the precursors for the inner epicuticle (IE) into the space. (From Locke, 1976.)

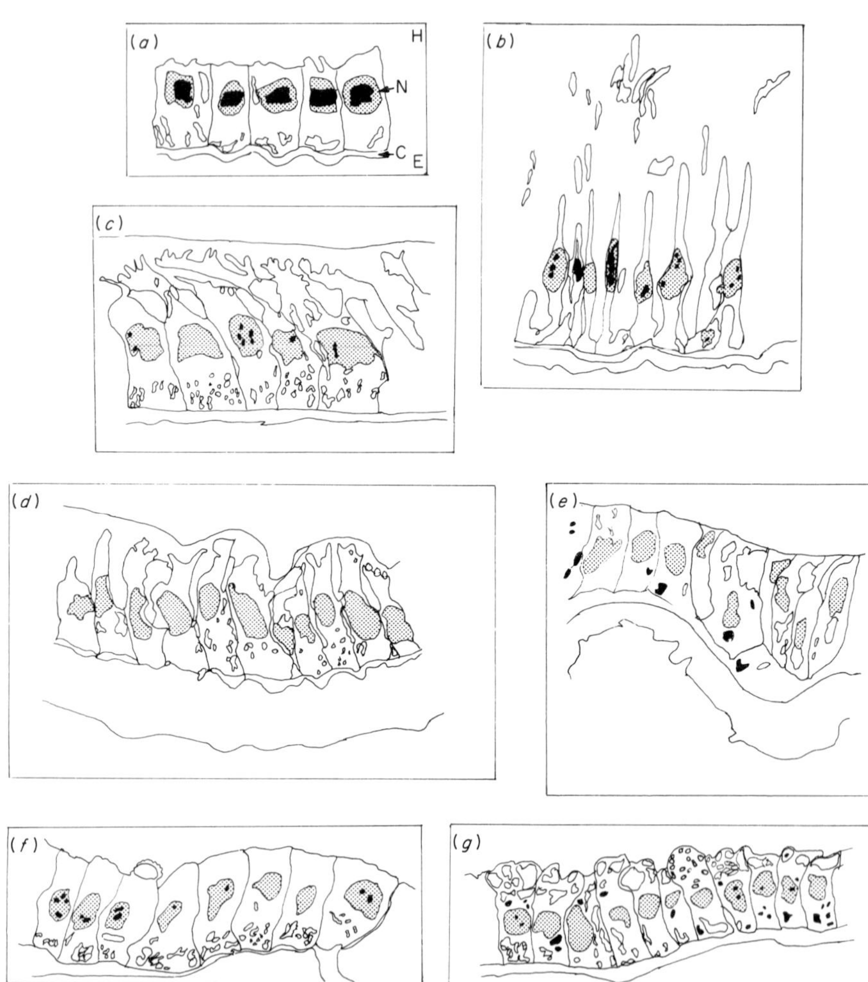

Fig. 16. Changes in the shape of *Hyalophora cecropia* larval–pupal epidermal cells before, during and after the secretion and resorption of moulting fluid. The cells change in shape from cuboidal at 3 days before the larval pupal ecdysis (LPE) (*a*), to short then tall columnar during secretion of moulting fluid at 30 h before the LPE (*b*), becoming short columnar following resorption of moulting fluid at 10 h before LPE (*c*), returning to the original cuboidal shape at ecdysis (*d*, *e*, *f*, *g*). Accompanying these changes in cell shape are decreases in lateral surface interactions (*b*–*d*), increases in total cell surface (*b*–*d*), changes in chromatin condensation (*a* contrasted with *b*–*s*), nuclear shape (*a*, *f*, *g* versus *b*–*e*), apical cell vacuolization (*a*–*b*, contrasted with *c*–*d*, and *e* contrasted with *f*–*g*), and basal cell vacuolization (*a*–*f* contrasted with *g*). Shaded areas in each cell represent the nucleus (N). Dark regions represent chromatin (if in the nucleus) or condensed material within the cytosol. H, haemolymph or basal surface of epidermis; E, exuvial or apical surface of epidermis; C, cuticle. (From Hikida and Jungreis, 1978, 1979.)

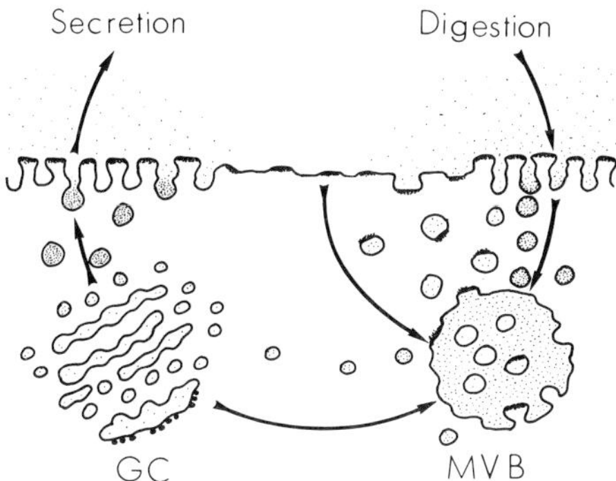

Fig. 17. The Golgi complex in secretion and digestion. The apical plasma membrane and the soluble components of the cuticle of *Calpodes ethlius* during the larval–pupal transformation are both turned over through secretion and pinocytosis. The Golgi complex (GC) has a role in secretion and in the formation of the lytic enzymes that are passed to the multivesicular bodies (MVB). (From Locke, 1976.)

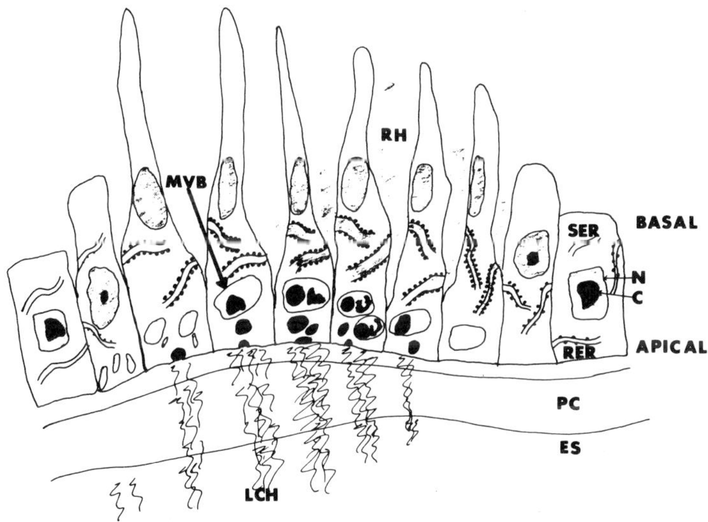

Fig. 18. Schematic representation of the resorption of moulting fluid and the larval cuticular hydrolysate (LCH) during the larval–pupal transformation of *Hyalophora cecropia.* Changes in cell shape during this period are described in Fig. 16. The cuboidal shaped epidermal cell contain little lysosomal structure such as multivesiculate bodies (MVB). As the cell becomes short and then tall columnar, the smooth endoplasmic reticulum (SER) becomes covered with ribosomes (RER). The larval cuticular hydrolysate enters the epidermal cells on the apical side via pinocytosis following diffusion across the pupal cuticle (PC) from the exuvial space (ES). Pinocytotic vesicles are incorporated into multivesiculate bodies, the contents hydrolysed and released and taken up by the rough endoplasmic reticulum. The RER begin to swell with newly synthesized material and release the products (RH) into the lateral inter-cellular spaces. Following resorption of the total larval cuticular hydrolysate, the epidermal cells become more cuboidal, lose much of the RER, experience a reduction in lysosomal activity and take on a more quiescent appearance. N, nucleus; C, chromatin. (From Hikida and Jungreis, 1978, 1979.)

fluid this flocculent material appears in the lateral extra-cellular spaces of the tall columnar epidermal cells in bodies or globules not delineated by membranes. This flocculent material is the presumed source of much protein, urate and "storage granules" deposited in fat body at this time (see citations in Wyatt and Pan, 1978). However, unlike the situation in *Calpodes*, little Golgi apparatus is present in the cytosol of *H. cecropia* or *M. sexta* prior to the larval–pupal ecdysis or initiation of pharate adult development (Greenstein, 1972; Sedlak and Gilbert, 1976; Hikida and Jungreis, 1978) an observation in *H. cecropia* which is not surprising in light of the fact that endo-cuticle formation normally requires up to 30 days post ecdysis (pupal) for completion.

4 Enzymes involved in cuticle synthesis and degradation

Hormonal control of the moulting process and the role of juvenile hormone in the decision of the epidermis to produce larval, pupal and/or adult type cuticle, as well as the histology, ultrastructure, and mode of synthesis of the cuticle have now been briefly described. The control of cuticle deposition and degradation is mediated hormonally and results from selective induction, activation, and/or de-activation of non-stage and stage specific enzymes. Some of the major enzymes involved in cuticle synthesis and degradation are listed in Table 1, while their metabolic inter-relationships are shown in Fig. 19. Unfortunately, information regarding the regulation of these enzymes during development is scant, and much work remains before control of these enzymes *in vivo* can be understood.

Several generalizations regarding synthesis and degradation can be made: 1. the capacity to produce cuticle is intact at all times in the life cycle, 2. many of the enzymes involved in cuticle synthesis remain inactive unless specifically activated (i.e. existing as pro-enzymes), 3. certain enzymes are laid down in the new cuticle to facilitate initial cuticle digestion or repair of damaged cuticle, 4. enzymes synthesized and released into the ecdysial space following apolysis can be in active or inactive forms, 5. the secretion of moulting fluid activates pro-enzymes present in moulting gel, 6. active enzymes present in moulting gel can not induce premature cuticle degradation, since their respective substrates are frequently the products of early cuticle degradation. Examples of some of these generalizations are readily found in the recent work of Bade and co-workers (1974, 1975, 1978, personal communication; Bade and Schoukimas, 1974; Bade and Stinson, 1976) on the control of chitinase activity in *M. sexta* during the larval–pupal transformation.

Highly active "endogenous" chitinase (i.e. tightly held chitinase which

TABLE 1 A tabulation of enzymes and their locations within the insect that are involved in cuticle synthesis, degradation and/or tanning

Enzyme	Tissue	Source	Citation
Catechol oxidase [E.C.1.10.3.1 (2-*O*-diphenol + $O_2 \rightarrow$ 2-0-quinone + 2 H_2O)]	cuticle	*Bombyx mori*	Kawase (1965)
		Calliphora erythrocephala	Karlson & Liebau (1961)
		Calliphora vicinia	Pau and Eagles cited in Neville (1975)
		Calpodes ethlius	Lai-Fook (1966)
		Drosophila sp	cited in Lai-Fook (1966)
		Drosophila virilis	Oshnishi (1954); Yamazaki (1969)
		Musca	cited in Lai-Fook (1966)
		Sarcophaga bullata	Lai-Fook (1966)
		Sarcophaga falculata	Dennell (1947)
		Spodoptera littoralis	Ishaaya (1972)
		Tenebrio molitor	cited in Lai-Fook (1966)
	eggs	*Melanoplus*	Bodine & Allen (1938)
	haemocytes	*Bombyx mori*	Kawase (1960)
		Calliphora erythrocephala	Karlson & Schweiger (1961)
		Drosophila sp	Rizki & Rizki (1959); Mitchell & Weber-Tracy (1965)
		Lucilia cuprina	Hackman & Goldberg (1967)
		Periplaneta americana	Mills *et al.* (1968)
Tyrosine hydroxylase [E.C.1.99.1.0 (tyrosine $\rightarrow$ *p*-tyrosine)]	cuticle	*Calliphora vicinia*	Pau *et al.*, cited in Neville (1975)
		Galleria mellonella	Retnakaran (1969)
	haemocytes	*Periplaneta americana*	Mills *et al.* (1968)
		Pieris brassicae	Post (1972)
Para-diphenyloxidase (= Laccase) [E.C.1.10.3.2. (2-*p*-diphenol + $O_2 \rightarrow$ 2-*p*-quinone + H_2O_2)]	cuticle	*Bombyx mori*	Yamazaki (1969)
		Drosophila virilis	Yamazaki (1969)
		Papillio xuthus	Yamazaki (1972)
	colleterial gland	*Periplaneta americana*	Whitehead *et al.* (1965b)

Table 1 (*contd*)

Enzyme	Tissue	Source	Citation
Polyphenolperoxidase [E.C.1.11.1.7 (reduced polyphenol + H_2O_2 → oxidized polyphenol + 2 H_2O)]	cuticle	*Calpodes ethlius*	Locke (1969b)
		Musca domestica	Hurst (1945)
		Tenebrio molitor	Hurst (1945)
Tyrosine peroxidase (= tyrosinase) [E.C.1.10.3.1 (*p*-tyrosine → 3,4-dihydroxyphenylalanine)]	cuticle	*Bombyx mori*	Kawase (1955)
		Schistocerca gregaria	Coles (1966); Anderson (1966)
	haemocytes	*Sarcophaga bullata*	Jones (1956)
Tyrosine decarboxylase [E.C.4.1.1.25 (*p*-tyrosine → tyramine)]	haemocytes	*Periplaneta americana*	Whitehead (1969)
Tyrosine aminotransferase [E.C.2.6.1.5 (L-tyrosine + 2 oxo-glutarate → L-glutamate + *p*-hydroxyphenylalanine)]	whole animals	*Calliphora erythrocephala*	Sekeris & Karlson (1962)
Dopa decarboxylase [E.C.4.1.1.26 (3,4-dihydroxyphenylalanine → 3-hydroxytyramine)]	epidermis	*Sarcophaga bullata*	Sekeris (1963)
	haemocytes	*Sarcophaga peregrina*	Whitehead (1970)
	whole body	*Calliphora erythrocephala*	Shaaya & Sekeris (1965)
Tyramine hydroxylase [E.C.1.99.1.0 (tyramine → 3-hydroxytyramine)]	haemocytes	*Periplaneta americana*	Jones (1962)
Acetyl-CoA acetyltransferase [E.C.2.3.1.9 (2 acetyl-CoA → CoA + acetoacetyl-CoA)]	whole body	*Calliphora erythrocephala*	Karlson & Ammon (1963)
β-glucosidase [E.C.3.2.1.21 (*N*-acetyldopamine-4-*O*-β-glucoside → *N*-acetyldopamine + glucoside)]	cuticle	*Tenebrio molitor*	Karlson cited in Wyatt (1968)
	epidermis	*Rhodnius prolixus*	Reynolds (1973)
	haemocytes	*Calliphora erythrocephala*	Sekeris (1964); Whitehead (1971a)

Table 1 (*contd*)

Enzyme	Tissue	Source	Citation
Chitin-UDP acetylamino deoxy-glucosyl transferase [E.C.2.4.1.16 (UDP-2-acetyl-amino-2-deoxy-D-glucose→chitin)]	whole body	*Prodenia eridania*	Porter & Jaworski (1965)
Chitinase [E.C.3.2.1.14 (hydrolysis of α1,4-acetylamino-2-deoxy-D-glucoside linkages in chitin and chitodextrin)]	cuticle	*Manduca sexta*	Bade (1974, 1975, pers. comm.)
	gut	*Periplaneta americana*	Powning & Irzykiewicz (1963)
	moulting fluid	*Bombyx mori*	Jeuniaux (1963); Kimura (1976)
	moulting fluid	*Manduca sexta*	Bade (1974, 1975, pers. comm.)
	moulting fluid	*Hyalophora cecropia*	Passonneau & Williams (1953)
Chitobiose [E.C.3.2.1.29 (chitobiose + H_2O→2-acetylamino-2-deoxy-D-glucose)]	moulting fluid	*Bombyx mori*	Jeuniaux (1963); Kimura (1976)
Dipeptidase [E.C.3.4.3.0.((β-alanyl-L-tyrosine→β-alanine + L-tyrosine)]	puparium	*Sarcophaga bullata*	Bodnaryk (1971a)
Proteinases [E.C.3.4.0.0 (protein →amino acids)]	moulting fluid	*Antheraea perynii*	Katzenellenbogen & Kafatos (1971a, b)
	moulting fluid	*Hyalophora cecropia*	Passonneau & Williams (1953)
	moulting fluid	*Manduca sexta*	Bade and Shoukimas (1974)
Esterases [E.C.3.1.0.0 (ester + H_2O →alcohol + carboxylic acid)]	epicuticle	*Calpodes ethlius*	Locke (1964)
	moulting fluid	*Antheraea pernyii*	Katzenellenbogen & Kafatos (1970, 1971c)

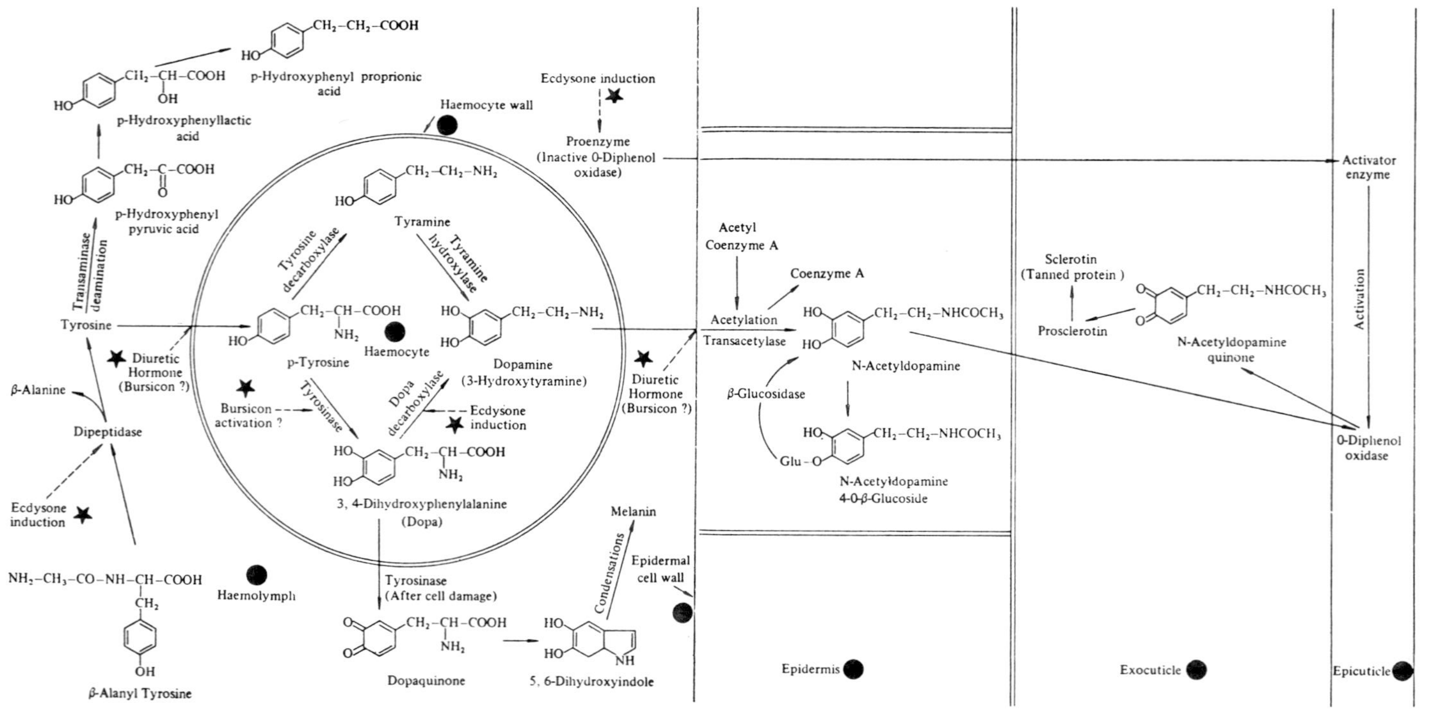

Fig. 19. A diagram relating the sequence of tanning metabolism reactions to their locations within the body. (From Neville, 1975.)

needs no added endogenous substrate to demonstrate enzyme activity) appears to be present in an inactive form (not a pro-enzyme) in cuticle of moulting *M. sexta* pharate pupae (Bade, 1974). This endogenous chitinase appeared to be activated by exposure of inactive cuticle to aliquots of larval–pupal moulting fluid (Babe and Stinson, 1976). Therefore, as a control mechanism to account for the initiation of cuticle digestion, activation of endogenous chitinase was proposed, since moulting fluid was required before any activity could be detected. During the latter part of the larval–pupal transformation, both unmasking of endogenous cuticle chitin and the seeming activation of endogenous chitinase occurs. Bade (1978) proposed that the function of a trypsin-like molecule present in moulting fluid (Bade and Shoukimas, 1974) is to alter the state of the chitin in the intact cuticle so that chitinase can attach to the chitin substrate and initiate its digestion.

Considerable endogenous chitinase activity can be measured in cuticle following the secretion of larval–pupal moulting fluid as well as in moulting fluid proper. Moreover, as the secretion and resorption of moulting fluid continues, the apparent activity of endogenous chitinase in the larval cuticle increases, suggesting the unmasking of activity already present in the cuticle. Unfortunately, this interpretation is incorrect, since the endogenous chitinase activity present in cuticle originates in the moulting fluid (Bade, 1978). Apparently, the endogenous chitinase secreted into moulting fluid preliminary to the larval–pupal ecdysis slowly diffuses into cuticle, thereby eliciting cuticle digestion.

A variety of chitin splitting enzymes can also be shown to be present in *M. sexta* moulting fluid. When moulting fluid is incubated for several hours with colloidal chitin, only 10% of the protein (and trypsin-like protease activity) is absorbed. However, closer examination reveals that over 50% of the chitinase activity is removed (Bade, 1978). Further attempts to selectively absorb the remaining moulting fluid chitinase activity meet with little success. Thus, at least two chitinases are present in moulting fluid (Bade, 1978). Based on these observations by Bade, it appears that the endogenous chitinase that one finds in cuticle in increasing activity preliminary to ecdysis is the moulting fluid chitinase that readily binds to chitin substrate.

A low grade exogenous chitinase activity can also be measured in new and inter-moult cuticle (Bade, 1978). This enzyme is directed entirely against exogenously supplied chitin substrate, since endogenous activity can not be elicited. Exogenous chitinase is laid down in cuticle at the time of cuticle synthesis, but in the absence of an exogenous supply of substrate, can not initiate cuticle degredation. The probable function of this enzyme lies in the repair of the inter-moult cuticle (see Lai-Fook, 1966).

In summary, endogenous cuticle chitin is not degraded by the low exogenous chitinase activity present in the inter-moult cuticle prior to the onset of moulting fluid secretion. Endogenous chitinase is secreted into the moulting space with the moulting fluid. Cuticle chitin becomes able to attack moulting fluid chitinase and to serve as substrate for it only after a trypsin-like protease has prepared the chitin substrate for enzyme attack. One of at least two chitinases present in moulting fluid then slowly diffuses into cuticle and is responsible for the endogenous chitinase activity associated with cuticle preliminary to the respective stage specific ecdyses. Exogenous chitinase is present in cuticle at all times and is most probably involved in the repair of damaged cuticle.

5 Moulting fluid: composition, secretion and resorption

Ecdysis and moulting is always preceded by the introduction of moulting fluid into the ecdysial space with its accompanying processes of cuticle digestion and hydrolysate resorption. The nature of moulting fluid, its mechanism of formation, and the role of the integumentary epithelium in the elaboration and resorption of this fluid during the larval–pupal transformation of *Hyalophora cecropia* and *Manduca sexta* are described in this section.

5.1 CONCENTRATION OF POTASSIUM IN MOULTING FLUID

In a report on the composition of moulting fluid collected from Cecropia silkmoths during the pupal–adult transformation, Passonneau and Williams (1953) qualitatively determined in both moulting fluid and haemolymph the presence of potassium, but failed to quantify the amount present. Since these authors noted significantly lower protein and significantly higher water contents, respectively, in moulting fluid than in haemolymph, they concluded that (*a*) moulting fluid is appreciably hypo-osmotic to haemolymph, and (*b*) moulting fluid is an "inert" solution containing little inorganic material. Some 20 years passed before the cationic and organic compounds of moulting fluid were quantitatively determined (Jungreis and Tojo, 1973; Jungreis, 1974, 1978b). Employing Cecropia silkmoths or synthetic diet fed tobacco hornworms (Riddiford, 1968; Yamamoto, 1969; Bell and Joachim, 1976), Jungreis and co-workers measured in *H. cecropia* or *M. sexta* pharate pupae the concentrations of potassium in integument, haemolymph and moulting fluid proper (Jungreis *et al.*, 1973; Jungreis and Tojo, 1973; Harvey *et al.*, 1975; Jungreis, 1974, 1978a, b; Jungreis, unpublished). In *H. cecropia*, potassium was present at concentrations of 149 ± 5 μ Eq/g and 132 ± 7 μ Eq/ml in intact integument and moulting fluid, respectively, while haemolymph potassium – determined simultaneously in the same animals – was 31 ± 3 μ Eq/ml. Thus, despite the accumulation of K^+ in the exuvial space, the concentration of potassium in haemolymph is maintained at the same level throughout all stages in development (Jungreis, Jatlow and Wyatt, 1973; Harvey *et al.*, 1975; Jungreis, 1978b; Tables 2–5). Moulting fluid is presumably a secretory product of the integumentary epidermal cells (see Locke and Krishnan, 1973; Jungreis and Harvey, 1975; Johnston and Jungreis, 1977; Jungreis and Hikida, 1978), although to date, its site or time of synthesis during the respective moults in these or other species has not been established with confidence. Reasoning that moulting fluid would be in ionic equilibrium with integumentary epidermal cells rather than with haemolymph, it was not unexpected that the concentrations of potassium in moulting fluid and integument were equivalent but dissimilar to that in haemolymph. However, information on the osmotic pressure of these fluids, which could provide insight into the mechanism of moulting fluid formation was at that time unavailable.

Using leaf fed *H. cecropia* silkmoths, Jungreis (1974) collected moulting fluid and haemolymph during the larval–pupal transformation (LPT) and measured both the concentration of potassium and the osmotic pressure in these fluids. The osmotic pressure of haemolymph was similar to that previously measured (Jungreis *et al.*, 1973), whereas moulting fluid – measured for the first time – had an osmotic pressure 38% greater than that of haemolymph (Table 3). Further, potassium occupies in moulting fluid a far larger fraction of the total osmotic pressure than it does in haemolymph with concentrations of potassium in the two fluids, of 176 ± 9 and $53 \pm 7\ \mu$Eq/ml,

TABLE 2 The concentration of potassium in haemolymph moulting fluid and integument of synthetic diet reared *H. cecropia* before, during and after the larval–pupal transformation (modified after Jungreis and Tojo, 1973; Jungreis *et al.*, 1973; Harvey *et al.*, 1975; Jungreis, 1978; Jungreis, unpublished); ( ), number of animals

Stage	Concentration Integument	mEq/l ± S.E. × Moulting fluid	Haemolymph
5th instar larvae	133 ± 5 (4)	–	34 ± 1 (4)
Initiation of spinning	133 ± 13 (4)	–	22 ± 0 (4)
Apolysis + 3 days	150 ± 6 (8)	–	35 ± 4 (8)
36–12 hours before larval–pupal ecdysis	149 ± 5 (4)	132 ± 7 (4)	31 ± 3 (4)
0–7 days after the larval–pupal ecdysis	120 ± 22 (5)	–	40 ± 4 (2)

TABLE 3 The inorganic ion composition and osmotic pressure of foliage and synthetic diet reared *H. cecropia* and synthetic diet reared *M. sexta* haemolymph and moulting fluid collected 36–12 h (*H. cecropia*) or 24–3 h (*M. sexta*) before the larval–pupal ecdysis

	Concentration (μEq/ml ± S.E. ×)			
	H. cecropia		*M. sexta*	
	Haemolymph	Moulting fluid	Haemolymph	Moulting fluid
K^+	52.7 ± 6.6 (3)	176 ± 8.7 (3)	38.5 ± 3.8 (16)	163.8 ± 11.3 (16)
Na^+	6.0 ± 0 (2)	1.5 ± 1.5 (2)	0.91 ± 0.28 (16)	0.87 ± 0.17 (16)
Ca^{++}	13.3 ± 0.7 (8)	14.1 ± 0.8 (8)	*ca.* 10	*ca.* 10
Mg^{++}	54.4 ± 5.6 (6)	15.5 ± 1.7 (6)	*ca.* 50	–
Cl^-	19.1 ± 1.7 (14)	24.8 ± 3.2 (14)	25.8 ± 1.5 (14)	6.7 ± 0.62
HCO_3^-	28.2 ± 6.13 (7)	85.7 ± 15.0 (7)	10.0 ± 2.63 (18)	90.3 ± 13.0 (18)
Osmotic pressure (mOsm)	330.5 ± 12.8 (6)	463.8 ± 13.7 (6)	319.2 ± 3.3 (15)	317.2 ± 6.1 (15)

*modified after Jungreis *et al.* (1973); Jungreis (1974, 1978b); unpublished

TABLE 4 The pH and concentrations of potassium, bicarbonate and chloride in haemolymph and moulting fluid of diet reared *M. sexta* (from Jungreis, 1978b)

	Haemolymph (mM)				Moulting fluid (mM)			
	pH	K^+	HCO_3^-	Cl^-	pH	K^+	HCO_3^-	Cl^-
Early	6.59 ± 0.03 (8)	39 ± 1 (6)	9 ± 3 (6)	31 ± 2 (3)	7.21 ± 0.05 (8)	70 ± 10 (6)	71 ± 28 (5)	7 ± 1 (3)
Middle	6.56 ± 0.01 (12)	33 ± 6 (5)	11 ± 6 (7)	22 ± 1 (3)	7.11 ± 0.02 (12)	140 ± 10 (5)	84 ± 20 (7)	5 ± 2 (4)
Late	6.54 ± 0.05 (5)	46 ± 4 (5)	10 ± 4 (5)	25 ± 2 (5)	7.21 ± 0.07 (5)	180 ± 14 (5)	100 ± 31 (5)	6 ± 1 (5)

TABLE 5 The pH and concentrations of potassium, chloride and bicarbonate in haemolymph and moulting fluid of diet and foliage reared *H. cecropia* (Jungreis, unpublished)

	Haemolymph				Moulting fluid			
	pH	K^+	HCO_3^-	Cl^-	pH	K^+	HCO_3^-	Cl^-
			LABORATORY REARED					
Early	6.69±0.03 (8)	33	5±2 (5)	19±3 (7)	7.28±0.03 (3)	130	24±11 (5)	35±7 (7)
Middle	6.68±0.03 (10)	33	5±2 (9)	19±1 (10)	7.39±0.03 (3)	130	29±10 (9)	31±3 (10)
Late	6.66±0.02 (5)	33	4±1 (5)	25±2 (9)	7.36±0.07 (3)	130	45±13 (5)	30±1 (8)
			FOLIAGE REARED					
Early	6.74 (1)	50	14 (1)	23 (1)	7.21 (1)	180	41.0 (1)	64 (1)
Middle	6.69±0.02 (3)	50	33±11 (3)	18±2 (3)	7.21±0.03 (3)	180	93±30 (3)	21±3 (3)
Late	6.68±0.01 (3)	50	28±10 (3)	25±6 (3)	7.09±0.04 (3)	180	104±20 (3)	20±1 (3)

respectively. During resorption of moulting fluid – when osmotic and ionic differences were still present between fluid compartments – appreciable quantities of pigments and products of the cuticular hydrolysate (see Bade, 1978) originating in the larval endocuticle and epidermis (but now present in moulting fluid) were being transported across the integumentary epithelium, and – after passing through the haemolymph – were taken across the midgut epithelium into the closed lumen of the midgut (Jungreis, 1974; unpublished). This constitutes evidence that selected movement of molecules between these fluid compartments was taking place at a time when an osmotic gradient in a direction opposite to pigment and hydrolysate movement was present (Jungreis, 1974). In laboratory reared *H. cecropia*, dissimilar levels of chloride exist between moulting fluids and haemolymph (see Table 5), again indicating restricted movement between these two fluid compartments. Tight junctions, a natural cytological barrier between epithelial cells, are restricted to chordates and are not found in arthropods (Flower and Filshie, 1975; McNutt and Weinstein, 1973; Noirot and Quennedey, 1974; Satir and Gilula, 1973; Staehelin, 1974). However, septate junctions are present in great abundance (Flower and Filshie, 1975) and Lord and DiBona (1977) have found that the septate junction has osmotic properties similar to those of chordate tight junctions. Thus, an effective barrier to ion movement between moulting fluid and haemolymph is present. However, in the absence of metabolic work, the osmotic pressure of moulting fluid can not be maintained at a level higher than that of haemolymph.

In *M. sexta* reared on synthetic diet, larval–pupal moulting fluid is slightly hypo-osmotic to haemolymph in early moulting fluid becoming iso-osmotic to haemolymph during formation of middle and late moulting fluid (Jungreis,

TABLE 6 The osmotic pressures of haemolymph and moulting fluid is synthetic diet reared *Manduca sexta* 12–5 h before the larval–pupal ecdysis. Equal volumes of fluid were pooled from 2–5 animals prior to analysis. Values represent the mean ± standard errors for 5 separate pooled samples (taken from Jungreis, 1978b)

	Osmotic pressure (mOsm)	
	Haemolymph	Moulting fluid
12 h before the larval–pupal ecdysis (Early)	317.4±9.8	288.8±8.1
8 h before the larval–pupal ecdysis (Middle)	308.4±9.8	331.1±18.9
5–3 h before the larval–pupal ecdysis (Late)	332.4±14.3	331.8±22.4

1978b) (Table 6). Like *H. cecropia*, the major components of *M. sexta* moulting fluid are potassium and bicarbonate (Table 3). Thus, the mechanisms for the secretion of moulting fluid in both *H. cecropia* and *M. sexta* though siimilar appear to be unique.

5.2 EVIDENCE FOR ACTIVE TRANSPORT OF POTASSIUM ACROSS THE PHARATE PUPAL INTEGUMENT "IN SITU"

The potential difference (PD) across the pharate pupal integument (PPI) of both *H. cecropia* and *M. sexta* have been measured *in situ* (Tables 7 and 8) (Jungreis and Harvey, 1975) and are very similar to the initial PD measured across isolated pieces of integument under *in vitro* conditions (see for example Schultz and Jungreis, 1977). The exuvial side of the integument was nearly always 5–10 mV positive (to the haemolymph side) in *H. cecropia* (Table 7) and 15–25 mV positive in *M. sexta* (Table 8) during those periods when moulting fluid is present in the exuvial space, whether it be during secretion or resorption of moulting fluid. Haemolymph potassium concentrations were always in the range of 35–50 mM while the respective paired concentrations in moulting fluid were 130–180 mM in both foliage and laboratory reared *H. cecropia* and *M. sexta* (Table 3; Jungreis and Tojo, 1973; Jungreis,

TABLE 7 The potential difference measured across the pharate pupal integument *in vivo* in selected body segments of Cecropia silkmoths at 60–16 h before the larval–pupal ecdysis (taken from Jungreis and Harvey, 1975)

	PD of exuvial side (+) with respect to haemolymph-side (−) (mV)			
	Abdominal segments			
Time before ecdysis of an individual (h)	Last	Fifth	Second	Second prothoracic segment
60	−20	−14	–	–
48	+7.6	−3.6[b]	−5.6[a]	+8.1
44	+8.9	−2.6[b]	+6.9	+8.4
40[b]	+8.8	+21.5	+12.0	+16.0
36[b]	+10.4	+5.6	+8.6	+12.6
30[b]	+5.6	+7.2	+9.4	+9.4
24[b]	+9.6	+0.8	+4.7	+6.6
24[b]	+6.8	+5.7	+7.8	+6.6
18–16	+2.1	+4.5	+4.6	+2.7

[a] Moulting gel rather than moulting fluid was present
[b] Secretion period

TABLE 8 Measurement of the *in situ* potential across the pharate pupal integument of *Manduca sexta* during secretion and resorption of moulting fluid. In one series of experiments, individual animals were measured, whereupon they were narcotized with CO_2 for 30 min and the potential across the integument measured in an adjacent segment of the abdomen. In a second series of experiments, animals were narcotized with CO_2 prior to the initial measurement, and then measured aerobically following recovery of the animal, 30 minutes later. Ten animals were measured in each group

	Potential (mV)			
	$O_2 \rightarrow CO_2$		$CO_2 \rightarrow O_2$	
Early	16.5 ±0.8	2.4 ±0.7	1.2 ±0.3	23.6 ±1.2
Middle	14.6 ±0.4	5.7 ±0.4	1.6 ±0.5	18.8 ±0.8
Late	17.7 ±0.5	5.6 ±0.2	2.0 ±0.5	24.4 ±1.5

1974, 1978b). Potassium equilibrium potentials, E_K, were calculated from these data (by assuming equivalent activity coefficients in both fluids) using the Nernst Equation:

$$E_K = \frac{-RT}{zF} \ln \frac{[K]_{exuvial}}{[K]_{haemolymph}}$$

where R is the gas constant, T is the absolute temperature, z is the valence of potassium, F is Faraday's constant, and K is the potassium concentration present in the respective fluid compartments. Solution of the Nernst equation reveals an E_K that is negative (i.e. opposite to that of the measured PD) and its magnitude is ca. -35 mV for both species regardless of diet. If the observed trans-epithelial PDs of $+5 - +25$ mV (Tables F, G) equal the sums of the respective K pump PDs + Nernst equilibrium PDs (=electrochemical PD = -35 mV), then the PD associated with the K pump must be $+40$ – $+60$ mV, values considerably lower than those associated with the K pump in the larval lepidopteran midgut ($+120$ – $+180$ mV, Blankemeyer, 1977), but still quite respectable. Clearly potassium is not in equilibrium across the PPI. This result indicates that potassium is either actively transported across the integument from haemolymph to exuvial space side during this period, or that its flux in this direction is coupled to that of another actively transported ionic species.

Comparison of the transport capacity as evidenced by the PD in *M. sexta* and *H. cecropia* reveals a significantly higher PD in *M. sexta* than in *H. cecropia* (Table 7 contrasted to Table 8). At constant tissue resistance, higher PDs reflect higher currents or greater rates of K transport. The integumentary resistance (measured *in vitro*) of *M. sexta* (350 Ω) is considerably less than that in *H. cecropia* (1200 Ω), such that at the the same trans-epithelial PD, *M. sexta's* integument would transport K at a greater rate than would *H. cecropia's* integument. Hence, based on trans-epithelial PDs the duration of moulting fluid secretion and resorption in *M. sexta* would be and is shorter than that observed in *H. cecropia*. This reasoning assumes that the rate at which the "old cuticle" is digested is dependent both upon the rate at which moulting fluid enters the exuvial space, and the rate at which the cuticular hydrolysate is reabsorbed.

5.3 EVIDENCE FOR ACTIVE TRANSPORT OF POTASSIUM ACROSS THE PHARATE PUPAL INTEGUMENT "IN VITRO"

The single most important technique in the study of epithelial transport (if one excludes isotopes) is via voltage clamping (called the short-circuit method when 0 mV is used as an end point) as introduced by Ussing and Zerahn (1951). The short-circuit current (I_{SC}) that one measures represents the sum of all the net fluxes measured simultaneously. In using this method, the electrical potential across the epithelial membrane is brought to zero by an external current. This electrical potential results from the passive movement of ions down their electrochemical gradients as well as from active transport of various charged molecules. Because electrochemical gradients will establish potential differences across semi-permeable membranes, the isolated preparation is bathed in solutions of identical ionic composition. Following voltage clamping of the PD at 0 mV, only actively transported cations or anions have a net movement across the epithelium.

The pharate pupal integuments of *H. cecropia* and *M. sexta* can be dissected free from the larval cuticle at 36–0 or 24–0 hours, respectively, before the larval–pupal ecdysis, when the former consists of a delicate cuticle (*H. cecropia*) or a tough cuticle (*M. sexta*) developing over a one-cell thick layer of pupal epidermal cells. Midgut, fat body and other adhering tissues are gently scraped from the haemolymph side of the integument and the larval cuticle separated carefully at the major points of attachment, the spiracular junctions. The isolated integument is then transferred as a flat sheet of epithelium to a glass or plexiglass chamber (see Wood, 1972), and tied in place with fine cotton twine. An area of 0.75–1.00 cm^2 is thus exposed to separated though identical bathing solutions, thereby eliminating diffusion based electro-chemical potentials. The potentials observed were present *in situ* and *in vitro* across the entire length of the pharate pupal integument (Tables 7, 8, 9). Measured potentials were normally under +20 mV – and frequently only +5 mV – and are reminiscent of those associated with insect Malpighian tubules and salivary glands (Maddrell, 1971; Berridge and Prince, 1972); and marginally to those of the gall bladder (Diamond *et al*., 1971). When pieces of integument isolated during the period from 36 to 3 hours before ecdysis were short-circuited (thereby eliminating a net flux for passively transported ions), a net potassium flux toward the exuvial side was always observed. However, this net flux was not always sufficient to account for the measured short-circuit current, especially during resorption of moulting fluid. The ratios of flux towards the exuvial side (influx) to flux towards the haemolymph-side (efflux) were calculated under both open-circuit and short-circuit conditions and compared to the values predicted by the flux ratio equation (see Ussing *et al*., 1960).

$$\frac{J_{K}{}^{+}\,\mathrm{in}}{J_{K}{}^{+}\,\mathrm{out}} = \frac{[K^{+}]\,\mathrm{in}}{[K^{+}]\,\mathrm{out}}\, e^{[(PD)zF/RT]} \tag{2}$$

where $J_{K}{}^{+}$ in is the uni-directional potassium flux toward the exuvial side, $J_{K}{}^{+}$ out is the uni-directional potassium flux toward the haemolymph-side, and PD is the potential difference. Following short-circuiting, the measured flux ratios exceeded calculated ratios by factors ranging from 42 to 2.4, whereas under open-circuit conditions they ranged from 64 to 5.4 (Table 9). Clearly potassium is being actively transported from haemolymph-side to exuvial side of the isolated integument. The net potassium flux at 30 hours before ecdysis accounted for up to 105% of the short-circuit current, although a fraction of the short-circuit current normally remained unassigned (Fig. 20, Table 9). At 16–24 and 2–1 hours before ecdysis (Fig. 21, Table 9),

TABLE 9 Unidirectional potassium fluxes across the integument isolated from the pharate pupa and bathed in oxygenated 32-K±-S-tris (pH 8.60). (The influx (J_{in}) from haemolymph to exuvial side was measured with ^{42}K and the efflux (J_{out}) was measured with ^{86}Rb under both open-circuit and short-circuit conditions. I_{sc} is given for the short-circuit experiments. Insects fed on foliage are designated by (F) and those fed on synthetic diet by (S). Measurements were made at from 20.5 to 26 °C) (taken from Jungreis and Harvey, 1975)

Time before Ecdysis of an individual (h)	Steady state PD (mV)	I_{sc} (μ-equiv./h)	%[a]	J_{in}/J_{out} Observed	J_{in}/J_{out} Predicted[b]
30 (S)	+6.0	0.212	105	4.33	1.00
24 (F)	+9.6	–	–	12.50	1.46
24 (F)	+8.4	–	–	5.41	1.39
24 (S)	+3.9	0.283	49	7.54	1.00
16 (F)	+14.0	–	–	41.80	1.73
16–12 (F)	+10.0	–	–	15.70	1.48
8–6 (F)	+12.0	–	–	6.76	1.60
8–6 (F)	+5.4	0.0727	83	3.65	1.00
6 (Ant., S)	+12.4	0.148	41	2.41	1.00
6 (Post., S)	+11.0	–	–	64.40	1.54
3 (F)	+6.8	0.195	75	5.02	1.00
2–1 (Ant., S)	+7.0	0.122	9	1.12	1.00
2–1 (Post., S)	+4.6	–	–	0.96	1.20

[a] Percent of the short-circuit current contributed by the net transport of potassium
[b] Predicted from the Nernst Equation based upon the observed steady state PD.

potassium flux ratios were close to predicted values (i.e. 1.0), and Jungreis and Harvey (1975) concluded that active potassium transport had ceased. (Data from Fig. 21 and Table 9 appear contradictory, with respect to active K transport at 16–12 h before ecdysis, but this can be attributed to initial difficulties in staging (see Appendices). The substantial I_{SC} present at those times when K^+ agreement is absent must be due to the active transport of an anion in a direction opposite that of potassium (which occurred during secretion of moulting fluid) since no other cation is present.

5.4 DECAY PROFILE IN "HYALOPHORA CECROPIA"

The shape of the short-circuit decay profile provides information regarding maximal transport capacity as well as the time sequence during which the transport capacity becomes lost *in vitro*.

Though frequently preceded by a transient negative phase, the exuvial side of the pharate pupal integument is always positive with respect to the haemolymph side (Figs. 20–24, Tables 7 and 8). Differences in the decay profile associated with secretion versus resorption of moulting fluid appear to be absent. This suggests that the active potassium transport taking place in the

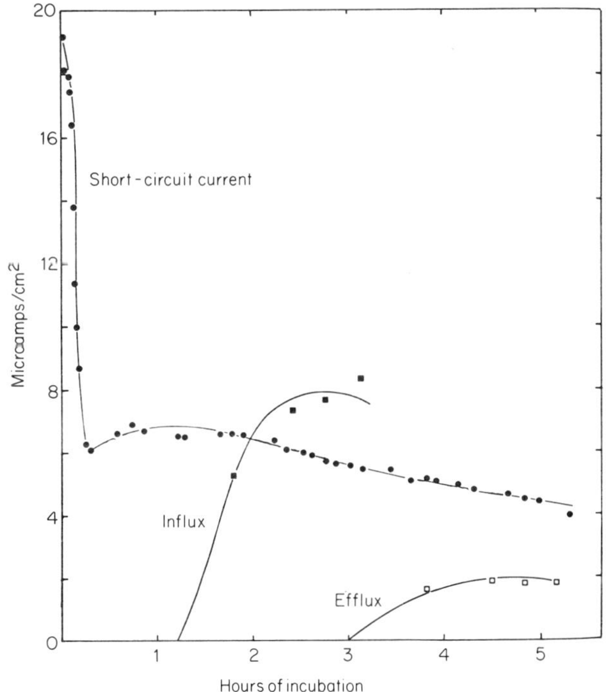

Fig. 20. Time course of the short-circuit current and unidirectional potassium fluxes across the pharate pupal integument of *Hyalophora cecropia* isolated during the secretion of moulting fluid at 30 h before the larval–pupal ecdysis. The integument was bathed on both sides with a solution containing 32 mM KCl, 5mM glucose, 200 mM sucrose and 5 mM tris-HCl (pH 8.6). The absence of an initial reversal of potential and the noticeable short-circuit decay after 1 h of incubation is typical of the integuments response in the absence of Ca^{++} or Mg^{++} in the bathing solution. The influx (haemolymph→exuvial side) ■ and efflux (exuvial→haemolymph side □ were measured sequentially with ^{86}Rb and are expressed in units of current (microamps/cm²). The net flux of potassium towards the exuvial side of the pharate pupal integument accounts for the entire short-circuit current (●) within experimental error. (From Jungreis and Harvey, 1975.)

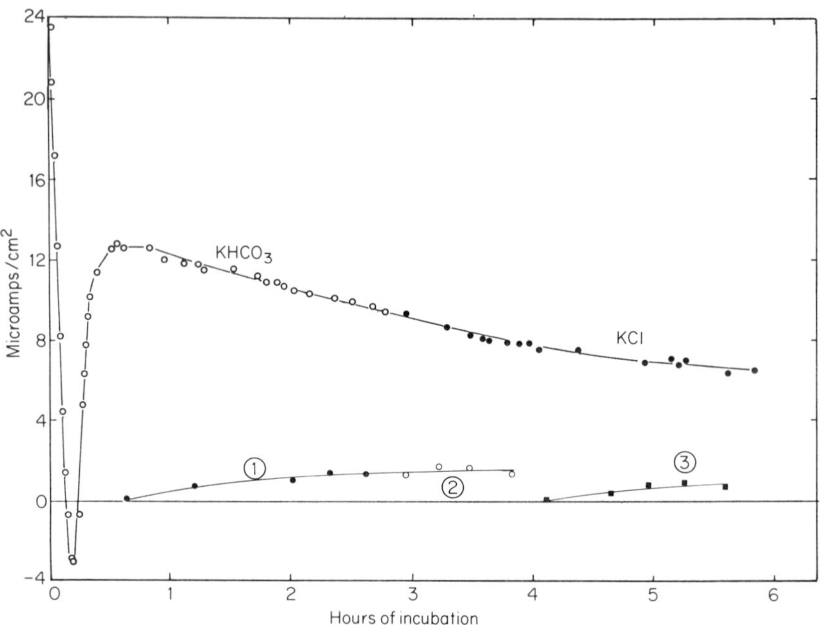

Fig. 21. Time course of the short-circuit current and unidirectional potassium fluxes across the pharate pupal integument of *Hyalophora cecropia* isolated during resorption of moulting fluid at 16–14 h before the larval–pupal ecdysis. The integument was initially bathed for 2.75 h in a solution containing 32 mM $KHCO_3$, 200 mM sucrose, and 5 mM tris-HCl (pH 8.6) (○), and then for 3 h in a solution where KCl was substituted for $KHCO_3$ (●). The short-circuit current was not affected when KCl was substituted for $KHCO_3$. The influx (haemolymph→exuvial side) in $KHCO_3$ (1) and in KCl (2), and the efflux (exuvial→haemolymph side) in KCl (3) were measured sequentially with ^{86}Rb and are expressed in units of current (microamps/cm^2). The net flux of potassium could account for no more than 10% of the short-circuit current within experimental error. It is concluded that active transport of potassium during the resorption of moulting fluid is not responsible for the observed short-circuit current.

living insect (Jungreis 1973; Jungreis and Harvey 1975) continues in the isolated preparation (Fig. 21).

The ability of the pharate papal integument to maintain a trans-epithelial potential *in vitro* is dependent upon the ionic composition of the bathing solution. When bathed in a saline containing *ca.* 30 mM KCl, 5 mM, $MgCl_2$, 5 mM glucose, 5 mM tris-HCl (pH 8.6) and 200 mM sucrose, the following short circuit or potential decay profile is obtained in *H. cecropia* (Figs 20–23): the potential (under 25 mV) and short-circuit current (under 25 microamps/cm^2) are initially positive, but both drop below zero within 10 minutes by "reversing" the net flux (presumably) with minimal values reached after about 20 minutes. During the second 20 minutes of incubation in this saline, the PD increases at a rate equal to its decline with a maximal

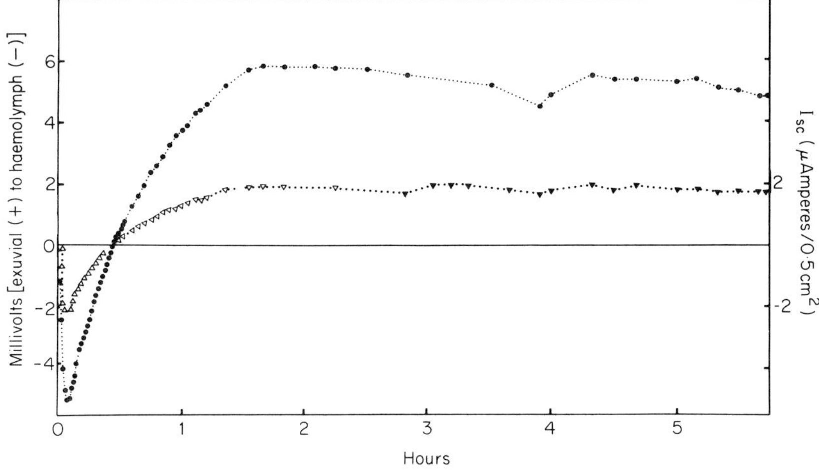

Fig. 22. Generalized response of the pharate pupal integument isolated from *Hyalophora cecropia* at 6 h before the larval–pupal ecdysis to open and short-circuit conditions. The integument was bathed on both sides with a solution containing 32 mM KCl, 5 mM $CaCl_2$, 5 mM $MgCl_2$, 200 mM sucrose and 5 mM tris-HCl (pH 8.6). The potential (PD) was measured continuously for the first 2.75 hours with the tissue short-circuited for 15–30 s every 4–15 min, whereupon it was continuously short-circuited for 3.25 h with the P.D. determined every 10–15 min under open circuit conditions. Note the typical initial potential or current reversal followed by a slow increase in these parameters until the maximal tissue response is reached after 1.75 (to 3) h. Little decay in these parameters is noted for ca. 5 h following attainment of the maximal parameter response. Short-circuit current discontinuously measured under open-circuit conditions (△). Short-circuit current measured continuously under short-circuit conditions ▲. PD across the pharate pupal integument (exuvial side positive to haemolymph side) measured continuously or discontinuously (●).

positive value (frequently exceeding the initial PD or short circuit current) reached only after 3 hours of incubation. This maximal potential is stable for at least 5 additional hours with up to 3 hours required before steady state potassium fluxes are achieved (influx or efflux).

A second decay profile is observed when calcium and magnesium are deleted from the bathing solution (Figs 20–21, 24). Under these experimental conditions, the complete potential or short-circuit current reversal requiring 3–5 hours in the presence of calcium and magnesium is limited to *ca.* 20 minutes with a marked decline in short-circuit current occurring thereafter.

The basic difference in decay profiles in the presence or absence of calcium and magnesium may reflect changes in extracellular spaces and epidermal cell vacuolization (see Hikida and Jungreis, 1978). The absence of divalent cations may promote the rapid decay of these compartmented cation or anion pools. Since the respective decay profiles occur during both secretion and resorption of moulting fluid in *H. cecropia*, the initial phases of the decay

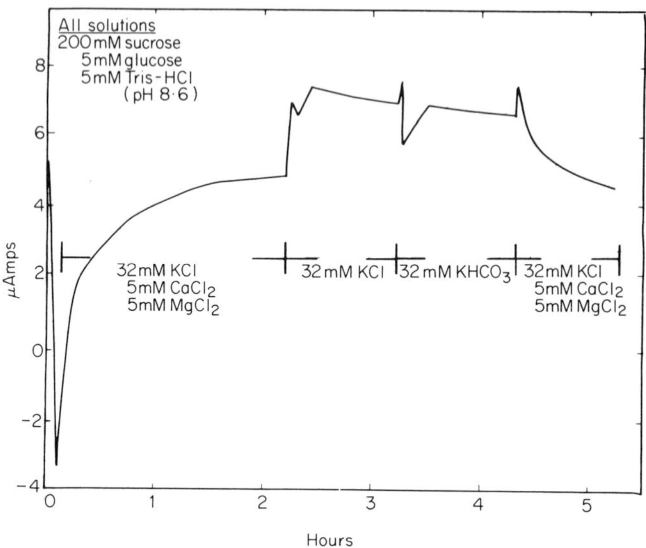

Fig. 23. The effects of calcium and magnesium, and the effects of replacing Cl^- with HCO_3^- on the time course of the short-circuit current measured across the pharate pupal integument (exuvial side positive to haemolymph side) of *Hyalophora cecropia* at 16–12 h before the larval–pupal ecdysis during resorption of moulting fluid. The current reversal and lengthy period required before the maximal current is attained are characteristic responses of the integument in solutions containing Ca^{++} and Mg^{++}. When Ca^{++} and Mg^{++} are removed from the bathing solution, the rapid increase in short-circuit current and the onset of a pronounced decay in that current are also characteristic responses of the pharate pupal integument during both secretion or resorption of moulting fluid or during the post-resorptive period *in vivo*.

profiles (i.e. those associated with potential reversals) are probably related to factors other than the potassium pump, whereas the latter portion of the decay profile is probably a reflection of changes in pump capacity.

The short-circuit current is affected by the concentration of potassium in the bathing solution, concentrations under 32 mM causing a drop in the I_{SC}, while those above 32 mM add little to the overall I_{SC} (Fig. 24). During both secretion and resorption, the anion accompanying actively transported potassium has little effect on the I_{SC} (Figs. 22 and 24), since the substitution of bicarbonate for chloride is without effect.

The potential (PD) and the short-circuit current maintained across the pharate pupal integument both *in situ* and *in vitro* is dependent upon the availability of oxygen and is therefore an energy dependent process (Fig. 25). When studied *in vitro*, only the exuvial side of the tissue is sensitive to the concentration of oxygen (Fig. 25). The changes in trans-epithelial potential occurring during oxygen deprivation are due to changes in I_{SC} and active potassium transport (see Table 8). One interesting feature of the response

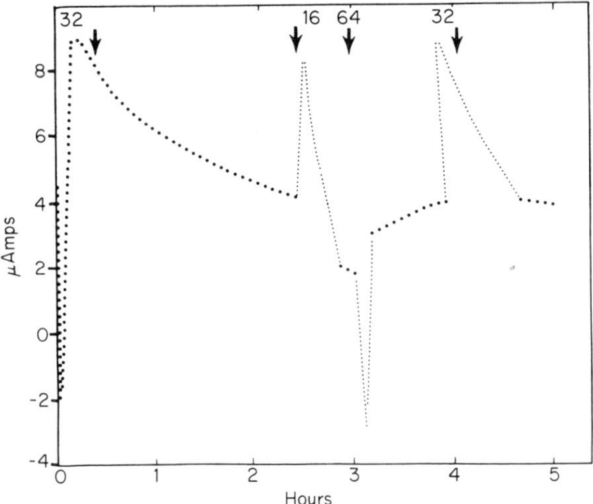

Fig. 24. The effect of the concentration of potasium in the bathing solution on the short-circuit current of the pharate pupal integument of *Hyalophora cecropia* measured at 36 h before the larval–pupal ecdysis during the secretion of moulting fluid. The basic bathing solution contained KCl, 200 mM sucrose and 5mM tris-HCl (pH 8.6). Calcium and magnesium were not present in the bathing solution and the characteristic decay profile exhibited by the integument in such solutions is readily observed. A decline in the potassium concentration from 32 to 16 mM halved the short-circuit current, whereas a doubling of the concentration of potassium from 32 to 64 mM failed to cause an increase in the short-circuit current. (From Jungreis, in preparation.)

of *H. cecropia's* integumentary epithelium *in vitro* to oxygen deprivation is the prolonged period needed to elicit a change in I_{SC} or trans-epithelial potential, namely 1–2 hours. This is in marked contrast to the response of other epithelia from this species such as the midgut, which rapidly loses its capacity to transport (as measured by I_{SC}) in less than five minutes (Harvey *et al.*, 1967; Wood, 1972; Blankemeyer, 1977).

The requirement of oxygen in the maintenance of I_{SC} and trans-epithelial potential implicate oxygen-sensitive energy-requiring processes in the active transport of potassium across the pharate pupal integument. However, this energy-dependent process is not dependent upon ouabain sensitive Na–K-ATPases, since this class of enzymes (see Keynes, 1969) is absent from epithelia of *H. cecropia* and *M. sexta* (Jungreis, 1977; Jungreis and Vaughan, 1977; Vaughan and Jungreis, 1977). Thus, addition of ouabain to the bathing medium is without effect on PD or I_{SC} (Fig. 26).

Potassium transport across the pharate pupal integument takes place in the absence of alkali metal co- or counter-transport, since agreement between I_{SC} and K^+ transport was observed when K^+ was the only cation in the incubation medium (Figs. 20–21). If choline or sodium are substituted for

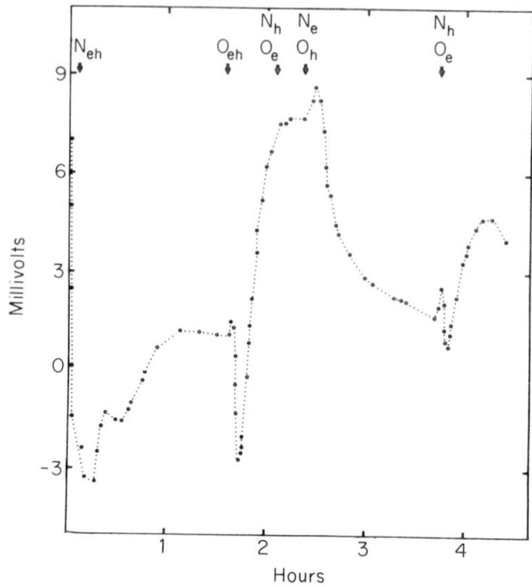

Fig. 25. Generalized response of the potential across the pharate pupal integument isolated from *Hyalophora cecropia* at 20 h before the larval–pupal ecdysis to oxygen deprivation. The integument was initially bathed on both sides with saline saturated with nitrogen containing 32 mM KCl, 5 mM $CaCl_2$, 200 mM sucrose and 5 mM tris-HCl (pH 8.6). A small potential of +1–3 mV (exuvial side positive to haemolymph) is regularly observed under anaerobic conditions. When the saline was oxygenated, the maximal potential was quickly reached. When nitrogen was re-introduced on the haemolymph side of the tissue, no effect on PD was observed, whereas re-introducing nitrogen on the exuvial side (haemolymph being oxygenated) initiated a marked loss in PD. The rate of PD decay under anoxic conditions is prolonged in *H. cecropia*, frequently requiring 2–4 h before the minimal PD (not = 0 mV) is recorded. N_{eh}, nitrogen saturated saline bathes both exuvial and haemolymph sides of the pharate pupal integument; O_{eh}, oxygen saturated saline bathes exuvial and haemolymph sides; N_hO_e, nitrogen bathes the haemolymph side, oxygen bathes the exuvial side; N_eO_h, nitrogen bathes the exuvial side, oxygen bathes the haemolymph side. (From Jungreis, in preparation.)

potassium in the bathing solution, then the ability to maintain a trans-epithelial potential drops to zero (Fig. 26). The response time for the loss in trans-epithelial potential following removal of potassium can be prolonged or rapid with differences probably arising from changes in the available extra-cellular potassium pools at the respective stages in development.

The pH of larval–pupal moulting fluid is only slightly alkaline (Table 5) and is *ca.* 0.5 pH units greater than that observed in haemolymph. When measured *in vitro*, only the exuvial side of the pharate pupal integument is sensitive to pH, (Fig. 27). Changes in pH on the haemolymph side between 3.0–11.5 having no effect on the measured I_{SC} or PD of the epithelium. The exuvial side of the pharate pupal integument is very sensitive to pH. When the pH of the bathing solution drops from 8.6 to 5.6, proportional but reversible changes in PD and I_{SC} are noted. However, when the pH on the exuvial

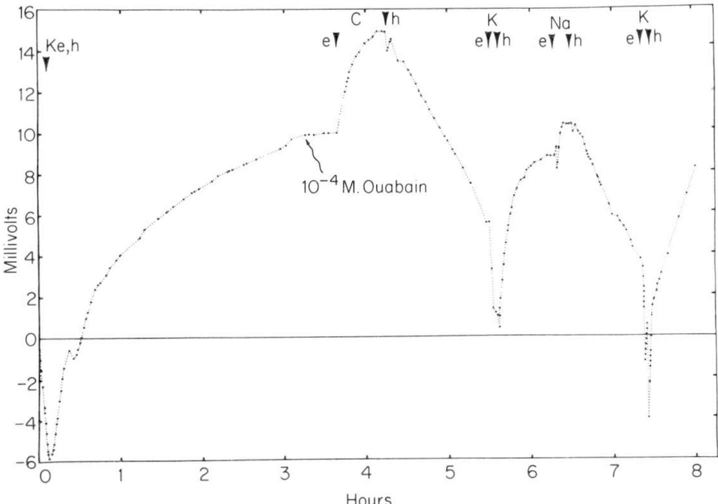

Fig. 26. The effects of ouabain, choline and sodium on the capacity of the pharate pupal integument of *Hyalophora cecropia* to maintain a PD (exuvial side positive to haemolymph) at 16–14 h before the larval–pupal ecdysis during resorption of moulting fluid. The bathing solution contained in all cases 5 mM $MgCl_2$, 5 mM $CaCl_2$, 200 mM sucrose and 5 mM tris-HCl (pH 8.6). Potassium was initially present in the solution (32 mM KCl) e, h bathing both sides of the integument ($K_{e,h}$). When the maximal potential was reached after 3 h of incubation, 10^{-4}M ouabain was added, but no change in PD was noted. The bathing solution was then changed to 32 mM choline chloride, changing in all cases the solution bathing the exuvial side before that bathing the haemolymph side ($C_{e,h}$). A slow reversible loss in PD reminiscent of the effect of anoxia was recorded for 1.5 h, after which 32 mM KCl was restored to the bathing solution ($K_{e.h.}$). The PD was restored to the steady state level within 15 m. Thereafter, 32 mM NaCl was substituted for 32 mM KCl ($Na_{e,h}$), and a reversible loss in PD comparable to that which followed substituting choline for potassium was noted. Restoration of potassium to the bathing solution again resulted in a restoration of PD. (From Jungreis, in preparation.)

side of the pharate pupal integument exceeds 10.0 even for one minute, an irreversible loss in transport capacity is noted (Fig. 27).

5.5 DECAY PROFILE IN "MANDUCA SEXTA"

The decay profile for the isolated integumentary epithelium of pharate pupal *Manduca sexta* is very different from that of *H. cecropia*. Trans-epithelial potentials measured *in situ* and short-circuit currents (Figs 27–31) are considerably larger than those observed in *H. cecropia*. However, trans-epithelial potentials measured under *in vitro* conditions appear to be less than those measured *in situ*. Unfortunately, individual animals have not been measured under both conditions, and the preceding observation may prove to be incorrect.

The maximal short-circuit current is normally not reached before 8 hours

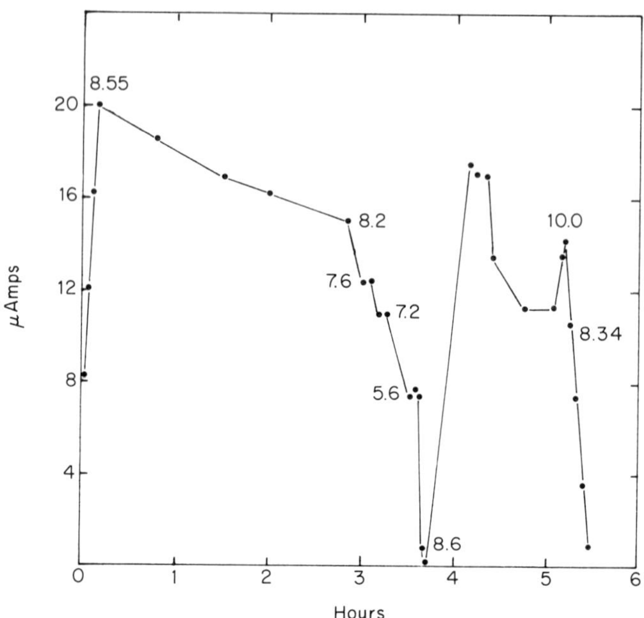

Fig. 27. Generalized effect of changes in pH of the bathing solution on maintenance of the short-circuit current across the pharate pupal integument (exuvial side positive to haemolymph side) of *Hyalophora cecropia* at 22–20 h before the larval–pupal ecdysis during secretion/resorption of moulting fluid. The bathing solution contains 32 mM KCl, 200 mM sucrose, 5 mM glucose and 5 mM tris-HCl (variable pH). Calcium and magnesium were not present in the bathing solution. Changes in pH were effective only on the exuvial side, the haemolymph side of the integument being refractory. When the pH was reduced from its initial values of 8.55 to 5.6, dramatic reversible losses in short-circuit current were noted. However, when the pH was raised to 10.0, the rapid decline in short-circuit current was irreversible. Numbers refer to the time and value of the pH of the bathing solution. (From Jungreis, in preparation.)

of incubation with times up to 12 hours not uncommon (Fig. 28). Whereas the maximal I_{SC} or PD in *H. cecropia* was sustained for more than 5 hours, that in *M. sexta* is not sustained, with a decline beginning almost immediately after the maximal value is reached. Initial potential reversals are observed only infrequently in *M. sexta*, and are of much shorter duration than those observed in *H. cecropia* (Figs 22 and 26). When calcium and magnesium are excluded from the bathing solution, the maximal short-circuit current is reached after only several hours of incubation (Fig. 29), with a marked decline in I_{SC} noted immediately. In this regard (i.e. rapid loss in transport capacity) *M. sexta* and *H. cecropia* are quite similar (compare Figs 21 and 29). Agreement between the short-circuit current and the direction of potassium transport has been determined during both secretion and resorption of pharate pupal moulting fluid (Jungreis, in preparation), but 100%

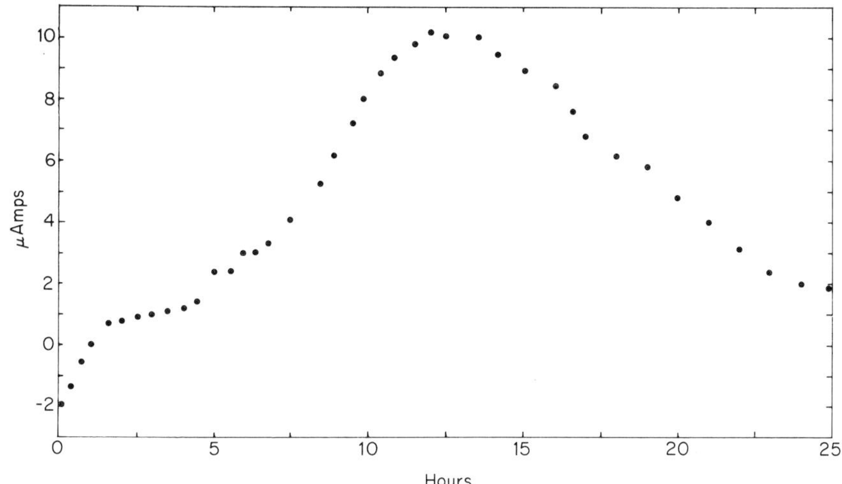

Fig. 28. Time course for the short-circuit current decay profile across the pharate pupal integument of *Manduca sexta* at 10 h before the larval–pupal ecdysis during resorption of moulting fluid. Although the initial short-circuit current was negative in contrast to *H. cecropia*, no initial current reversal is noted. Further, the maximal short-circuit current is not reached until 10–15 h of incubation, and no period of stable unchanging short-circuit current is present. The integument was bathed in oxygenated 32 mM KCl, 5 mM $MgCl_2$, 5 mM $CaCl_2$, 200 mM sucrose, 5 mM glucose and 30 mM tris-HCl (pH 8.6). (From Jungreis, in preparation.)

agreement has not been achieved. Substitution of bicarbonate for chloride in the bathing solution is without effect on the I_{SC} (unpublished observations).

The requirement of oxygen for maintenance of the short-circuit current is more pronounced in *M. sexta* than in *H. cecropia* (Fig. 30) with a complete loss in I_{SC} occurring within 5 minutes. The requirement for oxygen is just as evident during secretion as during resorption of moulting fluid. Elimination of calcium and magnesium from the bathing solutions is without effect on the rapid responsiveness of the integument of *M. sexta* to anaerobic conditions. As was noted in *H. cecropia*, ouabain sensitive Na-K-ATPases are not involved in maintenance of I_{SC} or trans-epithelial potential in *M. sexta* epithelia (Jungreis, 1977; Jungreis and Vaughan, 1977).

The effects of pH on the capacity of the pharate pupal integument to generate a short-circuit current were unexpected and different from those responses observed in *H. cecropia*. The haemolymph side of the integument remained refractory to changes in pH, consistent with observations in *H. cecropia*. In *M. sexta*, the exuvial side is also refractory at pHs under 9.0 (Fig. 31). However, if the pH on the exuvial side is raised to 10.0, even for one minute, an irreversible loss in transport capacity resulted. Thus, the exuvial side of the integumentary epithelium of *H. cecropia* and *M. sexta* both appear to have some molecular group that is sensitive to mid-alkaline

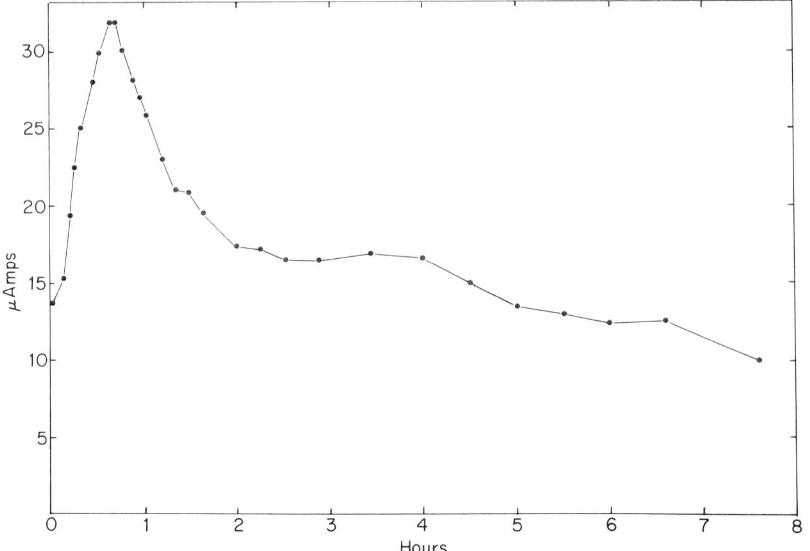

Fig. 29. Time course for the short-circuit current decay profile across the pharate pupal integument of *Manduca sexta* at 12 h before the larval-pupal ecdysis during resorption of moulting fluid. The integument was bathed in oxygenated saline identical to that described in the legend of Fig. 28 except that Ca^{++} and Mg^{++} were deleted. In contrast to the response of the integument when bathed in solutions containing Ca^{++} and Mg^{++}, deletion of these divalent cations from the bathing solution causes the short-circuit current to both reach its maximal value within 1 h, and to exhibit a biphasic decay profile thereafter, which is similar to that observed across the larval Lepidopteran midgut (see Schultz and Jungreis, 1977b), namely a fast decline requiring 30–45 min followed by a slow decline over a 6–10 h period.

pHs, for which changes in configuration result in changes in potassium transport capacity.

5.6 EVIDENCE FOR ACTIVE TRANSPORT OF BICARBONATE ACROSS THE PHARATE PUPAL INTEGUMENT

In his initial analysis of *H. cecropia* larval–pupal moulting fluid, Jungreis (1973, 1974) successfully analysed the major cationic components of both haemolymph and moulting fluid, but failed to identify the major anionic component(s), chloride being insufficient to achieve electroneutrality in moulting fluid.

In their analysis of active potassium transport across the pharate pupal integument, Jungreis and Harvey (1975) proposed that by analogy to the adult insect labial gland (Kafatos, 1968), integumentary epithelial potassium transport is accompanied by bicarbonate during the formation of larval–pupal moulting fluid. Data has now been collected which supports this proposal (Tables 4 and 5; Fig. 26; Jungreis, 1978b; Johnston and Jungreis, 1977).

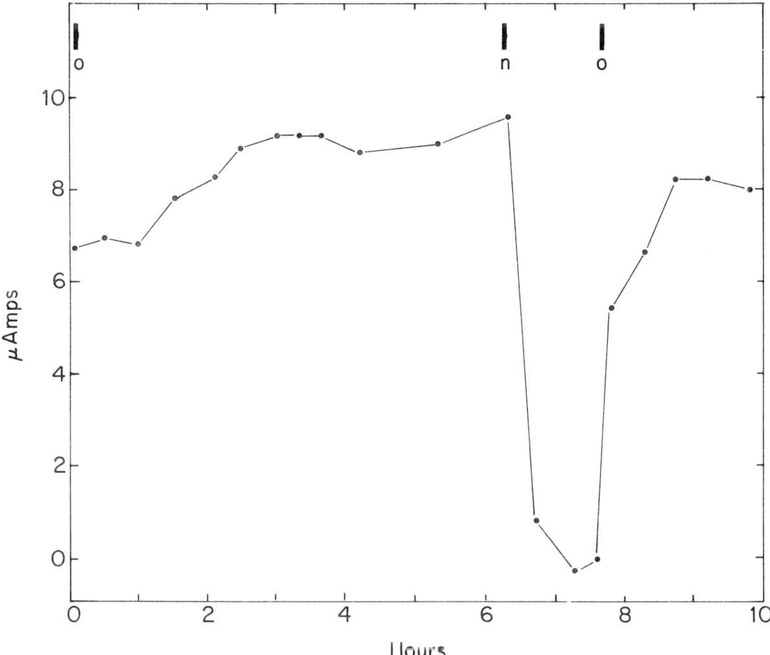

Fig. 30. Generalized relationship of the effect of anoxia on the short-circuit current (exuvial side positive to haemolymph side) maintained across the pharate pupal integument of *Manduca sexta* at 14 h before the larval–pupal ecdysis during secretion of moulting fluid. The bathing solution was identical to that described in the legend of Fig. 28. Note the slow increase in short-circuit current over (at least) a 7 h period that is characteristic of the pharate pupal integuments from *M. sexta*. When nitrogen (n) is substituted for oxygen (o) as the saturating gas, a rapid loss in short circuit current characteristics of the response of the larval-lepidopteran midgut (Wood, 1972; Blankemeyer, 1977) is noted. This response in *M. sexta* can be contrasted with that in *H. cecropia*, where the decline in short-circuit current following initiation of anoxia is extremely slow requiring hours rather than minutes to effect the minimal response. Note further that the short-circuit current across the integument of *M. sexta* drops to zero under anaerobic conditions, whereas that in *H. cecropia* is always appreciably above zero. The overshot in both PD and I_{SC} observed in *H. cecropia* following return of oxygen is absent in *M. sexta*. (From Jungreis, in preparation.)

The haemolymph cation composition of both *M. sexta* (Table 3; Jungreis, 1978b) and *H. cecropia* (Table 3; Jungreis, Jatlow and Wyatt, 1973; Jungreis, 1973, 1974) have been determined with reasonable precision in both diet- and foliage-reared animals. Analyses of the anion components though less exhaustive than those of the cations have now been completed in both species with regard to labile phosphates (Wyatt *et al.*, 1963; Jungreis, 1974; Jungreis *et al.*, 1976; Jungreis, 1978b), chloride (Jungreis, Jatlow and Wyatt, 1973; Jungreis, 1974, 1978b; Tables 3–5), bicarbonate (Jungreis, 1978b; Johnston and Jungreis, 1977; Tables 3–5), and amino acids (Jungreis, 1974, 1978b; Tables 10–11).

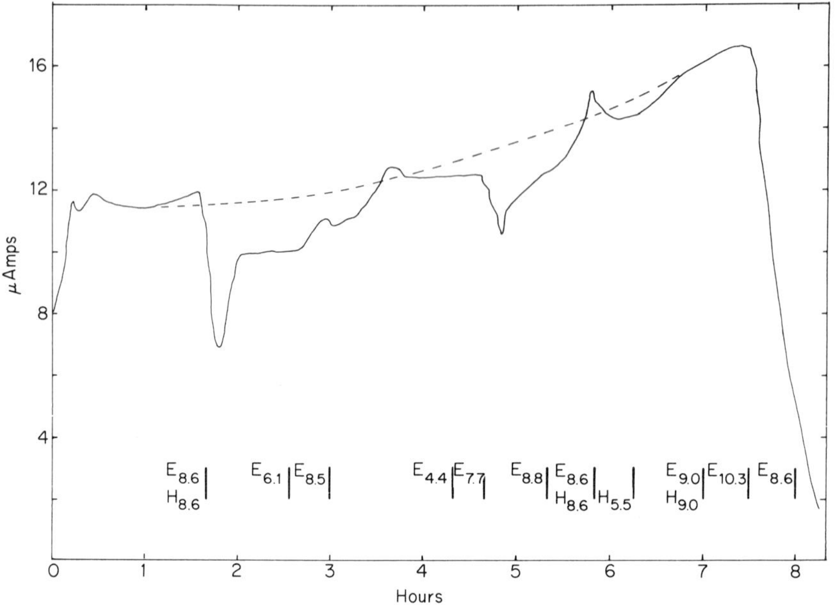

Fig. 31. Generalized response of the effect of pH on the short-circuit current across the pharate pupal integument of *Manduca sexta* at 14–12 h before the larval pupal ecdysis during secretion/resorption of moulting fluid. The bathing solution was identical to that listed in the legend of Fig. 28. The short-circuit current of the isolated integument of *M. sexta* appears to be refractory to reductions in pH as low as pH 4.4 on both haemolymph and exuvial sides, while the irreversible loss in transport capacity noted in *H. cecropia* following exposure of integument to pH's in excess of 9.8 persists in *M. sexta*. E, pH of the solution bathing the exuvial side of the integument; H, pH of the solution bathing the haemolymph side of the integument. When both E and H values are listed, solutions bathing the integument were replaced in contrast to having their pHs adjusted. The dashed line represents the projected short-circuit current had the bathing solution neither be changed nor the pH adjusted. (From Jungreis, in preparation.)

In *M. sexta*, the concentration of chloride in haemolymph is 4 times that in moulting fluid (Table 4), whereas the concentration of bicarbonate in haemolymph is only 10% of that of moulting fluid. Since the concentration of bicarbonate in *M. sexta* moulting fluid is *ca.* 100 mM, it is clearly the major anionic component and must accompany K across the integument. An increase in the concentration of haemolymph bicarbonate is noted prior to the formation of moulting fluid (Johnston and Jungreis, 1977; in preparation). If the quantity of bicarbonate is measured in haemolymph of feeding fifth instar larvae or throughout the larval–pupal transformation up to one day before the time of moulting fluid elaboration, then there is insufficient bicarbonate present even to account for the increase observed in haemolymph. Since during elaboration of moulting fluid there is insufficient bicarbonate to account for the observed increase in haemolymph, then bicarbonate

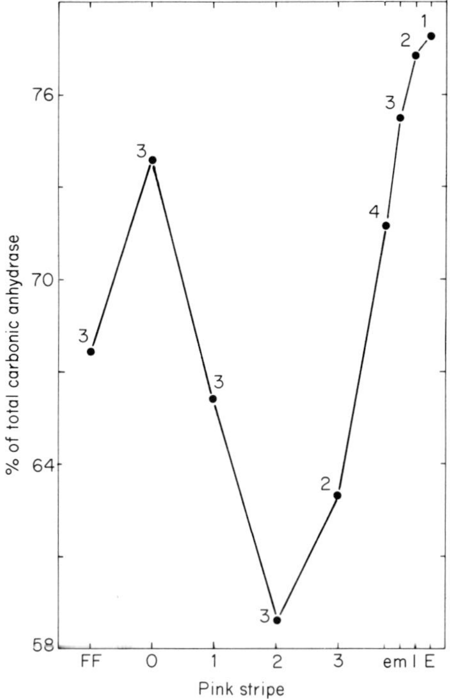

Fig. 32. The percent of the total carbonic anhydrase activity localized in integument during the larval–pupal transformation in the tobacco hornworm, *Manduca sexta*. The tissues or fluid measured were fat body, midgut, abdominal musculature, integumentary epidermis, larval cuticle, pharate pupal cuticle, moulting fluid and haemolymph. Integument normally retains the greatest portion of the total body capacity to enzymatically produce bicarbonate. The enzymatic activity of integument can be localized within the epidermal cells. An increase in the proportion of carbonic anhydrase localized in integument preliminary to and during the secretion and resorption of moulting fluid is initiated 2 days before the larval pupal ecdysis (= pink stripe + 2 days). During this period, the contribution of integument to the total activity increases from 58 to 78% (in newly ecdysed pupae). FF, feeding fifth instar larva; pink stripe, presence of ommochrome pigment along the dorsum following evacuation of the gut contents; days 0,1,2,3, are days after the onset of the pink stripe stage; e,m,l, refer to early, middle and late moulting fluid, respectively; E, day of the larval–pupal ecdysis. (From Johnston and Jungreis, 1978a; in preparation.)

present in haemolymph can not account for that appearing in moulting fluid. Furthermore, when measured as a function of carbonic anhydrase activity, the integument's capacity to produce bicarbonate exceeds that of all other body tissues and fluids combined (Fig. 32; Johnston and Jungreis, in preparation). These observations lend support to the hypothesis that bicarbonate appearing in moulting fluid is synthesized *de novo* in the integument. Apparent differences in the ability of *H. cecropia* to accumulate chloride in moulting fluid in synthetic diet reared (haemolymph/moulting fluid ratio =

0.65) but not foliage-reared animals (H/MF = 0.95) exist during both secretion and resorption of moulting fluid (Tables 3–5). The apparent exclusion of chloride from moulting fluid by the pharate pupal integument of *M. sexta* (H/MF – 4.3, Table 4) indicates that the integument serves as a barrier to free diffusion between haemolymph and moulting fluid.

Analysis of the contributions of bicarbonate and chloride to the anionic composition of *H. cecropia* moulting fluid reveals a major difference between diet- and foliage-reared animals. In foliage-reared animals, the concentrations of chloride in moulting fluid and haemolymph are virtually identical (20 mM) (Table 5), in agreement with values reported earlier by Jungreis (1974). The concentration of bicarbonate in moulting fluid is about 100 mM, four times that of the haemolymph concentration, and equal to that in *M. sexta* moulting fluid (Table 5 contrasted with Table 4). Further, bicarbonate present in *H. cecropia* moulting fluid is of sufficient magnitude to not only account for *ca.* 100% of the "missing" anionic components (see Tables 10 and 12), but also to confirm the proposal of Jungreis and Harvey (1975) that moulting fluid is primarily a potassium bicarbonate salt solution. In diet-reared *H. cecropia*, the concentration of chloride in moulting fluid is 50% greater than that in haemolymph (20 vs. 30 mM) (Table 5), while the concentration of bicarbonate in moulting fluid (30 mM), though 6 times that in haemolymph (5 mM), is still only 30% of that present in moulting fluid of foliage-reared animals. This apparent contradiction can be resolved by noting that the potassium concentration in foliage-reared (higher) vs. diet-reared (lower) moulting fluid differs by *ca.* 45 mM (Table 5). Summation of the bicarbonate-chloride concentrations in moulting fluid reveals a 45 mM differential between foliage- (higher) and diet-reared (lower) animals, a quantity of anions equal to the cation differential. The absence of reduced levels of chloride in moulting fluid from diet-reared animals suggests that these animals have a reduced capacity to produce bicarbonate relative to those reared on foliage.

5.7 OSMOTIC PRESSURE: CONTRIBUTION BY ORGANIC AND INORGANIC COMPONENTS

In both *M. sexta* and *H. cecropia* moulting fluid and in *H. cecropia* haemolymph, the sum of the contributions by the organic and inorganic components (Table 10–11) to the total osmotic pressure (Tables 3 and 6) equals the total osmotic pressure (Tables 12–13). The individual haemolymph component contributions from *M. sexta* when summed exceed the observed osmotic pressure by some 30%. However, if the presumed osmotic contribution from non-labile phosphates (i.e. total minus labile) is disregarded, then component parity with the measured osmotic pressure is achieved. Why *M.*

TABLE 10 The concentration of organic components in pharate pupal haemolymph and moulting fluid in foliage reared silkmoths *Hyalophora cecropia* (from Jungreis, 1974)

	Trehalose (μmoles/ml)	Phosphate (μmoles/ml) Total	Labile	Protein (%, W:V)	Polypeptides (%, W:V)	Amino acids (μmoles/ml)
Haemolymph	28.6 ± 2.3 (11)	76.1 ± 3.7 (11)	29 (11)	9.3 ± 0.4 (10)	2.3 ± 0.2 (10)	65.7 ± 5.1 (10)
Moulting fluid	2.1 ± 0.2 (10)	10.2 ± 4.1 (10)	0 (11)	2.8 ± 0.2 (10)	2.2 ± 0.1 (10)	78.1 ± 2.3 (10)

sexta differs from *H. cecropia* with regard to an osmotic contribution from non-labile phosphate is unclear, although the difference could be related to the fact that *H. cecropia* will be entering diapause, while *M. sexta* will continue developing into an adult following the larval–pupal ecdysis.

In *H. cecropia*, the contributions of alkali metals and bicarbonate represent only 60% of the measured osmotic pressure, regardless of diet, whereas they represent 80% of the osmotic pressure in *M. sexta*. The significance of this difference has not been determined.

5.8 EVIDENCE FOR THE PRESENCE OF A BARRIER TO DIFFUSION ACROSS THE PHARATE PUPAL INTEGUMENT BETWEEN MOULTING FLUID AND HAEMOLYMPH

Early workers studying moulting concluded that the movement of small organic molecules across the integumentary epidermis between moulting fluid and haemolymph occurs only by passive diffusion (Wigglesworth, 1933; Passoneau and Williams, 1953; Jeuniaux, 1958; Condoulis and Locke, 1966; Locke, 1966; Lensky *et al.*, 1970; Lensky and Rakover, 1972). This conclusion was never substantiated directly by determining that active processes were not involved. Rather it was inferred from the movement of small molecular weight components into or across the integumentary epithelium into moulting fluid.

In *M. sexta* and *H. cecropia*, measurement of the concentration of a variety of organic components reveals the presence of a restrictive barrier between moulting fluid and haemolymph (Tables 10–13; Lai-Fook, 1966; Flower and Filshie, 1975; Jungreis, 1974, 1978b). The concentrations of trehalose, glucose, total phosphate, labile and inorganic phosphate, total protein, polypeptides, amino acids, bicarbonate and chloride have been measured in haemolymph and moulting fluid during both secretion and resorption of moulting fluid (Tables 3–5, 10–13; Jungreis, 1973, 1974, 1978b; Jungreis *et al.*, 1975; Wyatt *et al.*, 1963). If an assignment of osmotic

TABLE 11 The concentration of organic components in pharate pupal haemolymph and moulting fluid in the tobacco hornworm *Manduca sexta* (from Jungreis, 1978b). Mean ± standard error. Number in parentheses is the number of pooled samples, each containing equal volumes from 2–5 animals

	Trehalose (μmoles/ml)	Phosphate (μmoles/ml)			Protein (%, W:V)	Polypeptides (%, W:V)	Amino acids (μmoles/ml)
		Total	Labile	Inorganic			
Early haemolymph	33.5 ± 2.2 (6)	103.7 ± 10.9 (6)	7.41 ± 0.75 (6)	1.31 ± 0.30 (6)	7.0 ± 0.5 (6)	1.8 ± 0.3 (6)	45.3 ± 5.2 (5)
moulting fluid	0.0 ± 0.0 (6)	32.7 ± 3.4 (6)	2.99 ± 0.41 (6)	2.20 ± 0.22 (6)	3.2 ± 0.1 (6)	3.9 ± 1.1 (6)	31.0 ± 2.7 (5)
Middle haemolymph	27.6 ± 1.6 (5)	103.9 ± 15.8 (5)	2.42 ± 0.75 (5)	0.83 ± 0.27 (5)	7.7 ± 0.6 (5)	1.5 ± 0.3 (5)	38.1 ± 1.4 (5)
moulting fluid	0.0 ± 0.0 (5)	37.8 ± 5.1 (5)	2.28 ± 0.60 (5)	2.21 ± 0.19 (5)	2.7 ± 0.3 (5)	5.3 ± 0.7 (5)	30.4 ± 1.8 (5)
Late haemolymph	32.0 ± 0.9 (5)	121.5 ± 21.1 (5)	0.72 ± 0.23 (5)	7.30 ± 0.1 (5)	1.4 ± 0.3 (5)	1.4 ± 0.3 (5)	38.9 ± 4.1 (5)
moulting fluid	0.0 ± 0.0 (5)	47.1 ± 8.0 (5)	1.58 ± 0.34 (5)	0.87 ± 0.24 (5)	3.3 ± 0.1 (5)	2.8 ± 0.6 (5)	41.7 ± 3.3 (5)

TABLE 12 The percent of the total osmotic pressure occupied by various constituents of haemolymph and moulting fluid from foliage reared *H. cecropia* (from Jungreis, 1974)

	Inorganic cations	Bicarbonate	Chloride	Labile phosphate	Proteins	Poly-peptides	Amino acids	Trehalose	Unde-termined
Haemolymph	38	8.5	4	13	2	5	20	9	0.5
Moulting fluid	45	18.5	4	0	0.5	3	17	0.5	11.5

TABLE 13 The percent of the total osmotic pressure occupied by various constituents in haemolymph and moulting fluid from *Manduca sexta* (from Jungreis, 1978b)

	Total osmotic pressure (%)									
	Inor-ganic cations	Bicar-bonate	Chloride	Labile phosphate	Proteins	Poly-peptides	Amino acids	Tre-halose	Unde-termined	Total P-labile
Early haemolymph	21.8	2.8	9.8	2.3	1.8	3.8	14.3	10.6	33	30.3
moulting fluid	58.9	24.6	2.4	1.0	0.9	9.0	10.7	0	−7.5	10.3
Middle haemolymph										
moulting fluid	45.3	25.4	1.5	0.7	0.7	10.7	9.2	0	6.5	10.7
Late haemolymph	22.9	3.0	7.5	0.3	1.8	2.0	11.7	9.6	40.4	36.3
moulting fluid	57.3	30.1	1.8	0.5	0.8	5.6	12.6	0	−8.7	13.7

equivalents is made to these organic components (see Jungreis, 1974, 1978), and if these contributions to the total osmotic pressure are added to those contributed by inorganic cations and anions, then virtually 100% of the osmotic pressure of the respective fluids can be assigned (Tables 12–13). The osmotic pressure of foliage-reared *H. cecropia* moulting fluid is appreciable hyper-osmotic to haemolymph (Table 3). Maintenance of this osmotic differential must restrict the movement of organic components between the two fluid compartments. The virtual absence of trehalose in moulting fluid (Tables 10–11), a sugar found to move passively across cell membranes (Jungreis and Wyatt, 1972), coupled with drastic differences in the concentrations of protein, polypeptides, labile phosphates and amino acids in the respective fluids of both species further supports the contention that unrestricted movement of solutes or water between the fluid compartments is unlikely to take place. Additional support comes from the work of Koeppe and Gilbert (1973), who found that immunologically unique proteins were selectively moved from haemolymph to the pupal cuticle. However, movement of the observed proteins, as well as the small molecular weight solutes described by other wokers can be attributed in pinocytotic activity (Locke, 1976; Hikida and Jungreis, 1978).

5.9 MODEL FOR ACTIVE ION MOVEMENTS DURING SECRETION AND RESORPTION OF MOULTING FLUID

Active transport of a specific cation or anion can be unequivocably demonstrated when 100% agreement is obtained between the short-circuit current and the net flux of that ionic species. In *H. cecropia*, secretion of moulting fluid takes place (at 25°C) at 36–24 hours before the larval–pupal ecdysis. When the isolated *H. cecropia* pharate pupal integument, dissected at 30 hours before the larval-pupal ecdysis, is bathed in calcium-magnesium-sodium free saline containing as the only cationic component potassium (and tris), potassium is actively transported from the haemolymph to exuvial sides of the integument with a flux ratio of 4.0 (Table 9). For this individual tissue, 105% of the short-circuit current could be attributed to actively transported potassium (Fig. 20).

Resorption of moulting fluid in *H. cecropia* occurs at 22–12 hours before the larval–pupal ecdysis. The contribution of actively transported potassium to the observed short-circuit current is only 9% at this time (Fig. 21), even though active transport of potassium could still be demonstrated *in vitro* (a flux ratio of 2.5 was observed (Table 9)). The observation that active potassium transport continued (at a reduced rate) in the direction of haemolymph towards exuvial side of the integument during resorption of moulting fluid, when coupled with the absence of other cations in the bathing solution, serves as evidence that the short-circuit current is being generated by the active transport of an ionic species of opposite charge to potassium in the opposite direction, namely exuvial towards haemolymph side of the integument. Chloride transport is not responsible for the observed short-circuit current for two reasons: (*a*) it represents only a small fraction of the total moulting fluid anion pool *in vivo* (especially in *M. sexta*—see Tables 4–5), and (*b*) substituting potassium bicarbonate for potassium chloride had no effect on the short-circuit current or rate of potassium transport (Figs 21 and 23). Since chloride and bicarbonate were the only two anions in the bathing solution when the quantity of actively transported potassium was noted to be insufficient to account for the short-circuit current, then the anion actively transported during resorption of moulting fluid that is responsible for the observed short-circuit current must be derived from an endogenous source within the epidermal cells of the integumentary epithelium.

Moulting fluid is essentially a potassium bicarbonate salt solution in foliage-reared *H. cecropia* and diet-reared *M. sexta*. The bicarbonate appearing in moulting fluid could either originate in haemolymph (unlikely, based on the data of Table 5) or endogenously within the epidermal cells themselves (Fig. 32). Since bicarbonate was not present in the bathing solution when 100% agreement between the short-circuit current and the net potassium flux was established (Jungreis and Harvey, 1975; Fig. 20), an endogenous origin for bicarbonate is favoured. Potassium uptake into the epidermal cells would then occur via a coupled exchange with a proton derived from the breakdown of carbonic acid: $CO_2 + H_2O \rightarrow H_2CO_3 \rightarrow H^+ + HCO_3^-$ In the intact animal, passive transport of bicarbonate during secretion of moulting fluid would account for the alkaline pH of moulting fluid (Tables 4–5).

The following model is proposed to account for the secretion and resorption of moulting fluid. Potassium is actively transported during the secretion of moulting fluid, accompanied by an anion derived either endogenously (bicarbonate) or from haemolymph (bicarbonate + chloride). At the basal surface of the epidermal cells (i.e. that in contact with haemolymph) there is a coupled exchange between a proton and a potassium ion. The pH of haemolymph therefore tends to become more acidic, while that of the epidermal cells and moulting fluid tend to become more alkaline during the secretion of moulting fluid. The rate of formation of bicarbonate by the epidermal cells is a function of both the effective carbonic anhydrase activity (Fig. 32; Johnston and Jungreis, 1977, 1979; in preparation), and the availability of carbon dioxide in the epidermal cells to regenerate carbonic acid. When the relative availability in haemolymph of carbon dioxide is low (as in the case of diet-reared *H. cecropia*), chloride derived from haemolymph accompanies actively transported potassium as a substitute for endogenously synthesized bicarbonate.

During resorption of moulting fluid, bicarbonate produced in the epidermal cells is actively transported out of the epithelial cells into haemolymph accompanied by potassium. At the apical surface of the epidermal cells, a proton produced in stoichiometric amounts with bicarbonate is exchanged for potassium in moulting fluid, thereby restoring the intracellular concentration of potassium. In moulting fluid, the proton combines with bicarbonate to form carbonic acid, which in turn dissociates into carbon dioxide and water. These compounds diffuse down a concentration gradient into the epidermal cells, thereby affecting resorption of the fluid volume from the exuvial space.

The limiting factor in both active potassium secretion and active bicarbonate resorption of moulting fluid may well be the effective activity of carbonic anhydrase within the epidermal cells. This conjecture can be supported by the observation that during the secretion of moulting fluid, when the concentration of potassium in the bathing solution is raised from 32 to 64 mM no increase in the short-circuit current is noted, whereas a reduction in the short-circuit current occurs when the concentration of potassium is reduced from 32 to 16 mM (Fig. 24). An alternate explanation of this last observation is that the potassium pump is saturated when the potassium concentration is about 32 mM. In any event, the effective carbonic anhydrase activity in the integumentary epithelium is at least 350-fold greater than that needed to account for the bicarbonate produced during moulting (Johnston and Jungreis, 1978a, 1979; in preparation).

6 Summary

A model for the secretion and resorption of moulting fluid requires the active transport of a cationic species (potassium) during the secretion of moulting fluid, and the active transport of an anionic species (bicarbonic) during moulting fluid resorption. The origin of the bicarbonate appearing in moulting fluid is the integumentary epithelium rather than haemolymph. Resorption of moulting fluid occurs via active transport of endogenously produced bicarbonate in the direction integumentary epithelium to haemolymph (accompanied by potassium moving passively), with a compensatory passive resorption into these cells of potassium bicarbonate from the moulting fluid. Moulting fluid is not the sol form of moulting gel and is isolated from haemolymph by the pharate pupal integument, which permits the establishment of an osmotic gradient between the two fluid compartments despite the continued diffusive movement of low molecular weight solutes between these compartments. Secretion and resorption of moulting fluid are thus two separate processes involving either exogenously supplied (potassium during secretion) or endogenously produced (bicarbonate during resorption) ions during the respective stages of moulting fluid elaboration. Specific hormonal involvement and changes in enzme activities during the elaboration, secretion and resorption of moulting fluid have yet to be correlated with the various facets of cuticle deposition and digestion.

7 Conclusion

Having surveyed the whole of moulting in insects, several as yet untried approaches to resolving the problems uncovered can now be presented. Virtually all the work done on moulting has been done on Lepidopterous or Dipterous insects. Do other insect orders solve the problems of moulting by similar or dissimilar mechanisms? More of the enzymes involved in cuticle synthesis and degradation need to be identified, their properties characterized with respect to changes in the internal environment, and their roles in the pre- and post-moulting process established. The relationship between the stage specific refractory nature of epidermal cells to hormones and the appropriate enzyme levels in these cells clearly needs to be determined. Specific hormone effects on the synthesis of individual cuticle components need to be studied, as does the role of moulting fluid in the initiation of cuticle digestion. Factors responsible for initiating the secretion and resorption of moulting fluid both *in situ* and *in vitro* need to be determined. Only after these and a host of other questions are answered will our understanding be sufficient to construct a unifying model of moulting in insects.

Acknowledgements

Technical assistance from James W. Johnston and Mila L. Arceneaux is gratefully acknowledged. I would also like to thank Professor Lynn M. Riddiford for numerous suggestions and considerable help in preparing the appendices, and Professor Michael Locke for generously supplying electron micrographs.

This paper was supported by NSF Grant PCM75–23456 and Biomedical Science Support Grant RR–07088.

References

Anderson, S. O. (1966). Covalent cross-links in the structural protein, resilin. *Acta Physiol. Scand.* (Suppl.), **66,** 1–81.

Anderson, S. O. (1976). Cuticular enzymes and sclerotization in insects. *In* "The Insect Integument" (Ed. H. R. Hepburn), pp. 121–144. Elsevier.

Bade, M. L. (1974). Localization of molting chitinase in insect cuticle. *Biochem. Biophys. Acta*, **372,** 474–477.

Bade, M. L. (1975). The pattern of appearance and disappearance of active molting chitinase in *Manduca* cuticle. The endogenous activity. *F.E.B.S. Letters*, **51,** 161–163.

Bade, M. L. (1978). Enzymatic breakdown of the chitin component in insect cuticle during the molt. (Submitted for publication.)

Bade, M. L. and Shoukimas, J. (1974). Neutral metal chelator-sensitive protease in insect molting fluid. *J. Insect Physiol.* **20,** 281–290.

Bade, M. L. and Stinson, A. (1976). Activation of cuticle chitinase: A probable new instance of activation by partial proteolysis. *In* "Proteolysis and Physiological Regulation" (Ed. D. W. Ribbons and K. Brew), **11,** p. 391. Miami Winter Symposium.

Barbier, R. (1967). Mise en envidence d'espaces intercellulaires importants dans l'hypoderme et de formations paracrystallines dan le liquide exuvial chez les larves de *Galleria mellonella* L. (Lepidoptere Pyralidae) lors de la mise en place de la cuticuline. *C. r. hebd. Seanc. Acad. Sci. Paris,* **264,** 2337–2340.

Barbier, R. (1971). Recherches sur la Morphologense Tegumentaire d'un Insecte Holometabole: *Galleria mellonella* L. (Lepidoptere Pyralidae). Theses de Docteur Es-Sciences Naturalles, Universite de Rennes, U.E.R. des Sciences Biologiques. Serie C., Ordre 122, Serie 42.

Barra, J. A. (1969). Tegument des Collemboles. Presence d'hemocytes a granules dans le liquide exuvial au cours de la mue. *C. r. hebd. Seanc. Acad. Sci. Paris,* **269,** 902–903.

Beament, J. W. L. and Treherne, J. E. (1967). "Insects and Physiology". American Elsevier, New York.

Bell, R. A. and Joachim, F. G. (1976). Techniques for rearing laboratory colonies of tobacco hornworm and pink bollworms. *Ann. Ent. Soc. Amer.* **69,** 363–373.

Berridge, M. J. and Prince, W. T. (1972). Transepithelial potential changes during stimulation of isolated salivary glands with 5-hydroxytryptamine and cyclic AMP. *J. exp. Biol.* **56,** 139–153.

Blankemeyer, J. T. (1977). The route of potassium transport in the silkmoth midgut. Ph.D. Dissertation, Temple University, Philadelphia, Pa.

Bodine, J. H. and Allen, T. H. (1938). Enzymes in ontogenesis (Orthoptera) IV. Natural and artificial conditions governing the action of tyrosinase. *J. Cell. Comp. Physiol.* **11,** 409–423.

Bodnaryk, R. P. (1971). Effect of exogenous moulting hormone (ecdysterone) on β-alanyl-L-tyrosine during puparium formation in the fleshfly *Sarcophaga bullata. Gen. Comp. Endocr.* **16,** 363–368.

Bollenbacher, W. E., Vedeckis, W. V., Gilbert, L. I. and O'Connor, J. D. (1975). Ecdysone titers and prothoracic gland activity during the larval–pupal development of *Manduca sexta. Dev. Biol.* **44,** 46–53.

Coles, C. G. (1966). Studies on resilin biosynthesis. *J. Insect Physiol.* **12,** 679–691.

Condoulis, W. V. and Locke, M. (1966). The desposition of endocuticle in an insect *Calpodes ethlius* Stoll, Lepidoptera, Hesperiidae. *J. Insect. Physiol.* **12,** 311–323.

Delachambrie, J. (1971). La formation des canaux cuticulaires chez l'adulte de *Tenebrio molitor* L. Etude ultrastructurale et remarques histochemiques. *Tissue & Cell* **3,** 499–520.

Dennell, R. (1947). A study of an insect cuticle; the formation of the puparium of *Sarcophaga falculata* Pad (Diptera). *Proc. R. Soc.* **B 134,** 79–110.

Diamond, J. M., Barry, P. H. and Wright, E. M. (1971). The route of transepithelial ion permeation in the gall bladder. *In* "Electrophysiology of Epithelial Cells" (Ed. G. Biebisch), pp. 23–33. Schattauer, Stuttgart.

Doane, W. W. (1973). Role of hormones in insect development. *In* "Developmental Systems of Insects" (Eds. S. J. Counce and C. H. Waddington), **2,** pp. 291–297. Academic Press, New York and London.

Ebeling, W. (1974). Permeability of insect cuticle. *In* "The Physiology of Insecta", 2nd Edn. (Ed. M. Rockstein), **6,** pp. 271–343. Academic Press, New York and London.

Fain, M. J. and Riddiford, L. M. (1975). Juvenile hormone titers in the hemolymph during late larval development of the tobacco hornworm, *Manduca sexta* (L.) *Biol. Bull., mar. biol. Lab., Woods Hole*, **149,** 506–521.

Fain, M. J. and Riddiford, L. M. (1977). Requirements for molting of the crochet epidermis of the tobacco hornworm larva *in vivo* and *in vitro*. *Wilhelm Roux's Archives*, **81,** 285–307.

Filshie, B. K. (1970). The fine structure and deposition of the larval cuticle of the sheep blowfly (*Lucilia cuprina*). *Tissue & Cell*, **2,** 479–489.

Flower, N. E. and Filshie, B. K. (1975). Junctional structures in the midgut cells of Lepidopteran caterpillars. *J. Cell. Sci.* **17,** 221–239.

Gibbs, D. and Riddiford, L. M. (1977). Prothoracicotropic hormone in *Manduca sexta*: Localization by a larval assay. *J. exp. Biol.* **66,** 255–266.

Gilbert, L. I. and King, D. S. (1973). Physiology of growth and development: Endocrine aspects. *In* "The Physiology of Insecta", 2nd Edn. (Ed. M. Rockstein), **1,** pp. 249–370. Academic Press, New York and London.

Goldsworthy, B. J. and Mordue, W. (1974). Review: Neurosecretory hormones in insects. *J. Endocrinol.* **60,** 529–558.

Greenstein, M. E. (1972). The ultrastructure of developing wings in the giant silkmoth, *Hyalophora cecropia*. I. Generalized epidermal cells. *J. Morph.* **136,** 1–22.

Hackman, R. H. (1964). The integument of insects. *In* "Physiology of Insecta" (Ed. M. Rockstein), **3,** pp. 471–506. Academic Press, New York and London.

Hackman, R. H. (1971). The integument of arthropods. *In* "Chemical Zoology" (Eds. M. Florkin and B. T. Scheer), **VI,** pp. 1–62. Academic Press, New York and London.

Hackman, R. H. (1974). Chemistry of the insect cuticle, *In* "The Physiology of Insecta" 2nd Edn. (Ed. M. Rockstein), **6,** pp. 215–270. Academic Press, New York and London.

Hackman, R. H. and Goldberg, M. (1967). The O-diphenoloxidases of fly larvae. *J. Insect Physiol.* **13,** 531–544.

Harvey, W. R., Haskell, J. A. and Zerahn, K. (1967). Active transport of potassium and oxygen consumption in the isolated midgut of *Hyalophora cecropia*. *J. exp. Biol.* **46,** 235–248.

Harvey, W. R., Wood, J. L., Quatrale, R. P. and Jungreis, A. M. (1975). Cation distributions across the larval and pupal midgut of *Hyalophora cecropia in vivo*. *J. exp. Biol.* **63,** 321–330.

Hepburn, H. R. (1976). "The Insect Integument". Elsevier.

Hikida, R. S. and Jungreis, A. M. (1978). Avesicular exocytosis in integumentary epithelial cells of *Hyalophora cecropia* during resorption of moulting fluid. (In press.)

Hikida, R. S. and Jungreis, A. M. (1979). Silkmoth integumentary epithelium during the larval–pupal transformation. (Submitted for publication.)

Hsiao, T. H. and Hsiao, C. (1977). Simultaneous determination of molting and juvenile hormone titers of the greater wax moth. *J. Insect. Physiol.* **23,** 89–93.

Hurst, H. (1945). Enzymatic activity as a factor in insect physiology and toxicology. *Nature, Lond.* **156,** 194–198.

Ishaaya, I. (1972). Studies of the haemolymph and cuticular phenoloxidase in *Spodoptera littoralis* larvae. *Insect Biochem.* **2,** 409–419.

Jenkin, P. M. and Hinton, H. E. (1966). Apolysis in arthropod moulting cycles. *Nature, Lond.* **211,** 871.

Jeuniaux, C. (1958). Resorption du liquide exuvial chez le ver a soie (*Bombyx mori*, L.). *Arch. Int. Physiol. Biochem.* **66,** 121–122.

Jeuniaux, C. (1963). "Chitine et Chitinolyse", pp. 181. Masson, Paris.

Johnston, J. W. and Jungreis, A. M. (1977). Microdiffusional determination of bicarbonate in hemolymph and moulting fluid of *Hyalophora cecropia* and *Manduca sexta*. *Am. Zool.* **17,** 61.

Johnston, J. W. and Jungreis, A. M. (1978). A micromethod for the analysis of bicarbonate. *Comp. Biochem. Physiol.* **61A,** 469–473.

Johnston, J. W. and Jungreis, A. M. (1978a). Comparative aspects of carbonic anhydrases in *Hyalophora cecropia. Am. Zool.* **18,** 640.

Johnston, J. W. and Jungreis, A. M. (1979). Comparative properties of mammalian and insect carbonic anhydrases: Effects of potassium and chloride on the rate of carbon dioxide hydration. *Comp. Biochem. Physiol.* **B.** (In press.)

Jones, J. C. (1956). The hemocytes of *Sarcophaga bullata* Parker. *J. Morph.* **99,** 233–257.

Jones, J. C. (1962). Current concepts concerning insect hemocytes. *Am. Zool.* **2,** 209–246.

Jungreis, A. M. (1973). Formation and composition of molting fluid in the silkmoth *Hyalophora cecropia. Am. Zool.* **13,** 270.

Jungreis, A. M. (1974). Physiology and composition of moulting fluid and midgut lumenal contents in the silkmoth *Hyalophora cecropia. J. Comp. Physiol.* **88,** 113–127.

Jungreis, A. M. (1977). Comparative aspects of invertebrate epithelial transport. *In* "Water Relations in Membrane Transport in Plants and Animals" (Eds. A. M. Jungreis, T. Hodges, A. Kleinzeller and S. G. Schultz), pp. 89–96. Academic Press, New York and London.

Jungreis, A. M. (1978a). Insect dormancy. *In* "Mechanisms of Dormancy and Developmental Arrest" (Ed. M. E. Clutter), pp. 47–112. Academic Press, New York and London.

Jungreis, A. M. (1978b). The composition of larval–pupal moulting fluid in the tobacco hornworm *Manduca sexta. J. Insect Physiol.* **24,** 65–73.

Jungreis, A. M., Dailey, J. C. and Hereth, M. L. (1975). alpha-Glycerol phosphatase and glycerol kinase activity in developing silkmoth tissues. *Amer. J. Physiol.* **229,** 1448–1454.

Jungreis, A. M. and Harvey, W. R. (1975). Role of active potassium transport by integumentary epithelium in secretion of larval-pupal moulting fluid during silkmoth development. *J. exp. Biol.* **62,** 357–366.

Jungreis, A. M., Jatlow, P. and Wyatt, G. R. (1973). Inorganic ion composition of hemolymph of the cecropia silkmoth: Changes with diet and ontogeny. *J. Insect Physiol.* **19,** 225–233.

Jungreis, A. M. and Tojo, S. (1973). Potassium and uric acid content in tissues of the silkmoth *Hyalophora cecropia. Amer. J. Physiol.* **224,** 21–226.

Jungreis, A. M. and Vaughan, G. L. (1977). Insensitivity of Lepidopteran tissues to ouabain. I. Absence of ouabain binding and Na^{+}-K^{+}-ATPases in laryal and adult midgut. *J. Insect Physiol.* **23,** 503–509.

Jungreis, A. M. and Wyatt. G. R. (1972). Sugar release and penetration in insect fat body: Relations to regulation of haemolymph trehalose in developing stages of *Hyalophora cecropia. Biol. Bull. mar. biol. Lab., Woods Hole,* **143,** 367–391.

Kafatos, F. C. (1968). The labial gland: a salt secreting organ of saturniid moths. *J. exp. Biol.* **48,** 435–453.

Karlson, P. and Ammon, H. (1963). Zum Tyrosinstoffwechsel der Insekten. XI. Biogenese und Schicksal det Acetylgruppe des N-acetyl-dopamins. *Hoppe-Seyler's Z. physiol. Chem.* **330,** 161–168.

Karlson, P. and Liebau, H. (1961). Zum Tyrosinstoffwechsel der Insekten. V. Reindarstellung, Kristiallisation und Substratspezifitat der O-Diphenoloxydase aus *Calliphora erythrocephala. Hoppe-Seyler's Z. physiol. Chem.* **326,** 135–143.

Karlson, P. and Schweiger, A. (1961). Zum Tyrosinstoffwechsel der Inseketen. IV. Das Phenoloxydase-System von *Calliphora* und seine Beeinflussung durth das Hormon Ecdyson. *Hoppe-Seyler's Z Physiol. Chem.* **323,** 199–210.

Karlson, P. and Sekeris, C. E. (1976). Control of tyrosine metabolism and cuticle sclerotization by ecdysone. *In* "The Insect Integument" (Ed. H. R. Hepburn), pp. 145–156. Elsevier.

Katzenellenbogen, B. S. and Kafatos, F. C. (1970). Some properties of silkmoth moulting gel and moulting fluid. *J. Insect Physiol.* **16,** 2241–2256.

Katzenellenbogen, B. S. and Kafatos, F. C. (1971a). Inactive proteinases in silkmoth moulting gel. *J. Insect Physiol.* **17,** 823–832.

Katzenellenbogen, B. S. and Kafatos, F. C. (1971b). Proteinases of silkmoth moulting fluid; Physical and catalytic properties. *J. Insect Physiol.* **17,** 775–800.

Katzenellenbogen, B. S. and Kafatos, F. C. (1971c). General esterases of silkmoth moulting fluid: Preliminary characteristics. *J. Insect Physiol.* **17,** 1139–1151.

Kawase, S. (1955). Tyrosinase in the integument of a marking mutant in the silkworm *Bombyx mori*. *Jap. J. Genet.* **31,** 284–292.

Kawase, S. (1960). Tyrosinase in the silk worm during the pupation period. *J. Insect Physiol.* **5,** 335–340.

Keynes, R. D. (1969). From frog skin to sheep rumen. A survey of transport of salts and water across multicellular structures. *Quart. Rev. Biophys.* **2,** 177–281.

Kimura. S. (1976). The chitinase system in the cuticle of the silkworm *Bombyx mori*. *Insect Biochem.* **6,** 479–482.

Koeppe, J. K. and Gilbert, L. I. (1973). Immunochemical evidence for the transport of haemolymph protein into the cuticle of *Manduca sexta*. *J. Insect Physiol.* **19,** 615–624.

Kunkel, J. G. (1975). Cockroach molting. I. Temporal organisation of events during the molting cycle of *Blatella germanica* (L.) *Biol. Bull. mar. biol. Lab.* **148,** 259–273.

Lai-Fook, J. (1966). The repair of wounds in the integument of insects. *J. Insect Physiol.* **12,** 195–226.

Laing, J. (1935). On the ptilinum of the blowfly (*Calliphora erythrocephala*). *Q. J. microsc. Sci.* **77,** 497–521.

Lensky, Y., Cohen, C. and Schneiderman, H. A. (1970). The origin, distribution and fate of the molting fluid proteins of the Cecropia silkworm. *Biol. Bull. mar. biol. Lab., Woods Hole*, **139,** 277–295.

Lensky, Y. and Rakover, Y. (1972). Resorption of moulting fluid proteins during the ecdysis of the honey bee. *Comp. Biochem. Physiol.* **B.41,** 521–531.

Locke, M. (1964). The structure and formation of the integument in insects. *In* "Physiology of Insecta" (Ed. M. Rockstein), **3,** pp. 379 470. Academic Press, New York and London.

Locke, M. (1966). The structure and formation of the cuticulin layer in the epicuticle of an insect, *Calpodes ethlius* (Lepidoptera, Hesperiidae). *J. Morph.* **118,** 461–494.

Locke, M. (1969a). The structure of an epidermal cell during the formation of the protein epicuticle and the uptake of molting fluid in an insect. *J. Morph.* **127,** 7–40.

Locke, M. (1969b). The localization of a peroxidase associated with hard cuticle formation in an insect, *Calpodes ethlius* Stoll, Lepidoptera, Hesperiidae, *Tissue & Cell*, **1,** 555–575.

Locke, M. (1970). The molt-intermolt cycle in the epidermis and other tissues of an insect *Calpodes ethlius*, Lepidoptera, Hesperiidae. *Tissue & Cell*, **2,** 197–223.

Locke, M. (1974). The structure and formation of the integument in insects. *In* "The Physiology of Insecta", 2nd Edn. (Ed. M. Rockstein), **6,** pp. 123–213. Academic Press, New York and London.

Locke, M. (1976). The role of plasma membrane plaques and golgi complex vesicles in cuticle deposition during the moult/intermoult cycle. *In* "The Insect Integument" (Ed. H. R. Hepburn), pp. 237–258. Elsevier.

Locke, M. and Krishnan, N. (1973). The formation of the ecdysial droplets and the ecdysial membrane in an insect. *Tissue & Cell*, **5,** 441–450.

Lord, B. A. P. and DiBona, D. R. (1976). Role of the septate junction in the regulation of paracellular transepithelial flow. *J. Cell. Biol.* **71,** 967–972.
Maddrell, S. H. P. (1971). The mechanisms of insect excretory systems. *Adv. Insect. Physiol.* **8,** 199–331. Academic Press, New York and London.
Malek, S. R. A. (1958). The origin and nature of the ecdysial membrane in *Schistocerca gregaria* (Forskäl). *J. Insect Physiol.* **2,** 298–312.
McNutt, N. S. and Weinstein, R. S. (1973). Junctional complexes. *Prog. Biophys. Mol. Biol.* **26,** 57–70.
Mills, R. R., Adrouny, S. and Fox, F. R. (1968). Correlation of phenoloxidase activity with ecdysis and tanning hormone release in the American cockroach. *J. Insect Physiol.* **14,** 603–611.
Mitchell, H. K. and Weber-Tracy, U. M. (1965). *Drosophila* phenol-oxidases. *Science,* **148,** 964–965.
Mitchell, H. K., Weber-Tracy, U. M. and Schaar, G. (1971). Aspects of cuticle formation in *Drosophila melanogaster*. *J. exp. Zool.* **176,** 429–444.
Mitsui, T. and Riddiford, L. M. (1976). Pupal cuticle formation by *Manduca sexta* epidermis *in vitro*: Patterns of ecdysone sensitivity. *Dev. Biol.* **54,** 172–186.
Mitsui, T. and Riddiford, L. M. (1978). Hormonal requirements for the laval–pupal transformation of the epidermis of *Manduca sexta in vitro*. *Dev. Biol.* **62,** 193–205.
Morgan, E. D. and Poole, C. F. (1977). Chemical control of insect moulting. *Comp. Biochem. Physiol.* **B.57,** 99–110.
Neville, A. C. (1975). Biology of the Arthropod Cuticle. *In* "Zoophysiology and Ecology" **4/5,** Springer-Verlag, New York.
Nijhout, H. F. (1975). A threshold size for metamorphosis in the tobacco hornworm, *Manduca sexta* (L.) *Biol. Bull. mar. biol. Lab., Woods Hole*, **149,** 214–225.
Nijhout, H. F. (1976). The role of ecdysone in pupation of *Manduca sexta*. *J. Insect Physiol.* **22,** 453–463.
Nijhout, H. F. and Williams, C. M. (1974a). Control of moulting and metamorphosis in the tobacco hornworm, *Manduca sexta* (L.): Growth of the last-instar larva and the decision to pupate. *J. exp. Biol.* **61,** 481–491.
Nijhout, H. F. and Williams, C. M. (1974b). Control of moulting and metamorphosis in the tobacco hornworm, *Manduca sexta* (L.): Cessation of juvenile hormone secretion as a trigger for pupation. *J. exp. Biol.* **61,** 493–501.
Noble-Nesbitt, J. (1963). The cuticle and associated structures of *Podura aquatica* at the moult. *Q. J. microsc. Sci.* **104,** 369–391.
Noble-Nesbitt, J. (1967). Aspects of the structure, formation and function of some insect cuticles. *In* "Insect and Physiology" (Eds. J. W. L. Beament and J. E. Treherne), pp. 3–16. American Elsevier, New York.
Noirot, C. and Quennedey, A. (1974). Fine structure of insect epidermal glands. *Ann. Rev. Entom.* **19,** 61–80.
Novak, V. J. A. and Slama, K. (1971). "Insect Endocrines". Pergamon Press.
Oshnishi, E. (1954). Tyrosinase in *Drosophila virilis*. *Annotnes Zool. Jap.* **27,** 33–39.
Passonneau, J. V. and Williams, C. M. (1953). The molting fluid of the Cecropia silkworm. *J. exp. Biol.* **30,** 545–560.
Porter, C. A. and Jaworski, E. G. (1965). Biosynthesis of chitin during various stages in the metamorphosis of *Prodenia eridania*. *J. Insect Physiol.* **11,** 1151–1160.
Post, L. C. (1972). Bursicon: its effect on tyrosine permeation into insect haemocytes. *Biochem. Biophys. Acta*, **290,** 424–428.
Powning, R. F. and Izrykiewica, H. (1963). A chitnase from the gut of the cockroach *Periplaneta americana*. *Nature, Lond.* **200,** 1128.
Retnakaran, A. (1969). Studies on the allosteric control of an enzyme involved in insect tanning. *Comp. Biochem. Physiol.* **29,** 965–974.

Reynolds, S. E. (1973). The plasticization of the abdominal cuticle in *Rhodnius*. Ph.D. Thesis, Univ. of Cambridge.

Richards, A. G. (1953). "The Integument of Arthropods". University of Minnesota Press, Minneapolis.

Riddiford, L. M. (1968). An artificial diet for Cecropia and other saturiid silkworms. *Science*, **160,** 1461–1462.

Riddiford, L. M. (1972). Juvenile hormone in relation to the larval–pupal transformation of the cecropia silkworm. *Biol. Bull. mar. biol. Lab., Woods Hole*, **142,** 310–325.

Riddiford, L. M. (1976). Juvenile hormone control of epidermal commitment *in vivo* and *in vitro*. *In* "The Juvenile Hormones" (Ed. L. I. Gilbert), pp. 198–219. Plenum Press, New York and London.

Riddiford, L. M. and Truman, J. W. (1978). Biochemistry of insect hormones and insect growth regulators. *In* "Biochemistry of Insects" (Ed. M. Rockstein), pp. 307–357. Academic Press, New York and London.

Rizki, M. T. M. and Rizki, R. M. (1959). Functional significance of the crystal cells in the larva of *Drosophila melanogaster*. *J. Biophys. Biochem. Cytol.* **5,** 235–240.

Satir, P. and Gilula, N. B. (1973). Junctional complexes in insects. *Ann. Rev. Entom.* **18,** 143–170.

Schmidt, E. L. (1956). Observations on the subcuticular layer in the insect integument. *J. Morph.* **99,** 211–226.

Schultz, T. W. and Jungreis, A. M. (1977). Origin of the short circuit decay profile and maintenance of the cation transport capacity of the larval Lepidopteran midgut *in vitro* and *in vivo*. *Tissue & Cell*, **9,** 255–272.

Sedlák, B. J. and Gilbert, L. I. (1976). Epidermal cell development during the pupal–adult metamorphosis of *Hyalophora cecropia*. *Tissue & Cell*, **8,** 637–648.

Sekeris, C. E. (1963). Zum Tyrosinstoffwechsel der Insekten, XII. Reinigung, Eigenschaften und Substratspezifität der Dopa-Decarboxylase. *Hoppe-Seyler's Z. physiol. Chem.* **332,** 70–78.

Sekeris, C. E. (1964). Sclerotization in the blowfly imago. *Science, Wash.* **144,** 419–420.

Sekeris, C. E. and Karlson, P. (1962). Zum Tyrosinstoffwechsel der Insekten, VII. Der katabolische Abbau des Tyrosins und die Biogenese der Sklerotisierungs-substanz. *Hoppe-Seyler's Z. physiol. Chem.* **329,** 210–221.

Shaaya, E. and Karlson, P. (1965). Der Ecdysontiter während der Insektenentwicklung II. Die postembryonale Entwickling der Schmeissfliege *Calliphora erythrocephala* Meig. *J. Insect Physiol.* **11,** 65–69.

Shaaya, E. and Sekeris, C. E. (1965). Ecdysone during insect development, III. Activities of some enzymes of tyrosine metabolism in comparison with ecdysone titer during the development of the blowfly, *Calliphora erythrocephala* Meig. *Gen. Comp. Endocrinol.* **5,** 35–39.

Shatz, L. (1952). The development and differentation of arthropod procuticle: staining. *Ann. Ent. Soc. Am.* **45,** 678–686.

Sláma, K., Romanuk, M. and Sorum, F. (1974). "Insect Hormones and Bioanalogues". Springer-Verlag, Berlin and New York.

Smith, D. S. (1968). "Insect Cells, Their Structure and Function". Oliver & Boyd, Edinburgh.

Staal, G. B. (1967). Plants as a source of insect hormones. *Proc. K. Ned. Akad. Wet.* **C 70,** 409–418.

Staehelin, A. L. (1974). Structure and function of intercellular junctions. *Int. Rev. Cytol.* **39,** 191–279.

Taylor, R. L. and Richards, A. G. (1965). Integumentary changes during moulting of arthropods with special reference to the subcuticle and ecdysial membrane. *J. Morph.* **116,** 1–22.

Truman, J. W. and Riddiford, L. M. (1974). Physiology of insect rhythms III. The temporal organization of the endocrine events underlying pupation of the tobacco hornworm. *J. exp. Biol.* **60,** 371–382.

Truman, J. W., Riddiford, L. M. and Safranek, L. (1973). Hormonal control of cuticle coloration in the tobacco hornworm, *Manduca sexta*: Basis of an ultrasensitive bioassay for juvenile hormone. *J. Insect Physiol.* **19,** 195–203.

Ussing, H. H., Kruhoffer, P., Thaysen, J. H. and Thorn, N. A. (1960). The alkali metals in biology. *In* "Handbuch der experimentellen Pharmakologie" **XIII,** pp. 124–188. Springer-Verlag, Berlin.

Ussing, H. H. and Zerahn, K. (1951). Active transport of sodium as the source of electric current in the short-circuited isolated for skin. *Acta Physiol. Scant.* **23,** 110–127.

Varjas, L., Paguia, P. and de Wilde, J. (1976). Juvenile hormone titers in penultimate and last instar larvae of *Pieris brassicae* and *Barathra brassicae*, in relation to the effect of juvennoid application. *Experimentia*, **32,** 249–251.

Vaughan, G. L. and Jungreis, A. M. (1977). Insensitivity of Lepidoptera tissues to ouabain. II. Physiological mechanisms for protection from cardiac glycosides. *J. Insect Physiol.* **23,** 585–589.

Verson, E. (1902). Observations on the structure of the exuvial glands and the formation of the exuvial fluid in insects. *Zool. Anz.* **25,** 652–654.

Wachter, S. (1930). The moulting of the silkworm and a histological study of the moulting gland. *Ann. Ent. Soc. Amer.* **23,** 381–389.

Weis-Fogh, T. (1970). Structure and formation of the insect cuticle. *Symp. R. ent. Soc., Lond.* **5,** 165–185.

Whitehead, D. L. (1969). New evidence for the control mechanism of sclerotization in insects. *Nature, Lond.* **224,** 721–723.

Whitehead, D. L. (1970). L-dopa decarboxylase in the haemocytes of Diptera. *F.E.B.S. Letters*, **7,** 263–266.

Whitehead, D. L. (1971). Some evidence for the likely mechanisms of action of the hormone that initiates sclerotization. *In* "Insect Endocrines" (Ed. V. J. A. Novak and K. Sláma), pp. 157–166. Pergamon Press, Oxford and New York.

Whitehead, D. L., Brunet, P. C. J. and Kent, P. W. (1965). Observations on the nature of the phenoloxidase system in the secretion of the left colleterial gland of *Periplaneta americana* (L.) *Nature, Lond.* **185,** 610.

Wigglesworth, V. B. (1933). The physiology of the cuticle and of ecdysis in *Rhodnius prolixus* (Triatomidae, Hemiptera) with special reference to the function of oenocytes and of the dermal glands. *Quart. J. Microsc. Sci.* **76,** 269–318.

Wigglesworth, V. B. (1959). "The Control of Growth and Form". Cornell University Press, New York.

Wigglesworth, V. B. (1970). "Insect Hormones", Oliver and Boyd, London.

Wigglesworth, V. B. (1972). "The Principles of Insect Physiology", 7th Edn. John Wiley & Sons, New York.

Wigglesworth, V. B. (1976). The distribution of lipid in the cuticle of *Rhodnius. In* "The Insect Integument" (Ed. H. R. Hepburn), pp. 89–106. Elsevier, Amsterdam.

de Wilde, J., Staal, G. B., deKort, C. A. D., deLoff, A. and Baard, G. (1968). Juvenile hormone titre in the haemolymph as a function of photoperiodic treatment in the adult Colorado beetle (*Leptinotarsa decemlinaeta* Say). *Proc. K. Ned. Akad. et.* **C 71,** 321–326.

Willis, J. H. (1974). Morphogenetic action of insect hormones. *Ann. Rev. Ent.* **19,** 97–115.

Wood, J. L. (1972). Some aspects of active potassium transport by the midgut of the silkworm, *Antheraea pernyi.* Ph.D. thesis, Cambridge University.

Wyatt, G. R. (1968). Biochemistry of insect metamorphosis. *In* "Metamorphosis" (Eds. W. Etkin and L. I. Gilbert), pp. 143–184. Appleton-Crofts, New York.
Wyatt, G. R. (1972). Insect Hormones, *In* "Biochemical Actions of Hormones" (Ed. G. Litwak), **2,** pp. 385–490. Academic Press, New York and London.
Wyatt, G. R. and Pan, M. L. (1978). Insect plasma proteins. *Ann. Rev. Biochem.* **47,** 779–817.
Wyatt, G. L., Kropf, R. B. and Carey, F. G. (1963). The chemistry of insect hemolymph. IV. Acid-soluble phosphates. *J. Insect Physiol.* **9,** 137–152.
Yamamoto, R. T. (1969). Mass rearing of the tobacco hornworm. II. Larval rearing and pupation. *J. Econ. Ent.* **62,** 1427–1431.
Yamazaki, H. I. (1969). The cuticular phenoloxidase in *Drosophila virilis. J. Insect Physiol.* **15,** 2203–2211.
Yamazaki, H. I. (1972). Cuticular phenoloxidase from the silkworm, *Bombyx mori*: properties, solubilization and purification. *Insect Biochhem.* **2,** 431–444.
Young, N. L. and Young, P. R. (1976). Biochemical control of moulting hormone titer in *Calliphora erythrocephala* during puparium formation. *Insect Biochem.* **6,** 169–177.
Zacharuk, R. Y. (1976). Structural changes of the cuticle associated with moulting. *In* "The Insect Integument" (Ed. M. R. Hepburn), pp. 299–321. Elsevier.
Zacharuk, R. Y. (1972). Fine structure of the cuticle, epidermis, and fat body of larval Elateridae (Coleoptera) and changes associated with molting. *Can. J. Zool.* **50,** 1463–1487.

Appendices

INTRODUCTION

The unavailability of synthetic diets for the larger species of Saturniid and Sphingid Lepidoptera has until recently precluded the routine study of these species at stages other than those of egg, diapause pupa and pharate adult. With the successful development of synthetic diets in the late 1960s (see House, 1967, 1974) and specifically those for the giant silkmoth *Hyalophora cecropia* (Riddiford, 1968) and the tobacco hornworm, *Manduca sexta* (Yamomoto, 1969; Bell and Joachim, 1976; Baumhover, *et al.*, 1977), it became feasible to study routinely the developmental changes which ensue during the larval–pupal transformation. The timing of prothoracicotropic hormone (PTTH = ecdysiotropin) and ecdysone release to initiate metamorphosis has been defined by ligation experiments for *H. cecropia* (Williams, 1952) and *M. sexta* (Truman and Riddiford, 1974; Nijhout and Riddiford, 1974), but it has not been until recently that workers have begun to study the cellular, physiological and biochemical changes which occur during that time (Bade and Wyatt, 1962; Fain and Riddiford, 1977; Harvey *et al.*, 1975; Haskell *et al.*, 1968; Judy and Gilbert, 1970; Jungreis, 1973, 1974, 1975a, b, 1978b; Jungreis and Harvey, 1975; Jungreis and Tojo, 1973; Jungreis and Vaughan, 1977; Jungreis and Wyatt, 1972; Jungreis *et al.*, 1973, 1974; Krishnakumaran *et al.*, 1967; Mitsui and Riddiford, 1976, 1978; Pan, 1971, 1977; Pan *et al.*, 1967; Riddiford, 1972, 1975; Wyatt and Pan, 1978). In spite of this proliferation of studies of this critical period, the published information (Bade and

Wyatt, 1962; Krishnakumaran *et al.*, 1967; Riddiford and Ajami, 1973; Pan, 1971; Harvey *et al.*, 1975; Williams-Boyce, 1978; Jungreis, 1978b; Riddiford, 1972; Williams, 1952; Nijhout and Williams, 1974; Truman and Riddiford, 1974) for staging these animals is incomplete.

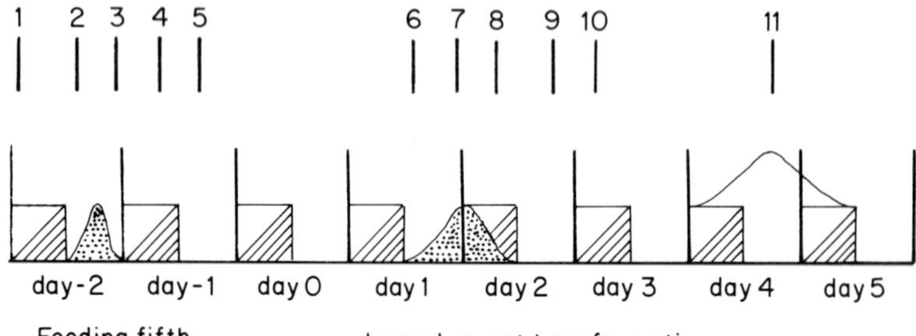

Fig. A. Timing of morphological events and hormone release in diet reared *Manduca sexta* during the late feeding fifth larval instar and the larval–pupal transformation at 25 °C under a 12L:12D photoperiod. The clear and cross-hatched periods denote photophase and scotophase, respectively. The beginning of each 24 h period is defined from the scotophase rather than from the photophase. Stippled areas denote the release of PTTH. 1. "Frosted" frass, 2. First release of PTTH, 3. First release of ecdysone, 4. Evacuation of the gut contents, 5. Initiation of wandering, 6. Second release of PTTH, 7. Ocellar retraction, end of wandering stage, 8. Second release of endysone; brain no longer necessary to complete the LPT, 9. Ocellar retraction completed in all pharate pupae, 10. Prothoracic glands no longer necessary, 11. Larval–pupal ecdysis. (Modified after Nijhout and Williams, 1974; Truman and Riddiford, 1974; Bollenbacher *et al.*, 1975.)

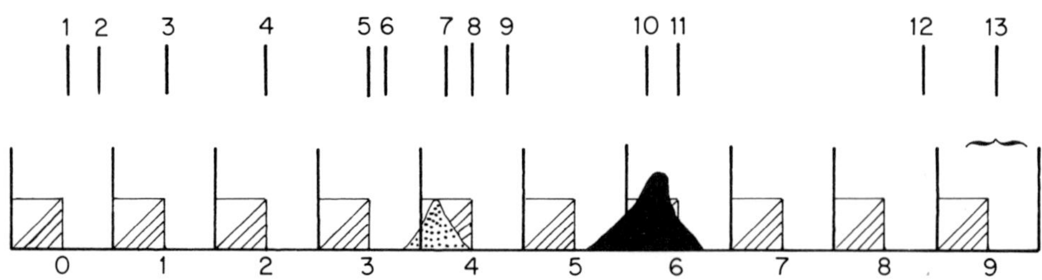

Fig. B. Timing of morphological events and hormone release in wild type, foliage reared *Hyalophora cecropia* during the larval–pupal transformation at 25 °C under a 12L : 12D photoperiod. The clear and cross-hatched periods denote photophase and scotophase, respectively. Stippled areas denote the single release of PTTH, while the shaded area denotes ecydsone release. 1. Evacuation of the gut by fifth instar larvae; viscera and integument are JH insensitive, 2. Onset of spinning, 3. Completion of the outer cocoon, 4. Completion of the inner cocoon, 5. Viscera become JH sensitive, 6. Animal seals cocoon if slit, 7. End of spinning (i.e. no further repair if cocoon is slit), 8. Brain no longer necessary to complete the LPT, 9. Begin ocellar retraction, 10. Ocellar retraction complete in all pharate pupae, 11. Prothoracic glands no longer necessary, 11. White prepupa, 12. Larval–pupal ecdysis (males earlier than females). (Modified after Riddiford, 1972; Riddiford, personal communication; Jungreis, unpublished observations.)

Criteria for staging *M. sexta* (Appendix I) and *H. cecropia* (Appendix II) between the larval and pupal stages in development are presented. Characteristics which can be observed without aid of the light microscope (see Williams, 1952; Riddiford, 1972) are emphasized, although histological, endocrinological and physiological criteria are also employed. The influence of photoperiod on developmental events, and the periods during which the pharate pupal integument has the capacity to secrete or resorb moulting fluid *in vitro* or *in vivo* are described (See Figs A and B).

APPENDIX I: STAGING CHARACTERISTICS OF DIET REARED "MANDUCASEXTA" DURING THE LARVAL–PUPAL TRANSFORMATION (25 °C)

Modified after Nijhout and Williams, 1974; Truman and Riddiford, 1974; Williams-Boyce, 1978; Jungreis, 1978b) Ecdysone and PTTH changes during the larval–pupal transformation are reported in Truman and Riddiford (1974), Nijhout and Williams (1974) and Bollenbacher *et al.* (1975)

Stage	Characteristics
Feeding fifth instar larva	Turquoise-blue feeding larvae have uniformly sized light coloured frass Integument uniformly dull, not shiny wet Absence of proleg swelling or puffiness Midgut distended with "dark" coloured contents Large salivary glands Fat body associated with considerable adhering haemolymph 1st release of PTTH
Purging fifth instar larva (1st 12 h)	Appearance of "frosted" frass Head–thorax have shiny wet appearance Prolegs puffy Midgut filled with lightly coloured undigested food Larva "firm" upon squeezing 1st release of ecdysone
Purged fifth instar larva (early pink stripe) (2nd 12 h)	Dorsal pink stripe (ommochrome pigment) Gut empty Initiation of wandering
Late pink stripe	Pink stripe dark Gut filled with clear gel-like fluid Wandering Negative geotaxic–very active
3 days before the larval–pupal ecdysis	Retraction of ocelli Dispersal of ommochrome pigment on dorsum Integument on head–thorax yellow-green

(*contd*)

Stage	Characteristics
	Forelegs and mandibles functional Prolegs cannot grasp Animal is motile Gut is initially filled with clear fluid Integument not completely separated from pupal cuticle Animal feels soft Salivary glands begin to decrease in size 2nd release of PTTH
2 days before the larval–pupal ecdysis	Ommochrome pigment absent Dorsal white ecdysial line appearing on thorax Dorsal aorta clear Integument yellow-green throughout Forelegs can not grasp Animal bends or twists, but is not motile Moulting gel appears in exuvial space between larval and pupal cuticles 2nd release of ecdysone
36–24 h before the larval–pupal ecdysis	White ecdysial line on dorsum very pronounced on metathoracic segments with tanning of last abdominal segments Cigar-shaped animal very firm Fat body visible through abdominal wall Dorsal aorta faint or not visible Animal unable to bend or twist Prolegs lightly tanned Prominent separation of larval and pupal cuticles with moulting gel Salivary glands greatly reduced in size, difficult to dissect free from other tissues
24–14 h before the larval–pupal ecdysis (early moulting fluid)	Ecdysial line visible on the dorsum Underlying cuticle lightly tanned in ventral region of last abdominal segments Body is turgid Larval cuticle "hard" to the touch Abdominal horn still prominent Little moulting fluid in exuvial space (20–50 μl) Tonofibrills between cuticle and underlying integumentary musculature intact Attempts to separate larval from underlying pupal cuticle results in haemolymph entry into the exuvial space
14–8 h before the larval–pupal ecdysis (middle moulting fluid)	Head tilted down *ca.* 45° relative to the plane of the body Symmetrical patches of pigment appear on the metathorax adjacent to the spiracles as bands 3 mm in width and 1 mm in height Abdominal horn worn down Bluish dorsal vessel readily observed Larval cuticle is pliable, but resists tearing

Stage	Characteristics
	Much moulting fluid is present in exuvial space (150–250 μl) Gut begins to accumulate dark coloured fluid
8–4 h before the larval–pupal ecydsis (late moulting fluid)	Head is tilted perpendicular to the plane of the body Animal soft to the touch Larval cuticle "bunched up" in posterior most segments and readily tears Symmetrical dark patches present on metathorax Little moulting fluid (0–50 μl) Gut fills with green-black fluid
Day of the larval–pupal ecdysis	Abdomen tanned Head–thorax initially light green and very soft to the touch, becoming tanned after 4–6 hours Proboscis distends and wings move down to cover thorax (2–6 hours) Cuticle is covered with wax and becomes non-pliable

Abbreviated criteria for staging at 25°C diet reared *Manduca sexta* between the feeding fifth and pupal stages in development using external morphological characteristics

Day	Stage	Time	Characteristic
3	Feeding fifth	2 days before pink stripe	4.5–7.5 g turquoise-blue larva, "frosted" frass
4	Feeding fifth	1 day before pink stripe	7.5–10 g turquoise blue larva gut purging, initiation of wandering
5	Pink stripe	5 days before larval–pupal ecdysis	Pink stripe (ommochrome pigment) along dorsum
6	Day 1	4 days before larval–pupal ecdysis	Animal is motile Initiation of ocellar retraction
7	Day 2	3 days before LPE	Animal bends or twists, but is not motile Ocellar retraction complete
8	Day 3	2 days before LPE	White ecdysial line on dorsum of thorax Animal unable to bend or twist
9	Day 4	1 day before LPE	Dorsal patches of pigment on metathorax on both sides of ecdysial line
10	Day 5	Day of larval–pupal ecdysis	Head–thorax light green becoming light brown Proboscis and wing sacs extended Abdomen more darkly tanned than thorax

APPENDIX II: FACTORS INFLUENCING THE TIME COURSE IN DEVELOPMENT OF "HYALOPHORA CECROPIA"

Genetics Consistent differences in timing of the larval–pupal transformation even under ostensibly the same photoperiod (17L : 7D) and temperature (25 °C) have been noted between my laboratory and that of Professor Lynn M. Riddiford. It appears that those animals which are offspring of field collected (*wild*) *Cecropia* grow larger (25–20 g on foliage, 17–25 g on synthetic diet) and develop faster than those from a laboratory population (*lab*) which has been continuously maintained on leaves or synthetic diet for 10 years (first in Professor G. R. Wyatt's laboratory, then in my laboratory) (16–25 g on foliage, 11–19 g on synthetic diet). The most important consistent difference in timing between the two has been in the time between the gut purge and the onset of spinning (2–7 hours for the *wild* larvae, mean = 4.2 (Lounibos, 1974); 8–12 hours for the *lab* strain) and in the time between the beginning of ocellar retraction and the larval–pupal ecdysis (−4.5–5 days for the *wild* larvae and 5–6 days for the *lab* strain). These differences may be due to slight variations in temperature and/or to real strain differences.
Photoperiod The times at which readily recognizable events occur during the larval–pupal transformation (LPT) are greatly influenced by photoperiod. Initiating of gut purging occurs about 0–4 hours after lights-on in larvae reared on synthetic diet under both short (12L : 12D) and long (17L : 7D) day photoperiods. A series of double ligations (between head and prothorax and between the first and second abdominal segments) of 10 larvae every 4 hours from early day 4 through day 6 (12L : 12D, 25 °C) confirmed (Riddiford, personal communication) the findings of Williams (1952). These ligations showed that in *Cecropia* PTTH release occurred at the beginning of day 4 (see Fig. B) causing ecdysone release and subsequently the initiation of ocellar retraction (larval–pupal apolysis) a day later. The prothoracic glands however were required for the complete pupation of the abdomen until early day 6 (0800 AZT). Pupal ecdysis itself is not indicated by the photoperiod but occurs at the end of the period of pharate pupal development (Truman, 1970); therefore, this ecdysis occurs over a wide range centred in the early part of the photophase on day 9 (*wild*) or 10 (*lab*).
Temperature In the descriptions of staging criteria that follow, the temperature in the laboratory was held constant at 25 °C. The sequence of events outlined below is characteristic of *lab H. cecropia*, with the differences between them and *wild* animals outlined above or specifically mentioned in the text that follows.

Stages and characteristics of developing Cecropia Silkmoths (See Fig. B)

Stage	Characteristics
Late, mature feeding fifth instar larva	The integument is opaque green (foliage reared=[F]) or blue (aqua) (reared on synthetic diet=[S]) with opaque cuticle over the dorsum. Orange [F] knobby pro- or meso-thoracic tubercles have a row of black pigment granules at their base, while the dorsally located yellow [F] or white [S] abdominal tubercles have only two spines. The lateral body wall shows an oily olive-green [F] or bluish-green [S] sheen as cessation of feeding approaches. Larvae readily use both legs and prolegs for grasping and climbing. Growth through the fifth instar is linear until it attains a plateau about 1 day before gut evacuation (Lounibos, 1974). "Frosted" frass (Nijhout and Williams, 1974) usually can be observed (more often in the *wild* larvae) at this time and is a good indicator that gut evacuation will follow in 1–2 days. Silk glands (posterior sections) are 0.25 to 0.75 mm in diameter at *ca.* 5 days before gut evacuation. 1.5 mm in diameter at gut evacuation, and can be detected in mature larvae by gently palping the lateral body wall following CO_2 but not cold anaesthesia. Both midgut and hindgut in [F] are greatly distended (2 cm in diameter) with leaf fragments, much fluid and numerous air bubbles. Larvae reared on [S] have little fluid and no air bubbles in their midguts which are only 1 cm in diameter. Haemolymph is bright green [F] or aqua [S] and melanizes readily upon exposure to air. Lancing or cutting the body wall results in turgor contractions and – if isolated from the body – curling of tissues containing muscle (i.e. integument and gut).
Day 0: Evacuation of gut contents	The larva wanders in search of a suitable place to initiate spinning of the cocoon. The lateral body wall is oily olive-green [F] or blue-green [S]. A slurry of midgut contents is evacuated over a 30–60 min period. Following evacuation, the anus is distended for up to 1 h during which approximately 0.5–2.0 ml of clear fluid of high potassium content are expelled. Evacuation results in a 25% loss in body weight (Jungreis and Tojo, 1973; Lounibos, 1974; Riddiford, unpublished observations). Initially flaccid and collapsed, the midgut becomes greatly distended with air within 3 hours after purging. The distended midgut has a diameter of 1 cm in both [F] and [S].
Day 0.5: Initiation of spinning	Spinning of the outer cocoon is initiated in *wild* strains by 4–6 hours (Lounibos, 1974) and in *lab* strains by 8–12 h after gut evacuation. This cocoon is thin, transparent, fragile and untanned. If the spinning process is disrupted, the animal will resume spinning. Legs and prolegs readily grasp and support the total body weight. The body length

(*contd*)

Stage	Characteristics
	has been reduced by *ca.* 1/4th relative to the Day 0 evacuated larval length, but a concomitant increase in girth is noted. The larva's pre-spinning length can be restored by manually elongating the animal. An additional 10–15% loss in body weight is noted (see Jungreis and Tojo, 1973). The midgut is fully distended with air.
Day 1.5: Spinning of the inner cocoon	The outer cocoon is completed and hardened. Tanning is initiated. Spinning of the inner cocoon begins. In *lab* animals only, when conditions for spinning are poor, disruption of spinning after completion of the outer cocoon prevents further spinning. (Under favourable conditions, spinning will resume.) In these disrupted *lab* animals, the larval–pupal transformation will then be lengthened for up to one month or more. An outline of the larva is readily seen within the cocoon prior to but not after tanning is complete. Legs and prolegs continue to have coordinated movements and will grasp objects, but are unable to support the body weight. An additional 10% loss in body weight is noted.
Day 3: Termination of Spinning	Both outer and inner cocoons are completed, and the cocoon is now fully tanned. The animal will repair cocoon if it is cut. The larva continues to grasp, but not hold, while the legs and prolegs lose the capacity to coordinate movements. The larva ceases to "walk". Heart contractions are detected through the dorsal thoracic cuticle, but cuticle over the abdominal segments is translucent. The midgut is distended with air, but less so than on the day of spinning.
Day 4.5: Larval–pupal apolysis (6–7 (*lab*) or 4.5–5.5 (*wild*) days before the larval–pupal ecdysis (LPE))	Early retraction of the larval ocellar pigment occurs (Kuhn and Piepho, 1936; Williams, 1952). An exuvial region or space (Verson, 1902; Wigglesworth, 1933) separates the larval from the pharate pupal cuticle. Though still prominent, legs and prolegs no longer grasp, but continue to respond to tactile stimuli. Cuticle over the dorsum, though translucent, is becoming more transparent. The now transparent silk gland is greatly reduced in diameter (*ca.* 0.5 mm) and begins to disintegrate into pieces 2–4 cm in length. Midgut is brownish and still "thick and muscular", although both epithelium (Judy and Gilbert, 1970) and musculature (Riddiford, 1972) are beginning to degenerate. A small air pocket is retained in the gut.
Day 5.5: (5–6 (*lab*) or 3.5–4.5 (*wild*) days before LPE)	Larval ocellar pigments are fully retracted. Irreversible longitudinal compression of the animal begins, and an outline of the barrel shaped "pre-pupa" of Williams (1952) can be discerned. The pharate pupa can not right itself when placed ventral side up. Not yet fully flattened,

Stage	Characteristics
	prolegs have ceased to function. Pupa type rotations of the abdomen are noted. Histolysis of the silk gland is now advanced, with the remaining tissue readily falling apart if touched. Haemolymph retains a greenish tinge, while the now green midgut is less muscular and opaque than at Day 4.5. Midgut (*ca.* 0.6 cm in diameter) contains only a small pocket of air.
Day 6.5: (4–5 (*lab*) days before LPE)	Forelegs respond feebly to tactile stimuli. The larval cuticle is only slightly translucent with heart pulsations easily observed through the dorsum of the underlying pharate pupal cuticle. Body length is slightly elongate relative to the pupal length. Midgut contains only a small pocket of air. Hindgut diverticulae are degenerating. Silk gland segments are transparent and have the appearance of a "melting-solid".
Day 6.5 (3–4 days before LPE)	The abdomen has the pupal barrel shape. Little spontaneous body movement is noted. Histolysis of the silk gland is now complete. Retension of appreciable quantities of silk gland will greatly prolong the period before the LPE. The larval cuticle is hard and virtually inseparable from the underlying pharate pupal cuticle, yet the cuticle over the dorsum is clear. Thoracic and abdominal tubercles point up prominently and fill with pale bluish, [S] opaque material. Apolysis is most pronounced in the head capsule and the last abdominal segments with the exuvial region difficult to detect visually in other body segments. Moulting gel as opposed to moulting fluid is present in the exuvial region (see Jungreis, 1974; Jungreis and Harvey, 1975). Longitudinal cutting of the body wall results neither in contraction nor in curling. Early degeneration of longitudinal and oblique integumentary musculature is noted.
Day 7.5: (3 (*lab*) days before LPE)	Larval abdominal tubercles are not bent parallel to the integument. The exuvial region is prominent in all regions of the body. The pharate pupa is more flaccid than on Day 6.5, presumably in response to marked degeneration of the abdominal longitudinal and oblique muscles. Incompletely formed pupal longitudinal muscles are present in the body wall. Moulting gel-sol rather than moulting fluid may be present in the most posterior segments of the abdomen.
Day 8–8.5: (60–48 h before LPE)	Thoracic tubercles are becoming transparent. Moulting fluid as opposed to moulting gel-sol is present in the most posterior abdominal segments, a similar transformation occurring in other proximal body segments by 48 h before

(*contd*)

Stage	Characteristics
	ecdysis. The loss of larval integumentary musculature is complete. Leaving the functional pupal longitudinal muscles. The larval-midgut epithelium is sloughed into the lumen of the pupal midgut, but midgut lumenal contents (Jungreis, 1974) do not accumulate at this time.
44–40 h before LPE	Moulting fluid, initially pale blue, accumulates in small pockets directly over the dorsum in the region formerly occupied by abdominal inter-segmental muscles. The exuvial region is filled with moulting sol and small quantities of moulting fluid. However, moulting fluid is not present in the thoracic tubercles. Tonofibrills extend from the epithelial tendon cells through the pupal cuticle and remain anchored to the larval cuticle.
36–30 h before ecdysis	The volume of moulting fluid greatly increases from about 50 μl, and is present in all body segments. Thoracic tubercles are filled with moulting fluid which is bluer (deep blue [F] and light blue [S]) than at 44–40 h before LPE, while moulting fluid in the most posterior segments is bright blue. Prolegs are greatly flattened into the body. If touched, abdomen shows a pupal rotation. Pharate pupae are very soft to the touch and the integument will break if pressure is exerted. The underlying pupal cuticle is becoming yellowish-white, especially in the posterior portion of the abdomen. Haemolymph is light green, whereas moulting fluid is deep blue. Connections persist between larval and pupal cuticles in the region of the spiracles, but tonofibrills between the epidermis and larval cuticle are no longer present.
30–34 h before LPE	Pharate pupae are blue-green in appearance with dorsal regions of the larval cuticle distended with the maximal quantity of moulting fluid that will accumulate (*ca.* 450 μl). The exuvial region separates larval from pupal cuticles except at the spiracles. The now soft larval cuticle can readily be dissected free from the pharate pupal cuticle in one piece. The lumen of the midgut fills with pigments and products derived from the epidermis and partially hydrolysed larval endocuticle.
12 h before LPE	The tubercles have collapsed with resorption of moulting fluid now virtually complete. The larval cuticle has a dry and crinkled appearance. Green or blue pigments are absent in both the larval and pharate pupal epidermis with the pupa now having a white appearance.
6–12 h before LPE	The larval cuticle becomes moist and crinkling of the cuticle disappears as it becomes flattened. Tanning of the underlying pharate pupal cuticle is noted in the posterior

Stage	Characteristics
	region of the abdomen. The larval cuticle is transparent and so fragile that it normally can be dissected from the pupal cuticle only in small pieces.
6 h before LPE	Crinkling of the larval cuticle reoccurs. The pharate pupa is very soft to the touch and frequently wiggles spontaneously. Tanning of the pharate pupal abdomen is complete.
1 or 2 h before LPE	Prominent white mid-lateral lines representing larval tracheae are noted. Tanning of the cuticle in the "thorax" is more pronounced, with the head region having a yellow appearance.
1 h before LPE	Pupal cuticle in the head region is green. Much wiggling is observed.
Day 9.5. (*wild*) or 10.5 (*lab*) – Day of larval–pupal ecdysis	The larval cuticle is shed. The cephalothorax is bright green, but will be tanned within 12 h. Formation of the endocuticle begins in earnest, but will not be completed for three to four weeks (see Locke, 1973). Fat body is the major tissue component in the pupa. The midgut is filled with a dark green fluid. Haemolymph is yellow-green [F] or pale yellow [S] in appearance.
Day of ecdysis + days	Pupal cuticle is dark brown. The rate of respiration is in excess of 20 μl g/h at 20 °C (see Schneiderman and Williams, 1954a). The midgut is fragile and filled with greenish-black fluid, whereas blood is bright yellow [F] or clear to pale yellow [S]. Fat body in [F] is yellow, ivory in [S]. Maximal volume of fluid is present in the lumen of the midgut.
Day of ecdysis + 4–10 weeks	The pupa is in metabolic diapause (respiration under 20 μl O_2 g/h at 25°C (see Williams, 1946; Schneiderman and Williams, 1954). The pupal endocuticle is fully formed.

Abbreviated criteria for staging at 25 °C *Hyalophora cecropia* between the larval and pupal stages in development using external morphological characteristics

Day	Stage	Time	Characteristic
–14–0	fifth larval instar	0–14 days before gut evacuation	Black pigment granules on dorsal thoracic tubercles
–4–0	mature 5th larval instar	0–4 days before gut evacuation	Ventro-lateral abdomen oily green in appearance

(*contd*)

Day	Stage	Time	Characteristic
0	gut evacuation	10–10.5 days before ecdysis	Gut contents evacuated, large air pocket in the gut, aimless wandering
Days after spinning			
0.5	0	9–10 days before ecdysis	Incomplete cocoon is thin, untanned, soft and transparent
1.5	+1 day	8–9 days before ecdysis	Tanned outer-cocoon is hard and nontransparent. Inner cocoon incomplete
2.5	+2 days	7–8 days before ecdysis	Cocoon completed. Legs and prolegs grasp, but body weight can no longer be supported
Days after larval–pupal apolysis			
4.5	0	6–7 days before ecdysis	Early retraction of larval ocellar pigment. Legs and prolegs no longer grasp
5.5	+1 day	5–6 days before ecdysis	Full retraction of larval ocellar pigment
6.5–7.0	+2–3 days	4–5 days before ecdysis	Prolegs non-functional, forelegs respond to tactile stimuli
7–7.5	+3–4 days	3–4 days before ecdysis	Abdomen barrel shaped
7.5–8	+4 days	3 days before ecdysis	Abdominal tubercles parallel to integument; moulting fluid in most posterior abdominal segment
8.5	+5 days	2 days before ecdysis	Moulting fluid accumulates in all parts of the exuvial region. Larval cuticle not soft
9.5	+6 days	1 day before ecdysis	Maximal volume of moulting fluid; larval cuticle very soft. Thoracic tubercles collapsed. Pharate pupal integument white
Days after the larval–pupal ecdysis			
10.5	0	day of ecdysis	Typical pupal appearance. Cephalothorax is very

Day	Stage	Time	Characteristic
			soft, fragile and incompletely tanned
	+1 day	day of ecdysis +1 day	Cuticle completely tanned. Body but not fragile
	+2–20 days	day of ecdysis +2–20 days	Respiratory rate of 50–20 μl O_2/g body wet weight-hour at 25 °C
	diapause pupa		Respiratory rate under 20 μl/g-h (25°C)

REFERENCES TO APPENDICES

Bade, M. L. and Wyatt, G. R. (1962). Metabolic conversions during pupation of the cecropia silkworm. 1. Deposition and utilization of nutrient reserves. *Biochem. J.* **83,** 470–478.

Baumhover, A. H., Cantelo, W. W., Hobgod, J. M., Knott, C. M. and Lam, J. J. (1977). An improved method for mass rearing the tobacco hornworm. Agriculture Research Service, U.S.D.A., Publication ARS–S–167.

Bell, R. A. and Joachim, F. G. (1976). Techniques for rearing laboratory colonies of tobacco hornworms and pink bollworms. *Ann. Ent. Soc. Am.* **69,** 363–373.

Fain, M. J. and Riddiford, L. M. (1977). Requirements for moulting of the crochet epidermis of the tobacco hornworm larva *in vivo* and *in vitro*. *Wilhelm Roux's Archives*, **181,** 285–307.

Harvey, W. R., Wood, J. L., Quatrale, R. P. and Jungreis, A. M. (1975). Cation distributions across the larval and pupal midgut of *Hyalophora cecropia in vivo*. *J. exp. Biol.* **63,** 321–330.

Haskell, J. A., Harvey, W. R. and Clark, R. M. (1968). Active transport by the Cecropia midgut. V. Loss of potassium transport during larval–pupal transformation. *J. exp. Biol.* **48,** 25–37.

House, H. L. (1967). Artificial diets for insects: A compilation of references with abstracts. *Inform. Bull. No. 5, Res. Inst., Can. Dept. Agr.* Belleville, Ontario.

House, H. L. (1974). Nutrition. *In* "Physiology of the Insecta", 2nd Ed. (Ed. M. Rockstein), pp. 1–62. Academic Press, New York and London.

Judy, K. J. and Gilbert, L. I. (1970). Histology of the alimentary canal during the metamorphosis of *Hyalophora cecropia* (L). *J. Morph.* **131,** 277–299.

Jungreis, A. M. (1973). Distribution of magnesium in tissues of the silkmoth *Hyalophora cecropia*. *Amer. J. Physiol.* **224,** 27–30.

Jungreis, A. M. (1974). Physiology and composition of moulting fluid and midgut lumenal contents in the silkmoth *Hyalophora cecropia*. *J. Comp. Physiol.* **88,** 113–127.

Jungreis, A. M. (1976a). Changes in fat body hexokinase activity during the larval–pupal transformation of the silkmoth *Hyalophora cecropia*. *Comp. Biochem. Physiol.* **B53,** 201–204.

Jungreis, A. M. (1976b). Regulation of *Hyalophora cecropia* fat body hexokinase by hexose phosphates common to the pathways of glycolysis, glycogen and trehalose synthesis. *Comp. Biochem. Physiol.* **B53,** 405–413.

Jungreis, A. M. (1977). Comparative aspects of invertebrate epithelial transport. *In* "Water Relations in Membrane Transport in Plants and Animals" (Ed. A. M. Jungreis, T. Hodges, A. Kleinzeller and S. G. Schultz), pp. 89–96. Academic Press, New York and London.

Jungreis, A. M. (1978a). The composition of larval–pupal moulting fluid in the tobacco hornworm, *Manduca sexta. J. Insect. Physiol.* **24,** 65–73.

Jungreis, A. M. (1978b). Insect dormancy. *In* "Mechanisms of Dormancy and Development Arrest" (Ed. M. E. Clutter), pp. 47–112. Academic Press, New York and London.

Jungreis, A. M., Dailey, J. C. and Hereth, M. L. (1975). α-Glycerol phosphatase and glycerol kinase activity in tissues of the silkmoth *Hyalophora cecropia* during the larval pupal transformation. *Amer. J. Physiol.* **229,** 1448–1454.

Jungreis, A. M. and Harvey, W. R. (1975). Role of active potassium transport by integumentary epithelium in secretion of larval–pupal moulting fluid during silkmoth development. *J. exp. Biol.* **62,** 357–366.

Jungreis, A. M., Jatlow, P. and Wyatt, G. R. (1973). Inorganic ion composition of hemolymph of the cecropia silkmoth: Changes with diet and ontogeny. *J. Insect Physiol.* **19,** 225–233.

Jungreis, A. M., Jatlow, P. and Wyatt, G. R. (1974). Regulation of trehalose synthesis in the silkmoth *Hyalophora cecropia*: The role of magnesium in body fat. *J. exp. Zool.* **187,** 41–46.

Jungreis, A. M. and Tojo, S. (1973). Potassium and uric acid content in tissues of the silkmoth *Hyalophora cecropia. Am. J. Physiol.* **224,** 21–26.

Jungreis, A. M. and Vaughan, G. L. (1977). Insensitivity of lepidopteran tissues to ouabain. I. Absence of ouabain binding and Na^+-K^+-ATPases in larval and adult midgut. *J. Insect Physiol.* **23,** 503–509.

Jungreis, A. M. and Wyatt, G. R. (1972). Sugar release and penetration in insect fat body: Relations to regulation of haemolymph trehalose in developing stages of *Hyalophora cecropia. Biol. Bull. mar. biol. Lab., Woods Hole,* **143,** 367–391.

Krishnakumaran, A., Berry, S. J., Oberlander, H. and Schneiderman, H. A. (1967). Nucleic acid synthesis during insect development. II. Control of DNA synthesis in the Cecropia silkworm and other saturniid moths. *J. Insect Physiol.* **13,** 1–57.

Kuhn, A. and Piepho, H. (1936). Ueber Hormanale Wirkungen bei der Verpuppung der Schmetterlinge. *Ges. Wiss. Göttingen, Nachr. Biol.* **2,** 141–154.

Locke, M. (1973) The structure and formation of the integument in insects. *In* "Physiology of the Insecta", 2nd Edn. (Ed. M. Rockstein), **VI,** pp. 124–213. Academic Press, New York and London.

Lounibos, L. P. (1974). Cocoon spinning behaviour of saturniid silkworms. Ph.D. Thesis, Harvard University.

Mitsui, T. and Riddiford, L. M. (1976). Pupal cuticle formation by *Manduca sexta* epidermis *in vitro*: Patterns of ecdysone sensitivity. *Dev. Biol.* **54,** 172–186.

Mitsui, T. and Riddiford, L. M. (1978). Hormonal requirements for the laval–pupal transformation of the epidermis of *Manduca sexta in vitro. Dev. Biol* **62,** 193–205.

Nijhout, H. F. and Williams, C. M. (1974). Control of moulting and metamorphosis in the tobacco hornworm, *Manduca sexta* (L.): Cessation of juvenile hormone secretion as a trigger for pupation. *J. exp. Biol.* **61,** 493–501.

Pan, M. L. (1971). The synthesis of vitellogenin in the Cecropia silkworm. *J. Insect Physiol.* **17,** 677–689.

Pan, M. L. (1977). Juvenile hormone and vitellogenin synthesis in the Cecropia silkworm. *Biol. Bull.* **153,** 336–345.

Pan, M. L., Bell, W. J. and Telfer, W. H. (1967). Vitellogenic blood protein synthesis by insect fat body. *Science, Wash.* **165,** 393–394.

Riddiford, L. M. (1968). An artificial diet for Cecropia and other saturniid silkmoths. *Science, Wash.* **160,** 1461–1462.

Riddiford, L. M. (1972). Juvenile hormone in relation to the larval–pupal transformation of the Cecropia silkworm. *Biol. Bull. mar. biol. Lab., Woods Hole,* **142,** 310–325.

Riddiford, L. M. (1975). Juvenile hormone induced delay of metamorphosis of the viscera of the Cecropia silkworm. *Biol. Bull. mar. biol. Lab., Woods Hole,* **148,** 429–439.

Riddiford, L. M. and Ajami, A. M. (1973). Juvenile hormone: Its assay and effects on pupae of Manduca sexta. *J. Insect Physiol.* **19,** 749–762.

Schneiderman, H. A. and Williams, C. M. (1954a). The physiology of insect diapause. VIII. Qualitative changes in the metabolism of the Cecropia silkworm during diapause and development; IX. The cytochrome oxidase system in relation to the diapause and development of the cecropia silkworm. *Biol. Bull. mar. biol. Lab., Woods Hole,* **106,** 210–229; 238–252.

Telfer, W. H. (1967). Cecropia. *In* "Methods in Developmental Biology" (Ed. F. H. Wilt and N. K. Wessells), pp. 173–182. Thomas Y. Crowell, New York.

Telfer, W. H. and Rutberg, L. D. (1960). The effects of blood protein depletion on the growth of oocytes in the Cecropia moth. *Biol. Bull. mar. biol. Lab., Woods Hole,* **118,** 185–210.

Truman, J. W. (1970). The control of ecdysis in silkmoths. Ph.D. Thesis, Harvard University.

Verson, E. (1902). Observations on the structure of the exuvial glands and formation of the exuvial fluid in insects. *Zool. Anz.* **25,** 652–654.

Wigglesworth, V. B. (1933). The physiology of the cuticle and of ecdysis in *Rhodnius prolixus* (Triatomidae, Hemiptera) with special reference to the function of oenocytes and of the dermal glands. *Q. J. microsc. Sci.* **76,** 269–318.

Williams, C. M. (1946). Physiology of insect diapause: the role of the brain in production and termination of pupal dormancy in the giant silkworm, *Platysamia cecropia. Biol. Bull. mar. biol. Lab., Woods Hole,* **90,** 234–243.

Williams, C. M. (1952). The physiology of insect diapause. IV. The brain and prothoracic glands as an endocrine system in the Cecropia silkworm. *Biol. Bull. mar. biol. Lab., Woods Hole,* **103,** 120–138.

Wyatt, G. R. and Pan, M. L. (1978). Insect plasma proteins. *Ann. Rev. Biochem.* **47,** 779–817.

Yamamoto, R. T. (1969). Mass rearing of the tobacco hornworm. II. Larval rearing and pupation. *J. Econ. Ent.* **62,** 1427–1431.

Morphology and Electrochemistry of Insect Muscle Fibre Membrane

Tom Piek and K. Djie Njio

Pharmacological Laboratory, University of Amsterdam, The Netherlands

1 Introduction

Insects have been adapted to terrestrial life since the earliest part of the upper carboniferous. Since that period they have developed a considerable degree of differentiation which has complicated attempts to describe physiological processes in insects in terms of general physiology. Also, comparison with other groups of animals, is often impossible. An example is the extraordinary diversity of feeding habits. This may be the cause of the diversity in ionic composition of the haemolymph which, particularly in phytophagous insects, is distinctly different from the "sea-water-like" composition of the haemolymph of marine invertebrates as well as of the blood of marine and terrestrial vertebrates. In this respect, insects may be unique, especially phytophagous species.

In view of the fact that the ionic composition of insect haemolymph may differ from what is generally accepted as a "normal" medium for excitable tissues, the question arises whether the environment, immediately adjacent to the plasma membrane of excitable tissues in insects, differs fundamentally in ionic composition from that in other animal groups. If the environment in direct contact with the outer margin of the plasma membrane of nerve and muscle fibres is assumed not to be fundamentally different from that surrounding the excitable tissues in other animals, a distinct barrier should exist separating the direct extracellular environment of nerve and muscle fibres from the haemolymph. In that case the special ionic composition of the haemolymph found in a number of insects requires specialized ion barriers, not only around the nervous system but also around the muscles. As regards the nervous system of insects the presence of ion barriers, as well as their morphological localization is relatively well known (*cf.* Treherne, 1976). Since the structure of the muscle fibre membrane is extremely complex it is conceivable that an ion barrier may also exist in this tissue (Piek 1974, 1975).

It is the special purpose of this chapter to describe the ultrastructural and electrochemical properties of the muscle fibre membrane as a basis for a discussion of the properties and localization of ion barriers.

2 Morphology of the membranes

The morphological structure of insect muscle bears a close resemblance to vertebrate skeletal muscle. Besides the mitochondria and the contractile fibres, a system of tubules and vesicles can be discerned. This sarcotubular system consists of a transversely orientated tubular system (T-system, TTS) and a longitudinally orientated sarcoplasmic reticulum (SR). These systems are involved in the generation of membrane potentials, and in the mechanism of muscle excitation, contraction and relaxation.

The membranes of the muscle fibres consist of three morphologically and probably also functionally distinct systems. These are the basal lamina, the plasma membrane including the membranes of the TTS, and the membranes of the SR.

The structure and functions of the contractile elements of insect muscle were reviewed by Elder (1975).

2.1 THE BASAL LAMINA (BASEMENT MEMBRANE)

In insects the muscle fibres are surrounded by a connective tissue sheath, often called basement membrane. For insect muscle fibres we prefer the synonym basal lamina, because of its laminal structure. The basal lamina of insect muscles vary considerably in thickness and structural detail. According to Edwards *et al.* (1956) the thickness is about 30 to 60 nm in high frequency muscles and about 50 to 90 nm in low frequency muscles. Bienz-Isler (1968a) measured a thickness of 28 nm in *Antherea* (= *Telea*) *pernyi* flight muscle and Huddart and Oates (1970) found 80–140 nm in leg muscles of *Carausius morosus* and *Locusta migratoria*. The basal lamina in *Apis mellifera* (Fig. 1) and *Galleria mellonella* is approximately 20 nm thick, in *Pieris brassicae* and *Philosamia cynthia ca.* 40 nm, in *Schistocerca gregaria ca.* 10 nm (Fig. 21), and in *L. migratoria* it reaches a thickness of about 150 nm (Fig. 2).

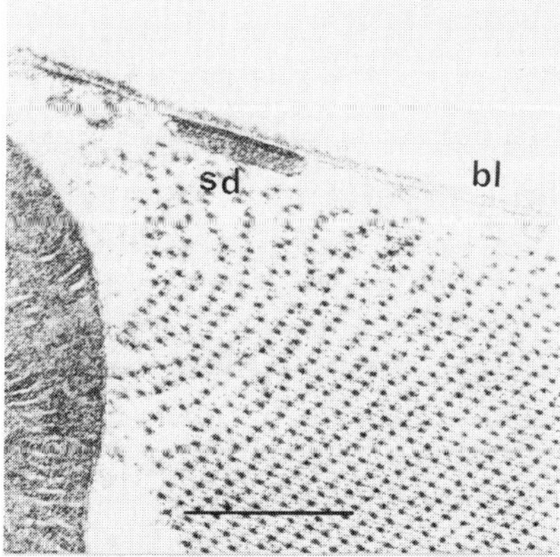

Fig. 1. Transverse section of a fibrillar flight muscle fibre of a worker honeybee *Apis mellifera*, showing part of a fibril and a mitochondrion. The fibre is limited by a plasma membrane, forming surface dyads (sd) with parts of the sarcoplasmic reticulum. The fibre is covered by a thin basal lamina (bl). Scale 0.5 μm.

The thickness measured from electron micrographs may greatly depend on the way the muscle has been treated before fixation (Clements and May, 1974, *cf.* section 6 and Figs 21 and 22).

In the fibrillar flight muscles of *Tenebrio molitor* the basal lamina consists of 3 distinct regions: an outermost component consisting of a diffuse or unorganized coarsly granular layer, an irregular dense sheath of about 25 to 30 nm and an inner sheath of 50 to 100 nm (Smith, 1960, 1961b). For the

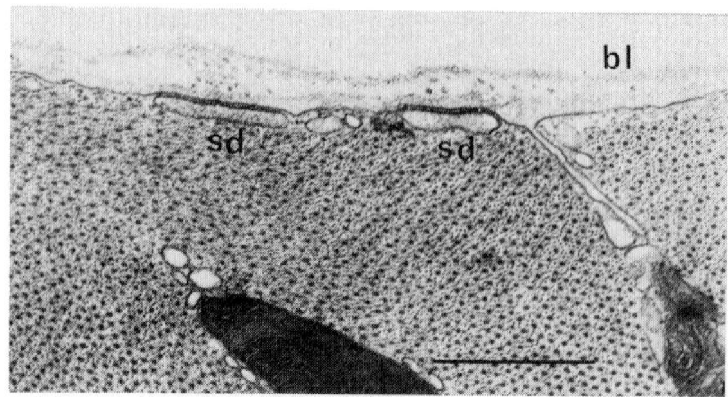

Fig. 2. Transverse section of a fibre of the extensor muscle of the tibia of *Locusta migratoria* showing the relatively thick basal lamina (bl) and surface dyads (sd) formed by the plasma membrane and the sarcoplasmic reticulum. Scale 0.5 μm.

fibrillar flight muscle of *Lethocerus* spp. Ashhurst (1967) mentioned a basal lamina consisting of amorphous material of about 50 nm thickness, covering the plasma membrane.

In *Locusta migratoria* (Ashhurst, 1959) and in *Periplaneta americana* (Ashhurst, 1961) the connective tissue surrounding the nervous tissue consists of collagen fibres embedded in a matrix of neutral mucopolysaccharides. In *Galleria mellonella* the histochemical properties of the neural lamella contrast markedly with that of most other insects, in that it contains a large amount of acid mucopolysaccharides in addition to neutral mucopolysaccharides and collagenous fibres (Ashhurst, 1964; Ashhurst and Richards, 1964).

Little is known of the exact structure and function of the basal lamina. One function obviously is that of a supporting tissue, binding fibres together. According to Ceasar and Edwards (1957) the lamina might protect cell units and their common interspace against too rapid changes in ion concentration. Ruska *et al.* (1958) suggested that the basal lamina may serve as a restrictive ion barrier and the plasma membrane as a selective ion barrier, and Bennett (1960) suggested that the basal lamina plays a role in ion permeability and in exchange phenomena. The work of Treherne (*cf.* Treherne, 1976), however, showed that relatively large molecules rapidly cross the neural lamina, and Ashhurst (1968) concluded that connective tissue in insects cannot function as a selective ion barrier. However, Clements and May (1974) have suggested that in the intact locust the basal lamina could serve as a barrier for glutamate (*cf.* section 6 and Figs 21 and 22). They observed that in muscles fixed by perfusion, the basal lamina is more compact than in completely isolated muscles fixed by immersion.

2.2 THE PLASMA MEMBRANE

The thin plasma membrane which surrounds the muscle fibre is invaginated at regular intervals (Figs 2 and 3) to form the transverse tubular system (TTS). Contrary to the basal lamina, the plasma membrane is uniform in thickness and structure. As in most cells the plasma membrane of the muscle fibre has a triple layered structure, consisting of two osmiophilic regions separated by a region of low density. Each region is about 2.5 nm thick, making a total thickness of about 7.5 nm (Smith, 1961b, 1968; Bienz-Isler, 1968a).

Occasionally the plasma membrane and the basal lamina lie in close apposition, forming desmosome-like structures (Figs 3 and 4). According to Elder (1975) these structures, the hemidesmosomes, are the attachment points of Z-line material to the plasma membrane and the basal lamina. However, not all hemidesmosomes coincide with Z-line attachment. Hemidesmosomes may also be found far from the Z-line structures especially in Lepidoptera where randomly distributed hemidesmosomes are frequently observed (Fig. 4).

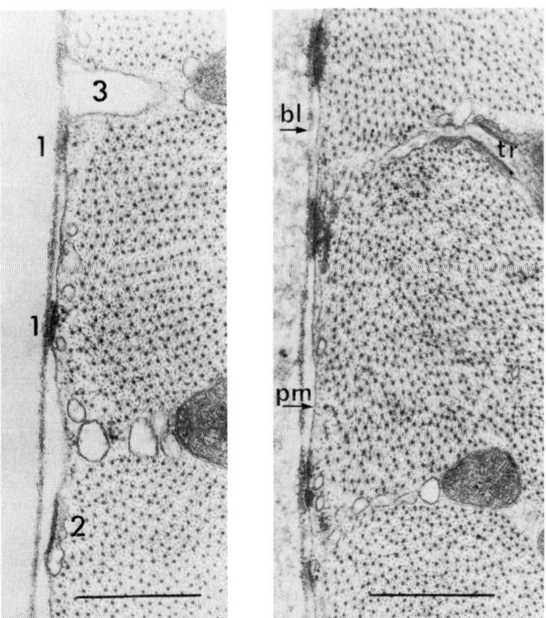

Fig. 3 (*left*). Transverse section of a sternopedal muscle fibre of *Actias selene* showing hemidesmosomes (1) connecting the basal lamina with the plasma membrane, surface dyads (2) connecting the plasma membrane with the sarcoplasmic reticulum, and an invagination (3) of the plasma membrane. Scale 0.5 μm.

Fig. 4 (*right*). Hemidesmosomes in the membrane of a sternopedal muscle fibre of *Actias selene*, showing the presence of electron dense structures in the plasma membrane (pm) and of dense material in the space between the plasma membrane and the basal lamina (bl). Note the triad (tr). Scale 0.5 μm.

According to Loewenstein (1970) desmosomes are places of increased ion permeability. Possibly also the hemidesmosomes are sites of low resistance pathways by which ion exchange may occur (*cf.* section 5.2).

Sometimes the surface plasma membrane lies in close apposition to an element of the sarcoplamic reticulum, forming surface dyads or sarcoplasmic dyads. Hagopian and Spiro (1967), investigating the femoral muscle of *Leucophaea maderae*, were the first to observe the surface dyads. Surface dyads were also observed by Cochrane *et al.* (1972) in the muscles of *Schistocerca gregaria* and by Piek *et al.* (1973) in *Philosamia cynthia*. Surface dyads have

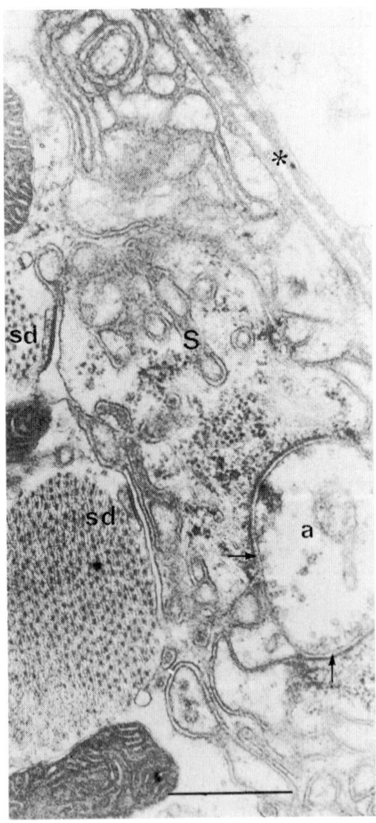

Fig. 5. Transverse section through a sternopedal muscle fibre of *Philosamia cynthia* at the level of the neuromuscular junction. The axon terminal (a) is embedded in a subsynaptic reticulum (S), the lumen of which appears to be continuous with the synaptic cleft (→), the lumen of the transverse tubular system, and with the outer medium of the fibre (*). Note the surface dyads (sd) in the part of the plasma membrane covered by the subsynaptic reticulum. Scale 0.5 μm.

also been seen in skeletal muscle fibres of *Locusta migratoria* (Orthoptera), *Actias selene* (Lepidoptera) and *Apis mellifera* (Hymenoptera) (Figs 1, 2, 3 and 5). The structure of surface dyads seems identical with that of the dyads present in the TTS (see below). Electron dense material, as visualized in electron micrographs, seems to originate from the SR-component and it almost spans the gap between the SR and the plasma membrane (Fig. 2).

In the obliquely striated muscles of annelids, modifications of the membrane resembling dyads occur at places where the peripheral vesicles of the SR make contact with the plasma membrane at the surface of the fibre (Torida *et al.*, 1975). Thus the presence of surface dyads is not restricted to insect muscles.

2.3 THE TRANSVERSE TUBULAR SYSTEM AND THE SARCOPLASMIC RETICULUM

Using the light microscope the presence of fine filaments forming longitudinal and transverse reticula have already been discerned in arthropod and vertebrate skeletal muscles by Ramon y Cajal (1890) with the Golgi-silver impregnation method. He was of the opinion that in arthropods this system was continuous with the tracheoles, penetrating into the muscle. Veratti (1902) found that the arrangement of the internal reticular apparatus differed characteristically between different muscles of the same species. For insects he correctly interpreted their nature as being quite distinct from the tracheal system.

The internal reticular structure has long been neglected, probably because the impregnation technique required for its demonstration is particularly capricious. Finally electron microscopy has confirmed the existence of the reticulum and has enabled a structural and topographical analysis of its components.

Bennett and Porter (1953), studying the breast muscle of the domestic fowl, were the first to observe a reticular structure with the electron microscope. It was distributed in repeating patterns related to the sarcomeres of the myofibrils. They identified this structure as the reticulum seen in light microscopy by the silver impregnation method of Ramon y Cajal (1890) and Veratti (1902). They introduced the term sarcoplasmic reticulum for the system of vesicles and tubules. They believed that the tubules were linked together to form lace-like sleeves around the myofibrils. The sarcoplasmic reticulum was thought to be analogous to the endoplasmic reticulum of other cells. Porter (1956) and Porter and Palade (1957) investigated the position of the SR in the sarcomere in the larvae of the amphibian *Amblystoma punctatum* and in various muscles of the rat. They were the first who used the term triad for the structure existing of two vesicles with an intermediate space, which was called the central element of the triad. The position of these triads was fixed for each muscle. It was also observed by Porter (1956) and Porter and Palade (1957) that the limiting membrane of the central element of a triad was thicker than that of the lateral vesicles, a first indication that the central element might not be continuous with the other two vesicles.

The term sarcoplasmic reticulum was thus originally used for both the transversely and longitudinally orientated system in the muscle. Anderson-Cedergren (1959), however, studying mouse skeletal muscle concluded that there were two different elements. She introduced the term T-system (TTS) for the transversely orientated tubules which were closely associated with the plasma membrane to form the central elements of triads. The name sarcoplasmic reticulum is now restricted to the components of the longitudinal system whose tubules and vesicles have no connection with the outer plasma membrane.

In fibrillar flight muscle of *Tenebrio molitor*, Smith (1961b) found an association of two vesicles, one from the TTS and the other from the SR, which he called a dyad. He considered the dyad analogous to the vertebrate triad. He also observed that the transverse and longitudinal systems were discontinuous.

It has already been mentioned that Porter (1956) and Porter and Palade (1957) found that the central element of the triad had a thicker membrane than the lateral elements, and thus might not be continuous with them. Evidence that in frog muscle the extracellular fluid has access to the central elements of the triads and not to the lateral elements was found by Huxley (1964) using the ferritin technique and electron microscopy, and at the same time by Endo (1964) using the fluorescent dye lissamine rhodamine B200. Experiments in which muscles of dissected insects were soaked in ferritin demonstrated that the extracellular fluid has access to the T-tubules (Smith, 1966b; Ashhurst, 1967; Beinbrech, 1972). Smith and Sacktor (1970) injected ferritin into the haemolymph of a living blowfly *Phormia regina*. This experiment also showed that the lumen of the tubules and the extracellular fluid are continuous in the intact living animal. The SR is not directly continuous with the extracellular medium.

Smith (1961b) observed that the membranes limiting the T-system have the same triple layered structure as the outer plasma membrane and the same thickness (7.5 nm). The membranes of the SR, however, are only about 5 nm thick and do not appear to show the three-layered structure of a unit membrane. Hoyle (1965) and O'Connor *et al.* (1965), however, observed that the SR membrane in *Periplaneta americana* and in *Schistocerca gregaria* has a triple layered structure of about 5.6 nm thickness. Hoyle (1965) observed an osmiophilic line of about 2 nm, apposing the TTS, a clear middle line of 2 nm, and an inner line of 1.6 nm. Continuity of the TTS membrane with the outer plasma membrane was found to be a regular occurrence in insect muscles (Smith, 1966a, b).

With the introduction of glutaraldehyde fixation, the plasma membrane could be clearly visualized and its continuity with the TTS could finally be established. Additional investigations on the development of the TTS during metamorphosis showed the formation of the TTS by invaginations of the plasma membrane (Bienz-Isler, 1968b; Beinbrech, 1972).

Electron microscopy has also revealed the fine structure of the dyad. The gap between SR and TTS components of the dyad is about 10 nm. From the surface of the SR membrane apposing the TTS, rows of electron-dense blocks, evenly spaced at about 30 nm intervals, project toward the T-system in both synchronous and asynchronous muscles (Fig. 6) (Smith, 1965, 1966b;

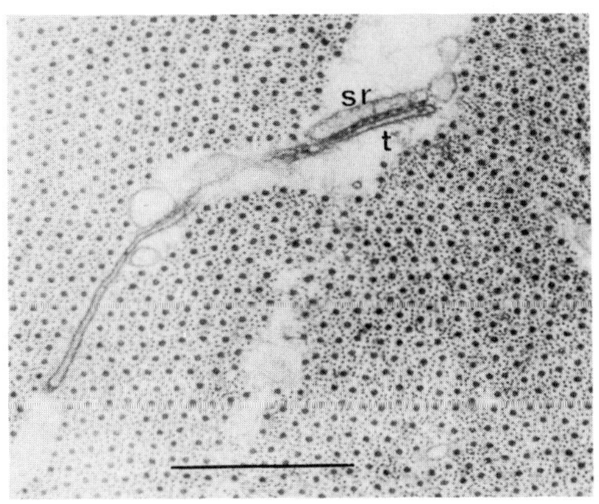

Fig. 6. A transverse tubule (t) forming a dyad with the sarcoplasmic reticulum (sr) in a fibre of the m. promotor coxae of *S. gregaria*. Note the electron-dense blocks in the space between the dyadic part of the tubule and the reticulum. Scale 0.5 μm.

Hagopian and Spiro, 1967; Smith and Sacktor, 1970; Anstee, 1971). The SR cisternae of the dyad may contain electron-dense material (Smith, 1965, 1966b; Ashhurst, 1967; Smith and Sacktor, 1970; Cochrane *et al.*, 1972). The T-tubules may contain electron-dense granules (Elder, 1975) (Fig. 7).

Pasquali-Ronchetti (1969) suggested that the elements of the SR of different sarcomeres were not interconnected. According to Cochrane *et al.* (1972), however, this view was based on incorrect observations and they conclude that the SR-systems of adjacent sarcomeres are interconnected.

In vertebrate muscle the TTS and SR develop simultaneously, and only in the last phase of their development are the triads formed. In insect muscle, a similar development was found by Bienz-Isler (1968b) in synchronous flight muscles of the lepidopteran *Antheraea pernyi*. Beinbrech (1972) studying

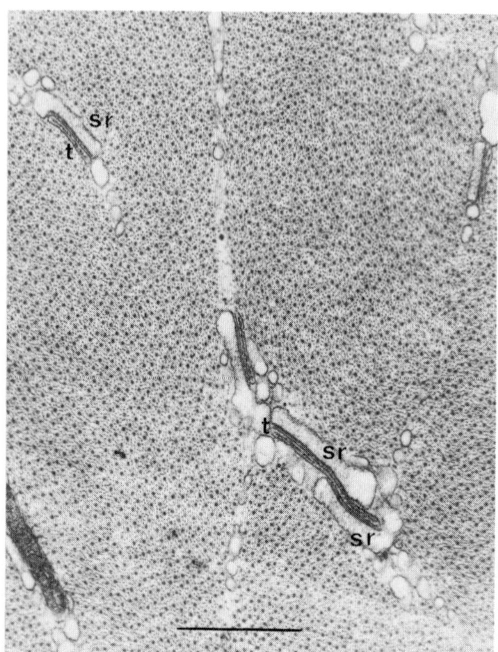

Fig. 7. Dyadic and triadic structures in a fibre of the metathoracic m. extensor tibiae of *L. migratoria.* Note the electron-dense material in the sarcoplasmic reticulum (sr) and the granules in the TTS (t). Scale 0.5 μm.

asynchronous flight muscles of *Phormia terrae-novae*, found that during metamorphosis the TTS is formed before the SR whereas the dyads are formed simultaneously with the SR.

In vertebrate muscles, there is a correlation between the speed of contraction and the extent of the SR (Porter, 1961; Fawcett and Revel, 1961; Revel, 1962). This correlation is also found in insect synchronous muscles (Cochrane *et al.*, 1972). Huddart and Oates (1970) stated that the speed of contraction was dependent not only on the extent of the SR but also on the number of dyads per unit area. They found that in the fast extensor tibia of the locust, *L. migratoria*, there are three times as many dyads per unit area than in the slower flexor tibiae of the stick insect (*C. morosus*). Although Huddart and Oates (1970) did not find a marked difference in the TTS of locust and stick insect muscles, they found that the SR was more developed in the locust than in the slower stick insect muscle.

In contrast to synchronous muscle, there is no correlation between the extent of the SR and the speed of contraction in the asynchronous (fibrillar) flight muscles, which contract with extremely high frequencies yet the SR is very much reduced (Smith, 1965; Smith and Sacktor, 1970). The TTS in these muscles is well developed, though somewhat irregular in distribution. Only in the fibrillar flight muscle of the water bugs *Lethocerus* spp. Ashhurst (1967) demonstrated a well developed SR and a reduced TTS. In all fibrillar flight muscles the SR is too sparse to account for the rapid removal of calcium ions which is necessary to account for the high contraction frequencies recorded in these muscles. Obviously the presence of a well developed SR is coupled to a high frequency excitation-contraction coupling system rather than to a high frequency oscillation of the myofibrils.

2.4 NEUROMUSCULAR JUNCTIONS

On physiological grounds, it has been supposed that specialized areas must be present on the muscle fibre, where the nerve stimulus is transmitted to the muscle fibre. Doyère (1840) investigating the nervous system of Tardigrada was the first to observe the junction between nerve and muscle fibre in invertebrates. What he actually saw was the branching of the nerve over the muscle fibre. Kühne (1871) named these structures Doyère's hillocks or cones. Doyère's hillock was thought to be analogous to the endplate in vertebrates. Sherrington (1897, *cf.* De Robertis, 1964) introduced the term synapse to describe the site where the transmission of impulses occurs. Arvanitaki (1942) defined a synapse as "surfaces of contact anatomically differentiated and functionally specialized for transmission of the liminal excitation from one element to the following in an irreciprocal direction".

Two types of neuromuscular junction were discerned in insects: the "endplate" type usually found in less differentiated muscles and the "non-endplate" type found in highly differentiated muscles. In the "endplate" type the branching nerves made contact on Doyère's hillock; in the "non-endplate" type the axon terminals were diffusely distributed over the muscle fibres (Mangold, 1905; Marcu, 1929; Auber, 1960). This classification is also used by Hámori (1963) in his electron microscope studies.

The distribution of neuromuscular junctions in insect muscle has been visualized with the light microscope (Kühne, 1871; Foettinger, 1880; Ramon y Cajal, 1890; Mangold, 1905; Macu, 1929). The structure of the synapses was revealed using the electron microscope (Edwards *et al.*, 1958a; Auber, 1960; Hámori, 1961).

In insects there are several ways in which an axon can make synaptic contact with the muscle fibre. The axon may lie in a groove in the surface of the muscle fibre as in *Vespa carolina* leg muscle (Edwards *et al.*, 1958a), or the axon may make synaptic contact with projections from the surface of the muscle, which take the form of pillars or sheets, as in locust retractor unguis muscle (Fig. 8). These axons are normally covered by glial cells, but

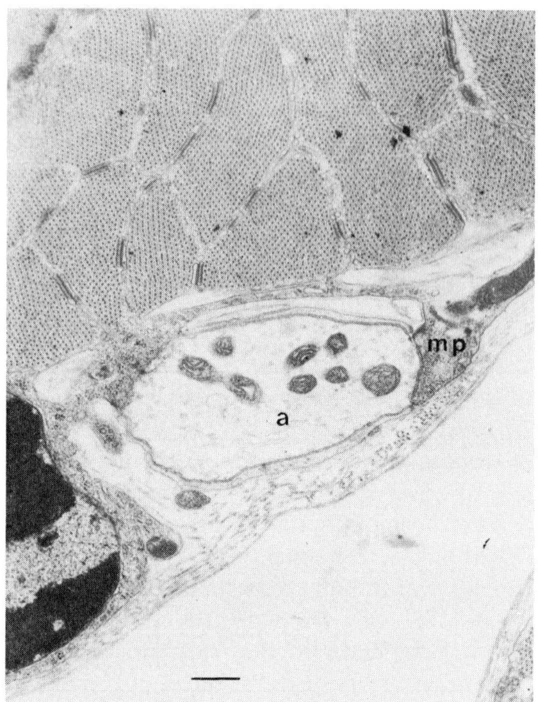

Fig. 8. Transverse section of the synaptic area of a muscle fibre (m. retractor unguis) of *S. gregaria* showing a muscular pillar (mp) making contact with the axon terminal (a). Note the numerous dyads. Scale 0.5 μm.

some axon terminals are "naked" and only separated from the haemolymph by the basal lamina as in *Calliphora erythrocephala* (Hardie, 1976). Finally, the axon can penetrate into the muscle, where it is completely surrounded by the muscle cell. The axon may lie either superficially as in *Drosophila melanogaster* flight muscle (Shafiq, 1964) or well away from the surface of the muscle as in *Tenebrio molitor* flight muscle (Smith, 1960). These axon terminals usually have no glial investment. The occurrence of the different types of neuromuscular junction in insects has been tabulated by Osborne (1970).

The axon terminal is characterized by the presence of numerous synaptic vesicles, about 20–40 nm in diameter. At the neuromuscular junctions, these vesicles are usually clustered in groups, lying close to the axon membrane. It is generally accepted that the synaptic vesicles contain transmitter substances.

In some crustaceans it was possible to distinguish between inhibitory and excitatory synapses since they contain different synaptic vesicles. The inhibitory axon terminals contain flattened vesicles, whereas the excitatory axon terminals contain spherical vesicles (Uchizono, 1966, 1967; Nadol and Darin de Lorenzo, 1968; Kosaka, 1969; Atwood *et al.*, 1972). According to Lund and Westrum (1966), Pensa and Ciccarelli (1968), and Bodian (1970) in vertebrates there are two populations of vesicles. The form of the vesicles of one population changes under the influence of the fixation fluid and the buffer used, while the other population is unaffected and remains spherical. Recently Tombes (1976) found spherical and flattened synaptic vesicles in nerves innervating the spermathecal muscles of weevils. We have tried several combinations of fixation fluids and buffers on the locust extensor tibia muscle, but were unable to note any difference in the form of the synaptic vesicles in axon terminals which obviously transmitted excitatory as well as inhibitory nerve impulses.

According to Atwood *et al.* (1969) it might be possible to distinguish between fast and slow axons by the diameter of the axon terminals. In the cockroach extensor tibia muscle the mean diameters of the slow and fast axon terminals were 1.2 and 2.0 μm respectively. Jahromi and Atwood (1969) suggested that the slow axons innervate muscles with long sarcomeres whereas the fast axons innervate muscle fibres with short sarcomeres. This, however, makes no sense for muscle fibres with dual innervation (excitatory and inhibitory).

In most cases, the postsynaptic muscle membrane has a thickened aspect. In *Tenebrio molitor* flight muscle this thickening is caused by intercalation of a layer of electron-dense material about 7.5 nm thick (Smith, 1960). In the cockroach extensor tibiae, the apparent thickening of the postsynaptic membrane is due to deposition of an electron-dense, granular material, in a 10–15 nm thick layer, leaving a relatively clear area of 5–10 nm. The dense layer is interrupted periodically every 18–20 nm by less dense "gaps" about 6 nm wide (Atwood *et al.*, 1969). In the synaptic region, the pre- and postsynaptic membranes retain their triple-layered appearance. In *Phormia regina* the postsynaptic membrane bears dense structures at intervals of about 11 nm projecting into the synaptic gap (Smith and Sactor, 1970). Between the presynaptic axonal membrane and the apposing postsynaptic muscle membrane lies the synaptic cleft. This cleft ranges in width from 7.5 nm in *T. molitor* flight muscle (Smith, 1960) to 25 nm in *Galleria mellonella* integumental muscle (Belton, 1969). In most cases, the synaptic cleft is either empty or filled with some electron-dense material. Septate desmosomes in the synaptic cleft were observed by Osborne (1967) in *Phormia terrae-novae* and by Hardie (1976) in *Calliphora erythrocephala* larvae.

In insects, the plasma membrane of the muscle fibre is invaginated in most of the synapses, to form a more or less complex membrane system, the subsynaptic reticulum or rete synapticum. In these muscle fibres the

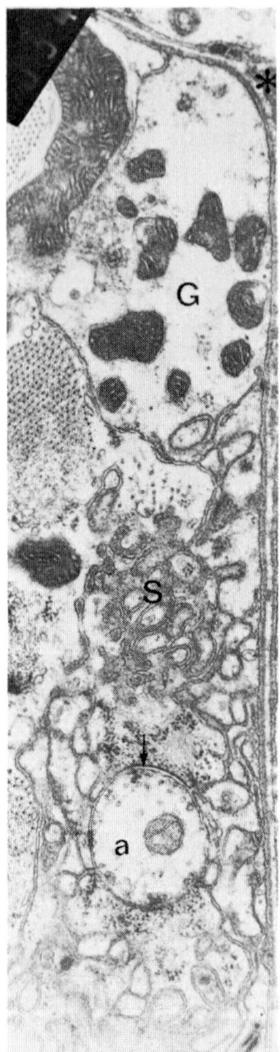

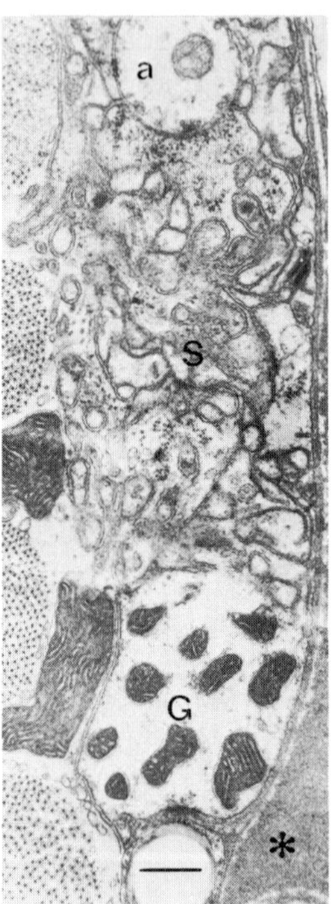

Fig. 9. Transverse sections of the synaptic region of the sternopedal muscle of *P. cynthia* showing the extensive subsynaptic reticulum (S). The reticulum formed by the muscle fibre and the glial cell (G) forms a seal of the axon terminal (a) leaving only a very long and small passage from the synaptic cleft (→) to the outer medium (*). Scale 0.5 μm. The two pictures represent overlapping parts of the same synapse.

tubules in the subsynaptic region are formed by the apposing membranes of the muscle and glial cells (Fig. 9). The spaces between the membranes are therefore extracellular. The function of the subsynaptic reticulum is not clear. It has been suggested that the subsynaptic reticulum has a function in propagating action-potentials. In some muscle fibres, however, the subsynaptic reticulum is completely absent, as in *Drosophila melanogaster* flight muscle (Shafiq, 1964), and in some synapses of the promotor coxae muscle of *S. gregaria* (Fig. 10).

In a number of insect muscles the sarcoplasm immediately beneath the

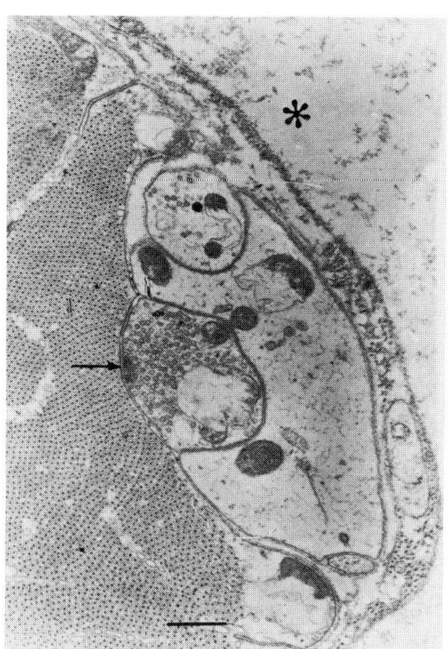

Fig. 10. Transverse section of the synaptic region of a muscle fibre (m. promotor coxae) of *S. gregaria* showing a neuromuscular junction without subsynaptic reticulum. Nevertheless the synaptic cleft (→) is connected with the outer medium (∗) via a narrow space. Scale 0.5 μm.

plasma membrane of the muscle fibre in the synaptic region, is filled with large numbers of granules, the aposynaptic granules. These are dense, osmiophilic, homogenous granules 5–15 nm in diameter. They are aggregated most closely in the region of the synapse and become more dispersed with increasing distance from the plasma membrane (Edwards *et al.*, 1958a, 1958b; Edwards, 1959; Smith, 1960; Belton 1969). Belton (1969) has suggested that these granules are glycogen.

In some insect muscles postsynaptic vesicles are found. The vesicles are 40 to 120 nm in diameter in *Tenebrio molitor* (Smith, 1960), 50 to 150 nm in *Aeshna* sp. (Smith, 1961c). In *Periplaneta americana* the postsynaptic vesicles and the presynaptic vesicles are similar in size i.e. about 40 nm (Atwood *et al.*, 1969). The postsynaptic vesicles have a more or less dense content and are thought to be the site of enzyme storage.

The sheath of glial cells that accompanies the axons on their course from the central nervous system to the neuromuscular junction is interrupted at the axon terminal thus allowing a close (synaptic) contact between the axonal membrane and the muscle plasma membrane (Smith, 1968). In most neuromuscular junctions, the confluence of the glial and muscle plasma membranes forms a perfect cover, leaving only a very narrow passage between the synaptic cleft and the haemolymph. In the moth *Philosamia cynthia*, the cleft between glial cell and muscle fibre and the synaptic cleft as well as the lumen of the TTS are interconnected by very short passage ways (Piek, 1974, 1975). The possible consequences of this structural organization are discussed in section 6.

3 Ionic composition

The plasma membrane of muscle fibres forms a boundary between the sarcoplasm and the direct environment of the fibre, and at the same time forms a connection between the two compartments by active and passive transport of molecules and ions. The plasma membrane is generally considered to regulate and maintain the interior environment of the fibre, and it is also considered to control, up to a certain degree, the exterior environment of the fibre in the direct vicinity of the plasma membrane (Piek, 1974, 1975).

Active and passive transport of ions causes an electrical current. Before the electrical phenomena can be explained in terms of ion fluxes, a description of the unequal distribution of ions among the compartments on both sides of the membrane is discussed.

3.1 THE OUTER MEDIUM OF THE MUSCLE

Insects possess an internal fluid called haemolymph. This fluid is often considered to be the direct outer medium for cells and fibres, but this may not be true at least for the nervous system (*cf.* Treherne, 1976), and for the muscle fibres (Piek, 1974, 1975). As an interchange between the intermediate extracellular fluid and the haemolymph must be present, the composition of the latter fluid may be important for an understanding of the ionic composition of the true extracellular environment of excitable tissues. A strong argument for this view is that substituting the haemolymph for an artificial saline solution in most cases results in changes of the electrochemical properties of the fibres.

The composition of insect haemolymph has been reviewed recently by Florkin and Jeuniaux (1974). We shall discuss the matter briefly; data on the ionic composition of the haemolymph of different insect orders being summarized in Table 1.

Florkin and Jeuniaux (1974) demonstrated that the sodium ion is the dominant cation in haemolymph of Apterygota and Exopterygota, with an exception of stick insects (Cheleutoptera), of which *Carausius morosus* possesses a haemolymph with magnesium as the dominant cation. With regard to ionic composition, the haemolymph of Endopterygota is less uniform. In Diptera and some of the Coleoptera sodium is the main cation. Lepidoptera, however, are characterized by a low concentration of sodium ions and relatively high concentrations of potassium and magnesium.

The dominant anion in insect haemolymph is chloride, which is found in Endopterygota in concentrations varying from 10 to 86 mmol l^{-1}, and in Exopterygota from 93 to 144 mmol l^{-1}. Phosphate is usually present in small amounts, but in Orthoptera the concentrations vary from 16 to 40 mmol l^{-1}. The third inorganic anion is bicarbonate, which has been demonstrated in the haemolymph of a few insects in concentrations of about 15 mmol l^{-1}. Organic anions are present in insect haemolymph in relatively high concentrations. Amino acids have been found in Exopterygota in concentrations varying from 14 to 30 mmol l^{-1} (Table 1), and in Endopterygota from 26 to 76 mmol l^{-1}. The relative concentration at which the different amino acids are present in the haemolymph varies with the different orders. Recently for example, Collett (1976) found in *Calliphora erythrocephala* haemolymph, 22 mmol l^{-1} free amino acids, of which 11.35 was proline. Citrate has been found in all insects. In Lepidoptera Tsuji (1909, *cf.* Levenbook and Hollis, 1961) found 49 mmol l^{-1} citrate in the haemolymph of *Bombyx mori*, but the highest value found by Levenbook and Hollis (1961) was 32 mmol l^{-1}.

Besides amino acids insect haemolymph contains other amphoteric ions, the most intriguing being phosphorylcholine, demonstrated in a number of different insect orders (*cf.* Bridges, 1972). Haemolymphs of certain Lepidoptera contain a high concentration, the highest value of about 20 mmol l^{-1} to be found in *Galleria mellonella.* No role has been suggested for phosphorylcholine (Bridges, 1972).

A special problem in interpreting the measured values of ion concentrations in haemolymph is the binding of ions to plasma proteins and to haemocytes (Bishop *et al.*, 1925; Tobias, 1948a; Clark and Graig, 1953; Barsa, 1954; Carrington and Tenny, 1959; Brady, 1967a, b; Plantevin, 1967; Weidler and Sieck, 1977). Carrington and Tenny (1959) using an ultrafiltration method demonstrated that 15–20% of the calcium and magnesium ions in the haemolymph of the moth *Telea* (=*Antherea*) *polyphemus* were bound to macromolecules, which did not pass through a collodium or cellophane dialysis membrane. No binding of either potassium or chloride was found by Carrington and Tenny (1959), but in *Galleria mellonella* Plantevin (1967) found some binding of sodium and potassium ions up to 9 and 2% respectively. Weidler and Sieck (1977), however, demonstrated by the method of ultrafiltration, a significant binding to macromolecules in the haemolymph of *P. americana* for sodium ions (22%), magnesium ions (25.5%), calcium ions (16.2%), chloride (10.3%), phosphate (26.9%), but not for potassium ions.

In the cockroach, *Periplaneta americana*, Tobias (1948a) measured the sodium, potassium and magnesium ion concentrations in both whole and

TABLE 1 Ions in the haemolymph of a number of insects from different orders. The concentrations are expressed in mmol l^{-1}, the extreme values between round brackets, the number of observations between square brackets

Order	Na^+	K^+	Ca^{2+}	Mg^{2+}	Cl^-	Phosphate	HCO_3^-	Citrate	Amino acids*	References†
EXOPTERYGOTA										
Odonata	149 [13] (104–179)	10 [12] (4–28)	7 [7] (2–6)	4 [5] (2–6)	110 [1]	4 [1]	15 [1]	1 [1]	23	7, 11, 15, 29
Dictyoptera	129 [2] (100–157)	12 [2] (8–15)	6 [6] (2–10)	7 [5] (3–11)	126 [2] (107–144)	1 [1]		1 [1]	14	1, 2, 11, 23, 31, 34
Cheleutoptera	14 [5] (9–21)	21 [5] (16–28)	6 [3] (4–8)	60 [3] (53–73)	98 [1]				20	7, 15, 28, 35
Orthoptera	91 [14] (22–234)	20 [14] (3–62)	7 [7] (1–14)	11 [7] (1–17)	97 [2] (93–101)	28 [2] (16–40)			35	4, 7, 11, 15, 25, 28, 30, 32
Heteroptera	106 [10]	15 [10]	10 [8]	6 [7]				2 [1]		7, 11, 22, 23, 28, 29
ENDOPTERYGOTA										
Diptera	106 [5] (40–206)	16 [15] (2–58)	6 [8] (4–10)	11 [6] (7–9)	26 [2] (15–36)	4 [1]	15 [1]	4 [10] (0–13)	26	7, 10, 15, 16, 19, 22, 23, 28
Lepidoptera	11 [58] (1–40)	41 [69] (10–96)	13 [66] (3–65)	31 [58] (7–52)	42 [9] (10–86)	4 [3] (3–6)		19 [9] (5–49)	76	3, 4, 5, 7, 8, 9, 11, 13, 14, 15, 16, 17, 18, 20, 21, 22, 27, 28, 30, 32, 33
Coleoptera	77 [25] (2–165)	28 [28] (4–65)	19 [18] (5–78)	48 [17] (5–99)	32 [2] (19–44)	4 [2] (3–5)		4 [4] (2–7)	71	7, 11, 14, 15, 16, 23, 24, 26, 28, 29
Hymenoptera	34 [13] (2–154)	39 [13] (19–61)	6 [10] (1–9)	8 [10] (1–12)	33 [1]			4 [3] (2–6)	70	6, 7, 15, 16, 23, 30

* Calculated from data in Florkin and Jeuniaux (1974), Table 5

† 1, van Asperen and van Esch (1954); 2, van Asperen and van Esch (1956); 3, Babers (1938); 4, Barsa (1954); 5, Bialaszewicz and Landau (1938); 6, Bishop *et al.* (1925); 7, Boné (1944); 8, Brecher (1929); 9, Carrington and Tenny (1959); 10, Chen and Friedman (1975); 11, Clark and Craig (1953); 12, Clark (1958); 13, Drilhon (1934); 14, Drilhon and Busnel (1943); 15, Duchâteau *et al.* (1953); 16, Florkin and Jeuniaux (1974); 17, Gese (1950); 18, Heller and Moklowska (1930); 19, Hevert (1974); 20, Huddart (1966b); 21, Kafatos (1968); 22, Levenbook (1950); 23, Levenbook and Hollis (1961); 24, Ludwig (1951); 25, Pepper *et al.* (1941); 26, Patterson (1956); 27, Plantevin (1967); 28, Ramsay (1953); 29, Sutcliffe (1962); 30, Sutcliffe (1963); 31, Tobias (1948a); 32, Tobias (1948b); 33, Tsuji (1909, cf. 23); 34, Weidler and Sieck (1977); 35, Wood (1957). Brought upto date after Piek, 1975

centrifuged haemolymph, and found a fall of 40% of these ion concentrations after centrifugation.

According to Brady (1967a, b) the haemolymph of the cockroach, *P. americana*, usually contains a variable number of haemocytes per μl. He plotted the potassium and sodium ion concentrations in the haemolymph of a number of individuals against the haemocyte number, and found a relation, which he considered to be an indication that potassium was present in a high concentration in haemocytes, representing more than 50% of the total potassium in the haemolymph.

Considering this evidence for ion binding and the possibility of changes in distribution between haemocytes and plasma during isolation of the haemolymph, the described ionic composition of insect haemolymph must be regarded as a rough approximation of the original free concentration.

3.2 THE MYOPLASM

The ionic composition of frog and rat skeletal muscle fibres was originally reviewed by Conway (1957), who calculated the values for the myoplasm by correcting for the intercellular space the data obtained from whole muscle. It was generally believed, in those days, that the ionic composition of the intercellular spaces must be the same as that of the extracellular medium of the whole muscle. However, it will become evident in the next section, that at least for insect muscle the idea can no longer be maintained that the ionic composition of the fluid immediately outside the muscle fibre membrane is identical to the medium surrounding the whole muscle (haemolymph). The data concerning ionic compositions of whole skeletal muscles of insects, given in Table 2, therefore is more or less unreliable as an indication of the ionic composition of the myoplasm, which in homogenates may be heavily contaminated with the "extracellular" fluid inside the TTS.

Despite the above criticism concerning the reliability of the experimental data summarized in Table 2, the differences between different insect groups are so extreme, that it is impossible to neglect them. The group consisting of the Dictyoptera, Orthoptera, Cheleutoptera, and Coleoptera may be compared with the Lepidoptera, especially with regard to the concentration of sodium and potassium ions in the haemolymph and myoplasm. For the Lepidoptera $[Na^+]_o/[Na^+]_i = 0.53$ (± 0.22 SEM, $n=7$), and for the rest of the insect groups $[Na^+]_o/[Na^+]_i = 2.86$ (± 0.64 SEM, $n=7$). For the Lepidoptera $[K^+]_o/[K^+]_i = 0.53$ (± 0.04, $n=7$), and for the rest of the groups $[K^+]_o/[K^+]_i = 0.18$ (± 0.06 SEM, $n=7$). In the above division the phytophagous Lepidoptera are compared with the other insect groups, which also include the phytophagous stick insect *Carausius morosus* (Cheleutoptera). As a

TABLE 2 Ionic composition of insect muscles in relation to the ionic composition of the haemolymph. The data on muscle are not corrected for extracellular space. A, adults; L, larvae.

Insect species (Order)	Phase	Inorganic constituents ($mmol\ l^{-1}$)*					Refs†
		Na^+	K^+	Ca^{2+}	Mg^{2+}	Cl^-	
Periplaneta americana (A)	Haemolymph	107	17	–	1.7	–	1
(Dictyoptera)	Skeletal muscle	46	112	–	7.4	–	
Periplaneta americana (A)	Haemolymph	111	13	–	–	96	2, 9
(Dictyoptera)	Skeletal muscle	27	110	–	–	10	
Locusta migratoria (A)	Haemolymph	103	11	–	–	94	2, 9, 10
(Orthoptera)	Skeletal muscle	19	124	–	–	13	
Schistocerca gregaria (A)	Haemolymph	83	4.6	3.2	8.6	75	3
(Orthoptera)	Skeletal muscle	61	49	2.3	4.5	33	
Romalea microptera (A)	Haemolymph	64	18	–	–	–	4
(Orthoptera)	Skeletal muscle	44	128	–	–	–	
Carausius morosus (A)	Haemolymph	15	18	–	–	–	2, 11
(Cheleutoptera)	Skeletal muscle	13	103	–	–	–	
Telea polyphemus (A)	Haemolymph	3	54	–	36	–	5
(Lepidoptera)	Skeletal muscle	18	77	–	–	–	
Telea polyphemus (A)	Haemolymph	3	41	3	29	68	6
(Lepidoptera)	Skeletal muscle	5	79	7	–	15	
Sphinx ligustri (A)	Haemolymph	4	50	4.9	36	61	6
(Lepidoptera)	Skeletal muscle	21	84	9.7	–	15	
Bombyx mori (A)	Haemolymph	9	41	7.7	42	68	6
(Lepidoptera)	Skeletal muscle	12	98	17	–	13	
Actias selene (A)	Haemolymph	9	47	8.7	26	75	6
(Lepidoptera)	Skeletal muscle	16	116	14	–	14	
Philosamia cynthia (A)	Haemolymph	24	39	6.4	5.0	25	3
(Lepidoptera)	Skeletal muscle	14	78	2.0	0.2	24	
Samia cecropia (A)	Haemolymph	3	54	–	–	–	7
(Lepidoptera)	Skeletal muscle	20.0	88	–	–	–	
Tenebrio molitor (L)	Haemolymph	76	37	–	–	130–163	8
(Coleoptera)	Skeletal muscle	18	72	–	–	–	

*Concentrations of ions in muscle are expressed as $mmol\ l^{-1}$ in tissue water (modified from Usherwood, 1969)

† 1, Tobias (1948a); 2, Wood (1963); 3, Piek (1975); 4, Tobias (1948b); 5, Carrington and Tenny (1959); 6, Huddart (1966b); 7, McCann (1965); 8, Belton and Grundfest (1962b); 9, Wood (1965); 10, Hoyle (1955); 11, Wood (1957)

phytophagous insect *C. morosus* has a very low sodium concentration in the haemolymph, but the value for $[K^+]_o/[K^+]_i$ is similar to that of the non-phytophagous insects.

According to the Nernst equation (*cf.* section 5.2.1), and based on the outer and inner concentrations mentioned in Table 2, for the non-Lepidoptera the mean equilibrium potential for potassium ions, $E_K = -44$mV (± 9 mV SEM, $n = 7$) and the equilibrium potential for sodium ions, $E_{Na} = +27$ mV (± 6 mV SEM, $n = 7$). For the Lepidoptera, however, $E_K = -16$ mV (± 2 mV SEM,

$n=7$) and $E_{Na} = -16$ mV (± 12 mV SEM, $n=7$). For the non-Lepidoptera the $E_K \approx -44$ mV should be more positive than the resting membrane potential ($E_m \approx -55$ mV). This difference is not very large, and may be caused by incorrect measurement of ionic concentrations, or the involvement of other ion species in determining membrane potential (E_m).

In non-phytophagous insects, the position of E_{Na} (ca. $+27$ mV) is consistent with the amplitude of active membrane responses. For the phytophagous insects (Lepidoptera and *C. morosus*), the gap between the predicted and the experimental values is extremely large. The position of E_{Na} (*ca.* -16 mV) is certainly not in agreement with the direction in which the active membrane response shifts. It is even too negative to explain the reversal potential of the *epsp* (*cf* section 5.3), but other ion species may be involved in the generation of *epsp*s and active membrane responses (*cf.* section 5.3). As regards the resting membrane potential, the discrepancy between the experimental values (Table 2) and the potentials calculated according to an extended constant-field equation (*cf.* section 5.2.1) remains unexplained (Piek, 1975).

A special problem with regard to the ionic composition of the myoplasm are hydrogen ions. The available data indicate that under normal conditions the flux of hydrogen ions through the muscle fibre membrane is very small compared with that of potassium or chloride ions. This is probably due to the very low concentration of hydrogen ions. Woodbury (1971) replaced chloride by glutamate and could show that in frog skeletal muscle fibres the membrane permeability for hydrogen ions is about 500 times higher than for chloride. If this is also true for insect muscle fibres, the product of $[H^+]$ and its relative permeation might be small compared with similar products for potassium or chloride ions, but not small enough to ignore the hydrogen ions completely.

Aicken and Thomas (1975), using recessed-tip internal-sensitive microelectrodes, measured internal pH-values of 7.21 ± 0.02 (SEM) in *Carcinus maenas* (Crustacea) extensor muscle, bathed in a saline with a pH of 7.5. During a period of 90 min, Hinke and Menard (1976) found the internal pH, measured with microelectrodes in muscles of *Balanus nubilus* (Crustacea) to be remarkably constant. Even though the resting membrane potential increased in this period from -75 mV to -81.5 mV and the external pH decreased from 7.69 to 7.62, the internal pH remained about 7.25.

As yet similar data are not available for insect muscle fibres.

3.3 THE LUMEN OF THE SARCOPLASMIC RETICULUM

The ionic composition of vertebrate sarcoplasmic reticulum (SR) is almost unknown, except for the presence of a high concentration of calcium ions, and a possible high concentration of chloride. Baylor and Oetliker (1975) suggested from birefringence studies of skeletal muscle of *Rana temporaria* that after a stimulus the whole SR-membrane underwent a depolarization of 100 mV or more. It is generally accepted that muscle contraction in vertebrates is elicited by a release of calcium ions from the SR. This release is probably coupled to the observed depolarization. Electron microscopically, calcium was found in the SR of frog muscle by Constantin *et al.* (1965) and by Komnick (1969), and in the SR of mouse cardiac muscle by Shiina and Mizuhira (1970). Dialescu and Popescu (1973) localized calcium in both SR and TTS of frog skeletal muscle. In our experiments with flight muscles of *Pieris brassicae*, Ca-precipitate was found in the Z-line region (Fig. 11a) both with oxalate (Constantin *et al.*, 1965) and with dihydroxytartrate (Shiina and Mizuhira, 1970).

The subject of the ionic composition of the SR is somewhat outside the scope of this review, and has been recently reviewed for vertebrates by Endo (1977). For insects there is evidence that calcium ions are involved in excitation-contraction coupling, and that the sarcoplasmic reticulum is capable of accumulating calcium ions (*cf.* Aidley, 1975).

3.4 THE LUMEN OF THE TRANSVERSE TUBULAR SYSTEM AND THE SYNAPTIC CLEFT

Studies in the last 20 years on the sarcotubular system of skeletal muscle fibres have led to the conclusion that the membranes of the transverse tubular system (TTS) are continuous with the "surface" plasma membrane (*cf.* section 2.3). Therefore, the lumen of the TTS is considered to be in continuity with the extracellular space. This continuity does not necessarily mean

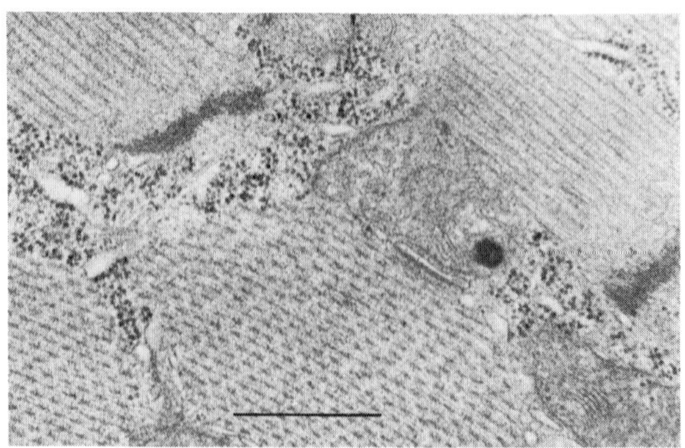

Fig. 11*a*. Distribution of the precipitate of Ca-oxalate in a flight muscle of *Pieris brassicae*. The precipitate is concentrated in the Z-line region. Scale 0.5 μm.

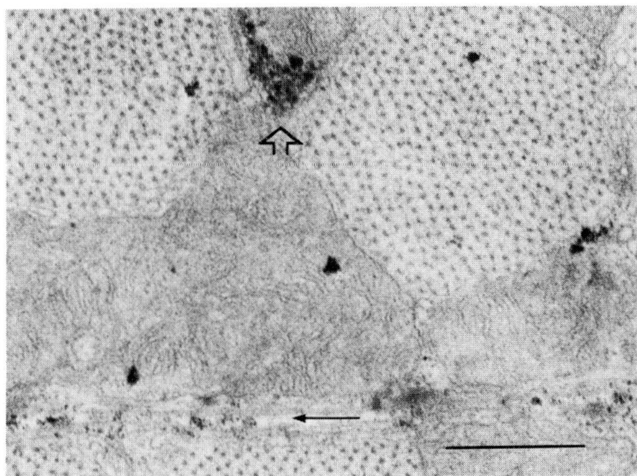

Fig. 11*b*. Distribution of the precipitate of a sodium antimonate complex in a flight muscle of *Pieris brassicae*. The precipitate is found in the intercellular space (→) and in the TTS-lumen (⇒). Scale 0.5 μm.

that the ionic composition in both compartments must be identical. Piek (1974, 1975) suggested that a highly active sodium–potassium exchange pump might be operating in the extensive transverse tubular membrane. This could lead to an accumulation of sodium chloride in the TTS-lumen.

In fact, the ionic composition of the TTS-lumen is a matter of conjecture. A quantitative evaluation of the chemical composition of the fluid inside the TTS is well beyond the range of current analytical techniques. For muscle fibres of Lepidoptera the prediction that the sodium chloride concentration inside the TTS-lumen is high, is based on two different observations: 1. the behaviour of inhibitory postsynaptic potentials (Piek, 1974, 1975) and 2, ultrastructural localization of sodium and potassium ions in the muscle (Njio and Piek, 1977).

1. Inhibitory postsynaptic potentials (*ipsp*s) in insect skeletal muscles are described in section 5.3. From these potentials it may be concluded that the equilibrium potential for chloride ions (E_{Cl}) must be about -60 mV. This equilibrium potential can only be attained if the ratio of [Cl^-] in the synaptic cleft over [Cl^-] in the myoplasm is about 10. In moth (*Philosamia cynthia*) muscle fibre, the synaptic cleft is separated from the outer medium by the space between muscle fibre and glial cell (*cf.* section 2.4). Therefore, the post-synaptic membrane can be considered as a part of the invaginated (non-surface) part of the plasma membrane. If the fluid in the synaptic cleft contains a high [Cl^-], this might also be the case for the rest of the invaginations, i.e. the TTS.

For muscle fibres of the Lepidoptera listed in Table 2, the ratios $[Cl]_o$ over

$[Cl^-]_i$ have an average of 2.4 ($\pm$ 1.7 SEM, $n=5$). This is far from the predicted value of about 10. The chloride concentration in the haemolymph may, however, not be relevant, since the ionic composition of the haemolymph may not be identical to that of the fluid adjacent to the outer edge of the plasma membrane as discussed earlier. Moreover, the experimental data concerning the $[Cl^-]_i$ must be corrected for contamination from fluid inside the TTS. If the volume of the TTS of *Philosamia cynthia* is taken as 5% of the total volume of the muscle fibre (Piek 1975) and taking $[Cl^-]_{homogenate} \approx 24$ mmol l^{-1} (Table 2) and assuming that $[Cl^-]_{TTS}/[Cl^-]_i \approx 10$, the corrected value for $[Cl^-]_i$ is given by the equation $5[Cl^-]_{TTS} + 95\,[Cl^-]_i \approx 2400$ mmol l^{-1}. Hence $[Cl^-]_i \approx 16.6$ mmol l^{-1} and $[Cl^-]_{TTS} \approx 166$ mmol l^{-1}. The 5% TTS-volume may be due to considerable swelling. In normal muscle fibres of insects, the TTS-volume is in the order of 1% of the total volume. As in these muscles, the chloride concentration is about 15 mmol l^{-1} (Table 2), the $[Cl^-]_i$ may be about 14 mmol l^{-1} and the $(Cl^-]_{TTS}$ about 140 mmol l^{-1}. These values agree with those described for swollen muscle fibres (Piek, 1975).

2. The electron microscopic precipitation techniques developed for the localization of sodium (Komnick, 1962) and potassium ions (Shiina and Mizuhira, 1970) may provide some indication of the presence of differences in ionic composition in different compartments inside and outside muscle fibres. Unfortunately, these techniques are far from quantitative. They can only discriminate between no precipitate, considerable precipitate, and a heavy precipitate. According to the intensity of the precipitation of sodium as a hexahydroxoantimonate (V)-complex, Njio and Piek (1977) concluded that muscle fibres of *Pieris brassicae* (Lepidoptera) fixed directly after the integument had been opened, probably contain more sodium in the TTS-lumen than outside the surface plasma membrane (Fig. 11b). Preliminary investigations with X-ray micro-analysis in the electron microscope indicate that most of the precipitate consists of sodium with a small amount of magnesium (Njio, unpublished results). Precipitates of potassium as a hexanitrocobaltate (III)-complex were found in the mitochondria, on the myofibrils and on the sarcoplasmic reticulum and the subsarcolemmal cisternae, but not in the lumen of the TTS.

The conclusion is that sodium ions, magnesium ions and chloride ions are present in the fluid inside the TTS, while sodium and chloride ions are present in higher concentrations than in the haemolymph. Since arguments are presented for a chloride concentration in the order of 150 mmol l^{-1}, the fluid inside the TTS-lumen may mainly contain sodium and magnesium chloride. It can also be concluded that the invaginated muscle fibre membrane in Lepidoptera, and possibly also in the other insects, creates its own external ionic environment in the TTS-lumen and in the synaptic cleft.

4 Permeability of the plasma membrane

By comparing the potassium ion and chloride conductances of amphibian muscle fibres with and without TTS, Eisenberg and Gage (1969) found little changes in the chloride conductance after detubulation. It was suggested (*cf.* Martonosi, 1972) that polyanion complexes are present inside the TTS, and that this may reduce the chloride concentration to such an extent as to explain the nearly zero chloride conductance in the TTS. In Amphibia, a sudden change in the external chloride concentration causes a rapid change in the muscle fibre membrane potential. However, a similar change in potassium ion concentration causes a much slower change in resting potential (Hodgkin and Horowicz, 1960). Moreover, potential changes due to changes in the external potassium concentration are asymmetric. Repolarization took place much slower than depolarization. The conclusion from these investigations is that the resting K^+-conductance is located in the TTS as well as in the surface membrane, whereas the Cl^--conductance is located mainly at the surface portion of the amphibian muscle fibre membrane (*cf.* Nakajima and Bastian, 1976). In rat diaphragm, however, the chloride conductance is mainly located in the TTS (Palade and Barchi, 1977).

Similar experiments with crayfish muscle fibres showed that their membrane potential changed rapidly when subjected to a sudden change in potassium concentration, and slowly to changes in chloride concentration (Orentlicher and Reuben, 1971). This indicates that the sites of chloride permeability may mainly be located within the TTS, whereas the potassium ion permeability may also, or exclusively, be located at the surface membrane. Reuben *et al.* (1964) found a swelling of the dyadic portion of the TTS after the fibres were exposed to solutions that induced an efflux of potassium chloride. The swelling could also be induced by an inward current applied through an intracellular microelectrode. Therefore, the swelling must be related to a translocation of chloride from the myoplasm to the transverse tubular lumen. The conclusion is that the movement of chloride can only account for swelling of the dyadic portion of the TTS, if the dyads are predominantly permeable to chloride (*cf.* Reuben *et al.*, 1976).

A summary of these investigations is that in Amphibia the TTS-lumen may contain little or scarcely any chloride, and the transverse tubular wall may not be very permeable to chloride, while in crustaceans the TTS-lumen may contain chloride and its wall may predominantly be permeable to chloride.

In the next section some arguments are presented in favour of the notion that for insects the distribution of selective permeabilities to potassium ions and chloride may be similar to that in crustaceans, and thus different from that in Amphibia.

4.1 CHLORIDE AND POTASSIUM IONS

Using skeletal muscle fibres of *Locusta migratoria* (Orthoptera) and *Periplaneta americana* (Dictyoptera), Wood (1965) replaced chloride by sulphate to varying degrees. This hardly affected the magnitude of the resting membrane potential and the internal chloride concentration. Wood concluded that the resting membrane in skeletal muscle fibres of these insects is poorly permeable for chloride. If, however, chloride would be translocated across the TTS-membrane producing a transient accumulation of chloride in the TTS-lumen, this high chloride concentration in the TTS-lumen would reduce considerably the effect of lowering chloride in the bathing saline (*cf.* section 3.4). In contrast to the findings by Wood (1965), Usherwood and Grundfest (1965), studying skeletal muscles of *Romalea microptera* (Orthoptera), found evidence for a high relative permeability for chloride, as compared with propionate and sulphate.

A sudden increase of the potassium ion concentration in the bathing fluid causes a quick fall in the resting membrane potential, as demonstrated for a skeletal muscle fibre of the grasshopper, the locust and the cockroach (Usherwood 1967a, 1969). During KCl-influx the resting membrane potential falls very rapidly to a value approximately intermediate between the predicted values for E_K and E_{Cl} (*cf.* section 5.3). Therefore, the values for the relative permeability of the membrane for potassium and chloride may be of similar magnitude. The ratio potassium-permeability over chloride permeability was 1.4 for the locust *Schistocerca gregaria* (Lea, 1972; Lea and Usher-

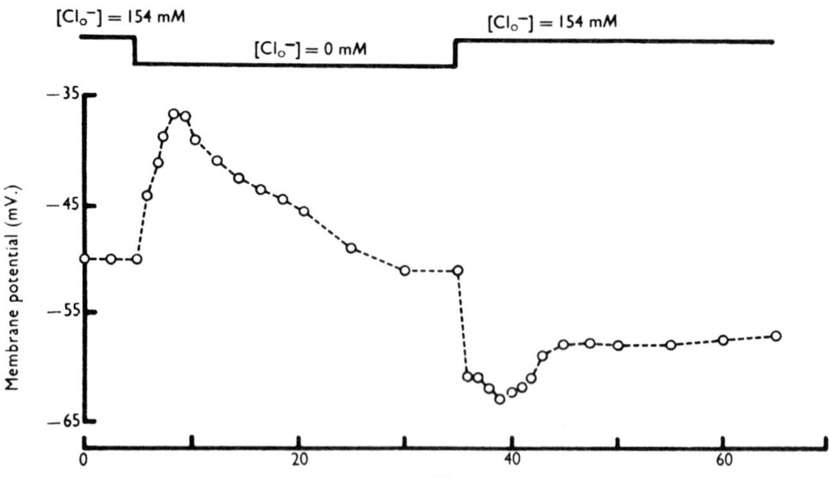

Fig. 12. Transient changes in membrane potential of a coxal adductor muscle fibre of *Schistocerca gregaria* caused by substituting of the chloride in the saline by sulphate. (From Lea and Usherwood, 1973b.)

wood, 1973b). Substitution of chloride by sulphate produced a transient depolarization with an average of 13 mV (Fig. 12). The long time lag of the chloride transients could be explained if the site of chloride conductance was the membrane of the transverse tubular system.

In muscle fibres of the retractor unguis of *S. gregaria*, loaded with KCl by increasing the external KCl concentration to 100 mmol l^{-1}, and subsequently treated for about 30 min with normal saline (10 mmol KCl per litre), the TTS component of the dyad, but not of the sarcoplasmic reticulum part, was markedly swollen (Cochrane and Elder, 1967). Swelling of the TTS also occurred when the external chloride concentration was reduced whilst maintaining the external potassium concentration constant. This indicates that the chloride permeability of insect muscle fibres may be located in the membrane of the TTS, as it probably is in crustaceans.

When locust, grasshopper or cockroach skeletal muscle fibres are loaded with KCl by increasing the potassium ion concentration, the membrane is quickly depolarized, but when a loaded muscle is returned to normal saline, the repolarization occurs much slower. Figure 13 shows the rapid depolarization by 100 mmol l^{-1} KCl and the subsequent slower repolarization in 10 mmol l^{-1} KCl in a grasshopper muscle fibre. The repolarization is characterized by an initial fast phase and a subsequent slow phase. This can be explained as follows. A sudden increase in $[K^+]_o$ depolarizes the membrane approximately according to the Nernst relation. The new membrane potential causes a change in the chloride permeation through the membrane, *i.e.* influx may be augmented. Restoring the $[K^+]_o$ to its initial value causes a repolarization of the membrane, however, to a limited extent. After an

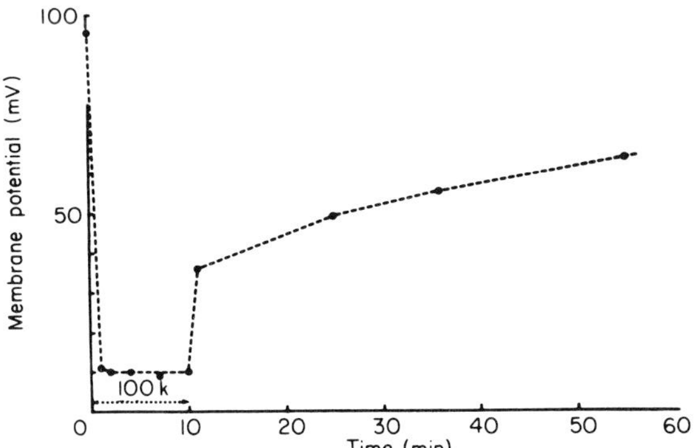

Fig. 13. Effect of changing the external potassium ion concentration, from 10 mmol l^{-1} to 100 mmol l^{-1}, and vice versa, at a constant chloride concentration (154 mmol l^{-1}), on the membrane potential of a *Romalea microptera* retractor unguis muscle fibre. The rate of depolarization is much greater than the rate of repolarization, the latter being characterized by an initial fast phase and a subsequent slow phase. (From Usherwood, 1969.)

initial shift in a negative direction, the partly repolarized membrane drives out the excess internal chloride. If the efflux of chloride can take place through the transverse tubular wall only, an accumulation of chloride in the TTS-lumen may prevent a further fast efflux of chloride and a further fast repolarization of the membrane. Thus the asymmetrical curve shown in Fig. 13 agrees with the view that the permeation site of chloride is mainly the TTS.

4.2 OTHER CATIONS

Changes in the external sodium ion concentration affect the resting membrane potential in Dictyoptera, Orthoptera, Cheleutoptera and Lepidoptera (Wood, 1961; Huddart, 1966c). The potential changes are small, and Usherwood (1969) suggested that the cause might be an effect on the sodium pump. Whatever the effect of sodium ions on the resting membrane potential may be, it is evident that the resting membrane is either not or only slightly permeable to sodium ions.

If sodium ions are replaced by ammonium ions, the fibres of the tibial muscle of *Telea polyphemus* (Lepidoptera) and the sternopedal muscle of *Philosamia cynthia* are depolarized (Rheuben, 1972; Piek, 1975). Caesium ions, having about the same hydrated ion diameter as ammonium or potassium ions, cause only a small depolarization in *P. cynthia*.

The resting membrane potential is only little affected by calcium or magnesium ions, as has been demonstrated in skeletal muscle of *P. cynthia* (Piek, 1975) and *Drosophila melanogaster* larvae (Jan and Jan, 1976a). Increasing the external calcium concentration for *D. melanogaster* resulted in a hyperpolarization. Such an effect is of "the wrong sign" to be due to a membrane permeability, and has been explained in terms of other actions of calcium on the membrane (Jan and Jan, 1976a, b).

4.3 OTHER ANIONS

McCann (1964) exchanged chloride for bromide, nitrate, acetate and sulphate in the saline suitable for preparations of heart muscle of *Hyalophora* (=*Philosamia*) *cecropia*, and found evidence for the following sequence of permeabilities: acetate > nitrate > bromide > chloride > sulphate. The membrane was depolarized in sulphate and hyperpolarized in acetate. The observation of McCann (1964) indicating that the insect muscle fibre membrane might be more permeable for acetate than for chloride, and a similar observation with propionate using muscle fibres of the lobster, *Homarus americanus* (Motokizawa, *et al*., 1969), at first sight appear to contradict the generally accepted view that acetate and propionate are impermeant ions

(Hagiwara *et al.*, 1964; Reuben *et al.*, 1964; Usherwood and Grundfest, 1965; Takeuchi and Takeuchi, 1966). To resolve this contradiction Piek *et al.* (1973) studied the effects of a number of monocarboxylates on resting potentials of skeletal muscle fibres of locusts (*S. gregaria* and *L. migratoria*) and a moth (*P. cynthia*). Exchange of most of the chloride in the saline for an equivalent amount of sulphate, or for divalent or trivalent carboxylates, resulted in a transient depolarization. Exchange of chloride for monocarboxylates with a $pK_a > 4.0$ always resulted in a reversible hyperpolarization; however, the latter was preceded by a short-lasting depolarization. These initial depolarizations are comparable with the transient depolarizations caused by di- or trivalent ions. When chloride was exchanged for monocarboxylates, the initial depolarization (i.e. the chloride transient) is interrupted by a large hyperpolarization (Fig. 14). Restoration of the chloride concentration resulted in a depolarizing potential shift back to the original value now preceded by a transient hyperpolarization (Fig. 14). These results could be explained by assuming a low permeability of the surface membrane and a relatively high permeability of the wall of the TTS for monocarboxylates. The results could, however, also be explained by assuming that the monocarboxylic acids penetrate the muscle fibre in the protonated, non-ionized, form.

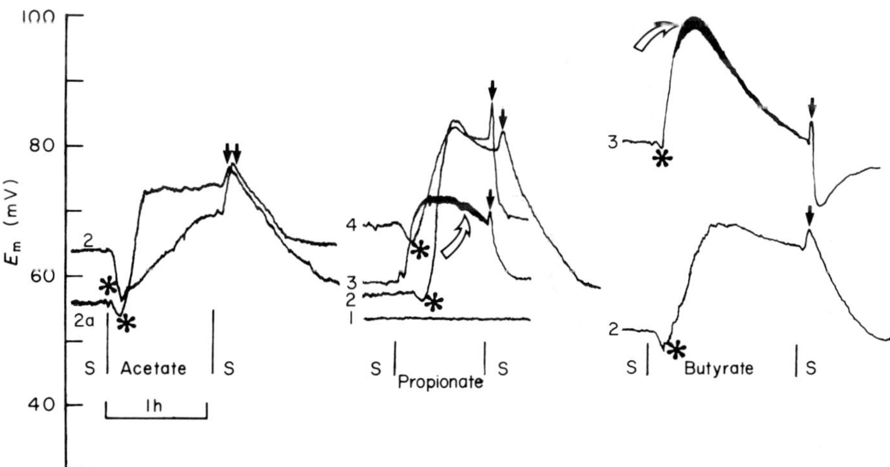

Fig. 14. Effect of exchange of chloride (66 mmol l^{-1} in *P. cynthia*, and 68 mmol l^{-1} in *S. gregaria* and *L. migratoria*) for an equivalent amount of acetate, propionate, or butyrate, on E_m of muscle fibres of the sternopedal muscle (2) and of the longitudinal flight muscle (2a) of *P. cynthia*, and on fibres of the retractor unguis muscle of the metathoracic leg of *S. gregaria* (3) and *L. migratoria* (4). The muscles were equilibrated in saline (s). Note the initial depolarization of the membrane (*) after exchanging chloride for monocarboxylates, and the initial hyperpolarization (→) on restoring the chloride concentration. In *S. gregaria* propionate and butyrate cause bursts of small potentials, here recorded with attenuation (⇒). The potential of sodium-glass-electrode against the reference electrode is also recorded (1). (From Piek *et al.*, 1973.)

TABLE 3 Relationship between the pK_a values and the activity of carboxylates tested on the rectractor unguis muscle of *S. gregaria* at pH 7 (from Piek *et al.*, 1973)

	pK_a	Activity*
Monocarboxylates		
Bicarbonate	6.4	51, 73, 43
Trimethylacetate	5.0	22
Propionate	4.9	24, 42, 35
p-Aminobenzoate	4.9	34, 25
Oenanthate	4.9	11, 12
Acetate	4.8	21, 23, 55
Butyrate	4.8	34, 36
Capronate	4.8	10, 10
γ-Hydroxybutyrate	4.7	13, 17
3,4,Dihydroxybenzoate	4.5	12
Benzoate	4.2	14, 11
Lactate	3.9	16, 10, 11
Glycolate	3.8	0, 0, 0
Formate	3.8	0, 8, 7, 6
α-Hydroxybutyrate	3.7	0, 0
Pyruvate	2.5	0, 0
Di- and trivalent carboxylates		
Malonate	2.8 (5.7)	0, 0
Succinate	4.2 (5.6)	0, 0
Phthalate	3.0 (5.3)	0, 0
Citrate	3.1 (4.7, 5.4)	0, 0
Amino acids		
Glycinate	2.4 (9.8)	0, 0
Alanate	2.3 (9.8)	0, 0
γ-Aminobutyrate	4.2 (10.4)	0, 0

* The activity is expressed as the hyperpolarization in a percentage of the initial value of membrane potential

To study the latter possibility, Piek *et al.* (1973) replaced most of the chloride in the saline of *S. gregaria* for a number of monocarboxylates of different pK_a-values (Table 3). A positive correlation was found between the hyperpolarizing activity of the acid and its pK_a-value. This indicates that monocarboxylic acids penetrate in their non-ionized form.

After entering the muscle fibres, the monocarboxylic acid will dissociate to an extent determined by the internal pH, and the pK_a of the acid involved. The released protons will decrease the internal pH, thereby inhibiting the dissociation and thus the penetration of the acid. In the first instance, the amount of carboxylic acid entering the fibre is determined by the pK_a.

The penetration of monocarboxylates causes a swelling of the muscle fibre (Piek *et al.*, 1973; Piek, 1975). Not only the myoplasm but also the TTS is swollen, particularly in the region of the dyad (Fig. 15). Since the treatment

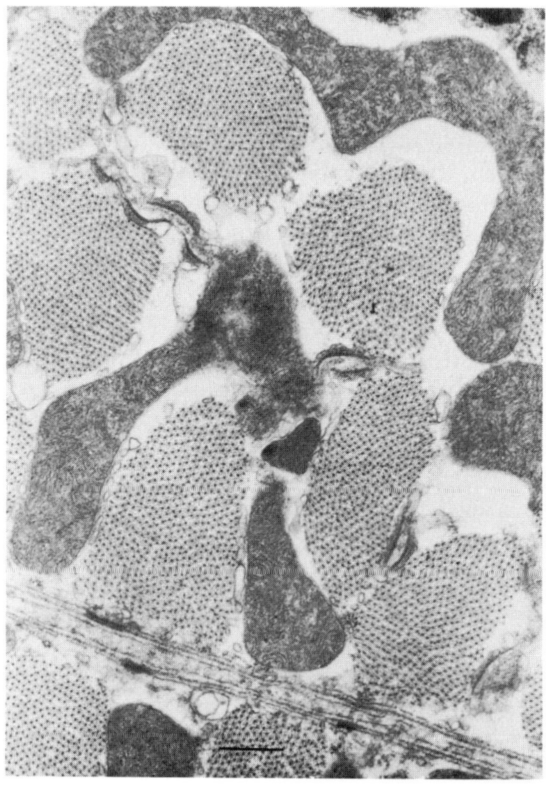

Fig. 15. Transverse section of parts of two fibres of the sternopedal muscle of *Philosamia cynthia* pretreated for 15 min with a saline in which 71 mmol l^{-1} chloride has been exchanged for 71 mmol l^{-1} bicarbonate. Note the swelling in the myoplasm and the dyadic portion of the TTS. Scale 0.5 μm.

of muscle fibres with monocarboxylates of a $pK_a > 4$ causes a reduction in intracellular chloride (Piek *et al.*, 1977) the swelling in the TTS may be caused by an efflux of chloride into the TTS. Although the resting membrane is hyperpolarized with a monocarboxylate (Motokizawa *et al.*, 1969; Piek, 1975), the inhibitory postsynaptic potential amplitude is increased, which can be explained by such a translocation of chloride.

4.4 HYDROGEN IONS

While the pH of the haemolymph or the bathing saline can be measured easily, the determination of the true hydrogen activity is not so easy. If the pH is calculated from the potential difference between a glass electrode and a reference electrode, the pH of an aqueous solution is found by interpolation between two or more values found using well defined buffer solutions. Insect haemolymph or saline, however, does not only differ from the standard buffer solutions in hydrogen ion activity, but also in the activity of other ions. Even if it is assumed that the glass-electrode is completely insensitive to ions other than hydrogen, the reference electrode will be sensitive to the other ions present in haemolymph or saline. Therefore, the pH-data obtained in this way may differ from the true $-\lg [H^+]$. The discrepancy probably does not exceed 0.3 pH-unit (Piek *et al.*, 1977).

Measurement of intracellular pH has presented many more technical difficulties. One method uses 5,5-dimethyl-2,4-oxazolidinedione (DMO) as an indicator, another method uses pH-sensitive glass microelectrodes. Recently Aickin and Thomas (1975) using pH-sensitive microelectrodes measured intracellular pH-values of 7.21 ± 0.02 (SEM) in *Carcinus maenas* (Crustacea) extensor muscles, bathed in a saline with a pH of 7.5. Hinke and Menard (1976) found in crustacean muscle bathed in a saline with a pH of 7.7 an internal pH of 7.3 with microelectrodes and with DMO. If hydrogen ions were distributed passively across the muscle fibre membrane, the internal pH should have been about 6.5 in these experiments. Hinke and Menard (1976) found that following prolonged equilibration the transmembrane hydrogen ion distribution varied with membrane potential, but not in accordance with a simple electrochemical equilibrium.

In insects the effect of changes in external pH on the resting potential of muscle fibres has been studied in *Schistocerca gregaria* (Orthoptera) by Washio (1971), in *Antherea polyphemus* (Lepidoptera) by Rheuben (1972), and in *Pieris brassicae* (Lepidoptera) by Piek *et al.* (1977). Washio (1971) found that membrane conductance was sensitive to changes in pH in standard (chloride) saline, and he observed that the conductance decreased with pH. This phenomenon cannot be caused by a direct contribution of hydrogen ions to the conductivity since in that case the conductance would decrease with increasing pH. This change in conductance was not found in a chloride-free (methylsulphate) saline. Washio (1971) concluded that chloride ions are involved in the pH sensitivity of the resting conductance, a notion which is in agreement with a similar phenomenon in frog muscle fibre (Hutter and Warner, 1967a, b). Rheuben (1972) and Piek *et al.* (1977) found an increase in conductance at a decreasing pH, and also a lower E_m (*i.e.* a more positive) at a lower pH. Rheuben (1972) noted that the pH sensitivity that she found might be caused by an effect on the $H_2CO_3 - HCO_3^-$ equilibrium, but this is hard to understand, since the saline used lacked bicarbonate in any appreciable concentration. Piek *et al.* (1977) found that if bicarbonate, or another monocarboxylate with a $pK_a > 4$, is present in the saline, muscle fibres of *P. brassicae* behave like a hydrogen electrode.

Woodbury (1971) studying frog muscle fibre, found a permeability ratio of hydrogen ions over chloride of about 500. If this is also true for insect muscle, the membrane current, which is normally carried by potassium and chloride ions, will also be carried by hydrogen ions in the presence of sufficient monocarboxylate. Woodbury (1971) emphasized that hydrogen ions must be transported by an active process, since it is well established that hydrogen ions are highly permeant and yet probably are not in electrochemical equilibrium (*cf.* Aickin and Thomas, 1975; Hinke and Menard, 1976). If such an active transport is mediated by a carrier, monocarboxylates might take part in such a putative carrier system (*cf.* section 5.2.3).

5 Electrical properties of the muscle fibre

Studies on the electrochemistry of muscle fibres was started at the end of the 18th century at the University of Bologna with the work of Galvani (1792), who observed that freshly isolated frog legs contract when short-circuited with two different metallic conductors. Volta concluded that these muscles were stimulated by an electric current originating from the contact between the two metals. It was not before Arrhenius' free-ion-theory that it was possible to understand how an electric current could flow from a metallic conductor of electrons via special contact surfaces, i.e. electrodes, through media containing ions. Biological tissues, however, are more complex than simple ionic solutions. Compartmentalization by membranes, and extreme infolding of membranes cause liquid junctions and membrane junctions between compartments of different ionic composition. The behaviour of ions in liquid junctions has been subjected to theoretical consideration by Planck (1890a, b). The task of formulating equations quantitatively describing observed membrane potentials has been undertaken by Goldman (1943), who considered multi-ionic potentials, which arise when mixtures of different ions are separated by an ion selective non-charged membrane.

5.1 LIQUID JUNCTIONAL POTENTIALS

Electron micrographs of insect muscle fibres (*cf.* section 3) show that the plasma membrane is highly invaginated (Smith 1961a, b, 1966a, b; Hagopian and Spiro, 1967; Pasquali-Ronchetti, 1970; Anstee, 1971; Piek *et al.*, 1973). Only a small part of the plasma membrane is in close contact with the lamina basalis. It has been estimated (Piek, 1974, 1975) that in the moth, *Philosamia cynthia*, 70% of the total surface of the muscle fibre plasma membrane is invaginated. As a result of this high degree of invagination, the major part of the membrane is in contact with the fluid present in the lumen of the invaginations, *i.e.* the TTS. Therefore, the fluid in the TTS forms an intermediate compartment between the major part of the plasma membrane and the extra-muscular fluid (Hodgkin and Horowicz, 1960).

A comparable situation is present in the nervous system (*cf.* Treherne and Pichon, 1972). The diffusion pathway from the haemolymph to the surface of the axoplasm consists of intercellular perineural channels and a complex network of channels formed by the glial cells. Pichon and Boistel (1967) found that the extra-axonal fluid, i.e. the fluid inside the sheath formed by the perineurium, had an average positive potential of 6.7 mV relative to the fluid in the bath. They explained this potential in terms of an unequal distribution of ions between the haemolymph and the extra-axonal fluid, and they suggested the presence of a relatively high sodium ion concentration in the extra-axonal fluid.

Treherne and Pichon (1972) presented the view that in insects the extra-axonal fluid is separated from the haemolymph by barriers formed by perineural septate desmosomes or by tight junctions. They suggested that insects, like vertebrates, but unlike crustaceans and other invertebrates, possess a blood-brain barrier. Such a barrier would greatly facilitate the processes responsible for increasing the sodium concentration in the extra-axonal space.

In the TTS of insect muscle fibres, structures like desmosomes or tight junctions are not present. Piek (1974, 1975), however, presented the view that a diffusion barrier might yet be present, the barrier now being formed by a liquid junctional potential in the fluid inside the TTS.

It must be born in mind that liquid junctional potentials are always multi-ionic potentials, and that a potential shift caused by concentration difference of one ion species is greatly diminished by transport of other ion species according to the new electrochemical potential. Therefore, the nature of liquid junctional potentials must be considered before discussing the probability of their existence in nerve and muscle.

When two electrolyte solutions "a" and "b" of different ionic composition are brought in contact, a potential difference can be measured across the junction assuming that all the ions are equally free to diffuse. This potential difference depends on the concentrations of ions and their relative mobilities and can be described, according to Henderson (1907, 1908), by the equation:

$$E_{\mathrm{L}}=\frac{RT\sum_{\mathrm{j}}(\mu_{\mathrm{j}}/z_{\mathrm{j}})(C_{\mathrm{j}}^{\mathrm{b}}-C_{\mathrm{j}}^{\mathrm{a}})}{F\sum_{\mathrm{j}}\mu_{\mathrm{j}}(C_{\mathrm{j}}^{\mathrm{b}}-C_{\mathrm{j}}^{\mathrm{a}})}\ln\frac{\sum_{\mathrm{j}}C_{\mathrm{j}}^{\mathrm{a}}\mu_{\mathrm{j}}}{\sum_{\mathrm{j}}C_{\mathrm{j}}^{\mathrm{b}}\mu_{\mathrm{j}}} \quad (1)$$

where R is the gas constant, T the absolute temperature, F Faraday's constant, C_j^a and C_j^b the concentrations of the ion species "j" in the two solutions "a" and "b", μ_j represents the mobility per valence unit and z_j the valence of the ion "j" and Σ_j represents the sum of all terms involved.

The use of this equation does not require any particular spatial arrangement of layers of solution at the boundary between the solutions, the only requirement being that the solutions should be a series of mixtures of the end solutions "a" and "b" (McInnes, 1961).

Considering the concentration differences observed between the extracellular axonal space, the synaptic cleft and the TTS-lumen on the one hand, and the haemolymph on the other hand, the liquid junctional potential difference between the former fluids and the haemolymph will be restricted to a few millivolts, as will be demonstrated below.

The osmotic pressure of haemolymph of insects, expressed in terms of freezing-point depression generally ranges from 0.5 to 0.9 °C (Sutcliffe, 1963). Therefore, the osmolarity will range from 250 to 500 mOsmol per litre. The higher values have been observed in Lepidoptera, in which the contribution of organic compounds to the osmotic pressure of haemolymph is much more pronounced than in Orthoptera or Dictyoptera (Sutcliffe, 1963). It seems reasonable to suppose that the contribution of ions (inorganic and organic ions) to the osmotic pressure of insect haemolymph will be in the order of 300 mOsmol per litre. In insect salines (*cf.* Huddart, 1971) the contribution of ions to the osmotic pressure varies from 270 to 350 mOsmol per litre, which agrees with the ion concentration in haemolymph. It is obvious that the fluid bordering the outer axonal membrane and the muscle fibre plasma membrane must be isotonic with the haemolymph. Since in liquid junctions, non-ionic molecules may add to the osmolarity independent of the electrochemical potential difference, their concentration in the extra-axonal and extramembranous fluid may be the same as in the haemolymph. Therefore, the maximal concentration of $NaCl + MgCl_2 + CaCl_2$ in the extramembranous fluid may be about 150 mmol l^{-1} NaCl or 100 mmol l^{-1} $MgCl_2$ or $CaCl_2$ or a mixture of NaCl, $MgCl_2$ and $CaCl_2$ in a concentration range varying from 100 to 150 mmoles per litre. Supposing that the extramembranous fluid contains about 150 mmol l^{-1} NaCl (*cf.* section 3.4), the theoretical value for the liquid junctional potential depends mainly on the sodium ion and chloride concentration in the haemolymph, as illustrated in the following examples.

1. For a "dictyopteran" haemolymph containing per litre: 100 mmol Na^+, 10 mmol K^+, 5 mmol Ca^{2+}, 5 mmol Mg^{2+}, 140 mmol Cl^- and 20 mmol organic anions the contribution of the ions to the total ionic osmolarity will be about 310 mOsmol per litre. Hence, the ionic osmolarity in the TTS must also be 310 mOsmol per litre.

If this is due exclusively to sodium chloride the TTS-lumen will contain

155 mmol NaCl per litre. Supposing that at 20 °C the transport numbers per valence unit (μ_j) are: 45 for Na^+, 67 for K^+, 53 for Ca^{2+}, 47 for Mg^{2+}, 68 for Cl^- and 30 as an approximation for the organic anions, according to Eqn (1) the liquid junctional potential would be 0.7 mV.

2. For a "lepidopteran" haemolymph containing per litre: 10 mmol Na^+, 40 mmol K^+, 10 mmol Ca^{2+}, 30 mmol Mg^{2+}, 40 mmol Cl^- and 90 mmol organic anions, the contribution of the ions to the total osmolarity will be about 220 mOsmol l^{-1}. Supposing again that the NaCl concentration in the TTS is maximal, i.e. 110 mmol l^{-1} NaCl, the theoretical liquid junctional potential will be 6.2 mV.

The above examples show that liquid junctional potentials are small compared with membrane potentials. Their presence, however, may be very important, since all ion species are distributed according to their electrochemical potentials, resulting in a situation in which the TTS-lumen may be relatively rich in anions and relatively poor in cations with the exception of the cations pumped from the myoplasm into the TTS-lumen.

5.2 RESTING MEMBRANE POTENTIALS

5.2.1 *Passive permeation*

If a membrane is permeable to a number of ion species for which the membrane has different permeabilities, a theoretical treatment of the correlation between the ionic composition on both sides of the membrane and the transmembrane potential difference is very complicated. If the membrane is considered to be homogeneous and its electric field is constant the theoretical treatment becomes less complicated. Goldman (1943) suggested in biological membranes the presence of a large number of dipoles, which are close to their isoelectric points, causing a more or less constant electric field inside the membrane. If this is true, the constant field concept may be a reasonable approximation.

Without the limiting conditions of equal ionic strength on both sides of the membrane, and univalent ions as the only ions involved, the so-called Goldman equation, or *constant-field-equation* can be derived from the general flux equation of Planck (1890a, b).

$$I_j = RT z_j \mu_j \frac{dC_j}{dx} + C_j \mu_j z_j^2 F \frac{dE}{dx} \tag{2}$$

In this equation I_j represents the flux from the inside of the membrane to the outside of ion "j" caused by two forces, a concentration gradient and an electric field, while E is the potential at a point x (x being the distance from the outer limit of the membrane towards the inner limit). Assuming a constant field in the membrane, the equation becomes a first-order differential equation,

$$I_j = RTz_j\mu\frac{dC_j}{dx} + \frac{C_j\mu_j z_j^2 FE}{t} \tag{3}$$

(Goldman 1943; Hodgkin and Katz 1949), in which t represents the thickness of the membrane, E is the potential on the inside of the membrane compared with the outside, and $(C_j)_o$ and $(C_j)_i$ are the activities of ion species "j" on the outside, and on the inside of the membrane. If potassium, sodium, magnesium, calcium, and hydrogen ions are taken into consideration, the constant-field equation can be described in an extended form (Piek, 1975) by

$$E_m = \frac{RT}{F}\ln\frac{-b+(b^2-4ac)^{1/2}}{2a} \tag{4}$$

in which, for example

$$a = [K^+]_i + \alpha[Cl^-]_o + \beta[Na^+]_i + 4\gamma[Mg^{2+}]_i + 4\delta[Ca^{2+}]_i + \varepsilon[H^+]_i$$

$$b = ([K^+]_i - [K^+]_o) - \alpha([Cl^-]_i - [Cl^-]_o + \beta([Na^+]_i - [Na^+]_o) + \varepsilon([H^+]_i - [H^+]_o)$$

$$c = -[K^+]_o - \alpha[Cl^-]_i - \beta[Na^+]_o - 4\gamma[Mg^{2+}]_o - 4\delta[Ca^{2+}]_o - \varepsilon[H^+]_o$$

Here α, β, γ, δ, and ε are the permeabilities of the membrane for chloride, sodium, magnesium, calcium and hydrogen ions respectively, relative to the permeability of the membrane for potassium ions.

Piek (1975) showed that even this extended constant field equation did not offer a reasonable description of the relation between the resting membrane potential (E_m) of a moth (*Philosamia cynthia*) skeletal muscle and the activities and relative permeabilities of a number of mono- and divalent ions in the bathing saline. A distinct deficit of more than 20 mV was left, the curve through the experimental points being negative compared with the theoretical curves.

Experimental data (Fig. 16) often show a sharp inflection in E_m versus lg $[K^+]_o$ curves, in the region of the natural $[K^+]_o$ (Hoyle, 1953; Wood, 1957; Belton and Grundfest, 1962; Huddart, 1966a; Piek, 1975; Deitmer, 1977). Two different explanations have been offered for this phenomenon. The sharp inflexion may result from a sudden change in relative permeability of the membrane for potassium ions. This might also explain the increase in E_m and the concurrent shift of the inflection to lower lg $[K^+]_o$ values in preparations equilibrated for a long period (5–6 h) in a low-potassium saline (Usherwood, 1967b). A different view has been presented by Huddart and Wood (1966), who suggested that in *Periplaneta americana* (Dictyoptera) and

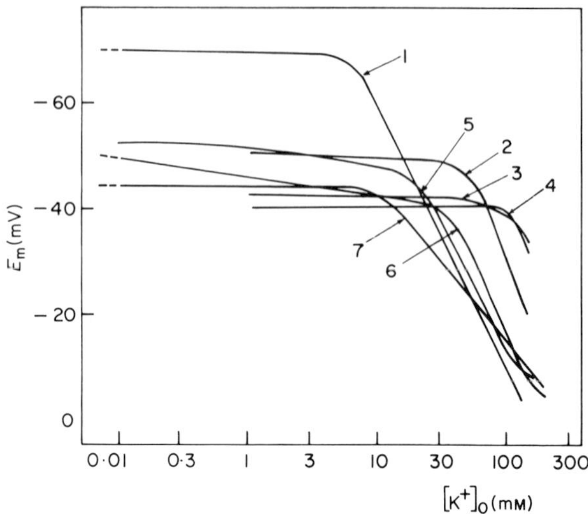

Fig. 16. Relation between the resting potential (E_m) and the external potassium concentration ($[K^+]_o$) of a number of insect skeletal muscle fibres, equilibrated for maximal 2 h in saline solution. 1. Flexor tibialis of *Schistocerca gregaria* (Orthoptera) (Hoyle, 1953); 2. longitudinal flight muscle of *Philosamia cynthia* (Lepidoptera); 3. sternopedal muscle of *P. cynthia* (Piek, 1975); 4. skeletal muscle from *Tenebrio molitor* larva (Coleoptera) (Belton and Grundfest, 1962b); 5. ventral longitudinal muscle of *P. cynthia* larva (Piek, 1975); 6. skeletal muscle of *Bombyx mori* (Lepidoptera) (Huddart, 1966a); 7. flexor tibialis muscle of *Carausius morosus* (Cheleutoptera) (Wood, 1957). The original experimental points have been omitted for the sake of clarity. (From Piek, 1975.)

in *Sphinx ligustri* (Lepidoptera) the potassium gradient across the membrane is maintained in part by an active process (*cf.* below).

5.2.2 *Electrogenic pumps*

If the resting membrane potential was determined to a considerable extent by electrogenic pump mechanisms, the addition of an inhibitor of metabolism should be followed by a fall in membrane potential. Huddart and Wood (1966) observed a slow decrease in membrane potential after treatment of the muscles of *P. americana* and *S. ligustri* with 2,4-dinitrophenol (DNP) or cyanide. Wareham *et al.* (1973) found a slow depolarization after administration of acetazolamide to bicarbonate buffered salines. McCann (1967), however, observed a distinct sudden fall in membrane potential of heart muscle fibres of *Samia cecropia*. A sudden fall was also demonstrated in skeletal muscles of *Philosamia cynthia* and *Schistocerca gregaria* after treatment with DNP (Piek *et al.*, 1973), however, only in preparations pretreated with propionate or bicarbonate, or with a saline in which potassium ions were

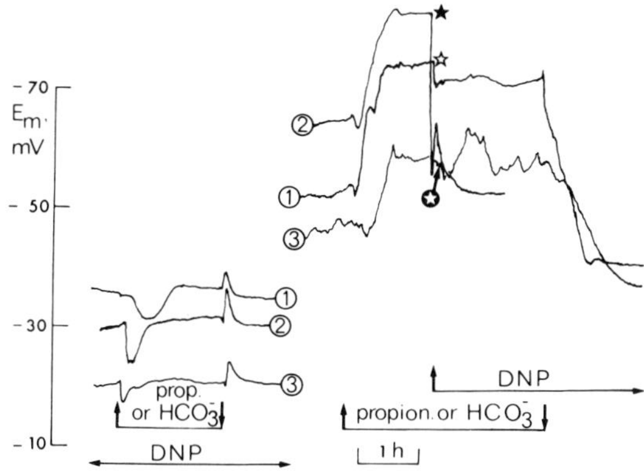

Fig. 17. Combined effects of propionate or bicarbonate and DNP. *Left.* Effects of propionate or bicarbonate on fibres with DNP. *Right.* Effects of DNP (added at *) on fibres pretreated with propionate or bicarbonate. 1. Retractor unguis of *S. gregaria* treated with bicarbonate; 2. Retractor unguis of *S. gregaria* treated with propionate; 3. Sternopedal muscle of *P. cynthia* treated with propionate. (From Piek *et al.*, 1973.)

omitted (Fig. 17). Therefore, the presence of a metabolic and electrogenic pump mechanism cannot be ruled out. Piek *et al.* (1973) demonstrated that DNP might also act as a proton carrier, thus competing with monocarboxylates. The ability to compete with DNP depends on the pK_a of the monocarboxylate. Only bicarbonate (pK_a=6.4) seems to compete successfully with DNP. Muscle fibres hyperpolarized with bicarbonate showed little or no depolarization with DNP (Fig. 17), and fibres first depolarized with DNP were sometimes hyperpolarized after exchange of chloride in the saline for bicarbonate, but never after exchange by other monocarboxylates.

An immediate and significant fall in E_m following exposure of the muscle to DNP is explained by a sudden inhibition of an electrogenic pump, and a slow and gradual decrease is often considered as an indication that the contribution of an electrogenic pump is insignificant (Huddart and Wood, 1966; Usherwood, 1969). An alternative view is that DNP first interferes with the passive transport of hydrogen ions and later on with the active transport of ions in general. If this is true, an immediate depolarization following treatment with DNP may indicate the presence of an important proton flux in the membrane, rather than the presence of an electrogenic pump mechanism. Piek *et al.* (1973) suggested that DNP should interrupt the electrogenic efflux of protons. An alternative concept including a facilitation of the influx of protons by DNP is discussed in the next section. The action of DNP on the muscle membrane is, however, far from clear, and must be elucidated by further research.

5.2.3 *Proton-coupled transport system*

It is generally assumed, that the transfer of reducing components between cytoplasm and mitochondria must occur indirectly by means of shuttles (cf. Crabtree and Newsholme, 1975). It is also generally assumed, that the transport of solutes through membranes is coupled to that of other compounds.

Around 1950 Lehninger (*cf.* Lehninger, 1975) provided experimental proof that the electron transport from nicotinamide adenine dinucleotide (NADH) to oxygen is the direct source of energy used for the phosphorylation of adenosine diphosphate (ADP), a coupling, which was prevented by 2,4-dinitrophenol (DNP). In 1953 Slater (*cf.* Lehninger, 1975) proposed that the coupling was realized by an intermediate reaction. Such a reaction has never been found, and Slater's chemical-coupling hypothesis provided no explanation for the fact that the mitochondrial membrane must be intact.

Around 1960 Mitchell (*cf.* Mitchell, 1973, 1976) proposed that an electrochemical gradient of hydrogen ions across the mitochondrial membrane serves as an intermediate coupling mechanism. According to Mitchell and Moyle (1965a, b) the active accumulation of monovalent or divalent cations by mitochondria is also accompanied by the ejection of protons into the cytoplasm, thus also providing a contribution to the hydrogen ion gradient across the membrane. Therefore, only an intact mitochondrial membrane is able to serve as a couple between the transport of solutes.

For the membrane of *Escherichia coli*, Mitchell (*cf.* Mitchell, 1973) proposed three types of transport, a symport, an antiport and a uniport. For ion transport systems an anion–cation symport, a cation–cation antiport and a weak acid uniport can be distinguished. In addition, some compounds may serve as intermediates in the coupling. According to Mitchell (1973) this deserves serious consideration because it corresponds to a system known in the proton-linked dicarboxylate uptake by mitochondria.

According to Simoni and Postma (1975), it is evident that bacteria generate a membrane potential, interior negative, of the order of 50 to 250 mV, and at the same time generate a pH gradient far from the electrochemical equilibrium. The presence of respiration-driven electrogenic proton extrusion has been established in a number of bacteria (*cf.* Simoni and Postma, 1975). A proton motive force thus created may be capable of driving ATP synthesis in the bacterial membrane.

A comparable ejection might be present in the plasma membrane of muscle fibres (*cf.* section 4.4). In analogy to the proton-linked transport of ions in the mitochondrial and bacterial membrane, it is suggested that in the insect muscle fibre plasma membrane, monocarboxylates, for example bicarbonate, could act as an intermediate compound for the proton-linked transport of,

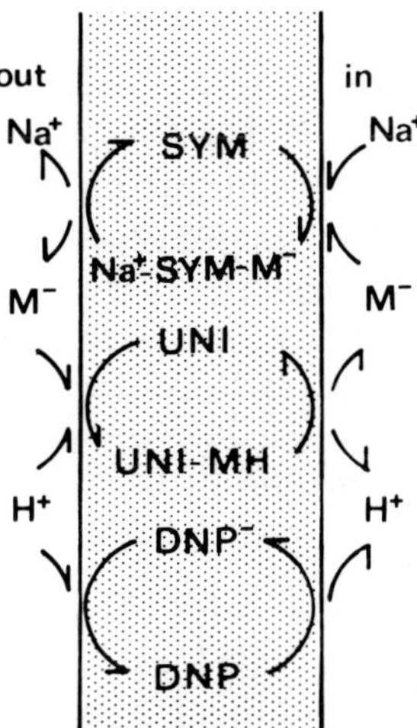

Fig. 18. Diagrammatic representation of the passive and the proton-transport-coupled permeation of ions across the plasma membrane of an insect muscle fibre. The internal hydrogen ion concentration is supposed to be displaced from its electrochemical equilibrium by a metabolic process (Woodbury, 1971; Mitchell, 1973, 1975), thus creating a protonmotive force for the translocation of other ions. Protons are transported down-hill from the outside to the inside as a uniport (UNI) with a monocarboxylic acid (MH). This proton-linked monocarboxylate (M^-) uptake enables the membrane to translocate other ions in the form of a symport (SYM). The role of the monocarboxylate as an intermediate compound might be to make the transport systems less dependent on each other. This system might also include a proton coupled transport of non-ionized compounds such as sugars (not included in the diagram). DNP permits the protons to flow into the fibre so uncoupling the driving force performed by the proton gradient.

for example, sodium ions (Fig. 18). The monocarboxylate could be transported back into the cytoplasm through a proton-coupled symport system.

As illustrated in Fig. 18, the presence of a specific proton conductor such as dinitrophenol (DNP) or azide should uncouple the system by permitting the recirculation of protons back across the membrane (Mitchell 1973).

The concept of proton coupled transport of ions in the plasma membrane of skeletal muscle fibres, and the role of monocarboxylates as an intermediate compound, still must be tested by crucial definitive experiments. At the moment the experimental data described in sections 4.3 and 4.4 are in agreement with the concept.

Propionate and other unbranched monocarboxylates like butyrate, pentanoate, hexanoate and octanoate, were found to affect active sodium transport in the toad (*Bufo marinus*) bladder (Hess *et al.*, 1975). It is conceivable that the function of monocarboxylates may not be restricted to the insect skeletal muscle fibre membrane, but may be of importance for membrane transport systems in general.

5.2.4 *Concluding remarks*

Different models are described to explain the existence and magnitude of resting membrane potentials. No special attention has been paid to theories based on ion-exchange or charged membranes. The reason for this is that equations for multi-ionic potential differences across ion-exchange membranes are very complex and therefore of no practical use, unless all ions involved are of the same valence. The charged membrane equations such as described by Wyllie (1954) are relatively simple and similar to the constant field equation of Goldman (1943). In this summary the theoretical discussion has been restricted to the passive translocation of ions described by the Goldman-equation and by the carrier-facilitated coupled transport theory.

It has been suggested by Wareham *et al.* (1973, 1974, 1975) that the resting membrane potential in muscle fibres of *Periplaneta americana*, and *Sphinx ligustri* contains a bicarbonate sensitive hyperpolarizing component, which may account for approximately one-third of the normal membrane potential. Piek *et al.* (1973) have found that this function of bicarbonate can be taken over by other monocarboxylates provided that the $pK_a > 4$.

At present the most attractive explanation of the resting membrane potential in insect muscle fibres seems to be as follows:

The resting membrane potential contains two components (*a*) one determined by the passive flux of potassium and chloride ions, and (*b*) one determined by a proton-coupled transport, in which a monocarboxylate acts as an intermediate compound.

5.3 EVOKED POTENTIALS

Insect skeletal muscle fibres respond to appropriate chemical or electrical stimuli with different types of transient potential changes, called evoked potentials. After a nerve impulse has reached the terminals, a local postsynaptic current is evoked in the postsynaptic membrane of the muscle fibre. The resulting postsynaptic potential (*psp*) may be either excitatory (*epsp*) or inhibitory (*ipsp*). If a muscle is depolarized by an *epsp* or by outward current via an intracellular microelectrode, the electrically excitable membrane may respond with a transient and graded potential change, the electrically excited response (*eer*).

5.3.1 *Postsynaptic potentials*

Fatt and Katz (1951) were the first to explain the generation of *epsp*s in vertebrate neuromuscular junctions as caused by the release of a transmitter substance, producing a transient non-selective increase in permeability of the membrane to ions, resulting in short circuiting of the postsynaptic membrane, In the sartorius muscle of *Rana temporaria* the reversal potential for the *epsp* ($E_{r(epsp)}$, see further in this section) was about −15 mV (Fatt and Katz 1951), in the sartorius muscle of *R. pipiens* −10 to −20 mV (Takeuchi and Takeuchi 1960), in the smooth muscle of the guinea-pig ileum about −9 mV (Bolton 1972). In sodium-deficient solution the $E_{r(epsp)}$ in frog skeletal muscle and guinea-pig ileum muscle became more negative, and in potassium-rich solution more positive, but the concentration of chloride did not influence the position of $E_{r(epsp)}$. Therefore, Takeuchi and Takeuchi (1960) and Bolton (1972) concluded that the transmitter substance makes the postsynaptic membrane more permeable to sodium and potassium ions, but not to chloride.

In the flexor tibialis muscle of *Locusta migratoria* (Orthoptera) the reversal potential is about zero (Del Castillo *et al*., 1953) and in *Philosamia cynthia* larvae (Lepidoptera), −6 to −21 mV (Yamaguchi *et al*., 1972). In *Telea polyphemus* (Lepidoptera) the existence of a nearly linear relation between the height of the evoked potential and the resting potential (E_m) demonstrates that also in this species the $E_{r(epsp)}$ is not far from zero potential (Huddart, 1971). When the membrane is depolarized by means of a stepwise increase in the external potassium ion concentration, the relation between evoked potential and E_m remains roughly linear, but $E_{r(epsp)}$ becomes distinctly positive (Huddart, 1971), indicating that K^+-activation is involved in the generation of the potential change.

Anwyl and Usherwood (1974) studied fibres of the metathoracic extensor tibiae muscle of *S. gregaria* with a voltage-clamp technique. In these experiments, the neurally-evoked responses were reduced to a small postsynaptic current by increasing the magnesium concentration to 40 mmol l^{-1}. Stimulation of the excitatory axon resulted in an inward current in the muscle fibre, the excitatory postsynaptic current (*epsc*). Iontophoresis of L-glutamate on the synaptic region of the muscle fibre produced a transient inward current, the glutamate current (*gc*). Reversal potentials were determined by clamping the membrane potential at different levels. A linear relationship was obtained for *epsc* and the clamped potential and for *gc* and the clamped potential (Fig. 19). The reversal potentials $E_{r(epsc)}$ and $E_{r(gc)}$, were about −3 mV. Perfusion with sodium-free (choline substituted) saline caused a 90–95% reduction in the amplitude of the glutamate current. In this saline the reversal potential varied between −10 and −15 mV. The reversal potentials were not affected by changing potassium or chloride concentration. Small glutamate currents recorded in sodium-free saline were increased in amplitude when the calcium concentration was increased. The results of the experiments of Anwyl and Usherwood (1974) suggest that the *epsc* and the glutamate current consist only of an inward current carried by sodium ions with calcium ions possibly making a small contribution.

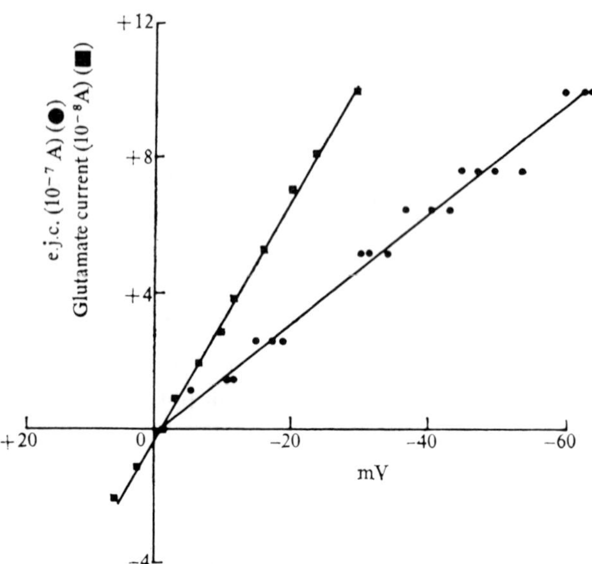

Fig. 19. Relationship between the membrane potential of a fibre of the metathoracic extensor tibiae muscle of *S. gregaria* clamped using the two microelectrode method, and the amplitude of the excitatory junctional current (e.j.c.) and the glutamate current. (From Anwyl and Usherwood, 1974.)

An indication that divalent cations may contribute to the generation of the *epsp* has already been obtained by Kusano and Grundfest (1967), who found, that in *Tenebrio molitor* (Coleoptera) skeletal muscle fibres the $E_{r(epsp)}$ in normal saline was close to zero potential, but high concentration of both Ca^{2+} or Mg^{2+} caused a shift of $E_{r(epsp)}$ in the positive direction. Kusano and Janiszewski (1976) also found a positive shift in $E_{r(epsp)}$ in *T. molitor* when the external calcium or magnesium ion concentration was increased, and they did not rule out the possibility that activation of the postsynaptic membrane for calcium ions could be involved in the generation of *epsp*s (*cf.* p. 239).

Jan and Jan (1976b) concluded that a change in the calcium ion concentration did not result in a change of the *epsp* in integumental muscle fibres of larvae of *Drosophila melanogaster*.

They further demonstrated that the contribution of chloride was not significant, that the contribution of potassium ions must be comparable to that of sodium ions, and that the contribution of magnesium ions was larger than that of potassium and sodium ions (*cf.* section 5.3.3).

5.3.2 *Electrically excited responses*

In myocardial fibres of *Hyalophora* (= *Philosamia*) *cecropia* (McCann, 1971) and in skeletal muscle fibres of locusts (Washio, 1972a, b) the amplitude of the *eer* increased with increasing $[Ca^{2+}]_o$. With a tenfold rise in $(Ca^{2+}]_o$ Washio (1972a) found an increase in *eer*-amplitude of about 29 mV. In Washio's experiments, however, all sodium ions had been replaced by tetraethylammonium ions. This might decrease the K^+-permeability, while the Na^+-current was drastically diminished by omitting sodium ions from the bathing fluid. Washio's experiments therefore did not indicate that the *eer* is caused by an increase in Ca^{2+}-permeability of the membrane. His experiments only showed that this might be the case when the contribution of other cations is excluded. Recently, a number of investigators have found, under conditions of high TEA-concentration, strong indication for Ca^{2+}-dependent all-or-none action potentials, which are not sensitive to sodium. These potentials were studied in the fly, *Sarcophaga bullata* (Patlak, 1976), in muscles of moth larvae, *Ephestia kuehniella* (Deitmer and Rathmayer, 1976), and in integumental muscles of larvae of the beetle, *Xylotrupes dichotomus* (Fukuda *et al.*, 1977). The resistance of the *eer* to high concentrations of tetrodotoxin indicates that Na^+-activation is not involved in the production of the *eer* (Patlak, 1976; Deitmer and Rathmayer, 1976; Fukuda *et al.*, 1977) (*cf.* p. 239).

When bathed in haemolymph or in a normal saline, insect muscle fibres usually respond to a depolarization with a graded response. This is supposed to be the result of two currents, an inward current of calcium ions and an outward current of potassium ions. If the potassium ion current is blocked by TEA, the calcium ion current becomes dominant.

5.3.3 *Theoretical considerations*

From sections 5.3.1 and 5.3.2 the conclusion can be drawn that divalent ions may contribute to the generation of *epsps*, glutamate potentials (Kusano and Grundfest, 1967; Anwyl and Usherwood, 1974; Kusano and Janiszewski, 1976; Jan and Jan, 1976) and the electrically excited responses (Washio, 1972a, 1972b; Patlak, 1976; Deitmer and Rathmayer, 1976) (*cf.* p. 239).

If the divalent cations Mg^{2+} and Ca^{2+} are taken into consideration the extended constant-field equation for the theoretical steady-state potential to which the *epsp* is directed, can be described as follows (Piek, 1975):

$$E_{epsp} = \frac{RT}{F} \ln \frac{-b+(b^2-4ac)^{1/2}}{2a} \tag{5}$$

in which a, b, and c are similar to the corresponding values in Eqn. (4), but in which the resting relative permeabilities α, β, γ and δ must be replaced by α', β', γ', and δ', representing the relative permeabilities during the rising phase of the *epsp*.

The E_{epsp} is often estimated by polarizing the resting membrane to different values by a block current. The reversal potential is the imposed potential at which activation of the postsynaptic membrane no longer results in a change in potential. The equation for this reversal potential (E_r) can be derived from the general flux Eqn (2), since E_r represents a steady state value present in the resting state. When I_j and μ_j are replaced by ΔI_j and $\Delta\mu_j$, P_j (the permeability coefficient of an ion species "j") becomes ΔP_j. If $\Delta I=0$, the equation for the reversal potential becomes

$$E_r = \frac{RT}{F}\ln\frac{-b+(b^2-4ac)^{1/2}}{2a} \tag{6}$$

in which, according to Piek (1975),

$$a=[K^+]_i+\frac{\Delta P_{Na}}{\Delta P_K}[Na^+]_i+4\frac{\Delta P_{Mg}}{\Delta P_K}[Mg^{2+}]_i+4\frac{\Delta P_{Ca}}{\Delta P_K}[Ca^{2+}]_i$$

$$b=[K^+]_i-[K^+]_o+\frac{\Delta P_{Na}}{\Delta P_K}([Na^+]_i-[Na^+]_o)$$

$$c=-[K^+]_o-\frac{\Delta P_{Na}}{\Delta P_K}[Na^+]_o-4\frac{\Delta P_{Mg}}{\Delta P_K}[Mg^{2+}]_o-4\frac{\Delta P_{Ca}}{\Delta P_K}[Ca^{2+}]_o$$

For *Drosophila melanogaster* Jan and Jan (1976b) suggested that the *epsp* is generated by an increase in P_K, P_{Na} and P_{Mg}. Assuming that for *D. melanogaster* the $[Mg^{2+}]_i$ is equal to 9.25 mmol l^{-1}, Jan and Jan (1976b) estimated that $\Delta P_{Na}/\Delta P_K=1.3$, and the $\Delta P_{Mg}/\Delta P_K=4.7$. The ignorance of the difference between β, β' and $\Delta P_{Na}/\Delta P_K$, or between γ, γ' and $\Delta P_{Mg}/\Delta P_K$, by Jan and Jan does not affect their results, but the reader is warned that their permeability factors are not equal to the factors for the resting membrane, and are also not equal to the factors during excitation, but represent only the change in permeability (ΔP) during the rising phase of the *epsp*.

5.3.4 *Peripheral inhibition*

Inhibitory postsynaptic potentials (*ipsp*s) have been recorded in skeletal muscles of Dictyoptera, Orthoptera, Cheleutoptera, Lepidoptera, Hymenoptera and Coleoptera (Usherwood and Grundfest, 1965; Ikeda and Boettiger, 1965a, b; Hoyle, 1966a, b; Bergman and Pearson, 1968; Miller, 1969; Iles and Pearson, 1969; Pearson and Bergman, 1969; Piek and Mantel, 1970a, b). In muscle fibres of locusts, the reversal potential of the *ipsp* ($E_{r(ipsp)}$) lies between −55 mV and −75 mV, the smaller values usually being obtained from fibres with low membrane potentials (Usherwood and Grundfest, 1965). According to these authors the $E_{r(ipsp)}$ is not changed by substituting potassium for sodium ions in the saline, provided the $E_{r(ipsp)}$ is determined soon after changing the external potassium concentration. That the major part of the current which flows during activation of the postsynaptic membrane of the inhibitory system is probably carried by chloride has been demonstrated for skeletal muscles of the locust *Schistocerca gregaria* by Usherwood and Grundfest (1962). Loading the fibre with potassium chloride by increasing the external potassium ion concentration and subsequently decreasing the potassium ion concentration down to the control value, reduces the hyperpolarizing *ipsp*, or even reverse it to a depolarizing response. This is probably caused by a shift of the equilibrium potential for chloride (E_{Cl}) to a more positive value, and possibly due to an uptake of chloride (Usherwood, 1967b).

If the *ipsp* is generated exclusively by an increase in P_{Cl} then

$$E_{r(ipsp)} = \frac{RT}{F} \ln \frac{[Cl^-]_i}{[Cl^-]_o} \tag{7}$$

Probably the function of peripheral inhibitory transmission is to stabilize the membrane potential by increasing the conductivity for chloride and thus more or less clamping the membrane to the E_{Cl}. If $E_{Cl} = E_m$ the observation of both depolarizing and hyperpolarizing *ipsps* may be caused by artifacts due to a shift of the resting and equilibrium potentials, either by exchanging the haemolymph for a saline or by penetration of the fibre by a microelectrode. In other words, *in vivo* the inhibitory activation may increase the conductivity of the membrane for chloride without altering E_m (Kuffler, 1960).

A demonstration of the effect of penetration by a microelectrode on the E_m and the *ipsp*-amplitude has been given by Huijbregts (personal communication). He impaled the most distal fibres of the extensor muscle of the tibia of *S. gregaria* (m. 135 c and m. 135 d, according to Snodgrass, 1929). The fibres were stimulated twice per second by an axon reflex from nerve 3c to nerve 3b (Pearson and Bergman, 1969). Immediately after penetration of the continuously stimulated muscle fibre, no *ipsps* were recorded (Fig. 20), but shortly afterwards the recorded resting potential decreased from −60 mV to −54 mV and then gradually to −44 mV. Later on the resting potential decreased stepwise to −36 mV. The decrease in E_m was accompanied by an increase in *ipsp*-amplitude, demonstrating the above view of Kuffler (1960). In the experiment of Fig. 20 the reversal potential for the *ipsp* may have been about −55 mV.

6 Ion barriers in the muscle fibre

For the locust *Locusta migratoria*, Hoyle (1953) suggested that the tracheolated membranes surrounding the various tissues could act as selective barriers between the haemolymph and the other tissues. Comparative studies

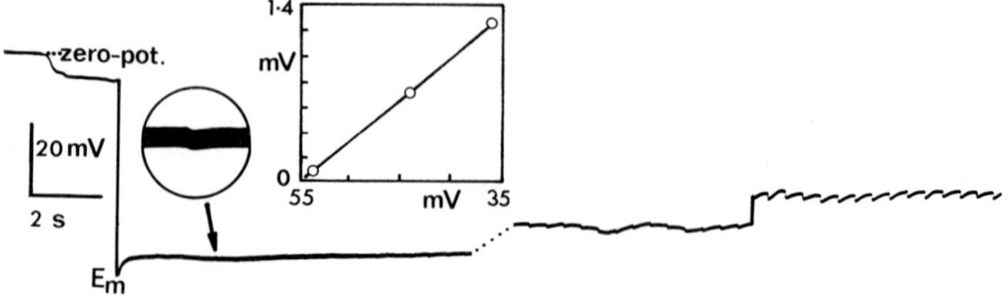

Fig. 20. Record of the membrane potential E_m against time of a tonic muscle fibre in the distal part of the metathoracic extensor tibiae muscle of *S. gregaria*. The record starts on the left with the recording microelectrode (filled with 2 mmol l^{-1} potassium citrate) in the extracellular position. After one second the microelectrode is brought in contact with the muscle fibre (small shift of the potential) and then the fibre is impaled by the microelectrode (shift to minus 60 mV). Within one second after impalement the initial membrane potential of −60 mV was reduced to −54 mV. Thereafter it decreased slowly to −44 mV and after some time the potential suddenly decreased to −36 mV. This decrease in E_m is correlated with an increase in the amplitude of inhibitory postsynaptic potentials (ipsps), evoked by indirect stimulation via an axon reflex twice per s. In the left part of the figure ($E_m = -54$ mV) the ipsps are scarcely visible (cf. left inset). The reversal potential for the ipsps is here about 55 mV. The right inset shows the relation between E_m and the size of the ipsp. (Huijbregts, unpublished results.)

(Thornton, 1963) indicated that a correlation exists between the sodium concentration in the haemolymph and the thickness of the sheath surrounding the central nervous system. In Sphingidae (Lepidoptera) the thickness of this layer of cells changes during life, being highly developed during the pupal stage, when the sodium concentration in the haemolymph is at its lowest. The fact that the sodium concentration in the whole central nervous system is significantly higher than in the haemolymph, provides some evidence that sodium ions are concentrated in the nervous system probably in the extracellular space (Thornton, 1963). The hypothesis, that the central nervous system of phytophagous insects may function in an extracellular medium with a relatively high sodium content has been confirmed by Treherne (1965a, b) for *Carausius morosus* (Cheleutoptera). According to Treherne and Pichon (1972), who reviewed the work on the ionic composition of the extraneuronal space, the "conventional" extraneuronal environment must be maintained by specialized mechanisms within the central nervous tissues. They argued that in insects a "blood-brain" barrier is present in the central nervous system.

The sheath of glial cells that accompanies the peripheral axons provides some evidence that the peripheral nervous system might also create its own extracellular environment. This, however, is probably not the case for all peripheral nerves. Neurosecretory neurons in *Carausius morosus*, for example, possess no fat body sheath, no perineurium and are only covered by a thin basal lamina (Orchard, 1976). Orchard suggested that there is no physical barrier between the neurosecretory nerve membrane and the haemolymph, but the observed discontinuously distributed cells present in the basal lamina may function as regulators. He demonstrated, however, that the action potentials in the neurosecretory axons are due to a current carried by calcium ions rather than by sodium ions. Magnesium ions do not seem to contribute as current carriers for the inward current. If this is the case, a barrier against sodium ions would not be necessary in these nerves.

The sheath of glial cells that accompanies the axons on their outward course from the central ganglia is interrupted at the axon terminal, allowing a close contact between the nerve terminal and the plasma membrane of the muscle fibre (Smith, 1968). This interruption is restricted to the synaptic part of the nerve terminal. The sheath does not disappear at the level of the neuromuscular junction; on the contrary, it forms a cover over the synaptic area (Edwards *et al.*, 1958b, Hamori, 1963). The confluence of the glial cell and plasma membrane of the muscle fibre often forms a perfect cover, leaving only a very narrow passage between the synaptic cleft and the extracellular fluid.

From differences in the sensitivity between completely isolated retractor unguis muscles and perfused femur preparations of *Schistocerca gregaria*, Clements and May (1974) concluded that the retractor unguis muscle within an intact femur contains a diffusion barrier which is damaged or destroyed by the mechanical disturbance caused by dissecting out the isolated muscle. They argued that if the diffusion barrier protects only the nerve endings, then one would expect the isolated preparation (with damaged diffusion barrier) to be more sensitive to synaptic blocking agents than the perfused femur preparations, and one would expect compounds acting on the non-synaptic membrane to affect both preparations equally. If, however, the diffusion barrier protects the whole muscle, the isolated preparation should show greater sensitivity than the perfused femur preparation to both classes of compounds. Compounds that are believed to act predominantly on the non-synaptic muscle fibre membrane include ryanodine (Usherwood, 1962) and ibotenic acid (Lea and Usherwood, 1973a, b) and also potassium ions. Ryanodine and potassium ions depressed the neurally evoked contractions to an almost identical extent in both types of retractor unguis preparations, ibotenic acid depressed the perfused femur preparations even more than the isolated preparations. These results are consistent with the concept that the diffusion barrier protects the nerve endings rather than the whole muscle fibre membrane.

The electron micrographs (Figs 21 and 22) of the basal lamina surrounding the retractor unguis muscle of *S. gregaria* fixed before dissection (compact structure) and after dissection (dispersed structure) suggest that the basal lamina constitutes the diffusion barrier (Clements and May, 1974).

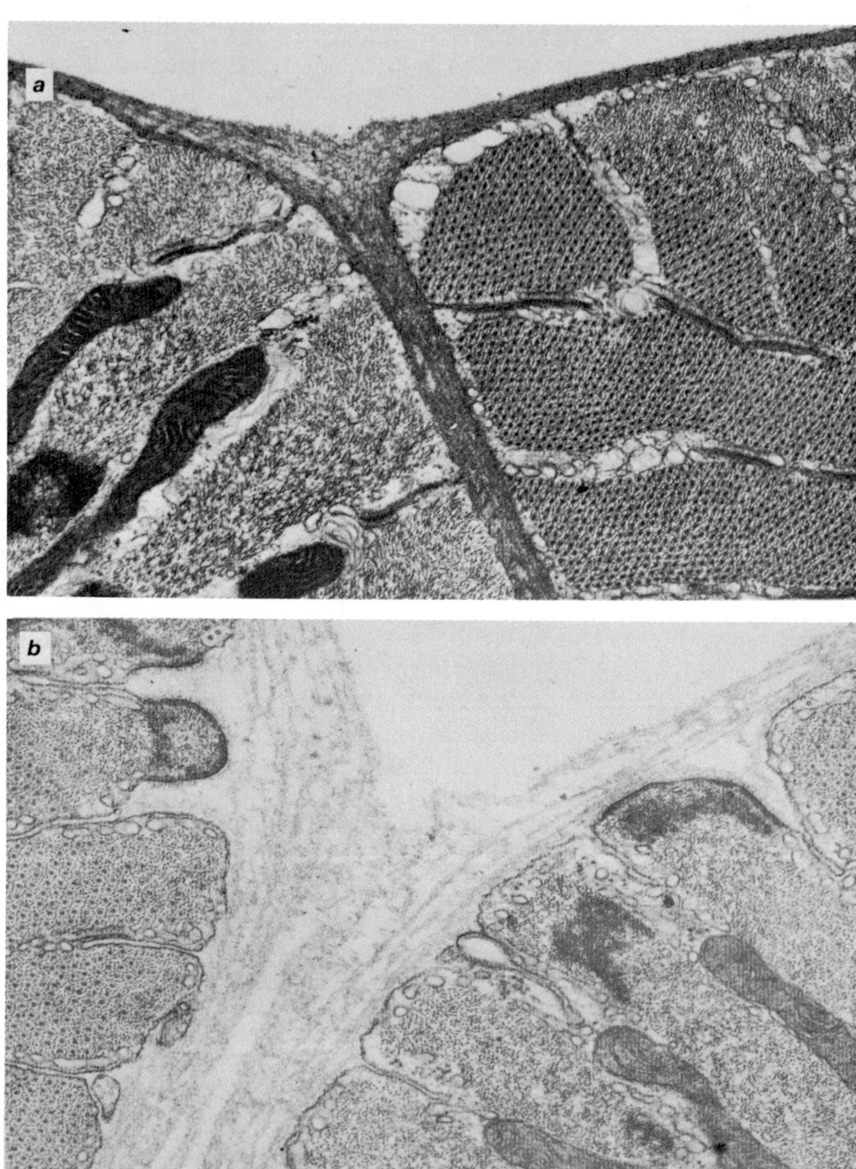

Fig. 21. Junction of two muscle fibres at the surface of the retractor unguis muscle of the rear leg of *S. gregaria*. (*a*) Perfused-femur preparation, fixed before dissection. The basal lamina sheath is compact. (*b*) Isolated preparation, fixed after dissection. The fibres have separated slightly, and the basal laminar sheath has an open appearance. Scale 0.5 μm. (From Clements and May, 1974.)

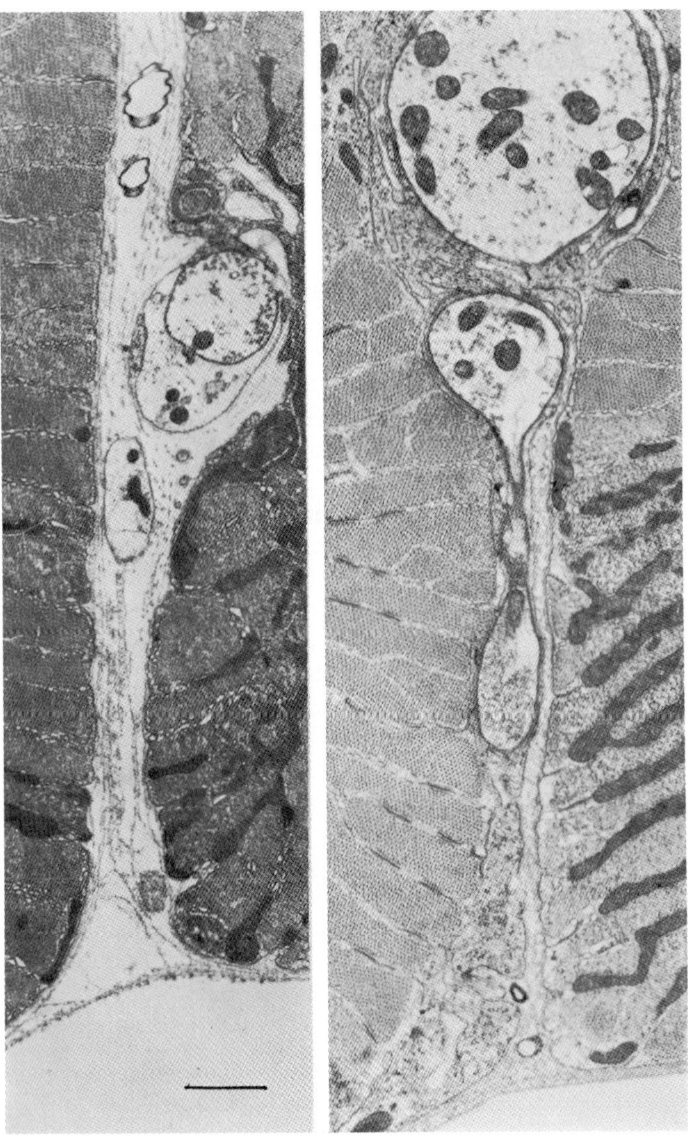

Fig. 22. (*right*) Junction of two muscle fibres in the retractor unguis of *S. gregaria* from a perfused-femur preparation. An axon terminal is situated between the two fibres. The axon terminal is isolated from the external medium of the muscle by the compact basal lamina. (From Clements and May, 1974.) (*left*) Junction of two fibres from an isolated preparation. An axon terminal is situated between the two fibres, and owing to the separation of the fibres and the opening up of the basal laminar sheath, the axon terminal is exposed to the outer medium. Scale 1 μm. (From Clements and May, 1974.)

Electron micrographs of insect muscle (Figs 2 to 9) show that the plasma membrane is highly invaginated. In crustaceans about 10% of the total plasma membrane is present in the cylindrical wall (here called the surface membrane) of the fibre (Falk and Fatt, 1964; Selverston, 1967), while 90% forms the wall of the transverse tubular system (TTS). In muscle fibres of *Philosamia cynthia* (Lepidoptera), about 70% of the total plasma membrane takes part in the formation of the TTS (Piek, 1974, 1975). In muscle fibres of *P. cynthia* and *Galleria mellonella* the synaptic regions are separated from the outer medium by a glial cell (Figs 5 and 9). Only very long and narrow channels connect the synaptic clefts with the outer environment (Belton, 1969; Piek, 1975). Therefore, the postsynaptic membrane may be considered as a part of the invaginated plasma membrane. Due to these structural peculiarities, the fluid inside the invaginated parts of the membrane (mainly the TTS) is the outer medium for the greater part of the plasma membrane.

The fact that some insect muscle fibres show inhibitory postsynaptic potentials (*ipsp*s) may indicate that the E_{Cl} across the inhibitory postsynaptic membrane is of the order of -60 mV, corresponding to a high chloride concentration in the lumen of the TTS (Piek, 1974, 1975). Electron microscopic identification of sodium ions indicates a relatively high sodium ion concentration in the lumen of the TTS of the butterfly *Pieris brassicae* (Njio and Piek, 1977, Fig. 11b). Moreover, the presence of magnesium ions is indicated by a precipitate of a magnesium–ammonium–phosphate complex.

The maintenance of a high sodium chloride concentration in the TTS-lumen could be explained by the presence of an ion pump (Fig. 23). The maintained high level of sodium chloride in the TTS-lumen may result in a steady liquid junctional potential at the boundary of the TTS-lumen and the haemolymph or the bathing saline. This liquid junctional potential would be caused by the transport number for chloride being about 1.5 times higher than that for sodium ions. It has been demonstrated (*cf.* section 5.1) that the liquid junctional potentials between two physiological saline solutions are in the order of a few millivolts. Therefore, their direct contribution to the measured potential difference between two electrodes, one extracellular, and one intracellular, can only be small. The junctional potential may, however, act as an ion barrier, inhibiting the influx of cations, other than Na^+ (and Mg^{2+}), and the efflux of anions (Piek, 1974, 1975).

The exact localization of the ion barrier is not known with certainty. Clements and May (1974) presented arguments that the barrier for glutamate and for drugs is restricted to the synaptic sites (Fig. 22) and may not be located in the whole lamina basalis surrounding the muscle fibre. Besides their experimental arguments for a restricted localization of the barrier, a theoretical argument against the presence of a barrier over the entire muscle fibre membrane can be put forward. If the whole lamina basalis would act

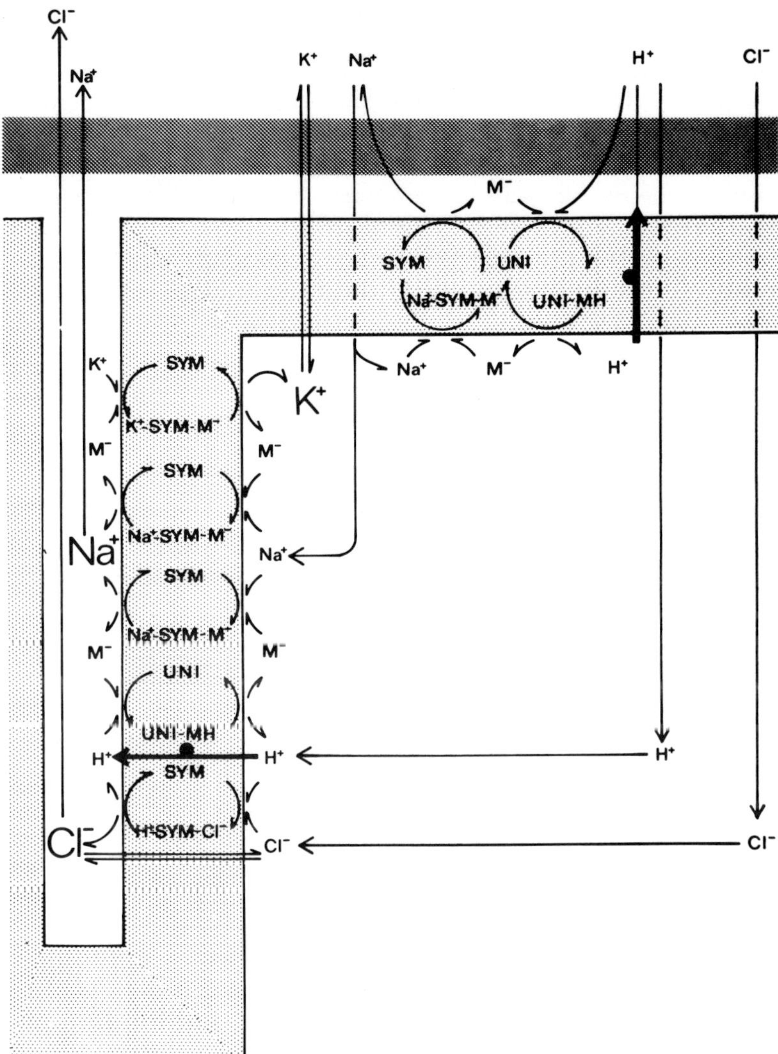

Fig. 23. Diagrammatic representation of the passive and the proton-transport-coupled permeation of ions across the tubular and the surface plasma membrane of an insect muscle fibre. The internal hydrogen ion concentration is thought to be displaced from its electrochemical equilibrium by a metabolic process. The passive permeation of potassium ions is suggested to occur predominantly via the surface membrane, the passive permeation of chloride, however, is suggested to occur via the transverse tubular wall. The transport of ions, coupled to the hydrogen ion transport by monocarboxylate (M^-) as an intermediate compound, is presented as in Fig. 18 (*cf.* legend of Fig. 18). In the TTS the transport of ions results in an accumulation of sodium chloride, the diffusion of which may cause a liquid junctional barrier, maintained by the ratio of the transport number for chloride over sodium ions being about 1.5.

as a barrier, active spots, involved in the regulation should be included in this barrier membrane. There is no morphological basis for such spots. The only special structures present in the basal lamina are the hemidesmosomes (Figs 3 and 4). If these hemidesmosomes should act as ion exchanging structures the transport should take place from the myoplasm to the haemolymph or *vice versa*, and not between the haemolymph and the fluid between the basal lamina and the plasma membrane. It is suggested, therefore, that a barrier exists only in those sites where invaginations are in contact with the haemolymph (Figs 2, 3 and 10). These invaginations may be formed by the transverse tubules as well as by the clefts between glial cells and muscle fibre membrane. The concept of Clements and May (1974) that in insect muscle the neuromuscular junction is protected by local barriers agrees with the concept of Piek (1974, 1975) that the muscle fibre membrane is protected partially by barriers located at the openings of the invaginations.

7 Concluding remarks

In insect muscles resting membrane potentials, postsynaptic potentials and active membrane responses presumably are determined mainly by ionic currents caused by ionic gradients across the plasma membrane and the relative permeability of the membrane for the various ions. In this respect, insects do not deviate from other animals. Arguments are presented, indicating that insect muscle fibres create their own external environment, in analogy to what has been demonstrated for the insect nervous system. This protection of excitable tissues in insects enables them to vary the ionic composition of the haemolymph. In phytophagous insects, in particular, the ionic composition of the haemolymph deviates extraordinarily from the conventional "sea-water-like" ionic environment of the excitable tissues of other animals.

Once the possibility had been created of varying the composition of the haemolymph, insects could have started to gradually vary the basis of their bio-electrical properties. They could have, for example, replaced sodium ions by calcium or magnesium ions, and this could be useful, since the food of phytophagous insects normally contains much more magnesium than sodium ions. Although experiments indicating a partial change in the nature of the ions carrying the membrane currents have been criticized, at present one cannot deny the existence of such a change.

Initial attempts to correlate electron microscopic findings with electrochemical phenomena have led to the concept of a highly active proton coupled ion transport system, also operating in the extensive transverse tubular membrane, which is permeable to chloride. This leads to a high concentration of sodium chloride in the transverse tubular lumen. The mechanism, which may be the base of the generation of the membrane potential is hypothetical, but the concept is supported by a number of observations reviewed in this paper.

Acknowledgement

The authors thank Professor C. van der Meer for the helpful discussions during the preparation of the manuscript.

References

Aickin, C. C. and Thomas, R. C. (1975). Micro-electrode measurement of internal pH of crab muscle fibres. *J. Physiol. Lond.* **252,** 803–815.

Aidley, D. J. (1975). Excitation-contraction coupling and mechanical properties. *In* "Insect muscle" (Ed. P. N. R. Usherwood). Academic Press, New York and London.

Anderson-Cedergren, E. (1959). Ultrastructure of motor endplate and sarcoplasmic components of mouse skeletal muscle fibre as revealed by three-dimensional reconstructions from serial sections. *J. Ultrastr. Res.* (suppl.) **1,** 1–191.

Anstee, J. H. (1971). An ultrastructural study of tettigoniid muscle in relation to function. *J. Insect Physiol.* **17,** 1983–1994.

Anwyl R. (1977a). Permeability of the post-synaptic membrane of an excitatory glutamate synapse to sodium and potassium. *J. Physiol. Lond.* **273,** 367–388.

Anwyl R. (1977b). The effect of foreign cations, pH, and pharmacological agents on the ionic permeability of an excitatory glutamate synapse. *J. Physiol.* **273,** 389–404.

Anwyl, R. and Usherwood, P. N. R. (1974). Voltage clamp studies of glutamate synapse. *Nature, Lond.* **252,** 591–593.

Arvanitaki, A. (1942). Effects evoked in an axon by the activity of a contiguous one. *J. Neurophysiol.* **5,** 89–108.

Ashhurst, D. E. (1959). The connective tissue sheath of locust nervous system: a histochemical study. *Q.J.microsc. Sci.* **100,** 401–412.

Ashhurst, D. E. (1961). A histochemical study of the connective tissue sheath of the nervous system of *Periplaneta americana. Q.J. microsc. Sci.* **102,** 455–461.

Ashhurst, D. E. (1964). Fibrillogenesis in the wax moth, *Galleria mellonella. Q.J. microsc. Sci.* **105,** 391–403.

Ashhurst, D. E. (1967). The fibrillar flight muscles of giant water-bugs: An electron microscope study. *J. Cell Sci.* **2,** 435–444.

Ashhurst, D. E. (1968). The connective tissues of insects. *Ann. Rev. Ent.* **13,** 45–74.

Ashhurst, D. E. and Richards, A. G. (1964). A study of the changes occurring in the connective tissue associated with the central nervous system during the pupal stage of the wax-moth, *Galleria mellonella. J. Morphol.* **114,** 225–236.

Atwood, H. L., Smyth, T. and Johnston, H. S. (1969). Neuromuscular synapses in the cockroach extensor tibiae muscle. *J. Insect. Physiol.* **15,** 529–535.

Atwood, H. L., Lang, F. and Morin, W. A. (1972). Synaptic vesicles: selective depletion in crayfish excitatory and inhibitory axons. *Science N.Y.* **176,** 1353–1355.

Auber, J. (1960). Observations sur l'innervation motrice des muscles des insectes. *Z. Zellforch.* **51,** 705–724.

Babers, F. H. (1938). An analysis of the blood of the sixth-instar southern armyworm (*Prodenia eridania*). *J. agric. Res.* **57,** 697–706.

Barsa, M. C. (1954). The behaviour of isolated hearts of the grasshopper, *Chortophaga viridifasciata*, and the moth, *Samia walkeri*, in solutions with different concentrations of sodium, potassium, calcium and magnesium. *J. gen. Physiol.* **38,** 79–92.

Baylor, S. M., and Oetliker, H. (1975). Birefringence experiments on isolated skeletal muscle fibres suggest a possible signal from the sarcoplasmic reticulum. *Nature, Lond.* **253,** 97—101.

Beinbrech, G. (1972). Elektronenmikroskopische Untersuchungen über die Entwicklung des Transversal-Systems und des sarkoplasmatischen Retikulums in asynchronen Insekten-Flugmuskeln. *Cytobiol.* **5,** (3), 448–462.

Belton, P. (1969). Innervation and neural excitation of ventral muscle fibres of the larva of the waxmoth, *Galleria mellonella. J. Insect Physiol.* **15,** 731–741.

Belton, H. and Grundfest, H. (1962). Potassium activation and K-spikes in muscle fibres of the mealworm larva (*Tenebrio molitor*). *Am. J. Physiol.* **203,** 588–594.

Bennett, H. S. (1960). The structure of striated muscle as seen by the electron microscope. *In* "The structure and function of muscle" (Ed. G. H. Bourne), **I,** pp. 137–181. Academic Press, New York and London.

Bennett, H. S. and Porter, K. R. (1953). An electron microscopic study of sectioned breast muscle of the domestic fowl. *Amer. J. Anat.* **93,** 61–106.

Bergman, S. J. and Pearson, K. G. (1968). Inhibition in cockroach muscle. *J. Physiol. Lond.* **195,** 22P.

Bialaszewicz, K. and Landau, C. (1938). Sur la composition de l'hémolymphe et sur les changements qu'elle subit au cours de la croissance et pendant la métamorphose. *Acta biol. exp.* **12,** 307–320.

Bienz-Isler, G. (1968a). Elektronenmikroskopische Untersuchungen über die imaginale Struktur der dorsolongitudinalen Flugmuskeln von *Antheraea pernyi* Guer. (Lepidoptera). *Acta anat.* **70,** 416–433.

Bienz-Isler, G. (1968b). Elektronenmikroskopische Untersuchungen über die Entwicklung der dorsolongitudinalen Flugmuskeln von *Antheraea pernyi* Guer. (Lepidoptera). *Acta anat.* **70,** 524–553.

Bishop, G. H., Briggs, A. P. and Ronzoni, E. (1925). Body fluids of the honey bee larva. II Chemical constituents of the blood, and their osmotic effects. *J. biol. Chem.* **66,** 77–88.

Bodian, D. (1970). An electron microscopic characterization of classes of synaptic vesicles by means of controlled aldehyde fixation. *J. Cell Biol.* **44,** 115–124.

Bolton, T. B. (1972). The depolarizing action of acetylcholine or carbachol in intestinal smooth muscle. *J. Physiol. Lond.* **220,** 647–671.

Boné, G. J. (1944). Le rapport sodium/potassium dans le liquide coelomique des insectes, 1: Ses relations avec le régime alimentaire. *Ann. Soc. roy. Zool. Belg.* **75,** 123–132.

Brady, J. (1967a). The relationship between blood ions and blood-cell density in insects. *J. exp. Biol.* **47,** 313–326.

Brady, J. (1967b). Haemocytes and the measurement of potassium in insect blood. *Nature, Lond.* **215,** 96–97.

Bridges, R. G. (1972). Choline metabolism in insects. *Adv. Insect Physiol.* **9,** 51–110.

Brecher, L. (1929). Die anorganischen Bestandteile des Schmetterlingspuppenblutes (*Sphinx pinastri, Pieris brassicae*). Veränderungen im Gehalt am anorganischen Bestandteile bei der Verpuppung (*Pieris brassicae*). *Biochem. Z.* **211,** 40–64.

Caesar, R. and Edwards, G. A. (1957). The physiologic significance of the basement membrane. *Anat. Rec.* **128,** 530.

Carrington, C. B. and Tenney, S. M. (1959). Chemical constituents of haemolymph and tissue in *Telea polyphemus* Cram. with particular reference to the question of ion binding. *J. Insect Physiol.* **3,** 402–413.

Chen, A. C. and Friedman, S. (1975). An isotonic saline for the adult blowfly *Phormia regina*, and its application to perfusion experiments. *J. Insect Physiol.* **21,** 529–536.

Clark, E. W. (1958). A review of literature on calcium and magnesium in insects. *Ann. ent. Soc. Am.* **51,** 142–154.

Clark, E. W. and Craig, R. (1953). The calcium and magnesium content in the haemolymph of certain insects. *Physiol. Zool.* **26,** 101–107.

Clements, A. N. and May, T. E. (1974). Studies on locust neuromuscular physiology in relation to glutamic acid. *J. exp. Biol.* **60,** 673–705.

Cochrane, D. G. and Elder, H. Y. (1967). Morphological changes in insect muscle during influx and efflux of potassium ions, chloride ions and water. *J. Physiol. Lond.* **191,** 30P–31P.

Cochrane, D. G., Elder, H. Y. and Usherwood, P. N. R. (1972). Physiology and ultrastructure of phasic and tonic skeletal muscle fibres in the locust, *Schistocerca gregaria. J. Cell Sci.* **10,** 419–441.

Collett, J. I. (1976). Some features of the regulation of the free amino acids in adult *Calliphora erythrocephala. J. Insect. Physiol.* **22,** 1395–1403.

Constantin, L. L., Franzini-Armstrong, C. and Podolsky, R. J. (1965) Localization of calcium accumulating structures in striated muscle fibres. *Science, N.Y.* **147,** 158–160.

Conway, E. J. (1957). Nature and significance of concentration relations of potassium and sodium ions in skeletal muscle. *Physiol. Rev.* **37,** 84–132.

Crabtree, B. and Newsholm, E. A. (1975). Comparative aspects of fuel utilization and metabolism by muscle. *In* "Insect muscle", (Ed. P. N. R. Usherwood). Academic Press, New York and London.

Deitmer, J. W. (1977). Electrical properties of skeletal muscle fibres of the flour moth *Ephestia kuehniella. J. Insect Physiol.* **23,** 33–38.

Deitmer, J. W. and Rathmayer, W. (1976). Calcium action potentials in larval muscle fibres of the moth *Ephestia kühniella Z.* (Lepidoptera). *J. comp. Physiol.* **112,** 123–132.

Del Castillo, J., Hoyle, G. and Machne, X. (1953). Neuromuscular transmission in a locust. *J. Physiol. Lond.* **121,** 539–547.

De Robertis, E. D. P. (1964). "Histophysiology of synapses and neurosecretion." Pergamon Press, Oxford.

Diculescu, J. and Popescu, L. M. (1973). Electron microscopic demonstration of calcium in mitochondria of the frog skeletal muscle in situ. *Exp. Cell Res.* **82,** 152–158.

Doyère, M. (1840) Mémoires sur les Tardigrades. *Ann. Sc. nat. série 2,* **14,** 269–361.

Drilhon, A. (1934) Sur le milieu intérieur des Lépidoptères. *C.R. Soc. Biol. Paris,* **115,** 1194–1195.

Drilhon, A. and Busnel, R. G. (1946). Documents sur le milieu intérieur d'un insecte aquatique pouvant servir a étayer une systématique biochemique. *Bull. Soc. Zool. Fr.* **71,** 185–188.

Duchâteau, G., Florkin, M. and Leclercq, J. (1953). Concentration de bases fixes et types de composition de la base totale de l'hémolymphe des insectes. *Arch. int. Physiol. Biochem.* **61,** 518–549.

Edwards, G. A. (1959). The fine structure of a multiterminal innervation of an insect muscle. *J. biophys. biochem. Cytol.* **5,** 241–244.
Edwards, G. A., Ruska, H., de Souza-Santos, P. and Vallejo-Freire, A. (1956). Comparative cytophysiology of striated muscle with special reference to the role of the endoplasmic reticulum. *J. biophys. biochem. Cytol.* (suppl), **2,** 143–157.
Edwards, G. A., Ruska, H. and de Harven, E. (1958a). Electron microscopy of peripheral nerves and neuromuscular junctions in the wasp leg. *J. biophys. biochem. Cytol.* **4,** 107–114.
Edwards, G. A., Ruska, H. and de Harven, E. (1958b). Neuromuscular junctions in flight and tymbal muscles of the cicade. *J. biophys. biochem. Cytol.* **4,** 251–256.
Eisenberg, R. S. and Gage, P. W. (1969). Ionic conductions of the surface and transtubular membranes of frog sartorius fibers. *J. gen. Physiol.* **53,** 279–297.
Elder, H. Y. (1975). Muscle structure. *In* "Insect muscle" (Ed. P. N. R. Usherwood). Academic Press, New York and London.
Endo, M. (1964). Entry of a dye into the sarcotubular system of muscle. *Nature, Lond.* **202,** 1115–1116.
Endo, M. (1977). Calcium release from the sarcoplasmic reticulum. *Physiol. Rev.* **57,** 71–108.
Falk, G. and Fatt, P. (1964). Linear electrical properties of striated muscle fibres observed with intracellular electrodes. *Proc. roy. Soc. Lond.* **B160,** 69–123.
Fatt, P. and Katz, B. (1951). An analysis of the end-plate potential recorded with an intracellular electrode. *J. Physiol. Lond.* **115,** 320–370.
Fawcett, D. W. and Revel, J. P. (1961). The sarcoplasmic reticulum of a fast-acting fish muscle. *J. biophys. biochem. Cytol.* (suppl.) **10,** 89–110.
Florkin, M. and Jeuniaux, C. (1974). Haemolymph: composition. *In* "The physiology of Insecta", (Ed. M. Rockstein), 2nd Edn. **5,** Academic Press, New York and London.
Foettinger, A. (1880). Sur les terminaisons des nerfs dans les muscles des insectes. *Arch. Biol.* **1,** 279–304.
Fukuda, J., Furuyama, S. and Kawa, K. (1977). Calcium dependent action potentials in skeletal muscle fibres of a beetle larva, *Xylotrupes dichotomus. J. Insect Physiol.* **23,** 367–374.
Fukuda, J. and Kawa, K. (1977). Initiation of Ca-spikes from an insect muscle fibre immmersed in a low pH saline solution containing carboxylic anions. *Life Sc.* **21,** 981–988.
Galvani, A. (1792). De viribus electricitatis in motu musculari. Thesis. *In* "Bononiensi Archigymnasio et Instituto Scientiarum Publici Professoris", Anatomici Emeriti, Academici Benedictini.
Gese, P. K. (1950). The concentration of certain inorganic constituents in the blood of the cynthia pupa *Samia walkeri* Felder and Felder. *Physiol. Zool.* **23,** 109–113.
Goldman, D. E. (1943). Potential, impedance, and rectification in membranes. *J. gen. Physiol.* **27,** 37–60.
Hagiwara, S., Chichibu, S. and Naka, K. (1964). The effects of various ions on resting and spike potentials of barnacle muscle fibres. *J. gen. Physiol.* **48,** 163–179.
Hagopian, M. and Spiro, D. (1967). The sarcoplasmic reticulum and its association with the T-system in an insect. *J. Cell Biol.* **32,** 535–545.
Hámori, J. (1961). Innervation of insect leg muscle. *Acta biol. Hung.* **12,** 219–230.
Hámori, J. (1963). Electron-microscope studies on neuromuscular junctions of end-plate type in insects. *Acta. biol. Acad. Sc. Hung.* **14,** 231–245.
Hardie, J. (1976). Motor innervation of the supercontracting longitudinal ventro-lateral muscles of the blowfly larva. *J. Insect Physiol.* **22,** 661–668.

Heller, J. and Moklowska, A. (1930). Über die Zusammensetsung des Raupenblutes bei *Deilephila euphorbiae* und deren Veränderungen im Verlauf der Metamorphose. *Bioch. Z.* **219,** 473–489.

Henderson, P. (1907). Zur Thermodynamiek der Flüssigkeitsketten. *Z. Physik. Chem.* **59,** 118–127.

Henderson, P. (1908). Zur Thermodynamik der Flüssigkeitsketten. *Z. Physik. Chem.* **63,** 325–345.

Hess, J. J., Taylor, A. and Maffly, R. H. (1975). On the effects of propionate and other short chain fatty acids on sodium transport by the toad bladder. *Biochim. Biophys. Acta,* **394,** 416–437.

Hevert, F. (1974). Zur Regulation der Ionenkonzentration von *Drosophila heydei,* Na^+, K^+, Ca^{2+} und Cl^-. *J. Insect Physiol.* **20,** 2225–2245.

Hinke, J. A. M. and Menard, M. R. (1976). Intracellular pH of single crustacean muscle fibres by the DMO and electrode methods during acid and alkaline conditions. *J. Physiol. Lond.* **262,** 533–552.

Hodgkin, A. L. and Horowicz, P. (1960). The effect of sudden change in ionic concentration on the membrane potential of single fibres. *J. Physiol. Lond.* **153,** 370–395.

Hodgkin, A. L. and Katz, B. (1949). The effect of sodium ions on the electrical activity of the giant axon of the squid. *J. Physiol. Lond.* **108,** 37–77.

Hoyle, G. (1953). Potassium ions and insect nerve muscle. *J. exp. Biol.* **30,** 121–135.

Hoyle, G. (1955). The effect of some common cations on neuromuscular transmission in insects. *J. Physiol. Lond.* **127,** 90–103.

Hoyle, G. (1965). Nature of the excitatory sarcoplasmic reticular junction. *Science, N.Y.* **149,** 70–72.

Hoyle, G. (1966a). An isolated insect ganglion-nerve-muscle preparation. *J. exp. Biol.* **44,** 413–427.

Hoyle, G. (1966b). Functioning of the inhibitory-conditioning axon innervating insect muscles. *J. exp. Biol.* **44,** 429–453.

Hoyle, G. (1969). Comparative aspects of muscle. *Ann. Rev. Physiol.* **31,** 43–84.

Huddart, H. (1966a). The effect of potassium ions on resting and action potentials in Lepidopteran muscle. *Comp. Biochem. Physiol.* **18,** 131–140.

Huddart, H. (1966b). Ionic composition of haemolymph and myoplasm in Lepidoptera in relation to their membrane potentials. *Arch. int.. Physiol. Biochim.* **74,** 603–613.

Huddart, H. (1966c). The effect of sodium ions on resting and action potentials in skeletal muscle fibres of *Bombyx mori* (L.). *Arch. int. Physiol. Biochim.* **74,** 592–602.

Huddart, H. (1971). Contraction of insect muscle. *Exp. Physiol. Biochem.* **4,** 219–288.

Huddart, H. and Oates, K. (1970). Ultrastructure of stick insect and locust skeletal muscle in relation to excitation-contraction coupling. *J. Insect Physiol.* **16,** 1467–1483.

Huddart, H. and Wood, D. W. (1966). The effect of DNP on the resting potential and ionic content of some insect skeletal muscle fibres. *Comp. Biochem. Physiol.* **18,** 681–688.

Hutter, O. F. and Warner, A. E. (1967a). The sensitivity of the chloride conductance of frog skeletal muscle. *J. Physiol. Lond.* **189,** 403–425.

Hutter, O. F. and Warner, A. E. (1967b). The effect of pH on the ^{36}Cl efflux from frog skeletal muscle. *J. Physiol. Lond.* **189,** 427–443.

Huxley, H. E. (1964). Evidence for continuity between the central elements of the triads and the extracellular space in frog sartorius muscle. *Nature, Lond.* **202,** 1067–1071.

Ikeda, K. and Boettiger, E. G. (1965a). Studies on the flight mechanism of insects-II.

The innervation and electrical activity of the fibrillar muscles of the bumble bee *Bombus*. *J. Insect Physiol.* **11,** 779–789.

Ikeda, K. and Boettiger, E. G. (1965b). Studies on the flight mechanism of insects-III. The innervation and electrical activity of the basalar fibrillar flight muscle of the beetle *Oryctes rhinoceros*. *J. Insect Physiol.* **11,** 791–802.

Iles, J. F. and Pearson, K. G. (1969). Triple inhibitory innervation of insect muscle. *J. Physiol. Lond.* **204,** 125P–126P.

Jahromi, S. S. and Atwood, H. L. (1969). Structural features of muscle fibres in the cockroach leg. *J. Insect Physiol.* **15,** 2255–2262.

Jan, L. Y. and Jan, Y. N. (1976a). Properties of the larval neuromuscular junction in *Drosophila melanogaster*. *J. Physiol. Lond.* **262,** 189–214.

Jan, L. Y. and Jan, Y. N. (1976b). L-glutamate as an excitatory transmitter at the *Drosophila* larval neuromuscular junction. *J. Physiol. Lond.* **262,** 215–236.

Kafatos, F. C. (1968). The labial gland: a salt-secreting organ of saturniid moths. *J. exp. Biol.* **48,** 435–453.

Komnick, H. (1962). Elektronenmikroskopische Lokalisation von Na^+ und Cl^- in Zellen und Geweben. *Protoplasma*, **55,** 414–418.

Komnick, H. (1969). Histochemische Calcium Lokalisation in der Skelettmuskulatur des Froches. *Histochemie*, **18,** 24–29.

Kosaka, K. (1969). Electrophysiological and electron microscopic studies on the neuromuscular junctions of the crayfish stretch receptor. *Jap. J. Physiol.* **19,** 160–175.

Kuffler, S. W. (1960). Excitation and inhibition in single nerve cells. *In* "The Harvey Lectures", 1958–1959, pp. 176–218. Academic Press, New York and London.

Kühne, W. (1871). Die Nervenendigung bei den wirbellosen Thieren. *In* "Stricker's Handbuch der Lehre von dem Geweben", Leipzig.

Kusano, K. and Grundfest, H. (1967). Ionic requirement for synaptic electrogenesis in neuromuscular transmission of mealworm larvae (*Tenebrio molitor*). *J. gen. Physiol.* **50,** 1092.

Kusano, K. and Janiszewski, L. (1976). Neuromuscular transmission in mealworm larvae (*Tenebrio molitor*). *In* "Electrobiology of nerve, synapse, and muscle". (Ed. V. P. Reuben, D. P. Pupura, M. V. L. Bennett and E. R. Kandel, Raven Press, New York.

Lea, T. J. (1972). Pharmacological and ionic properties in insect muscle fibre membranes. Thesis, University of Glasgow.

Lea, T. J. and Usherwood, P. N. R. (1973a). The site of action of ibotenic acid and the identification of two populations of glutamate receptors on insect muscle-fibres. *Comp. gen. Pharmacol.* **4,** 333–350.

Lea, T. J. and Usherwood, P. N. R. (1973b). Effect of ibotenic acid on chloride permeability of insect muscle-fibres. *Comp. gen. Pharmacol.* **4,** 351–363.

Lehninger, A. L. (1975). "Biochemistry. The molecular basis of cell structure and function". Worth Publ. Inc. New York.

Lettau, J., Foster, W. A., Harker, J. E. and Treherne, J. E. (1977). Diel changes in potassium activity in the haemolymph of the cockroach *Leucophaea maderae*. *J. exp. Biol.* **71,** 171–186.

Levenbook, L. (1950). The composition of horse bot-fly (*Gastrophilus intestinalis*) – larva blood. *Biochem. J.* **47,** 336–346.

Levenbook, L. and Hollis, V. W. (1961). Organic acid in insects – I. Citric acid. *J. Insect Physiol.* **6,** 52–61.

Loewenstein, W. R. (1970). Intercellular communication. *Sci. Amer.* **222** (5), 79–86.

Ludwig, D. (1951). Composition of the blood of Japanese beetle (*Popillia japonica* Newman) larvae. *Physiol. Zool.* **24,** 329–334.

Lund, R. D. and Westrum, L. E. (1966). Synaptic vesicle differences after primary formalin fixation. *J. Physiol. Lond.* **185,** 7P–9P.
MacInnes, D. A. (1961). "The principles of electrochemistry". Dover, New York.
McCann, F. V. (1964). The effect of anion substitution on bioelectric potentials in the moth heart. *Comp. Biochem. Physiol.* **13,** 179–188.
McCann, F. V. (1965). Unique properties of the moth myocardium *Ann. N.Y. Acad. Sci.* **127,** 84–99.
McCann, F. V. (1967). The effect of metabolic inhibitors on the moth heart. *Comp. Biochem. Physiol.* **20,** 399–409.
McCann, F. V. (1971). Calcium action potentials in insect myocardial fibres. *Comp. Biochem. Physiol.* **A40,** 353–357.
Mangold, E. (1905). Untersuchungen über die Endigung der Nerven in den quergestreiften Muskeln der Arthropoden. *Z. allg. Physiol.* **5,** 135–205.
Marcu, O. (1929). Nervenendigungen an den Muskelfasern von Insekten. *Anat. Anz.* **67,** 369–380.
Martonosi, A. (1972). Biochemical and clinical aspects of sarcoplasmic reticulum function. *In* "Current topics in membrane and transport", (Ed. F. Bronner and A. Kleinzeller). Academic Press, New York and London.
Miller, P. L. (1969). Inhibitory nerves to insect spiracles. *Nature, Lond.* **221,** 171–173.
Mitchell, P. (1973). Performance and conservation of osmotic work by proton-coupled solute porter systems. *J. Bionergetics,* **4,** 63–91.
Mitchell, P. (1975). Vectorial chemistry and the molecular mechanics of chemiosmotic coupling: Power transmission by proticity. *Biochem. Soc. Trans.* **4,** 399–430.
Mitchell, P. and Moyle, J. (1965a). Stoichiometry of proton translocation through the respiratory chain and adenosine triphosphatase systems of rat liver mitochondria. *Nature, Lond.* **208,** 147–151.
Mitchell, P. and Moyle, J. (1965b). Evidence discriminating between chemical and chemiosmotic mechanisms of electron transport phosphorylation. *Nature, Lond.* **208,** 1205–1206.
Motokizawa, F., Reuben, J. P. and Grundfest, H. (1969). Ionic permeability of the inhibitory postsynaptic membrane of lobster muscle fibres. *J. gen. Physiol.* **54,** 457–461.
Nadol, J. B. and Darin de Lorenzo, A. J. (1968). Observations on the abdominal stretch receptor and the fine structure of associated axo-dendritic synapses and neuromuscu lar junctions in *Homarus*. *J. comp. Neur.* **132,** 419–444.
Nakajima, S. and Bastian, J. (1976). Membrane properties of the transverse tubular system of amphibian skeletal muscle. *In* "Electrobiology of nerve, synapse, and muscle", (Ed. J. P. Reuben, D. P. Purpura, M. V. L. Bennett and E. R. Kandel). Raven Press, New York.
Njio, K. D. and Piek, T. (1977). The localization of sodium and potassium ions in a flight muscle of *Pieris brassicae*. *J. Insect Physiol.* **23,** 919–929.
O'Connor, A. K., O'Brien, R. D. and Salpeter, M. M. (1965). Pharmacology and fine structure of peripheral muscle innervation in the cockroach, *Periplaneta americana*. *J. Insect Physiol.* **11,** 1351–1358.
Orchard, I. (1976). Calcium dependent action potentials in a peripheral neurosecretory cell of the stick insect. *J. comp. Physiol.* **112,** 95–102.
Orentlicher, M. and Reuben, J. P. (1971). Localization of ionic conductances in crayfish muscle fibres. *J. Membr. Biol.* **4,** 209–226.
Osborne, M. P. (1967). The fine structure of neuromuscular junction in the segmental muscles of the blowfly larva. *J. Insect Physiol.* **13,** 827–833.

Osborne, M. P. (1970). Structure and function of neuromuscular junctions and stretch receptors. *In* "Insect Ultrastructure". Symp. roy. ent. Soc. (Ed. A. C. Neville), **5,** pp. 77–100. Blackwell, Oxford.

Palade, P. T. and Barchi, R. L. (1977). Characteristics of the chloride conductance in muscle fibres of the rat diaphragm. *J. gen. Physiol.* **69,** 325–342.

Pasquali-Ronchetti, I. (1969). The organization of the sarcoplasmic reticulum and T-system in the femoral muscle of the housefly, *Musca domestica. J. Cell Biol.* **40,** 269–273.

Patlak, J. B. (1976). The ionic basis for the action potential in the flight muscle of the fly, *Sarcophaga bullata. J. comp. Physiol.* **107,** 1–11.

Patterson, D. S. P. (1956). The accumulation of citrate in insect tissues. *Arch. int. Physiol. Biochim.* **64,** 681–683.

Pearson, K. G. and Bergman, S. J. (1969). Common inhibitory motoneurons in insects. *J. exp. Biol.* **50,** 445–471.

Pensa, P. and Ciccarelli, B. (1968). Ultrastructure of spinal cord synapses during strychnine intoxication. *Experientia,* **24,** 1025–1026.

Pepper, J. H., Donaldson, F. T. and Hastings, E. (1941). Buffering capacity and composition of the blood serum and regurgitated digestive juices of the mormon cricket (*Anabrus simplex* Hald). *Physiol. Zool.* **14,** 470–475.

Pichon, Y. and Boistel, J. (1967). Microelectrode study of the resting and action potentials of the cockroach giant axon with special reference to the rôle played by the nerve sheath. *J. exp. Biol.* **47,** 357–373.

Piek, T. (1974). Ion barriers in muscle fibres. *Arch. int. Physiol. Biochim.* **82,** 337–339.

Piek, T. (1975). Ionic properties. *In* "Insect muscle", (Ed. P. N. R. Usherwood). Academic Press, New York and London.

Piek, T. and Mantel, P. (1970a). A study of the different types of action potentials and miniature potentials in insect muscle. *Comp. Biochem. Physiol.* **34,** 935–951.

Piek, T. and Mantel, P. (1970b). The effect of the venom of *Microbracon hebetor* (Say) on the hyperpolarizing potentials in a skeletal muscle of *Philosamia cynthia* Hübn. *Comp. gen. Pharmacol.* **1,** 87–92.

Piek, T., Njio, K. D. and Mantel, P. (1973). Effects of anions and dinitrophenol on resting membrane potentials of insect muscle fibres. *J. Insect Physiol.* **19,** 2373–2392.

Piek, T., Mantel, P. and Wijsman, J. P. M. (1977). Effects of monocarboxylates and external pH on some parameters of insect muscle fibres. *J. Insect Physiol.* **23,** 773–777.

Planck, M. (1890a). Über die Erregung von Elektrizität und Wärme in Elektrolyten. *Ann. Physik Chem.* **39,** 161–186.

Planck, M. (1890b). Über die Potentialdifferenz zwischen zwei verdünnten lösungen binärer Elektrolyte. *Ann. Physik Chem.* **40,** 561–576.

Plantevin, G. (1967). Dosage de Na^+, K^+, Ca^{2+} et Mg^{2+} de l'hémolymphe de *Galleria mellonella* L. par spectrophotométrie de flamme. *J. Insect Physiol.* **13,** 1907–1920.

Porter, K. R. (1956). The sarcoplasmic reticulum in muscle cells of *Amblystoma* larvae. *J. biophys. biochem. Cytol.* (4 suppl) **2,** 163–170.

Porter, K. R. (1961). The sarcoplasmic reticulum. Its recent history and present status. *J. biophys. biochem. Cytol.* (suppl.) **10,** 219–226.

Porter, K. R. and Palade, G. E. (1957). Studies on the endoplasmic reticulum. III Its form and distribution in striated muscle cells. *J. biophys. biochem. Cytol.* **3,** 269–299.

Ramon y Cájal, S. (1890). Coloration par la méthode de Golgi de terminaisons des trachées et des nerfs dans les muscles des ailes des insectes. *Z. wiss. Mikr.* **7,** 332–342.

Ramsay, J. A. (1953). Active transport of potassium by malpighian tubules of insects. *J. exp. Biol.* **30,** 358–369.

Reuben, J. P., Girardier, L. and Grundfest, H. (1964). Water transfer and cell structure in isolated crayfish muscle fibres. *J. gen. Physiol.* **47,** 1141–1174.

Reuben, J. P., Eastwood, A. B., Zollman, J. R., Orentlicher, M. and Brandt, P. W. (1976). Permeability characteristics of the transverse tubular system: Determinants of the signal for contractile activation in crayfish. *In* "Electrobiology of nerve, synapse, and muscle", (Ed. J. P. Reuben, D. P. Purpura, M. V. L. Bennett and E. R. Kandel). Raven Press, New York.

Revel, J. P. (1962). The sarcoplasmic reticulum of the bat cricothyroid muscle. *J. Cell. Biol.* **12,** 571–588.

Rheuben, M. B. (1972). The resting potential of moth muscle fibre. *J. Physiol. Lond.* **225,** 529–554.

Ruska, H., Edwards, G. A. and Caesar, R. (1958). A concept of intracellular transmission of excitation by means of the endoplasmic reticulum. *Experientia*, **14,** 117–120.

Selverston, A. (1967). Structure and function of the transverse tubular system in crustacean muscle fibres. *Am. Zool.* **7,** 515–525.

Shafiq, S. A. (1964). An electron microscopical study of the innervation and sarcoplasmic reticulum of the fibrillar flight muscle of *Drosophila melanogaster*. *Q. J. micr. Sci.* **105** (1), 1–6.

Shiina, S. and Mizuhira, V. (1970). Subcellular localization of sodium and potassium ions in skeletal muscle. *Acta histochim. cytochim.* **3,** 74–79.

Shiina, S., Mizuhira, V., Uchida, K., Amakawa, T. and Tsuzi, K. (1968). Electronmicroscopic study on the distribution of sodium and potassium in heart muscle cell. *J. Electron Micr.* **17,** 267–268.

Simoni, R. D. and Postma, P. W. (1975). The energetics of bacterial active transport. *Ann. Rev. Biochem.* **44,** 523–554.

Smith, D. S. (1960). Innervation of the fibrillar flight muscle of an insect, *Tenebrio molitor* (Coleoptera). *J. biophys. biochem. Cytol.* **3,** 447–466.

Smith, D. S. (1961a). Reticular organizations within the striated muscle cell. An historical survey of light microscopic studies. *J. biophys. biochem. Cytol.* (suppl.), **10,** 61–87.

Smith, D. S. (1961b). The structure of insect fibrillar flight muscle. A study made with special reference to membrane systems of the fiber. *J. biophys. biochem. Cytol.* (suppl.), **10,** 123–158.

Smith, D. S. (1961c). The organization of the flight muscle in a dragonfly, *Aeshna* sp. (Odonata). *J. biophys. biochem. Cytol.* **11,** 119–145.

Smith, D. S. (1965). The organization of flight muscle in an Aphid *Megoura viciae* (Homoptera). *J. Cell Biol.* **27,** 379–393.

Smith, D. S. (1966a). The organization and function of the sarcoplasmic reticulum and T-system of muscle cells. *Progr. Biophys. molec. Biol.* **16,** 107–142.

Smith, D. S. (1966b). The organization of flight muscle fibers in the Odonata. *J. Cell Biol.* **28,** 109–126.

Smith, D. S. (1968). "Insect cells. Their structure and function". Oliver and Boyd, Edinburgh.

Smith, D. S. and Sacktor, B. (1970). Disposition of membranes and the entry of haemolymph-borne ferritin in flight muscle fibres of the fly *Phormia regina*. *Tissue and Cell*, (21) **2,** 355–374.

Snodgrass, R. E. (1929). The thoracic mechanism of the grasshopper and its antecedents. *Smithonian misc. Coll. Wash.* **82,** no. 2.

Sutcliffe, D. W. (1962). The composition of haemolymph in aquatic insects. *J. exp. Biol.* **39,** 325–344.
Sutcliffe, D. W. (1963). The chemical composition of haemolymph in insects and some other arthropods, in relation to their phylogeny. *Comp. Bioch. Physiol.* **9,** 121–135.
Takeuchi, A. and Takeuchi, N. (1960). On the permeability of end-plate membrane during the action of transmitter. *J. Physiol. Lond.* **154,** 52–67.
Takeuchi, A. and Takeuchi, N. (1966). On the permeability of the presynaptic terminal of the crayfish neuromuscular junction during synaptic inhibition and the action of γ-aminobutyric acid. *J. Physiol. Lond.* **183,** 433–449.
Thornton, J. W. (1963). The relationship of low sodium, and herbivorous diet of Sphingidae to the ionic composition of their haemolymph and the activity and histology of their central nervous system. Ph.D. Thesis, University of Washington.
Tobias, J. M. (1948a). Potassium, sodium, and water interchange in irritable tissues and haemolymph of an omnivorous insect *Periplaneta americana. J. cell. comp. Physiol.* **31,** 125–142.
Tobias, J. M. (1948b). The high potassium and low sodium in the body fluid and tissues of a phytophagous insect, the silkworm *Bombyx mori* and the change before pupation. *J. cell. comp. Physiol.* **31,** 143–148.
Tombes, A. S. (1976). Myoneural junctions and neurosecretory endings on spermathecal muscle fibres of two weevils *Sitophilus granarius* and *Hypera postica. J. Insect Physiol.* **22,** 1573–1580.
Torida, N., Kuriyama, H., Tashiro, N. and Ito, Y. (1975). Obliquely striated muscles. *Physiol. Rev.* **55,** 700–756.
Treherne, J. E. (1962). The distribution and exchange of some ions and molecules in the central nervous system of *Periplaneta americana* L. *J. exp. Biol.* **39,** 193–217.
Treherne, J. E. (1965a). Some preliminary observations on the effect of cations on conduction processes in the abdominal nerve cord of the stick insect, *Carausius morosus. J. exp. Biol.* **42,** 1–6.
Treherne, J. E. (1965b). The distribution and exchange of inorganic ions in the abdominal nerve cord of the stick insect, *Carausius morosus. J. exp. Biol.* **42,** 7–27.
Treherne, J. E. (1976) Extracellular cation regulation in the insect central nervous system. *In* "Perspectives in experimental biology", (Ed. P. Spencer Davies) **1,** Pergamon, Oxford.
Treherne, J. E. and Pichon, Y. (1972). The insect blood-brain barrier. *Adv. Insect Physiol.* **9,** 257–313.
Uchizono, K. (1966). Excitatory and inhibitory synapses in vertebrate and invertebrate animals. *Abstr. 6th. int. Congr. Electron Micr.*, pp. 431–432, Kyoto.
Uchizono, K. (1967). Inhibitory synapses on the stretch receptor neurone of the crayfish. *Nature, Lond.* **214,** 833–834.
Usherwood, P. N. R. (1962). The action of the alkaloid ryanodine on insect skeletal muscle. *Comp. Biochem. Physiol.* **6,** 181–199.
Usherwood, P. N. R. (1967a). Permeability of insect muscle fibres to potassium and chloride ions. *J. Physiol. Lond.* **191,** 29P–30P.
Usherwood, P. N. R. (1967b). Insect neuromuscular mechanisms. *Am. Zool.* **7,** 553–582.
Usherwood, P. N. R. (1969). Electrochemistry of insect muscle. *Adv. Insect Physiol.* **6,** 205–278.
Usherwood, P. N. R. and Grundfest, H. (1965). Peripheral inhibition in skeletal muscle in insects. *J. Neurophysiol.* **28,** 497–518.

Van Asperen, K. and Van Esch, I. (1954). A simple microtitration method for the determination of calcium and magnesium in the haemolymph of insects. *Nature, Lond.* **174,** 927.

Van Asperen, K. and Van Esch, I. (1956) The chemical composition of the haemolymph in *Periplaneta americana.* With special reference to the mineral constituents. *Arch. Neerl. Zool.* **11,** 342–360.

Veratti, E. (1902) Ricerche sulla fine struttura della fibra muscolare striata. *Mem. Ist. Lombardo Cl. Sc. mat. nat.* **19** (3), 87. (Engl. tr. in: *J. biophys. biochem. Cytol.* (suppl.) **10,** 3–59, 1961).

Wareham, A. C., Duncan, C. J. and Bowler, K. (1973). Bicarbonate ions and the resting potential of cockroach muscle: implications for the development of suitable saline media. *Comp. Biochem. Physiol.* **A45,** 239–246.

Wareham, A. C., Duncan, C. J. and Bowler, K. (1974). The resting potential of cockroach muscle membrane. *Comp. Biochem. Physiol.* **A48,** 765–797.

Wareham, A. C., Duncan, C. J. and Bowler, K. (1975). The resting potential of the muscle membrane of moths. *Comp. Biochem. Physiol.* **A52,** 295–298.

Washio, H. (1971). The chloride conductance and its pH sensitivity in insect muscle. *Can. J. Physiol. Pharmacol.* **49,** 1012–1014.

Washio, H. (1972a). The ionic requirements for the initiation of action potentials in insect muscle fibers. *J. gen. Physiol.* **59,** 121–134.

Washio, H. (1972b). Calcium inward currents in insect muscle fibers. *Can. J. Physiol. Pharmacol.* **50,** 1114–1116.

Weidler, D. J. and Sieck, G. C. (1977). A study of ion binding in the hemolymph of *Periplaneta americana. Comp. Biochem. Physiol.* **A56,** 11–14.

Wood, D. W. (1957). The effect of ions upon neuromuscular transmission in a herbivorous insect. *J. Physiol. Lond.* **138,** 119–139.

Wood, D. W. (1961). The effect of sodium ions on the resting and action potentials of locust and cockroach muscle fibres. *Comp. Biochem. Physiol.* **4,** 42–46.

Wood, D. W. (1963). The sodium and potassium composition of some insect skeletal muscle fibres in relation to their membrane potentials. *Comp. Biochem. Physiol.* **9,** 151–159.

Wood, D. W. (1965). The relationship between chloride ions and resting potential in skeletal muscle fibres of the locust and cockroach. *Comp. Biochem. Physiol.* **15,** 303–312.

Woodbury, J. W. (1971). Fluxes of H^+ and HCO_3^- across frog skeletal muscle cell membrane. *In* "Homeostasis of the brain", (Ed. K. B. Siesjö and S. C. Sørensen). Munksgaard, Kopenhagen.

Wyllie, M. R. J. (1954). Ion-exchange membranes. I. Equation for the multi-ionic potential. *J. Phys. Chem.* **58,** 67–73.

Yamaguchi, H., Lockshin, R. A. and Woodward, D. J. (1972). The intersegmental muscles of silkmoth: ionic components of activity. *J. Insect Physiol.* **18,** 243–258.

Yamamoto, D. and Fukami, J-I. (1977). Ionic requirements for non-synaptic electrogenesis in the muscle fibres of a lepidopterous insect. *J. exp. Biol.* **70,** 41–47.

NOTE ADDED IN PROOF

Whilst this article was in press several relevant papers appeared concerning, the ionic composition of the haemolymph, excitatory postsynaptic potential (*epsp*) and the electrically excited response (*eer*) in insect skeletal muscle.

In the cockroach *Leucophaea maderae*, Lettau *et al.* (1977) found a diel rhythm in the potassium ion activity of the haemolymph. These changes are of an order which could affect the excitability of nerve and muscle or trigger circadian activity.

Using voltage clamp Anwyl (1977a) found the reversal potential in the extensor tibiae of the locust to be 3–4 m V positive, and $\Delta P_{Na}/\Delta P_K = 0.9$. In fibres with a dual excitating innervation the slow and fast excitatory currents have equal reversal potentials (Anwyl, 1977b).

Yamamoto and Fukami (1977) reported that in segmental muscle fibres of *Galleria mellonella* larvae the graded *eer* was converted into an all-or-none response in the presence of 10 mmol/l tetraethylammonium. The amplitudes of these spikes increased with increasing $[Ca^{2+}]_o$, but not with increasing $[Na^+]_o$ or $[Mg^{2+}]_o$. This agrees with earlier observations reviewed in 5.3.2.

The idea that in insect muscle during the *eer* the current is mainly carried by calcium ions is not supported by the passive transport theory as described by the constant field equation. Such a current cannot be simply explained in terms of a potential shift towards the equilibrium potential for calcium ions. Since the activity of the internal calcium ions is considered to be extremely low, in the constant field equation for the E_{epsp} (*cf.* section 5.3.3) the term $P_{Ca}[Ca^{2+}]_i$ could only compete with the term $P_K[K^+]_i$ if during *eer* the value for P_{Ca} was a million times higher than that for P_K, a situation which might be created with tetraethylammonium, but is highly unlikely under physiological conditions.

The *eer* is also generated during an *epsp*, at the moment P_{Na} equals P_K. Regarding the relatively low $[Ca^{2+}]$ it seems improbable that a passive calcium current could be considered to act as a serious competitor with the sodium and potassium current.

Recently Fukuda and Kawa (1977) reported that the skeletal muscle fibres of the beetle larva *Xylotrupes dichotomus*, normally showing graded responses to depolarizing current pulses, became capable of generating all-or-none reponses when, at a pH = 6.0, acetate was added to the saline in a concentration range of 8 to 64 mmol/l. In addition to acetate they found that other monocarboxylates could also be effective. This observation shows a similarity to the results described in sections 4.3 and 4.4. It would be interesting to study the possible coupling between the calcium ion transport during *eer*, and the proton transport that may be mediated by monocarboxylic acids (*cf.* section 5.2.3).

Theories of Pattern Formation in Insect Neural Development

John Palka

Department of Zoology, University of Washington, Seattle, USA

1 Introduction

This essay represents the efforts of a neurobiologist to understand theories formulated by developmental biologists for explaining the formation of patterns, and to evaluate their application to the analysis of neural development.

I will not weigh the evidence for and against the theories themselves – that task belongs to the realm of developmental biology. Rather, in Part I, I will state a number of theories as clearly as possible, together with a few of the observations which led to their formulation. In Part II I will review some of the empirical findings concerning the development of various parts of the insect nervous system, with special emphasis on sensory systems and only an appreciative nod to motor neurons. Following the description of the development of each neural subsystem is a brief summary in which possible applications of the theories of pattern formation to that particular tissue are highlighted. The essay closes with some speculations on how the various ideas and findings scattered throughout its many pages might be interrelated.

2 Part I

Collected in Part I are accounts of some half dozen theories common in the literature on insect development. These accounts are by no means complete, for a thorough review of any one of them would occupy the full length of this review. I have tried in each case to state the essential ideas of the theory and to give a sufficiently detailed description of some of the relevant experiments or observations to make the theory come alive, at least to some extent, for the reader who has no prior familiarity with the subject. Most of the empirical evidence presented in Part I is based on experiments done on the tissue which has been found most convenient by the developmental biologists who study pattern formation, namely the integument.

2.1 COMPARTMENTS

A striking empirical observation has led to the formulation of a rather detailed theory of how the bodies of adult insects might arise from discrete building blocks called *compartments* (García-Bellido, Ripoll and Morata, 1973, 1976; Crick and Lawrence, 1975; Lawrence and Morata, 1976).

The observation is based on the technique of marking a single cell and all of its progeny by genetic means. The resulting marked clone constitutes a pictorial history of a single cell and its progeny up to the time the observations are made, a history recorded in *spatial* terms. For example, if marking a single cell in the embryo resulted in the formation of several discrete patches of marked cells in the adult, substantial cell migration would be implied, for how else could the progeny of a single cell become separated from each other? In fact, continuous clones are the usual outcome of cell marking in insects, and migration in insect development is believed to occur mainly over short distances.

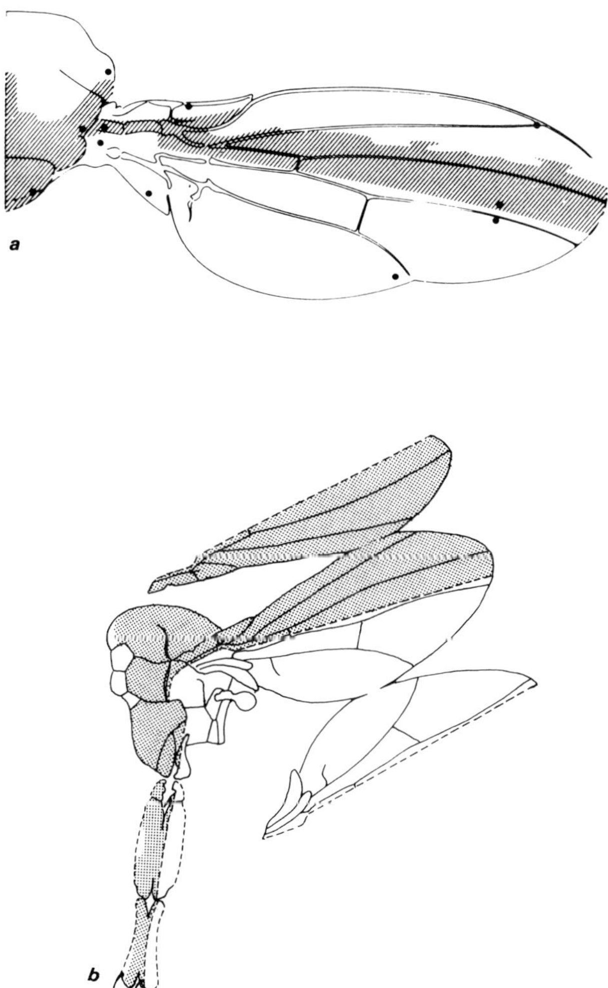

Fig. 1. Marked clones illustrating compartments. The clones (shaded) are exceptionally large because the flies were mutant for one of the Minute genes which caused their cells to divide at a slower rate than usual, while the cells of the clones were wild type for this locus, divided at the usual rate, and achieved relatively greater numbers. (*a*) A clone on the dorsal side of the wing and thorax (García-Bellido *et al.*, 1976). (*b*) A clone including both the dorsal and ventral anterior compartments of the wing, the anterior thorax and the anterior leg (Steiner, 1976). In both cases note the smoothness of the lines separating the two compartments.

Let us consider the most outstanding result of such experiments. It is observed that when single cells of *Drosophila* are marked even as early as the blastoderm stage of embryonic development and the wings of adults are studied, marked clones never cross a particular line on the wing just anterior to but not coinciding with the fourth longitudinal vein (Fig. 1). A clone in the anterior part of the wing can grow as far back as this line, and a clone in the posterior region can grow forward to meet it, but in wild type flies the line is never crossed. No structural barrier has been detected along this anterior–posterior (A–P) dividing line, and in fact the line is only demonstrable by an experimental procedure which marks clones. Because of the clonal separation of the developmental histories of the anterior and posterior wing regions, resulting in a sharp boundary between them and no detectable intermixing, the regions are called the anterior and posterior *compartments.*

Anterior and posterior compartments are not limited to the wing. They have been traced on the central regions of the mesothorax and on the mesothoracic leg, so that the entire mesothoracic segment is divided into anterior and posterior compartments (Fig. 1b). The pro- and metathorax show similar divisions (e.g. Steiner, 1976).

Within a single segment, compartmentalization appears to be a progressive matter. Clearly, if the egg itself were marked with an appropriate marker, every cell in the adult would be marked. At the other extreme, if a cell were only marked just prior to its last division, at most two marked cells would be found in the adult. It is observed, as would be expected, that clones induced later in development are generally smaller than clones induced earlier. But the exact behaviour of clones induced at intermediate stages has proven to be very interesting (Fig. 2). Early induced clones may run from the wing, along the thorax and out the corresponding leg while still remaining within

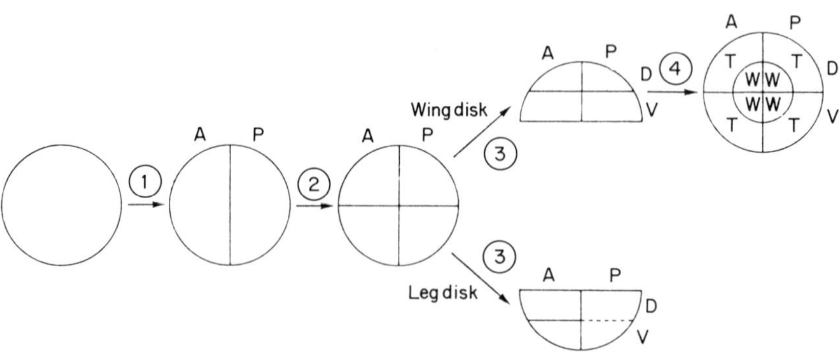

Fig. 2. The sequence of compartment formation in the mesothorax as described in the text (Morata and Lawrence, 1977).

the A or P compartment of the mesothoracic segment, as shown in Fig. 1b. Clones induced later will respect not only the fundamental segmental and A–P restrictions, but in addition will be limited according to disc: wing and dorsal thorax from the wing disc, leg and ventral thorax from the leg disc. Still later, clones will fail to cross from wing to thorax at all, and furthermore will not cross from the dorsal to the ventral surface of the wing proper. In the last described subdivision, clones on the wing itself will respect a proximo-distal dividing line. Thus, the territory which the clone derived from a single marked cell may occupy becomes progressively smaller as development proceeds. Each compartment present at a given time in development appears to subdivide into exactly two compartments at a later time (Fig. 2).

Another important feature of compartments is that at the earliest time that they can be marked they consist of several cells; at no time does there appear to be a single cell which is the progenitor of a whole compartment. This is inferred from the observation that in wild type flies a single clone never fills a whole compartment, *i.e.* when a single cell is marked some cells are always found in its compartment which are not marked and which must, therefore, have descended from a different mother cell. A rough estimate of the number of founder cells of a compartment has been proposed from the observation that early induced clones typically occupy 10–20% of any given compartment; thus, there may have been 5–10 founder cells.

If the clones induced in the anterior compartments of many different wings are mapped, their shapes and locations are found *not* to fall into clear classes. It could be, for example, that founder cells are arranged in a single row and each one buds off a row of cells lengthwise along the wing. If this were true, we should find just 5–10 classes of strip-shaped clones. Actual maps appear *not* to fit this expectation, leading to the important conclusion that the progeny of a given founder cell do not play a fixed role in morphogenesis. Indeed, one cannot speak of a "given" founder cell, because there is no evidence that that particular cell could be recognized in the next individual on any criterion that we know of. The founder cells of a compartment appear to have interchangeable roles, but all of their progeny remain within that compartment, and conversely, the compartment is not invaded by cells from any other source. For this reason, Crick and Lawrence (1975) have characterized the compartment as a polyclone – all the cells are directly descended from a *group* of founder cells.

Far reaching inferences about how the formation of compartments might be related to the information encoded in the genome have been made (García-Bellido, 1975). The immediate basis for these theories is the behaviour of several so-called homeotic mutants, in which a transformation of segments or parts of segments occurs which causes the appearance in a particular location of structures that are appropriate to a different location. For example, the mutant *bithorax* (*bx*) causes the anterior part of the haltere, the appendage of the dorsal metathorax, to be transformed into anterior wing, the appendage of the dorsal mesothorax (Fig. 3a; Lewis, 1963). The mutation *postbithorax* (*pbx*) causes posterior haltere to be replaced by posterior wing. It is found that the line which forms the posterior limit of wing tissue in *bx* is just that which in normal wings separates the anterior from the posterior compartment (compare Figs 3a and b). The same is true for the anterior

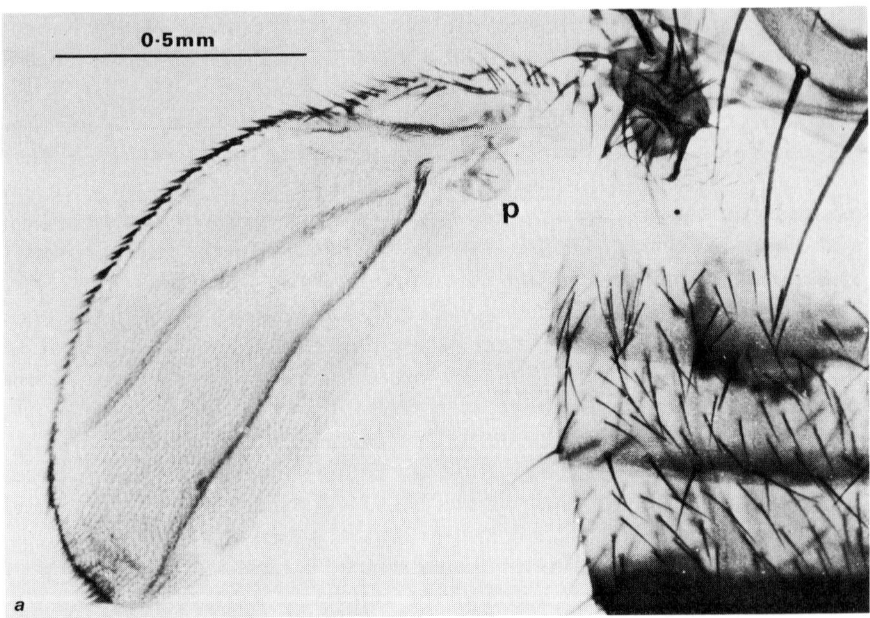

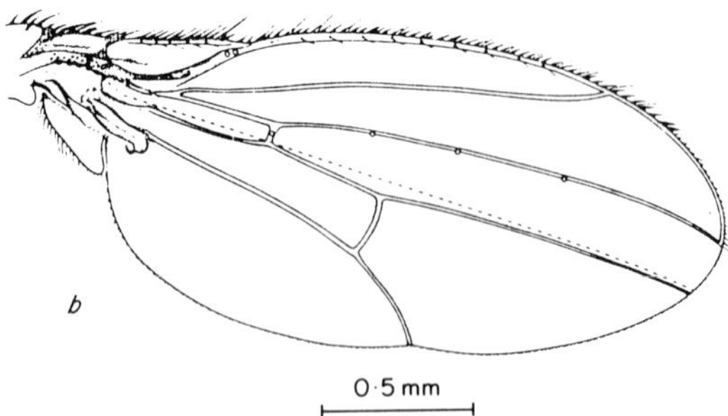

Fig. 3. The *bithorax* phenotype. In (*a*) the actual morphology can be seen: the posterior tissue (p) is still haltere, but the anterior tissue has formed wing. (*b*) Shows the A–P compartment line determined as in Fig. 1. The dividing line between wing and haltere tissue follows the A–P boundary very closely (Crick and Lawrence, 1975).

limit of wing tissues in *pbx*. The two mutants can be combined and the resulting flies develop a second pair of wings which resemble the normal wings in virtually every detail except for being slightly smaller. These mutations affect the legs and the thorax in similar ways, so that *bx pbx* flies actually develop a second mesothorax with appropriate dorsal and ventral appendages (Lewis, 1963).

Inasmuch as the difference between anterior mesothorax and anterior metathorax lies in the activity of a single gene, *bx*, García-Bellido calls this gene a *selector gene*. It is supposed to select between meso- and metathoracic quality. Mutations generally cause the alteration, diminution or loss of a gene product, so presumably the wild type function of *bx* is to prevent the anterior compartment of the third thoracic segment from assuming the structure of the mesothorax; when wild type function is lost, haltere tissue transforms into wing tissue, etc. The same argument is applied to *pbx*. In general, selector genes are supposed to be those genes that produce the difference between any two compartments which are descended from a single mother compartment. Every time a compartment splits into two, another digit is added to a binary code word (Kauffman, 1973, 1975) which describes each of the new compartments. Thus, the anterior compartment of the wing divides into the anterior-dorsal and the anterior-ventral compartment. Each of these divides in turn into anterior–dorsal–proximal and anterior–dorsal–distal, plus anterior–ventral–proximal and anterior–ventral–distal. In this progressive fashion, smaller and smaller building blocks form, each under the influence of the activity of some small set of selector genes which control the presumably large number of structural genes which are necessary for the building of cells and cell products. (For an interesting hypothesis of how compartments might arise, see Kauffman, 1977, and Kauffman *et al.*, 1978.)

2.2 POSITIONAL INFORMATION

The idea that the *location* of a particular cell in the body governs its differentiation is an old one in developmental biology. In the past decade, however, this basic idea has attracted new attention, largely because of its more explicit formulation under the name of *postional information* by Wolpert (1969, 1971).

The central idea of positional information is perhaps best summarized by contrasting it with other ways in which the differentiation of a cell might be controlled. An extreme alternative would be that the pedigree of a cell – the exact sequence of cell divisions which led to its production – is what counts. Another alternative is that structural patterns are produced by a chain of inductive processes – cell C_1 induces "twoness" in adjacent cell C_2, which induces "threeness" in cell C_3, etc. This, too, has been proposed in some cases: one, the induction of bracts by bristle organs, is discussed later in Part I; another, the retina, is discussed in Part II.

In contrast, the hypothesis of positional information proposes that animals are built of regions with boundaries, and that location between boundaries is a continuous variable encoded in some physical–chemical form that can be recognized by a given cell. The cell responds to the information that, e.g. it is at grid point Right 5/Anterior 7 by differentiating, let us say, a glandular secretion. If that particular cell were removed and another one from the same region substituted, it too would produce a glandular secretion. A cell from some other part of the body, however, if transposed to that same location, might differentiate a hair rather than a gland. Thus, there are two consecutive processes operating within the cell: recognition of its position, and the response to that positional information. In a population of cells, the outcome of these two processes is a pattern, be it the markings on the wing of a butterfly or the particular bristle pattern which characterizes the foreleg of a male of *Drosophila* or, in principle, the pattern of vibrissae on a rodent's snout.

A number of phenomena are nicely compatible with such a view. For example, it is observed that when clones of homeotically transformed cells are formed on *Drosophilia* appendages, the structures they develop are systematically related to their location. An anteriorly situated clone of wing tissue on the haltere forms anterior wing structures, not posterior ones. Clones of

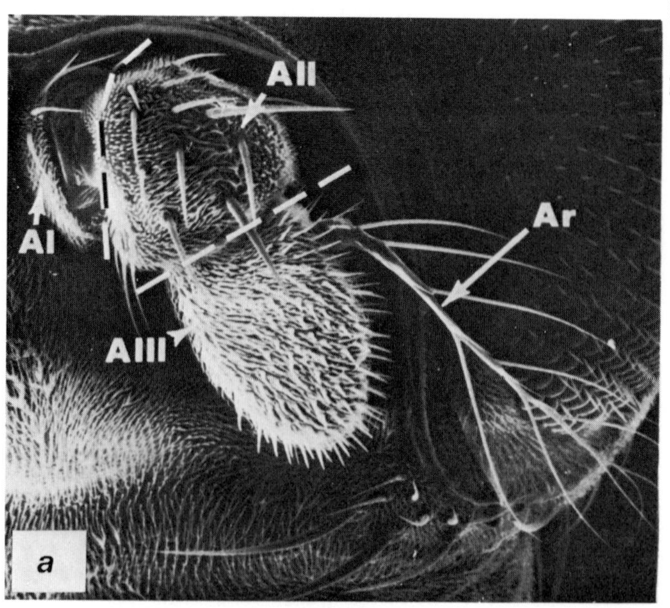

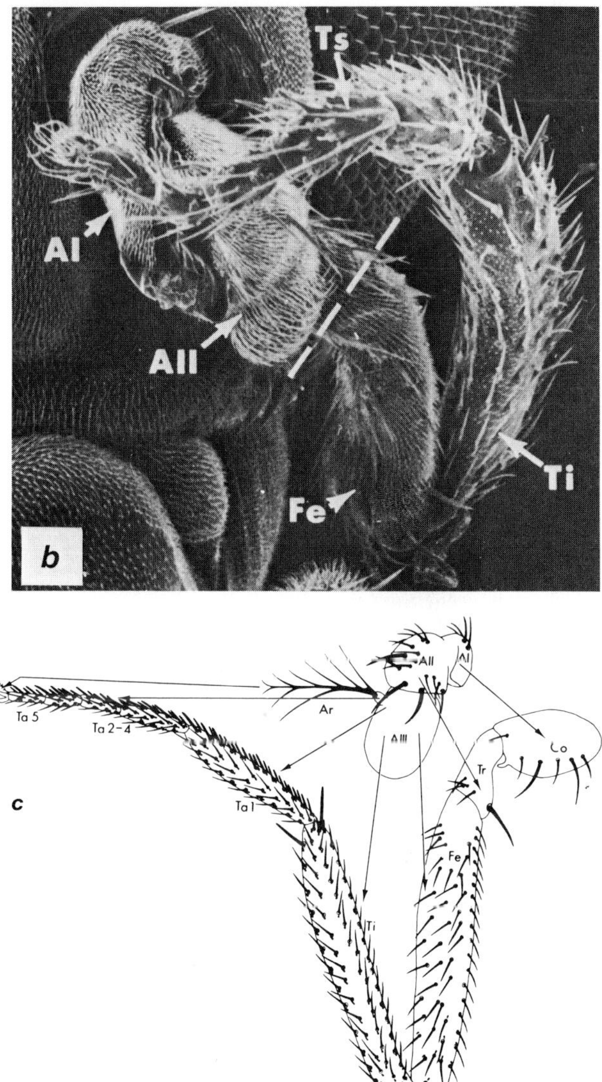

Fig. 4. The *Antennapedia* phenotype. (*a*) Scanning electron micrograph of a wild type antenna; the three antennal segments are marked AI, AII and AIII; Ar is the arista. (*b*) Scanning electron micrograph of the transformed appendage; Fe, Ti and Ts mark femur, tibia and tarsus respectively. The dashed lines in (*a*) and (*b*) show where cuts were made to produce experimental degeneration of the afferent axons (see Fig. 32). (Stocker *et al*., 1976.) (*c*) A diagram showing the homologies of leg and antennal regions determined by mapping which leg structures appear in which regions of the antenna in phenotypically mosaic appendages. Such mosaics are frequent; the mutation is said to show variable expressivity (Postlethwait and Schneiderman, 1971).

leg tissue on the tip of the antenna form tarsal claws, not any of the more proximal leg structures (Fig. 4). Many examples could be adduced for this generalization.

Wolpert has suggested (1969) that the physical–chemical nature of positional information may well be universal in the same sense that the genetic code is universal. The morphogenetic response to particular positional values would vary from appendage to appendage and species to species, of course, but butterfly wing tissue should be able to recognize and respond in orderly fashion to hummingbird leg positional information.

The hypothesis of positional information embodies no assumptions about what the nature of the code might be. Historically, the first guess was a system of gradients (Wolpert, 1969, 1971), but the more recent polar coordinate model (French *et al.*, 1976) provides an alternative; both are discussed below. Similarly, the hypothesis does not explain what makes a leg cell different from an antennal cell in the first place – that difference is assumed.

2.3 GRADIENTS

The first explicit formulation of a gradient hypothesis for insect epidermis was that of Locke (1959). Its essential observations were as follows. A small

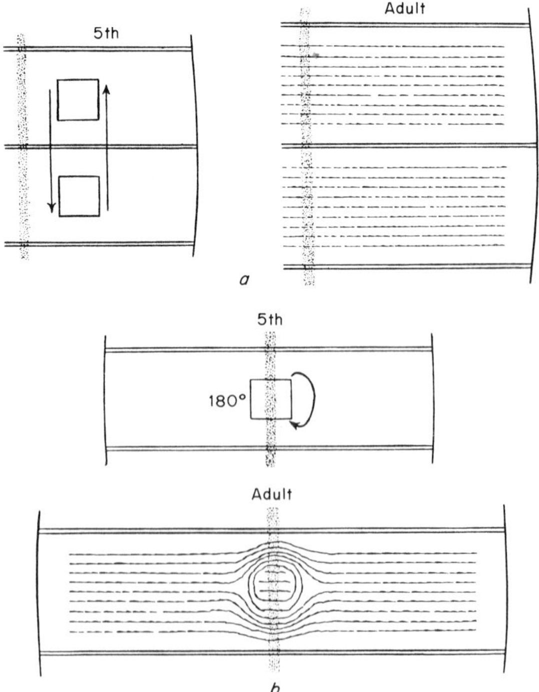

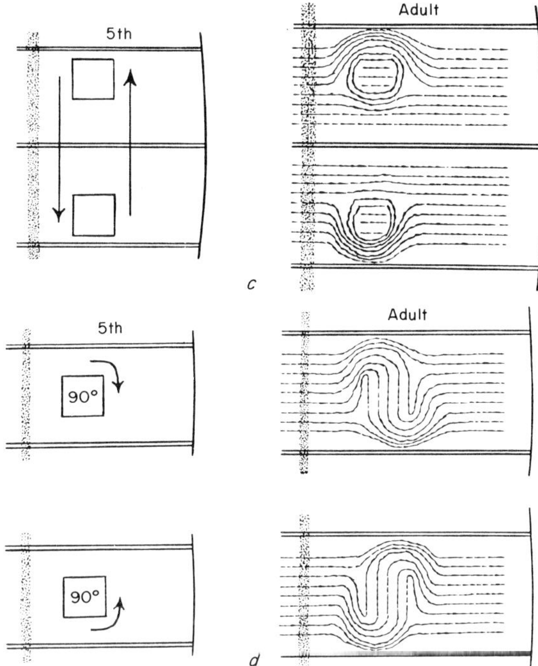

Fig. 5. The effects of transposing or rotating pieces of larval abdominal integument upon the cuticular ripple pattern of adult *Rhodnius*. The double lines indicate the segment margins, the stippled bar the midline. Note that the drawings of the ripple patterns are simplified, and in particular (*b*) does not show that the ripples within the island form two concentric patterns, one anterior and the other posterior. These appear in Locke's photographs and descriptions and are shown in the simulations of Fig. 8 (Locke, 1959).

square of cuticle from the abdomen of a larval *Rhodnius* was transplanted in the same orientation and (*a*) to the same antero-posterior level of the same segment, or (*b*) the same level of an adjacent segment. In these operations, the pattern of transverse ripples which sculptures the surface of the cuticle of the adult insect was undisturbed (Fig. 5a). If, however, the square of larval cuticle was (*a*) rotated by 180° before reimplantation into its original site, or (*b*) grafted to a different antero-posterior level, the graft looked as if it were isolated and the ripple lines of the host were deflected to pass around its edges (Figs 5b, c). If the square was rotated by 90° before reimplantation, an S-shaped pattern was produced in which the ripples of the host and graft became confluent at the original lateral margins of the graft (Fig. 5d). From these and other systematic experiments Locke inferred the presence of a segmentally repeated, axially oriented gradient whose effect was to make cells of a given antero-posterior level in the segment seek contact preferentially with other cells of the same level. Locke then (1960) presented some additional evidence for the role of intersegmental membranes in separating the gradient of one segment from that of the adjacent ones.

It may be useful at this point to make a clear distinction, already recognized by Locke, between evidence for *polarity* in a piece of tissue and evidence for the presence of a *gradient*. If rotation of a piece of tissue results in some morphogenetic effect (as in Figs 5b and d), this could be either because the tissue or each cell within it has information about direction (for example because the dorsal pole of each cell is different from the ventral pole) or

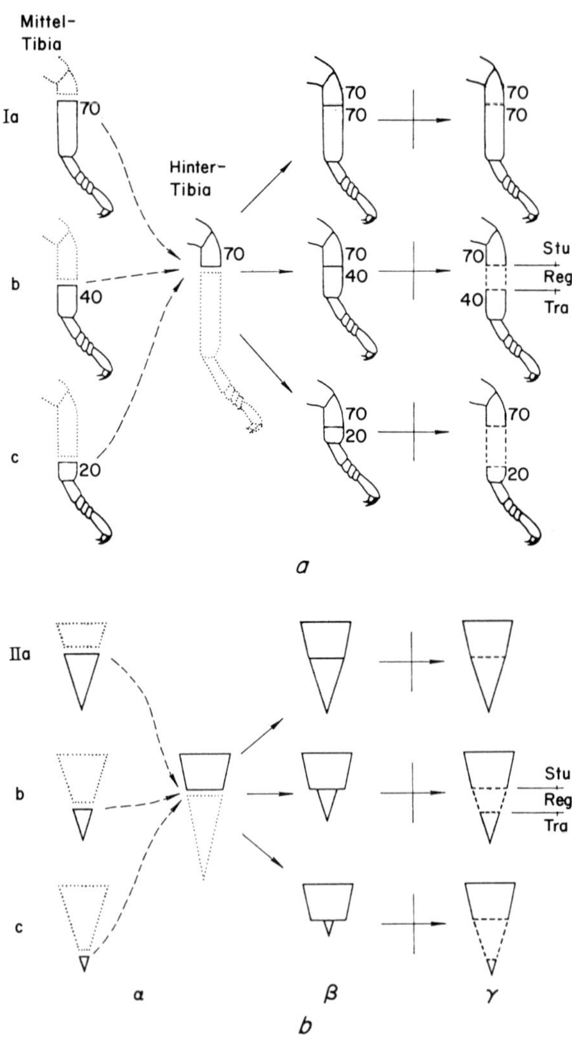

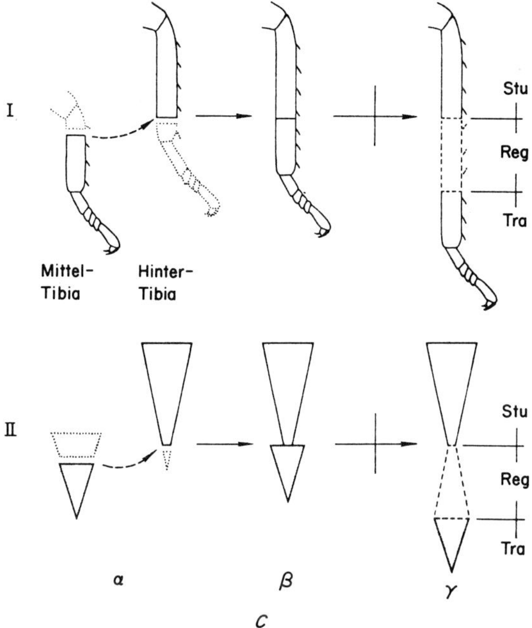

Fig. 6. The effects of grafting together leg segments of different lengths. (*a*) Experimental results. The numbers indicate the level of the tibia at which a cut was made, solid outlines tissue which was retained and dotted lines tissue which was removed. The initial appearance of the grafts is shown in the third column, their final appearance in the last column. Intercalary regeneration compensates for the missing levels in b and c. (*b*) The interpretation of these experiments in terms of a longitudinal gradient. (*c*) Long tibial fragments grafted to long stumps, and the interpretation of the results in terms of a gradient (Bohn, 1970a).

because each cell is associated with some factor whose value or nature is different at every level from the most dorsal to the most ventral (or along any other axis). An additional experiment is required to distinguish between these two alternatives – for example, a translocation experiment such as Figs 5a and c. In this case the orientation of the graft relative to the host is not altered, only its location is changed. If polarity were the only important factor, this procedure should produce no morphogenetic effect, but if a gradient is involved, an effect is expected. The distinction between the two is clearly important in principle, but can be technically difficult to establish when working with very small pieces of tissue.

Gradients have been supposed to exist not only in the various segments along the main axis of the body, but also in the proximo-distal segments of body appendages. Strong evidence for this supposition comes from the work of Bohn (1970a, b; 1971). If a particular leg segment is transected and the distal stump grafted to a different leg but at the same fractional level of either the same or a different segment, the tissue simply heals and a new, compound appendage is established (Fig. 6a). If a short distal piece is grafted onto a short proximal stump, intercalary regeneration (regeneration between the two original pieces of tissue) occurs until the length of the segment is restored to normal. The interpretation of these experiments in terms of gradients is shown in Fig. 6b: intercalary regeneration occurs to the extent required to re-establish the slope of the gradient.

An alternate explanation of such experiments, however, is that the segment has some other way of measuring its length, intercalary regeneration occurs until normal length is restored, and the hypothesis of a gradient fits the results only coincidentally. Therefore, the following experiment has been taken as particularly strong evidence for the gradient hypothesis. If a long distal piece is grafted onto a long proximal stump, so that the segment and the whole leg are made longer than normal, intercalary regeneration still occurs (Fig. 6c). What is more, the polarity of the regenerate tissue is opposite to that of the leg – its bristles and spines point towards the body rather than away from it. The interpretation of this striking result in terms of gradients is shown in Fig. 6c: the operation introduces as discontinuity in the gradient and intercalary regeneration occurs until the apposed high-value and low-value regions are connected by a gradient of normal slope. This is precisely the same explanation as was given for the cases of Fig. 6a, except that the geometry of the situation requires a reversal of polarity within the regenerate segment.

Experiments of this type, while they provide persuasive evidence of the existence of gradients in abdominal segments, leg segments, and by extension any discrete region of the body, offer no direct indication of the nature of the gradient. At least two different kinds of gradients have been proposed and vigorously defended: (*a*) gradients of diffusible materials, and (*b*) gradients of a non-diffusible cellular property, its adhesiveness. These are not mutually exclusive ideas – for example, a diffusion gradient present at one time during development could result in the establishment of a gradient of adhesiveness which provides the immediate forces influencing pattern formation and reconstitution later, and particularly during the postembryonic stages when experiments are usually done. The discussion of the next two sections, therefore, concentrates on presenting both ideas in the experimental context which led to their formulation.

2.4 DIFFUSION GRADIENTS

One possible physical–chemical embodiment of positional information is a gradient of a diffusible substance, produced at a geographically discrete source and destroyed at a sink. If the concentration of the substance could be detected by cells lying between source and sink, the position of those cells could be specified. For more detailed accounts of diffusion-based models than is given here, see the reviews by Locke (1967) and Lawrence (1973).

A diffusible substance with a morphogenetic influence (morphogen) is qualitatively suggested, for example, by the behaviour of cuticular hairs in the vicinity of an interruption of the intersegmental membrane in the abdomen of *Oncopeltus* (Fig. 7), as described by Lawrence (1966a). The hair pattern

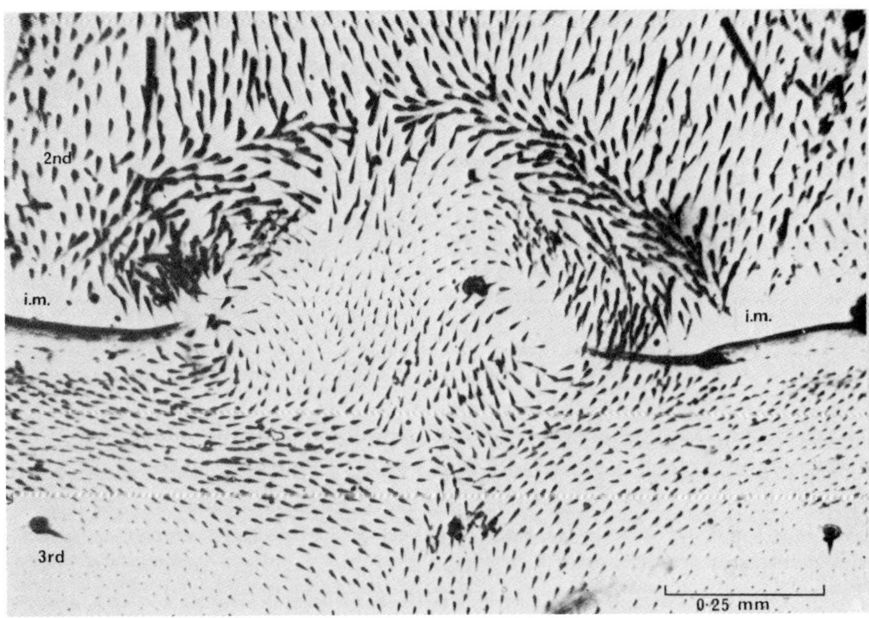

Fig. 7. The bristle pattern in *Oncopeltus* around a naturally occurring interruption in the intersegmental membrane (Lawrence, 1966a).

strongly suggests the spreading of a hair orienting effect from one segment into the adjacent one. Similarly, Piepho (1955) had found that implanting small pieces of intersegmental membrane into the middle of a segment in the moth *Galleria* produced an orderly and widely distributed effect on hair orientation.

Lawrence (1966a) showed that a gradient model which assumed the diffusion of some substance having an influence on surface anatomy could formally account both for his results on *Oncopeltus* and those of Piepho on *Galleria*, and also for the results of Locke (1959) (Fig. 5) on *Rhodnius*, which had originally been interpreted in terms closer to the concept of differential adhesiveness (see below). The same conclusion was reached by Stumpf (1965a, b; 1966a, b; 1967), and opposed by Locke (1966a, b) in a spirited

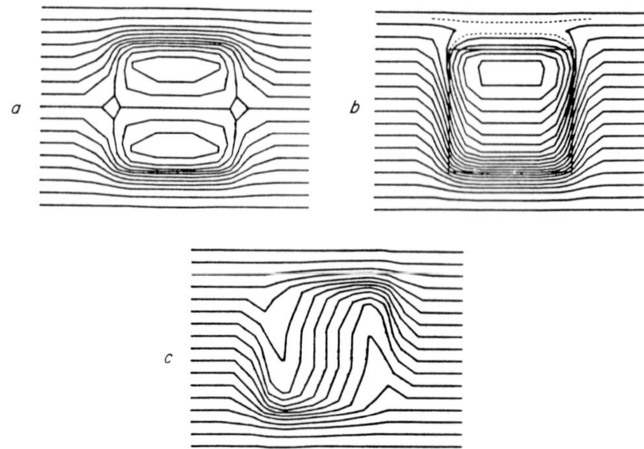

Fig. 8. The predicted ripple patterns of adult *Rhodnius* based on a model assuming a linear gradient of a diffusible morphogen between the segment boundaries; the experimental manipulations are the same as those of Fig. 5: (*a*) 180° rotation, (*b*) Translocation to a more posterior level, (*c*) 90° rotation (Stumpf, 1967).

exchange. The degree to which Stumpf's calculations matched Locke's results can be seen by comparing Figs 8 and 5.

Substantial detailed modelling of diffusion-based gradient models was conducted by Lawrence, Crick and Munro (1972), and the results were tested against the effect on the pattern of cuticular ripples in *Rhodnius* of translocation or rotation of pieces of cuticle. Three types of models were considered: 1. The anterior and posterior margins of a segment actively maintain different levels of a morphogen, and the resulting gradient between them is linear, the concept developed particularly by Stumpf. 2. A gradient is established within the segment, but in addition each cell exhibits a homeostatic tendency to maintain its own, previously established internal concentration of the morphogen. Thus, it responds to transplantation down the gradient where its morphogen diffuses away by producing more than before, and to transplantation up the gradient by producing less and becoming a sink for morphogen from surrounding cells. At each cell division the cell's target concentration of morphogen is reset to the actual concentration present in the cell at that time. If the cells of a graft do not divide, their target concentrations remain constant. 3. Cells actively transport morphogen to their neighbours against the local slope of the gradient, a suggestion made by Lawrence (1966a). Models 2 and 3 share the feature of stability over time. In model 1, the cells of a small graft in a large field should succumb to the morphogen concentrations generated within that field, which does not happen. Quantitative variation of parameters in these three classes of models showed that while class

1 models showed generally the right behaviour, class 2 models produced the closest match to the ripple patterns resulting from a variety of translocation and rotation operations. Some experimental evidence was given for the role of cell divisions.

2.5 GRADIENTS OF ADHESIVENESS

Recently, Nardi and Kafatos (1976a, b) have drawn renewed attention to the possible role of a gradient in adhesiveness similar to but more explicit than that originally postulated by Locke. Their model derives from results of grafting experiments on the wing of the moth *Manduca* and emphasizes phenomena of a somewhat different kind than we have reviewed thus far.

The adhesiveness gradient model proposes that: 1. Surface adhesiveness is a cellular property which is non-diffusible and is conserved in the face of surgical experiments. 2. There is a gradient in cell adhesiveness in the moth wing, the adhesiveness of proximal cells being the greatest. The gradient is non-linear, falling off more steeply proximally than it does distally. Thus, the difference in adhesiveness between any two cell populations increases with increasing separation of their location in the gradient. Semiquantitative predictions from these assumptions are set out by analogy with the behaviour of mixtures of immiscible liquids.

One class of predictions relates to effects on the size and shape of a piece of graft tissue of its source and its final location. For example, it is predicted that square graft pieces translocated either up or down the gradient should tend to become rounded. If the cells of the graft are more adhesive than those of the surrounding cells, they will move so as to maximize contact with each other. If the surrounding cells are the more adhesive, they will force out the cells of the graft. A graft should also become smaller and smaller with time, for precisely the same reasons. In fact, the forces acting at the boundary can be thought of as a pressure. The effects on shape and size should be greater as the difference between the site from which a piece of tissue is taken and the site in which it is implanted increases. A sampling of data illustrating these trends is given in Fig. 9.

In addition to changes in shape and size of reoriented or translocated graft tissue, changes in its internal pattern occur. The scales which form the patterns on moth wings are arranged approximately parallel to the veins, thus proximo-distally on the wing. Figure 10 shows the result of 180° rotation of a square piece in a proximal (*a*) and distal (*b*) region of the wing. The rosette configuration seen in (*a*) also occurs when proximal tissue is transplanted to a more distal site. It usually does not appear when distal tissue is transplanted into a proximal location. It is also less and less evident after

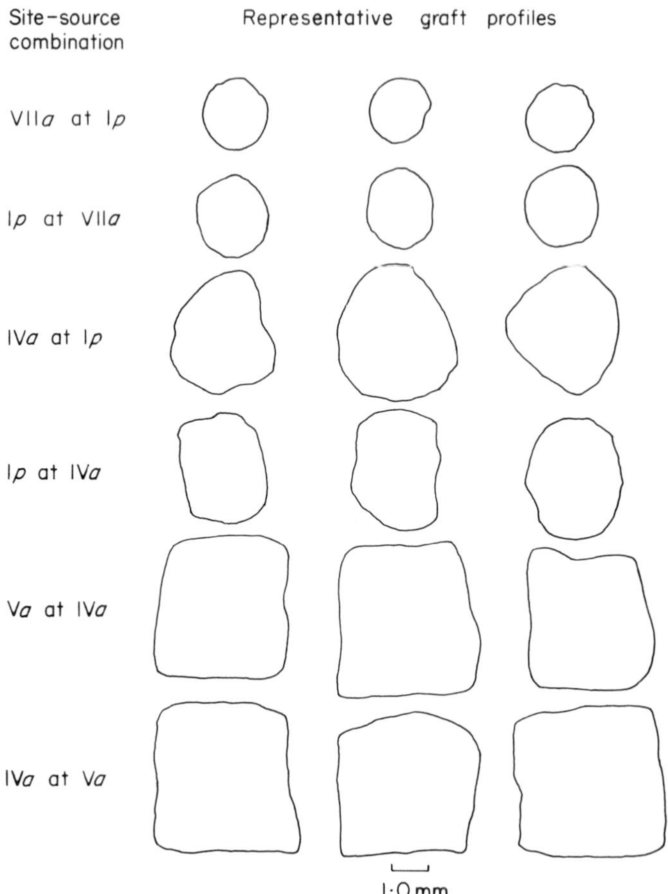

Fig. 9. The relationship of the size and shape of a graft of wing epidermis in *Manduca* to the separation between the source of the graft tissue and its site of implantation. The wing regions were numbered with I being proximal and VII distal, *a* anterior and *p* posterior. The original pieces were all squares of approximately the same size. When the graft was moved only a short distance, its appearance did not change much; when moved a long distance in either direction, it became smaller and rounder (Nardi and Kafatos, 1976a).

180° rotation as one proceeds towards the tip of the wing, as can be seen by comparing (*a*) and (*b*).

The explanation offered by Nardi and Kafatos for these observations is as follows. The highly adhesive proximal cells will produce the expected rounded graft shape, clearly seen in (*a*). But because within the graft tissue itself the more proximal cells are more adhesive than the more distal ones, they will tend to migrate into the interior of the graft while the somewhat

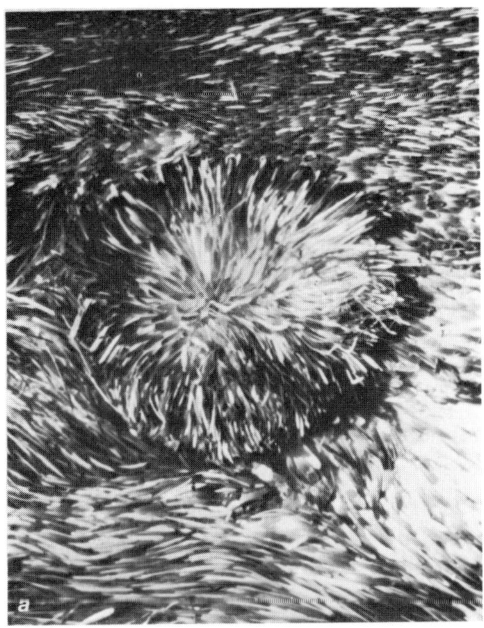

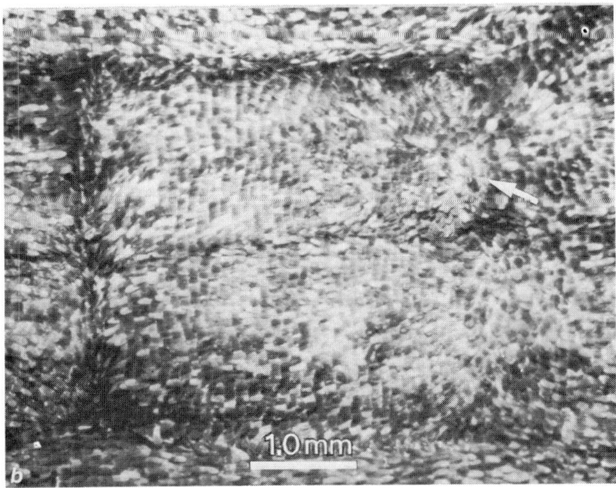

Fig. 10. Effects of 180° rotation of square pieces of epidermis in *Manduca*. Rotation and reimplantation in a proximal location (*a*) produces contraction and rounding of the graft and a rosette configuration of the scales. In a distal location (*b*) the reversed polarity of the graft is evident but the original size and shape are maintained and only a small rosette is formed (arrow). (Nardi and Kafatos, 1976a.)

less adhesive distal ones will tend to accumulate at the surface, at the interface between graft and host tissue. The end result of these movements is a more or less radial or rosette shaped pattern of scales. The same arguments apply to cell migrations initiated by 180° rotation. The effect is much less marked distally (Fig. 10b), where grafts neither round up as much as they do proximally, nor form such conspicuous rosettes. This is explained by supposing that the slope of the adhesiveness gradient here is shallower than in proximal regions.

In these experiments, as in all others dealing with gradients or other embodiments of positional information, the spatial organization of the mechanisms which ultimately produce a pattern is inferred from rather indirect experiments. A gradient of a factor having a direct influence on morphogenesis in insects has not been demonstrated. In fact, a gradient in adhesiveness has not been directly measured. But there are many other observations which indicate that cells from different regions of a given appendage, or from different appendages, probably do differ in adhesiveness (e.g. García-Bellido, 1966). There is also very direct evidence that cell migration does occur as a result of graft rotation.

As is described in the next section, if a limb is transected and grafted back to the original site following rotation, it will de-rotate during subsequent moults. The evidence that this occurs at the graft junction is very strong

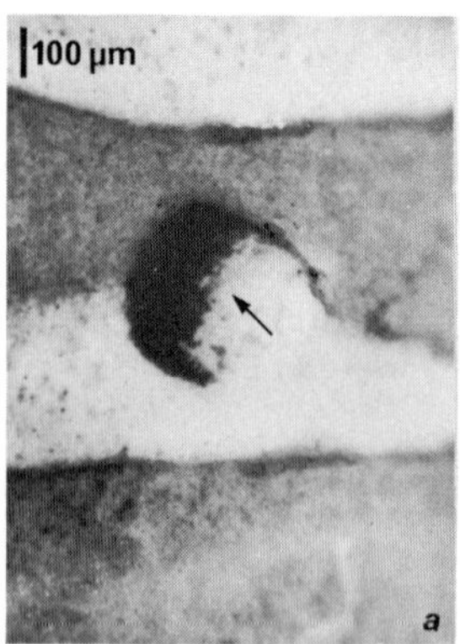

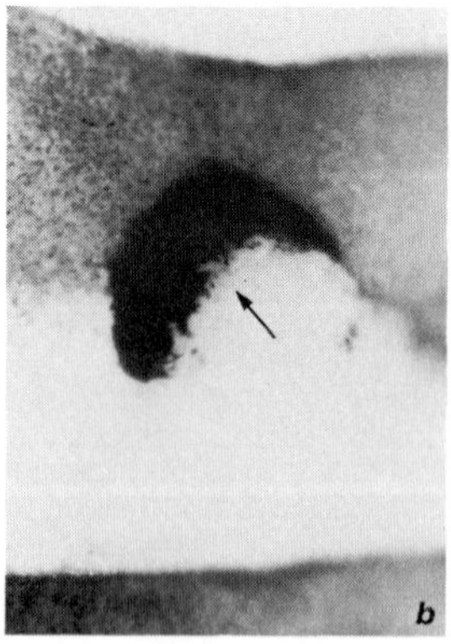

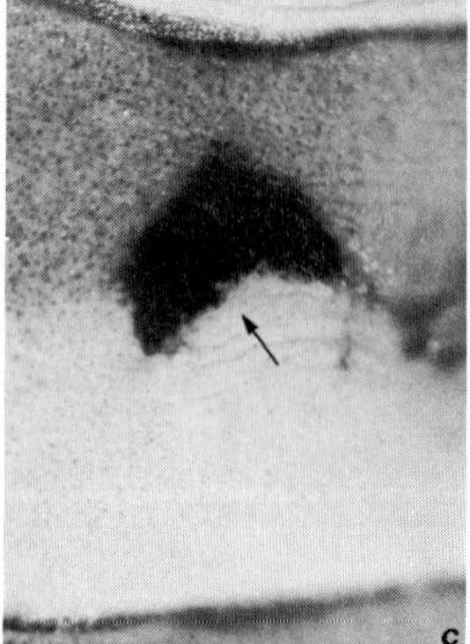

Fig. 11. Physical de-rotation of a graft in an abdominal segment of *Dysdercus*. The anterior half of the host's segment is grey (because of a colour mutation), the posterior half is white. The anterior half of the graft is red (black in the photograph) because it was taken from a wild type animal, while the posterior half is white. The graft was originally implanted with a 90° rotation in the 3rd larval instar. (*a*) shows its appearance early in the 4th instar, (*b*) late in the 4th instar and (*c*) shortly before metamorphosis. The cuticle is transparent, so the pictures reveal the underlying cells; the arrows point to an irregularity in the red–white border which can be followed throughout the sequence (Nübler-Jung, 1974).

(Bohn, 1965). Similarly, there are reports of the de-rotation of grafts of abdominal cuticle (Bohn, 1974; Lawrence, 1974; Nübler-Jung, 1974). In the last of these, the experimental material (*Dysdercus*, Heteroptera) was especially well suited to cytological observation, because the cuticle is transparent so that the pigmented epidermal cells underneath can be seen. Time lapse films showed the patterns of cell migration in the graft tissue which led to a reorientation of the whole graft (Fig. 11).

In a recent detailed study, Nübler-Jung (1977) has returned yet again to the types of experiments which led Locke (1959, 1960) to postulate a segmental gradient. Her work, not summarized here because of length limitations, re-emphasizes the observation of physical de-rotation of previously rotated grafts, and interprets the guiding influences for this de-rotation in terms of the gradient of adhesiveness set forth explicitly by Nardi and Kafatos (1976a, b). In addition, she presents evidence for intercalary regeneration in places where tissues of different axial origins are apposed. This would be expected both from a gradient model as envisioned by Bohn (see above) and from a polar coordinate model as postulated by French *et al.* (see below).

2.6 POLAR COORDINATE MODELS

A number of experimental observations suggest that cells around the circumference of an appendage such as a leg, a cercus or an antenna respond to positional information which tells them where around the circumference they are located. For example, if the tibia of a cockroach is transected in the middle and the distal part is grafted back to the stump but with a 90° rotation, the graft will be found to de-rotate back to its original correct orientation (Bohn, 1965; Fig. 12a). Usually 2 or 3 moults are required for the re-establishment of the proper alignment between the two parts. If a similar operation is performed, but instead of grafting the distal part of the leg back to its original stump it is grafted to a similar stump on the opposite side of the body, de-rotation does not occur. Instead, two supernumerary legs form at the graft junction. Their locations are predictable: in the case of simple left-right exchange, they form on the anterior and posterior faces of the leg; if the graft is also inverted dorso-ventrally, the supernumeraries form dorsally and ventrally (Bohn, 1965; Figs 12b, c).

Phenomena of this type clearly have something to do with position-dependent properties of cells. Bohn supposed that there are gradient systems in the leg not only in the proximo-distal axis, as described in the previous section, but also from dorsal to ventral and anterior to posterior. Cell position around the circumference would then be specified by two numbers, one for each of the two gradients present in the transverse plane of the leg.

More recently a different set of coordinates has been proposed (French

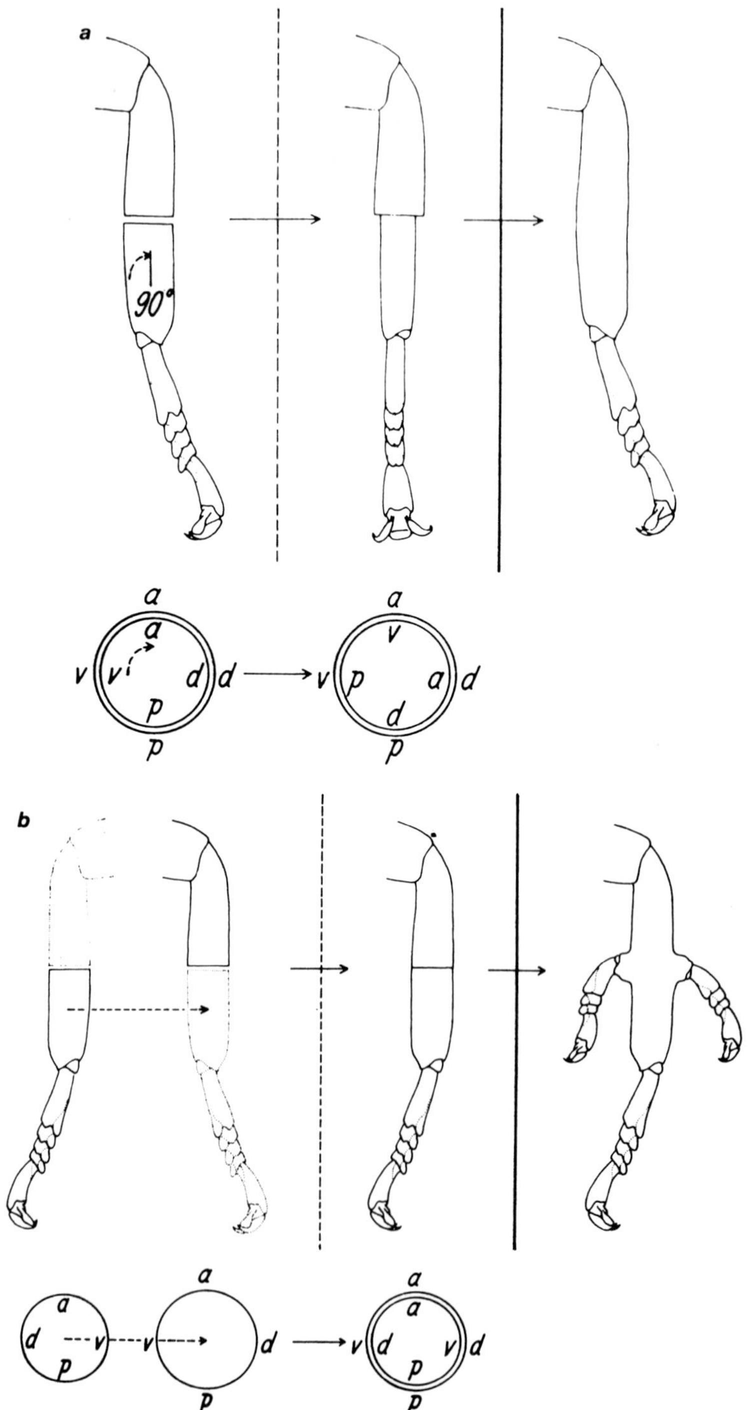
a
90°
a
a
v
v
d
d
p
p
a
v
v
p
a
d
d
p
b
a
d
v
p
v
a
d
p
a
a
v
d
v
d
p
p

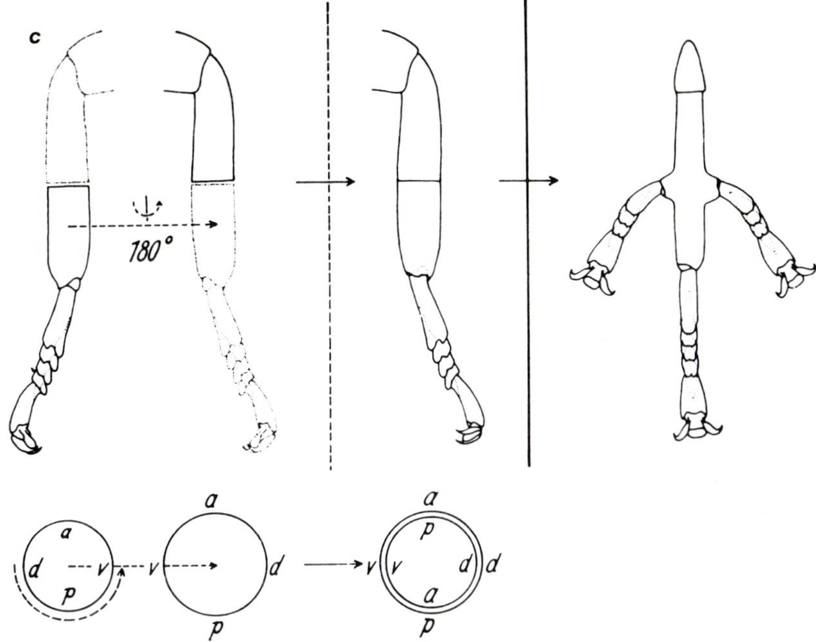

Fig. 12. Effects of leg rotation and exchange in *Leucophaea*. (*a*) Rotation without exchange is followed by de-rotation. (*b*) and (*c*) Supernumerary legs are formed following left–right exchange in locations where graft and host show maximum misalignment (Bohn, 1965).

and Bullière, 1975a, b) and a rather explicit model developed which describes and predicts phenomena seen not only in cockroach legs but also in *Drosophila* imaginal discs and in the regenerating legs of amphibians (French *et al.*, 1976; Bryant *et al.*, 1977).

The original observation leading to this model was as follows. If a thin strip of cuticle from one face of a cockroach femur is taken out and implanted in a different circumferential location in another leg, intercalary regeneration occurs at the graft margins (French and Bullière, 1975a). The structures which form are always those which establish the proper circumferential *order* between the two cut surfaces which the operation has brought into contact. Thus, in Fig. 13 where locations around the circumference have been assigned the numbers 1–12, inserting region 4, 5, 6, 7 between regions 8 and 9 has stimulated the growth of two intermediate strips of tissue: region 8 between face 7 of the graft and face 9 of the host; and regions 5, 6 and 7 between face 4 of the graft and face 8 of the host. Regeneration leads to a filling in of gaps by the shortest possible route. Furthermore, this finding holds irrespective of the position of either the graft or the implantation site.

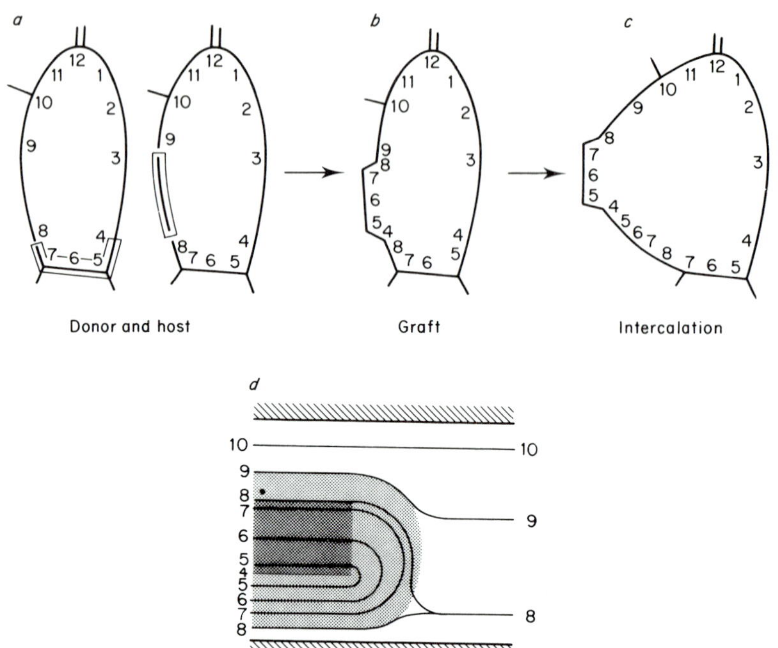

Fig. 13. Effect of grafting a longitudinal strip of femur tissue in *Blabera* into a new location: intercalary regeneration occurs as described in the text. The schematic cross-sections of the leg show some of the bristles used as positional markers. (*d*) shows the way graft and host tissues interact at the end of the strip (Bryant *et al.*, 1977).

It is this last observation that permits the differentiation of a polar coordinate model from a gradient model. A gradient must have a high end and a low end. In the polar coordinate model, however, there is no discontinuity around the circumference, just as there is no discontinuity on the face of a clock. We use the numbers 1–12 to identify the hours of the day, but the difference between 12 and 1 o'clock is just like the difference between 3 and 4 o'clock – just one hour of time which flows uninterruptedly.

In principle this is a clear-cut distinction between gradient and polar coordinate models of positonal information, but in practice difficulties arise. Not all regions of a limb circumference are quite equipotent, so it has been suggested that the numbers on the face of this "clock" need not be equally spaced (French *et al.*, 1976). However, once this is permitted the operational difference between the two classes of models necessarily becomes blurred.

Regeneration by the shortest distance is one of the two principal postulates of the polar coordinate model. The other is distal regeneration from any complete circumference. The application of this rule permits the prediction of the location of supernumerary regenerates. As described above, the empirical

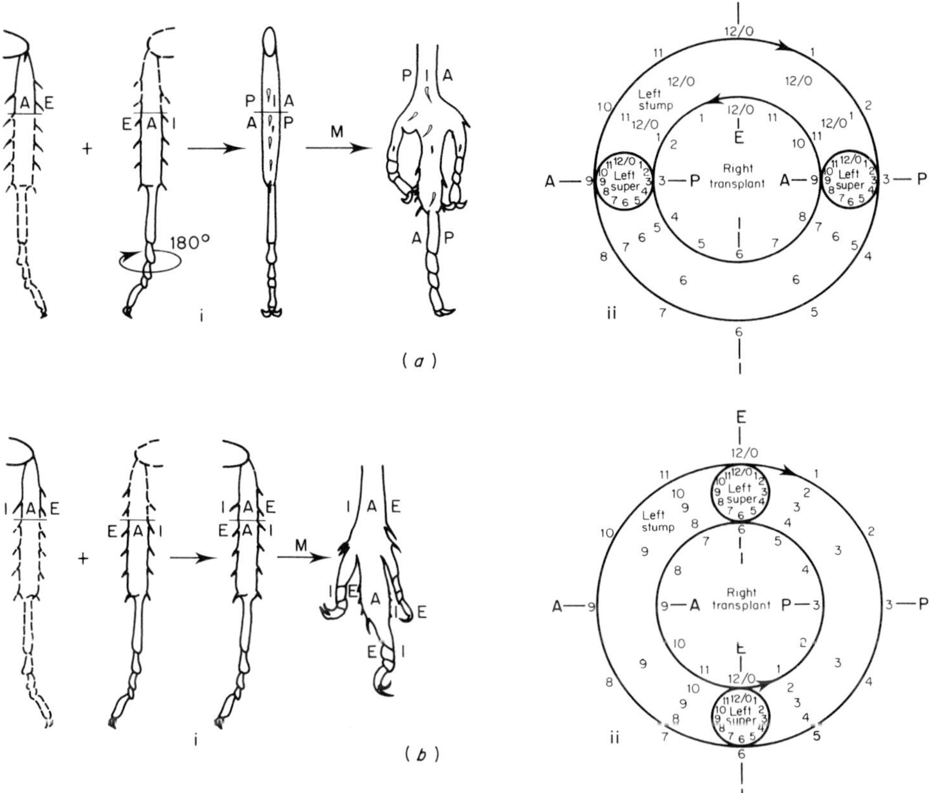

Fig. 14. The predictions of the polar coordinate model for occurrence and location of supernumerary appendages in exchange operations with and without 180° rotation. I, E, A and P refer to internal, external, anterior and posterior faces of the femur respectively. Further details in the text (French *et al.*, 1976).

observation is that supernumerary regenerates are formed when tissues which normally lie on opposite faces of the leg are brought together by the graft. When a right leg is grafted to a left stump, dorsal and ventral regions in host and graft are properly matched, but anterior and posterior faces are opposed. In Fig. 14a, 12/0 matches 12/0 dorsally and 6 matches 6 ventrally, so no regeneration occurs at these points. As one approaches the horizontal plane, however, regeneration by the shortest distance does occur between the graft (inner circle) and host (outer circle). Regeneration at the point where the difference between the apposed clock numbers of graft and host is the greatest requires the formation of a complete series of numbers, a complete circumference. It is here that supernumerary limbs form, as they should if distal growth occurs from a complete circumference. Of course, the same rule predicts that regeneration should occur following simple amputation, as indeed it does.

The limbs of holometabolous insects develop from imaginal discs. It is very gratifying, therefore, that the results of a wide variety of surgical experiments on the imaginal discs of *Drosophila* are in accord with the polar coordinate model. The details are reviewed by French *et al.*, (1976).

2.7 INDUCTION

Induction is the term applied to an interaction between two tissues – the inducing tissue and the tissue which is competent to respond – which results in the responding tissue acquiring some property or form which it would not acquire without the proximity of the inducing tissue. The term can equally well be applied to a similar interaction between two cells; the scale of the process does not matter.

One of the clearest examples of induction in insect embryogenesis is the influence of ectoderm on the development of mesodermal structures in the lacewing, *Chrysopa*; the following description is based on the account in Kühn (1965). In Fig. 15 are summarized the effects of removing either mesoderm or ectoderm during the formation of the embryonic ventral groove. Parts a_1, a_2 and a_3 show the normal course of development: the medial plate (hatched) will give rise to mesodermal structures such as fat body, various groups of muscles and the midgut; the lateral plate will give rise to the epidermis and the tracheae. If all or part of the mesoderm is removed by cautery with a hot needle prior to its inward migration through the ventral groove (parts b, c and d), the corresponding sets of mesodermal structures fail to form but the overlying epidermis forms normally. It develops its usual histological structure though its shape is distorted, apparently because the internal organs which would normally stretch it out are lacking. However, if ectodermal regions are destroyed (parts e, f and g), the result is quite different. Not only is the epidermis corresponding to the lesion lost, but the underlying mesodermal structures also fail to form. Apparently there is a non-reciprocal interaction between ecto- and mesoderm at this stage of development: ectoderm can differentiate normally in the absence of mesoderm, but mesoderm requires the presence of ectoderm. Ectoderm is the inducer and mesoderm is competent to respond to it. A number of other phenomena interpreted as induction are summarized by Counce (1972).

An example of particular interest to neurobiologists because it occurs on a cellular scale is provided by Tobler's (1966) analysis of the developmental interactions between bristle organs and their associated bracts in *Drosophila*. As is described in greater detail in Part II, insect bristles are typically

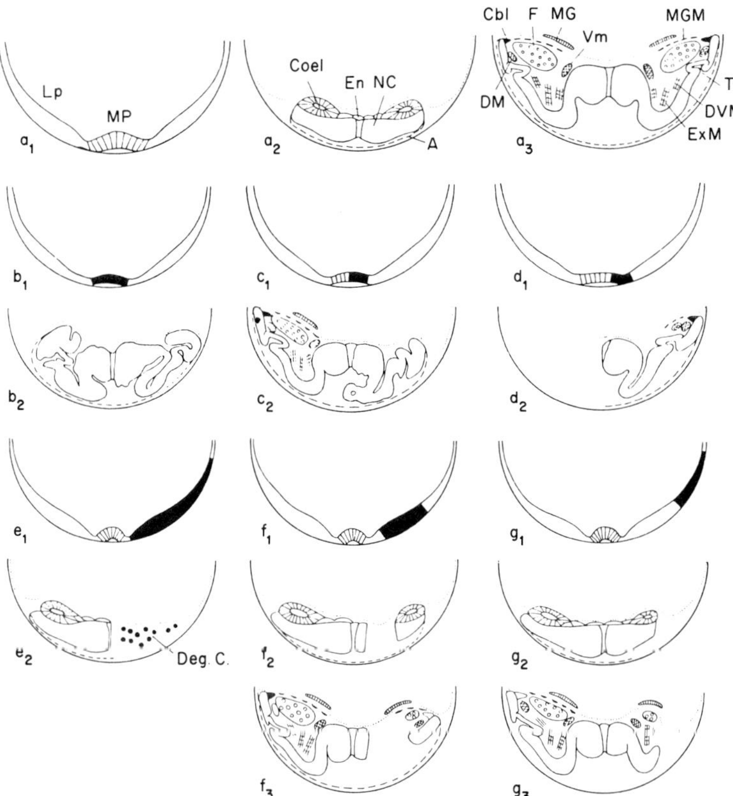

Fig. 15. Induction of mesodermal structures by ectoderm in *Chrysopa*. Details in text. A, amnion; NC, nerve cord; Cbl, cardioblasts, Coel-coelom; Deg. C., degenerating cells; DM, dorsal muscles; DVM, dorso-ventral muscles; En, endoderm; ExM, muscles of extremities; F, fat body; MG, midgut rudiment; MGM, midgut musculature; MP, middle plate; LP, lateral plate; Tr, tracheal invagination, Vm, ventral muscles (Kühn, 1965).

formed by four cells all descended from a single stem cell: a nerve and a sheath cell and, important for this discussion, a trichogen cell which forms the bristle shaft and a tormogen cell which forms the socket. The bract is a small cuticular projection formed by an epidermal cell adjacent to the bristle socket. Bracts occur in some but not all parts of the body of a fly; for example, they are found on distal but not proximal leg segments, and in proximal but not distal wing regions.

Tobler mixed the cells from parts of two different, genetically marked, leg discs: one disc was taken from a fly homozygous for *yellow*, a mutation which makes the bristles and bracts be a pale yellow rather than a brownish colour; the other was from a fly marked with both *ebony*, which makes bristles and bracts black, and with *multiple wing hair* which causes a subdivision of bracts and of the common, tiny cuticular projections, the hairs. The disc tissue was macerated with fine tungsten needles, mixed, and injected into the abdomen of a third instar larva. When the larva metamorphosed into an adult, the injected imaginal disc tissue likewise metamorphosed and formed cuticle appropriate to the leg regions corresponding to the parts of the discs which had been used in a given experiment.

The essential observation was that a number of instances were found in which a bristle organ and its associated bract were of different genotype (Fig. 16). Bristles which should have had bracts did have them and no isolated

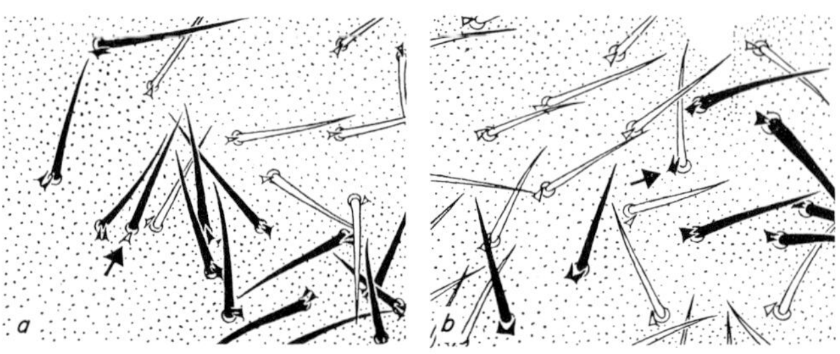

Fig. 16. Induction of bracts by bristle organs in *Drosophila*. Micromosaics of dissociated and reaggregated leg disc tissue in which the bract and bristle are sometimes of different genotype (arrows). Details in text. (Tobler, 1966.)

bracts were found. Since cells which were not in contact with a bristle never formed bracts, one would suspect that bract formation was induced. Since the bract forming cell often came from a different disc than the bristle forming cells, it must have been true that the induction took place subsequent to the mixing of cells from the two discs.

A number of other observations are consistent with this hypothesis. For example, two different chemical agents which interfere with cell division (Mitomycin C and nitrogen mustard) cause the formation of bristle organs lacking sockets (Tobler, 1969; Tobler and Pfluger, 1970; Tobler and Maier, 1970). These socketless bristles invariably lack bracts, whereas all normal, socketed bristles in the same animals which should have bracts do have them. There is a perfect correlation between fully formed bristle organs and the formation of associated bracts, and between incomplete bristle organs and the lack of bracts.

Since bristle organs lacking sockets also lack bracts, one might suppose that it is the tormogen cell that acts as the inducer. However, there are two mutants, *Hairless* and *shaven-depilate*, in which sockets without bristle shafts are formed. These too lack bracts. Thus, a complete bristle organ appears to be required for the induction of a bract, and the malfunctioning of either the tormogen or the trichogen cell (or perhaps the mother cell from which they both arise) results in a failure of induction. The role of the other two cells of the bristle organ, the nerve and sheath cells, in this process is not known.

Thus, induction can occur between two apposed tissues in a region-specific manner, between very small groups of cells and, by extension, between single cells.

3 Part II

The goal of Part II is to describe at least some of what is known about the development of a number of neural subsystems. Almost all of them are sensory systems, for several reasons. First, sensory systems begin at the integument, and the theories of Part I have been based largely on experiments on the integument. Second, in sensory systems we are able to follow the relationships between sets of cells through two or three well defined layers, which is not yet possible for motor systems. And finally, the fact that I myself have worked almost exclusively with sensory systems may have had some influence.

Several threads run through the various sections of Part II. From the section on the retina through to the section on abdominal segments there is an increasing emphasis on the presence and characteristics of uniquely identified neurons; these are taken up further in the synthesis. There are repeated discussions of what are loosely called supportive interactions between cells, especially cells of different layers, and the last section but one (on cerci and their associated interneurons) shows with especial directness how slight an interference with one layer can produce a structural and/or functional alteration in the next layer. There are a number of scattered indications that timing of morphogenetic events may be important in the formation of accurate neural networks.

The various subsystems have many other features in common. But since the goal of this entire essay is to consider the extent to which the theories of Part I help us interpret the findings of Part II, I have written a summary at the end of each section which highlights this particular relationship. A more general discussion is reserved for the synthesis. For a different view of insect neural development, see Bentley (1973; 1975).

3.1 DEVELOPMENT OF THE RETINA

The development of the eyes and optic lobes of insects is the subject of a voluminous literature and many reviews (e.g. Meinertzhagen, 1973, 1975;

Shelton, 1976). In the following sections I will present a few results, mostly from recent papers, which find possible interpretations among the theories of Part I of this essay.

3.1.1. *The origin of cells making up the retina*

Early in development of the eye (e.g. in the embryo of hemimetabolous insects, the pupa of Lepidoptera, or the imaginal discs of advanced Diptera) a small group of cells assembles into clusters resembling ommatidia. This

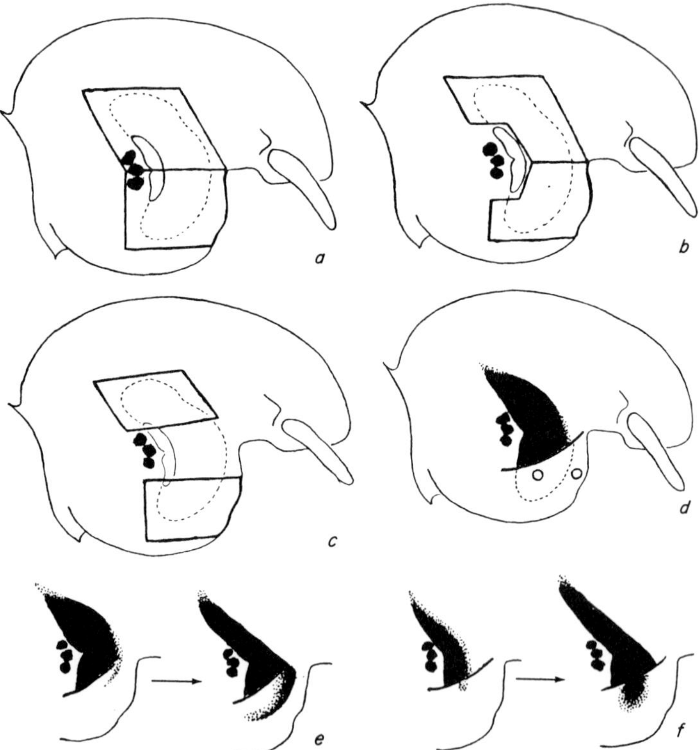

Fig. 17. Operations on the prospective eye field of *Aedes*. In (*a*) (*b*) and (*c*) are shown the designs of three related experiments: prospective eye field with the differentiation centre (just anterior to the three larval ocelli shown in black), without the differentiation centre, and with only small fragments of it was cut out and divided in half as shown by the polygonal outlines, and subsequently implanted into the internal haemolymph space of the head capsule. (*a*) and (*c*) differentiated eyes, (*b*) did not. (*d*), (*e*) and (*f*) show the effects of implanting barriers into the eye field. The long strip of non-eye epidermis forms a complete block in (*d*). The partial blocks in (*e*) and (*f*) result in waves of differentiation spreading in unusual directions across the remainder of the eye field (White, 1961).

early site or "differentiation centre" (White, 1961) is the locus from which differentiation spreads, most often in an anterior progresssion as in *Drosophila* (e.g. Ready *et al.*, 1976), or first in a small posterior wave and then in a major anterior wave as in *Ephestia* (e.g. Nardi, 1977). A number of studies have provided evidence that as the area of retina grows, competent cells in front of the anterior margin of the growing eye are in some sense recruited to become eye.

A particularly elegant demonstration of the course of development beginning with the differentiation centre has been provided by White (1961; 1963) in the mosquito. The differentiation centre is recognizable in the first instar larva as a small area of thickened epidermis, the optic placode, located at the posterior margin of the region which will ultimately become the compound eye, the prospective eye field. The optic placode grows slowly during the first three instars as a wave of mitoses, producing a thickened epithelium, sweeps gradually forward. At the beginning of the fourth (last) instar, the optic placode still occupies only a small portion of the prospective eye field, but the differentiation of retinal cells has started, again in the most posterior region. By the end of the fourth instar the entire prospective eye field has been converted to optic placode, and differentation, recognized by the formation of pigment, has caught up with it.

The following experimental results were obtained with early fourth instar larvae (Fig. 17). Pieces of prospective eye field including the optic placode, when transplanted into the head capsule of a host larva, differentiated into retina (a). Similar pieces lacking the optic placode failed to differentiate (b). If most of the placode was removed from a graft piece, differentiation still proceeded; 20–30 placode cells appeared to be sufficient to initiate eye formation (c). A piece of head epidermis from a region other than the prospective eye field implanted into the path of the wave of placode formation and subsequent differentiation stopped both processes (d). If the barrier was incomplete, the mitotic and differentiative waves passed around it even if this meant that they had to travel in an unusual direction in the eye field beyond the barrier (e, f). Only living epidermis formed a barrier; cuticle from the same region but with the epidermal cells killed and removed was ineffective. A piece of epidermis from the prospective eye field implanted into the same site became incorporated into the eye.

Evidently, the advancing edge of the optic placode has some special organizing capacity which causes the conversion of adjacent competent head epidermis into additional placode. Shelton and Lawrence (1974) took advantage of eye colour mutants of *Oncopeltus* to show unambiguously that cells which were not yet eye cells at the time of grafting would become eye cells under the influence of their host. Mosaic eyes were produced, containing both host and donor tissue (Fig. 18). The same clear result has been obtained in

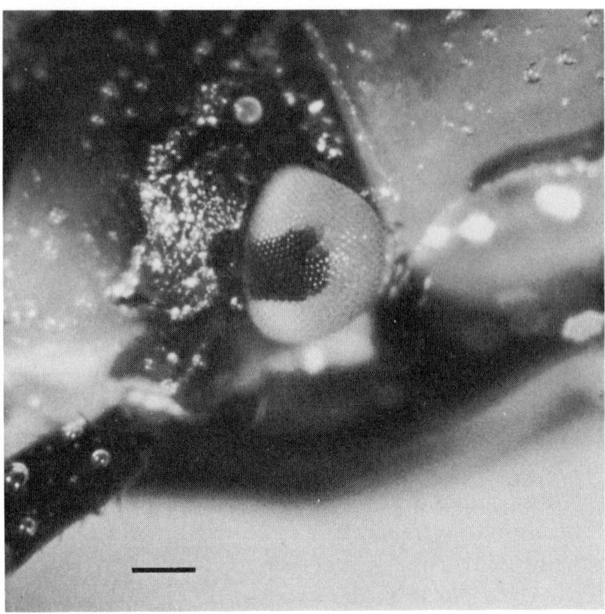

Fig. 18. Prospective eye field epidermis of a red eyed donor implanted into the eye field of a white eyed host in *Oncopeltus*. The tissue was incorporated and formed a mosaic eye (bar is 0.2 mm). (Shelton and Lawrence, 1974.)

cockroaches (Shelton *et al.*, 1977). However, not all epidermis can be induced to become eye. In mosquitoes, and also in dragonflies (Mouze, 1975), only epidermis from the prospective eye field (the region of the head capsule which will normally become occupied by eye tissue) appears to be competent to respond to the influence of the advancing front of eye formation. However, in *Oncopeltus* (Green and Lawrence, 1975) and *Ephestia* (Nardi, 1977) non-eye epidermis from regions close to the eye will also respond, but more distant epidermis will not. In cockroaches it has been claimed (Hyde, 1972) and denied (Shelton *et al.*, 1977) that prothoracic epidermis will respond; in dragonflies (Mouze, 1975) it does not. Meinertzhagen (1973) has raised the interesting possibility that the exact location within the prothorax from which graft tissue is taken might influence its competence to respond.

3.1.2 *Assembly of ommatidia*

Since there are usually 8 retinula cells in insect ommatidia, and they are arranged in a very regular pattern, it was tempting to suppose that they arise by three successive divisions from a single mother cell and thus constitute a clone (e.g. Kühn, 1965). However, recent studies demonstrate that this is not so.

When the border of a graft such as is illustrated in Fig. 18 is examined carefully, it becomes apparent that the cells which make up any one ommatidium can come from both the host and the graft. All combinations of graft and host retinula and pigment cells can be found (Shelton and Lawrence, 1974; Shelton, *et al.*, 1977). In *Drosophila*, clones of unpigmented cells have been induced in wild-type eyes, and the same result obtained (Ready *et al.*, 1976; Fig. 19). A single ommatidium can be mosaic for retinula cells, or

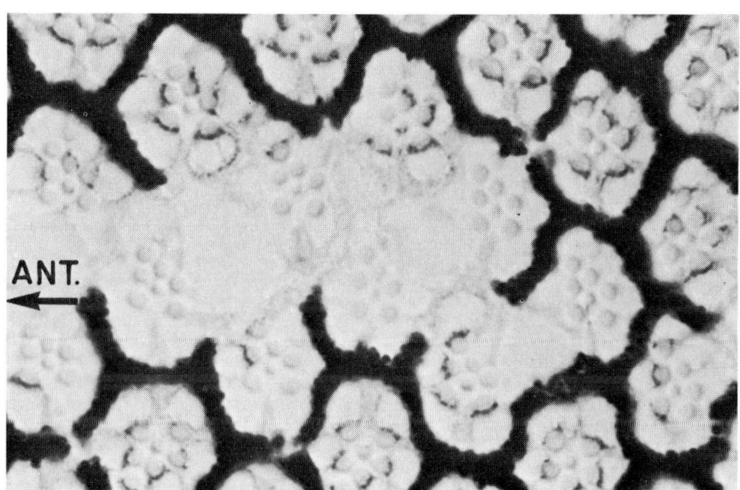

Fig. 19. A *white* clone in a wild type retina of *Drosophila*. The mutation causes the loss of pigment granules in both the primary pigment cells (heavy black borders) and the retinula cells (dark crescents bordering the rhabdomeres). Examination of individual ommatidia shows that any combination of mutant and wild type cells can be present (Ready *et al.*, 1976).

pigment cells, or both; no cell in an ommatidium has an obligatory clonal relationship to any other cell. The inference that the cells making up a single mature ommatidium are *not* related by cell lineage, but instead are collected from a pool of precursor cells, seems inescapable.

If cell lineage is not the mechanism for assembling the 8 retinula cells of an ommatidium or the associated pigment cells, what is? Ready *et al.* (1976) suggest a process of step-wise induction at the advancing border of the retina, in which the already differentiated elements, both retinula cells and pigment cells, would communicate instructions to the more recently recruited cells as they make contact with them, somewhat as new elements are added to the faces of a crystal as it grows. Some specific suggestions about possible mechanisms of communication by surface molecules are made. This is a graphic and attractive analogy, especially in view of the apparent importance of the advancing *front* of eye recruitment. It does not suggest how the process might start, of course, only how it might continue after the "seed crystal" was already organized.

However, it is not clear that the population of cells available for recruitment into ommatidia is homogeneous. Cell proliferation in the eye disc of *Drosophila* occurs in two successive waves (Ready *et al.*, 1976; Campos-Ortega and Gateff, 1976; Campos-Ortega and Hofbauer, 1977). Each retinula cell of the ommatidium has a particular location and other anatomi-

Fig. 20. A clone crossing the retinal equator in *Drosophila*. The equator can be found by drawing a zig-zag line between the arrows, separating ommatidia whose asymmetric rhabdomere patterns point in opposite directions. Small circles represent retinula cells, the large circles immediately around them the primary pigment cells (two per ommatidium), and the ovals the secondary pigment cells. The clone is found primarily below the equator, but a number of cells on the dorsal side are also marked (Ready *et al.*, 1976).

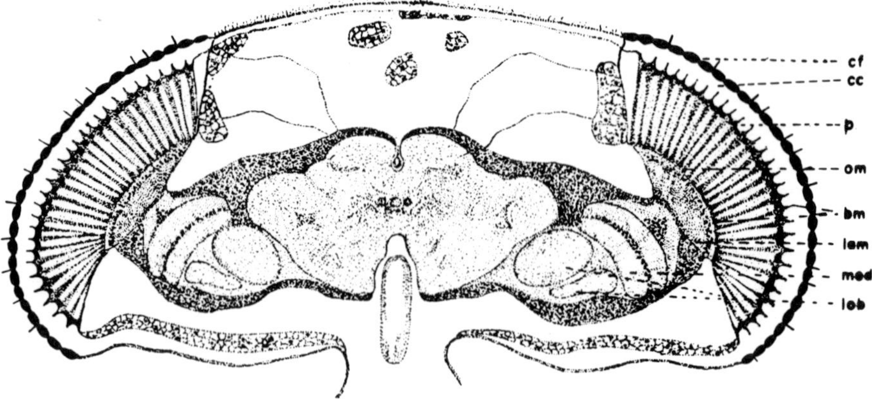

Fig. 21. The visual system and brain of *Drosophila*, seen in approximately frontal section. The corneal facets (cf), layer of crystalline cones (cc), ommatidia (om), lamina (lam), medulla (med) and lobula (lob) are indicated (Richards and Furrow, 1925).

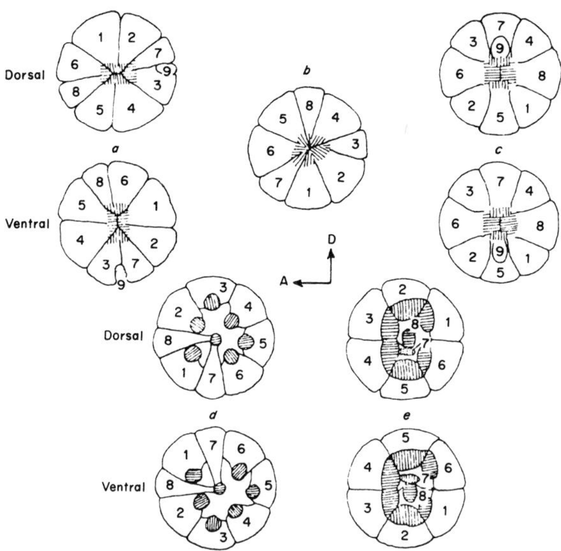

Fig. 22. The rhabdomere arrangement of a variety of insect species. Fused rhabdome eyes: (*a*) bee (*Apis*), (*b*) locust (*Schistocerca*), (*c*) butterfly (*Pieris*). Open rhabdome eyes: (*d*) fly (e.g. *Calliphora*), (*e*) waterbug (*Lethocerus*). In (*a*) and (*e*) the patterns in the dorsal and ventral halves of the eye are different, and both are shown (Meinertzhagen, 1975).

cal characteristics on the basis of which it can be assigned a number (see Figs 22d and 23b). It is found that cells 2, 3, 4, 5 and 8 arise and cluster together during the first mitotic wave, and cells 1, 6 and 7 mainly arise during the second and are added to the already established pre-ommatidial cell clusters. Campos-Ortega and his collaborators have offered some statistical evidence that in addition to this temporal sequence of acquisition of cells by the forming ommatidia, there may be a partitioning of the recruitment pool by cell class. Their working hypothesis is that a single retinula mother cell gives rise to one cell 7, one cell 8, and one set of cells 1–6; these last cells are interchangeable and acquire particular positions in an ommatidium according to the time at which they are incorporated. The whole set of eight cells can form a single ommatidium, but as we have seen they need not and may be distributed among several nearby ommatidia. There are formidable difficulties in testing this interesting hypothesis, but for the present it seems plausible and not ruled out by any of the available evidence.

3.1.3 *Polarity in the eye field and the retina*

Lawrence and Shelton (1975) have studied the polarity properties of the retina by grafting small pieces of genetically marked donor tissue in various

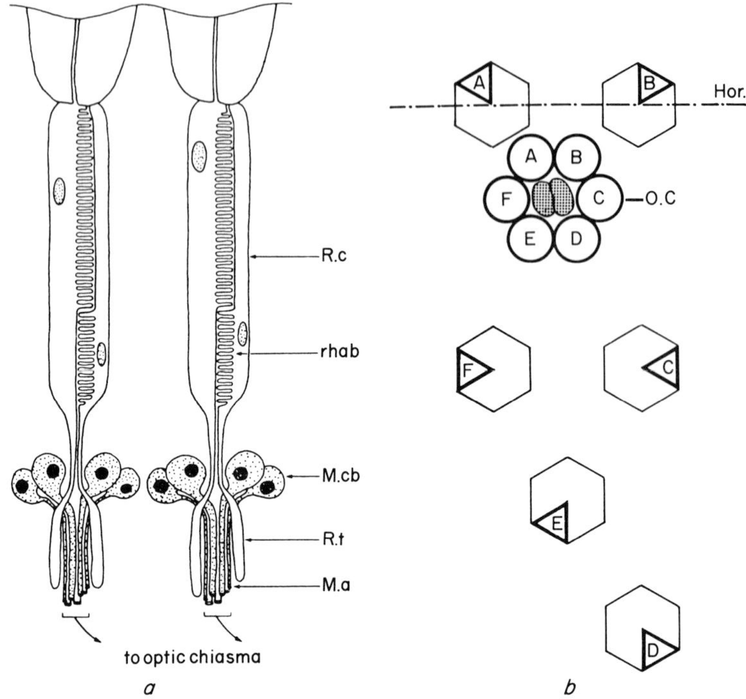

Fig. 23. The retina-to-lamina projection. (*a*) A fused rhabdome eye (the dragonfly *Sympetrum*). All of the short retinula cell axons terminate in a single cartridge, though not all at the same level. Retinula cells (R.c.), rhabdomeres (rhab), monopolar cell bodies (M.cb), retinula cell axon terminals (R.t.) and monopolar cell axons (M.a) are indicated. The cartridge contains four monopolar cells, two with large and two with small axons (Armett-Kibel, *et al.*, 1977). (*b*) An open rhabdome eye (the fly *Lucilia*). The optic cartridge (O.C.) is shown with two large central monopolar cell processes surrounded by six retinula cell terminals, each from a different ommatidium. The relative positions of the contributing ommatidia, and the origin and termination of each retinula axon, are indicated by letters. The ommatidium most directly overlying the cartridge sends out long retinula axons which do not synapse in the lamina. The projection pattern reverses at the equator, just as the pattern of rhabdomeres within the ommatidium does (Trujillo-Cenóz and Melamed, 1966).

orientations relative to the host. The grafts were taken from prospective eye field which still looked like ordinary head epidermis. The polarity of each ommatidium could be recognized because of the characteristic polarized pattern formed by the rhabdomeres (see, for example, Fig. 22).

The results were somewhat difficult to interpret because there was some reorientation of ommatidia along the graft margin even in control grafts, and because rotated graft tissue did not grow flat but showed a marked tendency to rise into "towers" within which ommatidial orientation was difficult to specify. With these limitations, the following constitute the most important observations: (*a*) In large grafts, at least some areas appeared to conserve their original polarity. (*b*) At the graft margins, some ommatidia with an orientation intermediate between donor and host appeared. (*c*) Within an ommatidium, each retinula cell had a polarity appropriate for the whole set; whatever influences were affecting polarity did so for the entire group of retinula cells within one ommatidium. (*d*) This polarity was not necessarily the same as the polarity of the primary pigment cells of the same ommatidium; the disparity between the two could be as much as 90°, the maximum detectable by the available morphological criteria. Disparities between the retinula cells and the pigment cells were apparently not frequent, but they have not been reported at all in normal tissue and would therefore seem to be significant.

In spite of the difficulty of the experiment and the variability of the results, it seems reasonable to conclude that the epidermis of the prospective eye field has a polarity prior to its recruitment into eye tissue, and that this polarity can survive the recruitment process.

3.1.4 *Are there compartments in the developing retina?*

The rhabdomeres within each dipteran ommatidium are arranged in an asymmetric pattern (Figs 19, 20, 22). The polarity of this pattern is constant over the dorsal half of the eye, and is reversed in the ventral half. The border separating the two regions of mirror-image symmetry is very straight and accurate, and has been studied in quantitative detail (e.g. Horridge and Meinertzhagen, 1970). Is this border a line of clonal restriction, and are the dorsal and ventral regions two separate compartments?

Ready *et al.* (1976) have provided a clear answer to this question, and it is no (Fig. 20). Clones do cross the equator. There is a tendency for clones to be elongate and to run parallel to the equator, so that if the eye is only examined externally they may appear to be confined to either the dorsal or the ventral half. But when the retina is sectioned, so that the location of the equator can be determined precisely, it becomes clear that the equator is not a line of strict clonal restriction such as the A–P compartment line of the wing. The number of marked cells crossing the equator to the side opposite the clone's centre of gravity is small, but crossing is a regular feature of clones in this location.

3.1.5 *A gradient of adhesiveness?*

Nardi (1977) made several observations which suggested a variation of adhesiveness over the eye. For example, anterior tissue translocated backwards towards the differentiation centre formed a smooth surface not obviously separated from the surrounding host tissue, even though it usually showed only poor differentiation into ommatidia. In contrast, posterior tissue moved to an anterior location tended to bulge from the surface so that a deep grove separated the two, even though the formation of ommatidia in the graft was excellent. In addition, grafts rotated 180° tended to assume a circular form, whereas grafts in the normal orientation became wedge-shaped.

Lawrence and Shelton (1975) remarked upon the tendency of rotated grafts to elevate into "towers". The same was apparently true in a number of the graft combinations performed by Mouze (1975). In all these cases, mechanical effects were associated with either rotation or translocation, not with the grafting procedure itself. They would be expected from the model of Nardi and Kafatos (1976) based on adhesiveness differences, but the tests required to argue for a *gradient* of adhesiveness are technically difficult to perform in such a small piece of tissue.

3.1.6 *Summary*

The retina, more than any other piece of nervous tissue in insects, has been studied using the methods of developmental biology and by means of experimental designs intended to test the ideas of developmental biology.

There appears to be no reason to doubt that the retina develops and responds to experimental interference much like any spatially discrete region of epidermis. The phenomenology of *induction* is conspicuous in its growth and differentiation. It has an empirically defined *polarity*. I have not found any compelling evidence for *gradients* (which is difficult to obtain for technical reasons), but none of the data argue against their presence. The retina does not appear to be built out of *compartments*. The cells of an ommatidium need not have a common clonal origin, but other important roles of *cell lineage* are not excluded. *Timing* may also be an important factor in the formation of the marvellously precise structure of the retina.

The cells of the retina differentiate from ordinary epidermal cells, so it is perhaps not surprising that in many respects they behave like them. Perhaps their first opportunity to express special nerve-cell-like attributes is in the lamina, the neuropile in which most of their axons synapse and otherwise interact with other nerve cells. Whether the familiar characteristics of epidermal cell behaviour appear again in this new setting is explored in the next section.

3.2 THE OPTIC LOBES

The anatomy of the optic lobes and of their relationship with the retina has been the object of intense study during the past 15 years. Excellent reviews are available, for example Trujillo-Cenóz (1972), Braitenberg and Strausfeld (1973), the symposium edited by Zettler and Weiler (1975) and Strausfeld (1976). An extensive comparative study has been published by Meinertzhagen (1976). The development of the optic lobes has also been reviewed extensively, notably by Edwards (1969) and Meinertzhagen (1973). I will first summarize some of the highlights of normal development, and then review some of the experimental findings concerning the relationship of the optic lobes and the retina.

3.2.1 *Basic anatomy*

The retina and the three optic ganglia (lamina, medulla and lobula) (Fig. 21) are all constructed of repeating structural units upon which the spatial features of the visual world are projected in precise topographic order. The ommatidia of most species of insects contain a set of eight retinula cells. All but two of them synapse in the underlying lamina; the two different cells have long axons which pass through the lamina to synapse in the medulla. Ommatidia fall into one of two functional and structural categories: fused rhabdome and open rhabdome (Fig. 22). In the first group, the microvilli of the receptors which bear the receptor pigment are aggregated into a single central structure, the rhabdome. Because of refractive index differences between the rhabdome and the surrounding retinula cell cytoplasm, light is internally reflected and can be trapped by the pigment of each of the retinula cells of that ommatidium. Thus, the entire ommatidium has a single optical axis and all its cells look in the same direction. In the open rhabdome group, the rhabdomeres (sets of microvilli produced by the individual retinula cells) are spatially separated and optically uncoupled, each one of them looking out at a slightly different angle through the common lens. However, particular retinula cells of nearby ommatidia look out along the same optical axis. Thus, as in fused rhabdome eyes, a set of receptor cells views along a single optical axis; the difference is that they belong to a group of ommatidia instead of a single one.

This difference in optical construction has structural correlates in the pattern of interconnection of the retina with the lamina. In fused rhabdome eyes all of the retinula cell axons from a single ommatidum (except the two long-axon ones) synapse with second-order neurons in a single cartridge, the repeating structural element of the lamina (Fig. 23a). In open rhabdome eyes, the short axons coming from adjacent ommatidia are recombined so that a single lamina cartridge receives synapses from all the cells looking out along a single optical axis, which, of couse, requires that each retinula cell in a given ommatidium should send its axon to a different cartridge (see Fig. 23b). In both cases, adjacent lamina cartridges view in adjacent directions in visual space.

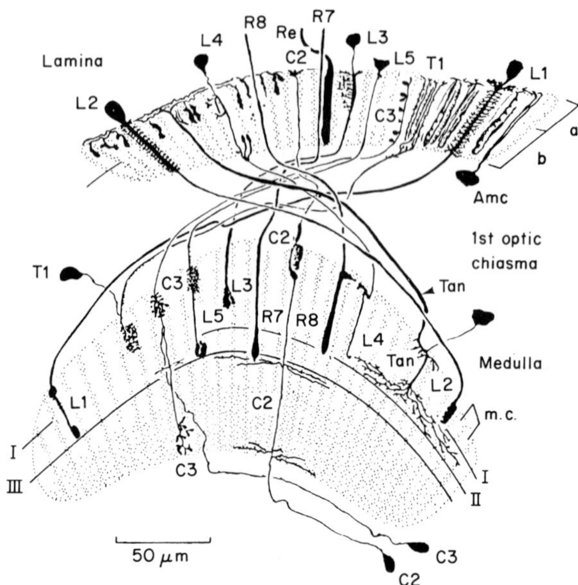

Fig. 24. The lamina-to-medulla projection (the fly *Musca*). Examples of the two large monopolar cells (L1 and L2) are shown crossing in the chiasma, as are the smaller monopolar cells (L3, L4 and L5). The short axon retinula cells (Re) terminate in the lamina, the long axon ones (R7 and R8) pass through the lamina and chiasma to terminate in the medulla. A sampling of centrifugal cells (C2 and C3), tangential cells (Tan and T1) and amacrine cells (Amc) is also shown (Braitenberg and Strausfeld, 1973).

This projection of visual space upon neural space is repeated at the next stage of neural processing as well (Fig. 24). The axons of lamina cartridges project to medulla cartridges in a precise order, but one that incorporates an antero-posterior chiasma. The long-axon retinula cells which did not synapse in the lamina join the appropriate axon bundles so that their connections in the medulla form the same map of visual space as those of the retina–lamina–medulla pathway.

The anatomy of subsequent levels is more complex and has not been subjected to experimental analysis, so it will not be described here.

3.2.2 *Formation of the optic lobes*

As described in the previous section, the retina develops by a process of recruitment or induction in the epidermis of the head capsule. The optic lobes, on the other hand, develop from zones of proliferation associated with the central nervous system. I give here a brief account emphasizing the post-embryonic development of the lamina and medulla, based largely on the studies of Nordlander and Edwards (1969a, b; reviewed in Edwards, 1969), Meinertzhagen (1973), and Mouze (1974).

Two separate zones of proliferation are found, the Inner Optic Anlage (IOA) which gives rise to the lobula, and the Outer Optic Anlage (OOA) which produces the cells of the medulla and the lamina. The same basic pattern of divisions is found in both: large neuroblast cells divide asymmetrically, giving a new neuroblast situated peripherally and a ganglion mother cell situated centrally. The ganglion mother cell divides at least once and sometimes several times to produce neurones which are arranged in columnar clusters. Thus, the oldest cells of any region are found closest to the corresponding neuropile and furthest from the neuroblasts.

The medulla and the lamina form from a single set of neuroblasts, the OOA, whose cells divide on two orthogonally oriented faces of the anlage (Fig. 25). Ganglion cells budded off parallel to the retina form the lamina; its oldest region is posterior and its youngest region is anterior, adjacent to the OOA. Ganglion cells budded off medially form the medulla; its oldest

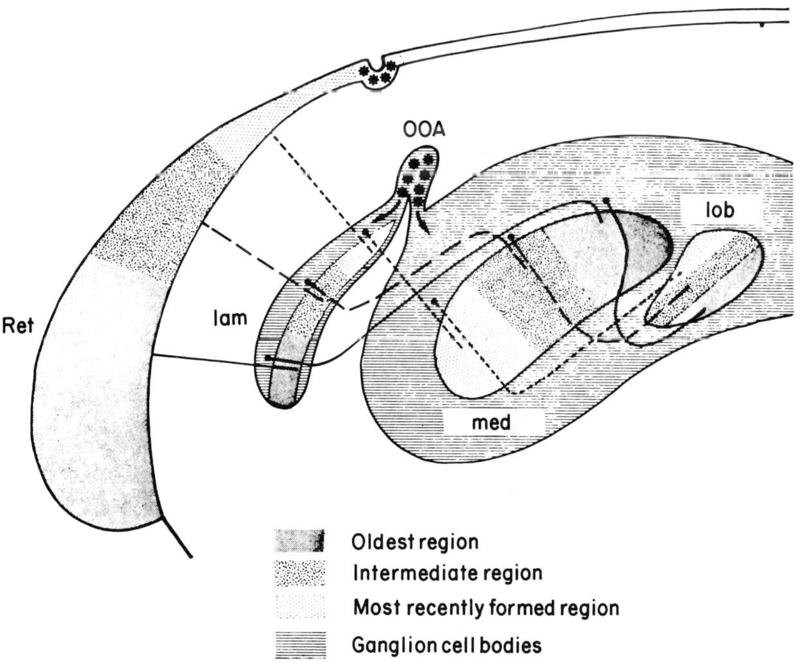

Fig. 25. The pattern of growth of the retina and the lamina (the dragonfly *Aeshna* or *Anax*). The asterisks indicate the advancing front of the retina and the Outer Optic Anlage (OOA). The stippling indicates the regions of retina and lamina which are of similar age. The retina (Ret), lamina (lam), medulla (med), and lobula (lob) are indicated (after Mouze, 1974).

region is the most medial but, following an approximately 90° rotation, becomes anterior.

From this description it is apparent that the oldest region of the lamina underlies the oldest region of the retina. Axons growing inwards from the earliest differentiating ommatidia encounter the first-formed lamina ganglion cells ready to receive them. Similarly, the second order axons originating in these earliest cartridges connect with the oldest region of the medulla, which is originally deepest (most central) but gradually swings forward. Thus, the oldest, most posterior, lamina axons are carried to the most anterior location in the medulla. The youngest lamina axons originate most anteriorly and synapse with the youngest region of medulla which is most posterior. Thus, an orderly chiasma forms between the two neuropiles. There is a second chiasma between the medulla and the lobula, but it will not be described here. The major reviews may be consulted for details.

3.2.3 *Do the optic lobes influence the development of the eye?*

We have seen that there is a close temporal and spatial correlation between the development of the eye and the associated optic lobes in insects. We next ask, are there important functional relationships implicit in this correlation – are there indications of inductive or trophic effects which might make the development of one of these topographically ordered structures dependent on the other? The older experimental literature has been reviewed by Bodenstein (1953), Pflugfelder (1958), Edwards (1969), Meinertzhagen (1973) and Mouze (1974).

We have already seen, in the summary of the studies of White (1961; 1963) that a piece of head epidermis containing the eye differentiation centre and the prospective eye field will, when implanted into the head capsule of a host larva, proceed to differentiate the entire retina. This result is typical of a number of studies which point to the independence of retinal differentiation from the presence of an underlying optic lobe, even though the two structures normally grow in close time synchrony. A recent example is the study of Mouze (1974), in which the following evidence appears: (*a*) Eyes transplanted to the abdominal integument with no accompanying optic lobe tissue but with at least a portion of the prospective eye field included recruit new ommatidia as usual, and these differentiate normally as far as could be established with the light microscope. (*b*) If a portion of the prospective eye field and the advancing eye is undercut and a thin plastic barrier is inserted which prevents the formation of connections between the retina and the lamina, recruitment and differentiation of the retina in the undercut region again proceed normally. (*c*) In both situations, however, there are clear retrograde effects of sectioning the axons of already differentiated retinula cells. Not only is there evidence of degenerative changes in the retinula cells, but other components of the ommatidia including the dioptric apparatus are also affected.

In my view, most of the more recent evidence points in the direction of retinal independence. Some of the earlier literature, however, indicated that the optic lobes do exert some degree of trophic or inductive influence upon the retina; an entry to this literature is provided by the reviews mentioned above. It is a difficult literature to evaluate. For example, Pflugfelder (1947) makes the plausible suggestion that while the retina is independent in its growth and differentiation in later, postembryonic stages, it requires the presence of the optic lobes to get started in the embryo. However, the embryos with which he worked were not clearly staged, and a systematic comparison of the behaviour of implants with and without attached optic lobes at the same stages of development was not provided. His conclusion may be perfectly correct – there is, after all, no other specific suggestion as to the origin of the differentiation centre – but the evidence is meagre. Other early papers have similar limitations.

3.2.4 *The effect of the retina upon the development of the optic lobes*

Whatever the details turn out to be, it is apparent that the relationship of the retina and the lamina is not a symmetric one, at least in postembryonic stages. For example, in a classic paper Kopeć (1922) reported that removing the optic lobes or the entire brain of a moth had no significant effect on the retina; neither did transplantation of the eye rudiment to the abdomen. However, removal of the eye rudiment prevented the development of the outermost layers of the optic lobe (the lamina, in modern terminology), and resulted in a deformation of the inner ganglia (medulla and lobula). In cases in which the retina regenerated, the effects upon the optic lobes were not seen.

Further indications of the dependence of the lamina upon the retina came from studies on mutants of *Drosophila* which were deficient in numbers of ommatidia even to the point of complete loss of one eye (e.g. Krafka, 1924; Richards and Furrows, 1925; Power, 1943). All these authors agreed that loss of retinal tissue had a profound effect on the lamina and progressively less on the medulla and lobula (Fig. 26). In *eyeless-2* individuals completely lacking facets, Power found a complete, 100% loss of the lamina, 84.4% loss of the medulla, and 58.7% and 57.1% loss in the two parts of the lobula complex. The loss of medulla tissue in individuals which were lacking only one eye was confined to the eyeless side. Furthermore, in many *Bar* individuals only a discrete region in the anterior part of the eye was missing. This deficiency was correlated with the loss of a posterior piece of the medulla,

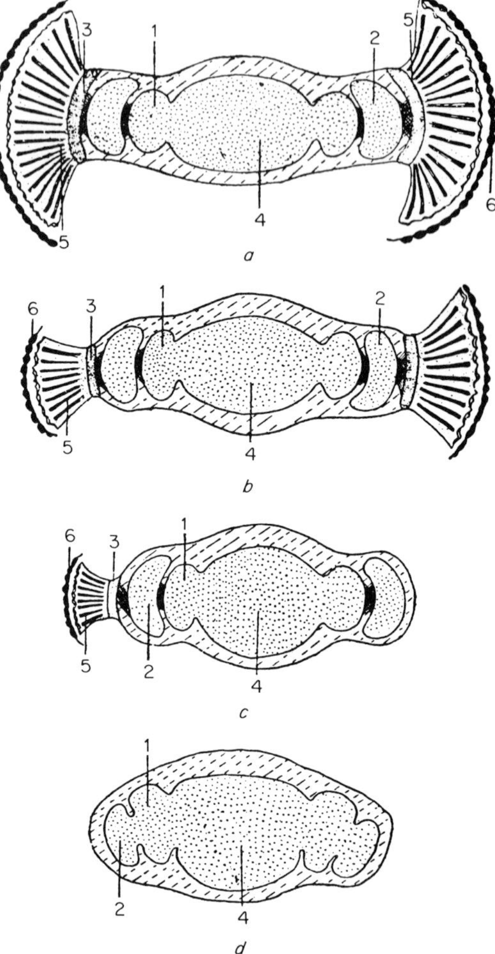

Fig. 26. The effect of loss of facets upon the size of the optic lobes in *eyeless* mutants of *Drosophila*. There is a progressive reduction in the size of the eyes shown from top to bottom, and a corresponding reduction in the lamina (3). The effects on the medulla (2) and lobula (1) cannot be seen. The corneal facets (6), retina (5) and basal lamina (4) are also shown (Richards and Furrow, 1925).

the very region to which anterior ommatidia project via the lamina and the external chiasma (Fig. 27). Finally, Power studied 10 *eyeless-2* individuals in which, for unknown reasons, the external chiasma was lacking and the lamina (underlying a normal eye) was therefore disconnected from the medulla. In these cases the loss of medulla tissue was as great as in individuals which were lacking an eye.

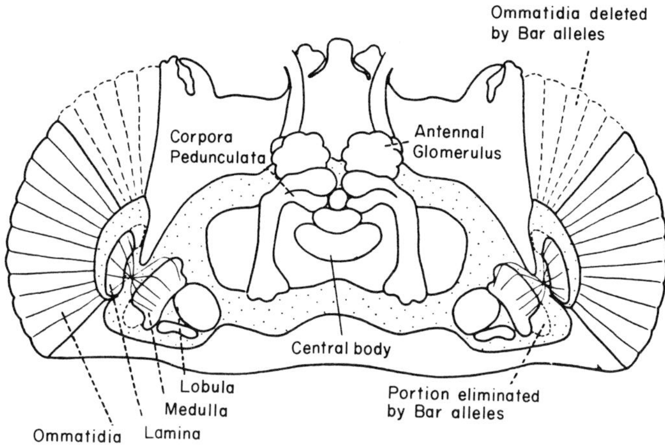

Fig. 27. The localized effect upon the lamina and medulla of the loss of particular ommatidia in the *Bar* mutant of *Drosophila*. The lost ommatidia, lamina and medulla are shown with dashed lines (after Power, 1943).

This careful study contains the basis for several important conclusions: 1. Either the development or the survival of the lamina depends profoundly on the existence of the retina. 2. The axons of the lamina similarly support the medulla. 3. These supportive effects are local: they do not extend to the other side of the brain, or to a neuropile which is not contacted by centripetal axons, or to a region of the neuropile which is selectively deprived of incoming axons.

In a modern version of this experiment, Meyerowitz and Kankel (1978) have produced clones of two mutations which produce marked abnormality of the retina, *rough* and *glassy*. As would be expected from Power's study, they found that the lamina was profoundly disorganized in the region to which the affected retinula cells projected, and that the effect was propagated to the medulla. In very small clones, affecting less than 1% of the eye surface, the correlation between retinal and laminal abnormality was not perfect. In all the larger clones, however, and also in mosaic visual systems produced by gynandromorph techniques, the disorganizing effect of the mutation was clearly propagated from the retina to optic lobe tissue. Power already had strong circumstantial evidence that the loss of optic lobe tissue which he saw was not due to a simultaneous (pleiotropic) effect of his mutations, and the mosaic techniques employed by Meyerowitz and Kankel, which assured that the optic lobes and most of the eye were phenotypically wild type and only small patches of retina were mutant, remove any lingering doubts on this point. Meyerowitz and Kankel also studied the dominant mutant *Glued* which produces severe abnormalities in both the retina and the optic lobes.

When clones of wild-type retina were produced in otherwise mutant animals, the under-lying lamina, composed of mutant cells, nevertheless assumed a normal histological appearance.

Inasmuch as the optic lobes of animals with *rough* and *glassy* clones were heterozygous and phenotypically wild type, we can also conclude that the phenotype of the optic lobes does not somehow spread outwards and control the phenotype of the retina. The retinae of *Glued* animals, whose optic lobes were mutant, formed regions of apparently perfect wild type construction in wild type clones, which even rescued the histological phenotype of the corresponding laminae from disorder. Again, we see that the morphogenetic interaction between retina and lamina is unidirectional, from outside to inside.

3.2.5 *Growth of axons from retina to lamina*

As described above, retinal differentiation proceeds from posterior to anterior. Axon bundles from the developing ommatidia grow inwards along the leading edge of an already-formed bridge of axons and glial cells connecting retina and lamina. What about the axon bundle from the very first ommatidium? Does it have any axons to follow into the lamina? The answer is yes. Trujillo-Cenóz and Melamed (1973) and Meinertzhagen (1973) have both studied the stalk of axons connecting the eye disc of larval flies with the optic lobe anlagen. These axons originate in an extra-ocular cluster of photoreceptors present in the larva anterior to the eye disc; they penetrate the disc on their way to the brain. When the compound eye starts to differentiate the axons of the first-formed retinula cells grow along the surface of this original larval axon bundle into the simultaneously developing lamina. Thus, all of the retinula cell axons of the compound eye follow pre-existing axons to the region in which they are to ramify and synapse.

Axons from fused rhabdome ommatidia travel as discrete bundles, and all the axons of a given bundle terminate in a single cartridge except for the long axons of cells 7 and 8 which continue into the medulla. But the six short axons from an open rhabdome ommatidium must each terminate in a different cartridge. Trujillo-Cenóz and Melamed (1973), and Meinertzhagen (1973) give an elegant account of how this projection pattern develops. Within the lamina, the ending of each short axon retinula cell (cell 1–6) swells into a growth cone from which many slender filopodia emerge in all directions. As they lengthen they intertwine and even invaginate into neighbouring growth cones without any obvious topographic pattern. Gradually a thick lateral extension appears, with its own growth cone; whether this extension was once a slender filopodium is not clear. The new lateral branch becomes dominant, the old growth cone recedes, and finally the new one turns inwards

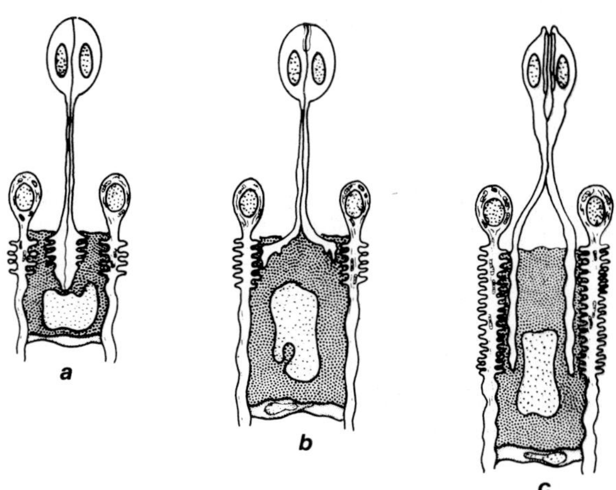

Fig. 28. Development of the interweaving pattern of retina-to-lamina connections in the fly *Phaenicia*. (*a*) Early (48 h after pupariation) all the short retinula cell axons from a single ommatidium terminate together in complex growth cones surrounded by glial cells (stippled); the rhabdomeres have not yet formed. (*b*) Between 50 and 80 h after pupariation stout lateral processes appear and contact adjacent monopolar cell processes; the rhabdomeres of the ommatidium are in contact. (*c*) In late pupae (90–100 h) the retinula cell axons have grown inwards and the interweaving pattern of projection has been established; the rhabdomeres have separated from each other (Trujillo-Cenóz and Melamed, 1973).

and follows along the second order monopolar neurons (Fig. 28). In this way a transformation of a fused rhabdome type of projection to a single cartridge into the interweaving projection characteristic of open rhabdome eyes is accomplished. The ommatidium itself goes through a similar transformation with approximately the same time course. Originally the rhabdomeres are all in contact, and only secondarily do they separate around a central space.

This description of the transformation process gives no hint of its astonishing accuracy. For example, Horridge and Meinertzhagen (1970) traced the paths of about 650 axons from 120 ommatidia in *Calliphora*, and found that every single one terminated in the correct cartridge. Meinertzhagen (1972) traced the paths of an additional 500 axons from 144 ommatidia in the region of the equator. He found 17 errors, of which 15 were associated with a minor dislocation of the equator of the retina defined, as usual, by the reversal of the asymmetric rhabdomere pattern. The errors were confined to single axons and were not propagated across the eye.

The growing retinula cell axons of the fused rhabdome eye of the locust behave similarly (Shelton, 1976). Within the lamina they form expanded growth cones whose filopodia extend sideways and contact those belonging to neighbouring axon bundles. But in this case, no major lateral process develops, and following the period of "exploration" the filopodia are retracted and the retinula cell axons from one ommatidium remain confined to a single cartridge.

Not only is there a precise topographic order in the projection of the whole set of ommatidia upon the set of cartridges, but every individual retinula cell projects to a particular location around the circumference of its appropriate cartridge. (This is illustrated in Fig. 23b and details are given in the general reviews listed above.) However, in the rare case when an axon enters an inappropriate cartridge, it can assume any circumferential position (Meinertzhagen, 1972). The mechanism establishing the usual precise order is therefore not a rigid, exclusive one. Quite possibly it is related to the fact that axons from a single ommatidium travel together in a single bundle in which their original topological relationships are preserved; the entire bundle may twist, but its axons do not interweave (e.g. Meinertzhagen, 1976).

3.2.6 *Summary*

It seems evident that as soon as the first step from the periphery of the nervous system into the centre is taken, the structural complexities which characterize the system become a dominant theme and hints about the developmental programmes employed in its construction become very difficult to extract. I will summarize the principal themes that I have recognized under two headings: themes directly suggested by the observations, and possible interpretations in the light of the theories of Part I.

1 The retina and the optic lobes develop from the different precursor cells in different locations and by different mechanisms. In the optic lobes there is no evidence for a wave of *induction* such as we saw in the retina; rather, the future cells of the various optic lobes are budded off from proliferation centres or anlagen.

2 One mechanism which seems very likely to play an important role in assuring an orderly projection of the retina upon the optic lobes is *timing*. At any given moment during development, the newest region of the retina is forming contacts with the newest region of the lamina, and this in turn with the newest region of the medulla.

3 While there is no evidence for the lamina inducing any nearby cells to become lamina also, there is clear evidence for some sort of supportive effect of the retina upon the lamina. The details are not clear, though in some cases the main effect seems to be on lamina cell differentiation (perhaps even survival) rather than on their initial proliferation.

4 There are suggestions in the literature of an inductive influence of optic lobe tissue upon the initial formation of the differentiation centre of the retina, but the experimental evidence is weak. I find the evidence to the contrary more convincing. Certainly in postembryonic stages of hemimetabo-

lous insects the retina develops perfectly well in the total absence of nearby optic lobe tissue.

5 *Cell-to-cell contact* seems to be a dominant feature of the growth of retinula cell axons to their proper destinations in the lamina. They grow initially along axons laid down much earlier in development, and as the retina grows the newer axons follow along the surface of already existing axon bundles. Within a single bundle originating in an ommatidium, axons maintain ordered contact with each other. When they enter the lamina they form growth cones whose many filopodia give the appearance of exploring their cellular environment.

6 At least in the case of open rhabdome eyes, *information about the location* of a given retinula cell in the ommatidium must be present at the tip of its axon, otherwise the lateral sprout which ultimately transfers the ending from its original location to a particular nearby cartridge might form in any direction.

7 However, the mechanism by which retinula cell bodies acquire their position-associated characteristics is not clear. A clonal origin for all the cells of a single ommatidium is ruled out, but an influence of cell lineage upon the position which a given cell from the precursor pool will acquire in the ommatidium to which it is recruited remains a strong possibility. The alternative possibility, that retinula cells acquire their characteristics purely by virtue of the positions in which they find themselves, is also not excluded.

8 Any mechanism which depends upon a rigid point-to-point recognition between retinula cells and lamina cells would seem to be excluded, since erroneous connections do form. The fact that almost all projection errors are associated with anomalies in the retina, even seemingly minor ones like the dislocation of the equator by one ommatidium, again suggests that the intimate local circumstances in which a cell body finds itself influence the growth of its axon.

9 The fact that errors of axonal trajectory were not propagated across the eye argues against any simple contact guidance mechanism for determining the fine details of neural projection patterns.

What of our developmental theories, then? The retina is apparently not subdivided into *compartments*. Compartments in central nervous tissue have been searched for but none have been found as yet, though their presence cannot be excluded (Kankel, personal communication). The projection of retinula cell axons is certainly associated with the position of their cell bodies within ommatidia. Some retinula cells are suspected of acquiring their position on the basis of their *lineage*. Others, however, seem to originate in a common pool, perhaps at least partly on the basis of *timing*. It is possible that here we have a case where *positional information* exerts a direct influence on the pattern of synaptic interconnections. The nature of this positional information is obscure. It would seem unlikely that a *diffusion gradient* through extracellular spaces could be involved, since the system would have to give instructions to cells which are adjacent to each other and arranged in a circular sequence. Rather, some form of direct cell-to-cell communication would seem to be more likely. The application of *polar co-ordinate models* to visual systems has not, to my knowledge, been explored. The geometry of ommatidia certainly invites such an exploration. A *gradient of adhesiveness* seems unlikely to play a role in the establishment of divergent connections such as those of the dipteran retina-to-lamina projection. However, it is a reasonable candidate for the guidance of axons from the level of the retina to the level of the lamina, particularly in view of the evidence that new axon bundles always follow old ones. There is no evidence for *induction* within any single region of the optic lobe, but the relationship between the retina and the lamina meets the minimum description of induction given in Part I: the normal form of the lamina is only expressed in the presence of adjacent, normal retina. This dependence of one level upon the preceding one extends at least into the medulla and perhaps also into the lobula.

3.3 ANTENNAE AND THEIR PROJECTION AREAS

3.3.1 *Anatomy of antennae*

As Fig. 29a shows, there is marked variation in the detailed anatomy of the antennae among the various insect groups (e.g. Schneider, 1964; Kaissling, 1971). Three principal segments are found: the proximal *scape*, which contains the only antennal muscles; the *pedicel*, which contains a complex mechanosensory apparatus, Johnston's organ; and the distal *flagellum*, which may be smooth, branched or annulated and may carry as many as 100 000 sensilla of several different types, innervated by up to a quarter of a million sensory neurons (Fig. 29b). In addition to their distinctive internal structures, the scape and pedicel also bear sensilla on their outer surface. Roughly 80% of the antennal sensilla are believed to be olfactory, 15% contact chemosensory but to include a mechanosensory neuron, and 5% purely mechanosensory (Schafer, 1973 and Schafer and Sanchez, 1973, for cockroaches).

3.3.2 *Anatomy of antennal lobes*

The anatomy of the antennal projections has been described a number of times. The paper of Power (1946) on *Drosophila* is a classic. Among the modern studies, those of Boeckh, Sandri and Akert (1970) on *Calliphora* and *Periplaneta*, Pareto (1972) on bees, Strausfeld (1976) on *Musca*, and Ernst

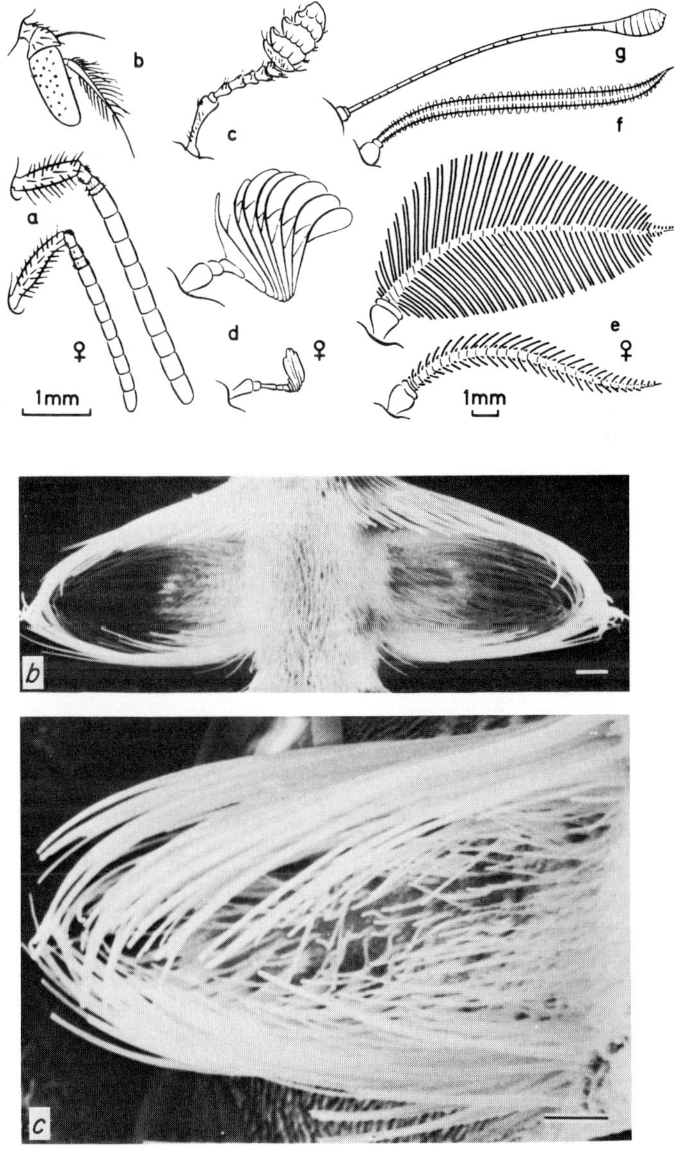

Fig. 29. Anatomy of the antenna. (*a*) Examples from a variety of insects: a, honeybee (*Apis*); b, fly (*Sarcophaga*); c, carrion beetle (*Necrophorus*); d, scarabid beetle (*Rhopaea*); e, saturniid moth (*Antheraea*); f, hawk moth (*Pergesa*); g, butterfly (*Vanessa*). In some cases male and female antennae are shown separately (Kaissling, 1971). (*b*) Scanning electron micrograph of one antennal annulus in the moth *Manduca* in which long trichoid sensilla are conspicuous (bar is 50 μm). (*c*) The same at higher magnification (bar is 50 μm) (Sanes and Hildebrand, 1976a).

et al. (1977) on *Periplaneta* and *Locusta* give many details on synaptic relationships as well as overall projection patterns.

The chemosensory (both air-borne and contact) and mechanosensory neurons of the antennae project to two separate areas of the deutocerebrum of the brain. The chemosensory fibres terminate in the antennal lobe, a region of neuropile which bulges out of the anterior and ventral part of the brain and is characterized by the presence of one to many glomeruli, areas of branching in which synapses occur (Fig. 30). The less darkly staining fibrous

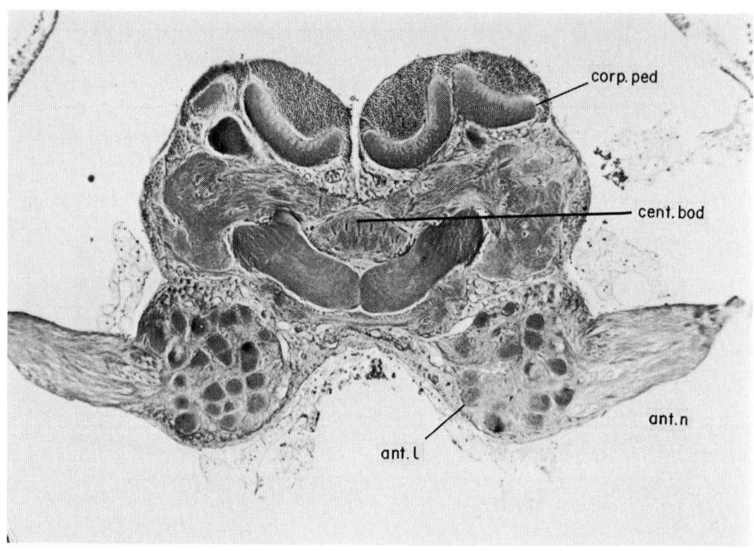

Fig. 30. Frontal section through the brain of a cockroach (*Periplaneta*) showing the entry of the antennal nerves and the glomerular structure of the antennal lobes. The corpora pedunculata and central body are also marked (Weiss, 1974).

areas in which the glomeruli are embedded consist of neuronal processes not bearing synapses. The mechanosensory fibres bypass the antennal lobe and terminate in a less sharply delineated area posterior to it known as the posterior antennal centre. It has been stated (Strausfeld, 1976) that chemosensory and mechanosensory fibres reaching the brain from other parts of the head and body, even as far away as the legs, similarly terminate in these modality-specific regions of the deutocerebrum.

3.3.3 *Development of the antenna*

The antennae of adult hemimetabolous insects develop directly from their miniature versions in the first instar nymph. We may take the slender, annulate antenna of the cockroach, recently studied by Schafer (1971, 1973; Schafer and Sanchez, 1973) as an example. In the first instar, each antenna of *Leucophaea* has approximately 46 annuli; its total surface area is about 5.2 mm^2; and it bears roughly 3700 sensilla. In the adult the number of annuli averages 118, the total surface area 43.8 mm^2, and the number of sensilla 33 700. The tip of the antenna wears away during the life of the animal and several annuli are also lost at each moult, so the amount of antennal tissue which forms during postembryonic development is even greater than these figures indicate.

How does the antenna actually grow? The first annulus of the flagellum is different from the rest, in that it becomes subdivided into 4 to 14 new annuli during each larval instar. These, of course, expand at the time of moulting. In addition, the next several annuli each divide into two during every instar, further increasing the total number. Thus, the addition of annuli occurs entirely at the base of the antenna, and the oldest annuli are always the most distal ones. As the animal matures, the dimensions of each annulus also increase, and the number of sensilla it bears increases. At the adult moult, a particularly large number of chemoreceptors is added to the surface of male, but not female, antennae; this is believed to be correlated with the males' sensitivity to female pheromones.

The antennae of holometabolous insects develop from imaginal discs. Postlethwait and Schneiderman (1971a) have given evidence that there is a line of clonal restriction between the antenna and the remainder of the head derived from the same disc. The histological aspects of the process have been described at least for *Calliphora* (Schoeller, 1964) and the moth *Manduca* (Sanes and Hildebrand, 1976a, b). The details are not pertinent to this essay, except to note that the differentiation of receptor cells described for moths is a highly synchronous process. Not only does it occur just during the brief period of metamorphosis, but within that period it is restricted to between 25 and 60 hours following pupal ecdysis. This stands in sharp contrast to the situation in cockroaches, where maturation of the animal may take more than a year and receptor cells are added in large numbers throughout this period.

As is the case with other sensilla, the sensilla of the antennna are formed by four clonally related types of cells: trichogen, secreting the seta or hair shaft; tormogen, forming the socket; nerve; and glia (for comparison, see Fig. 38). At various stages there may be more than one cell of each type, and in particular, most of the antennal sensilla have more than one nerve cell. Contrary to early claims based on light microscopy, each sensory cell sends a separate axon to the brain. The details are elegantly described by Sanes and Hildebrand (1976a, b).

3.3.4 *Growth of antennal nerves to the brain*

Sanes and Hildebrand (1975) describe the presence of a relatively small number of neurons near the tip of the developing antenna directly after pupation, well before the highly synchronous development of the antennal receptors (Fig. 31a). The axons of these cells are aggregated into two nerves, just as are those of the adult antenna (Fig. 31b). It seems entirely reasonable

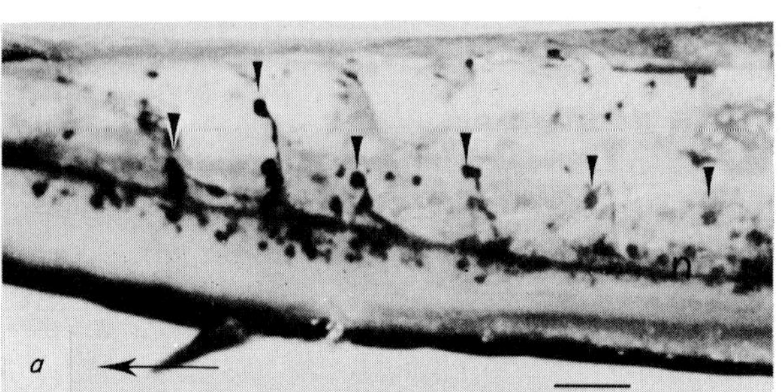

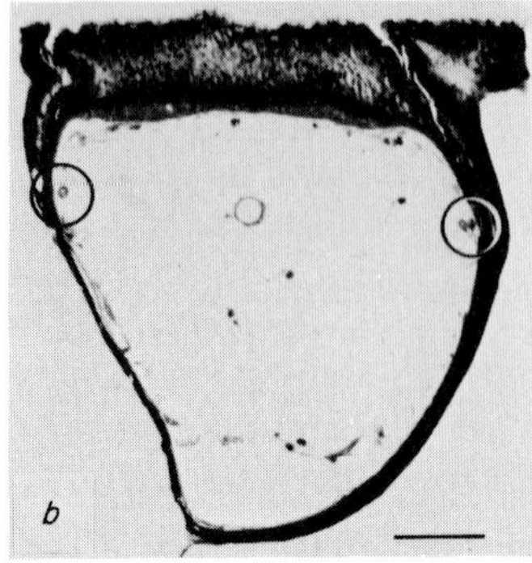

Fig. 31. Pioneer fibres in the antenna of *Manduca*. (*a*) Whole mount stained with methylene blue, showing the cell bodies of the pioneer fibres near the tip of the antenna, their axons collecting into a nerve. (*b*) Cross-section near the tip of the antenna, the two nerves formed by the pioneer fibres are circled (bars are 100 μm). (Sanes and Hildebrand, 1975.)

to suppose that the axons of the "ordinary" receptor cells follow these early-formed nerves to the brain, just as is supposed to be the case for retinula cell axons (see previous section).

The time, place and mode of origin of these early neurons is not known, but some cells staining with methylene blue and presumed to be nerve cells have been seen in the late imaginal disc.

The larvae of *Manduca* also have small antennae, and the discs are present just at their base (Sanes and Hildebrand, 1976a). The relationship of the axons of the larval antennal receptors to those of the developing adult antenna is not known but they are not essential to normal development, since removal of a larval antenna and the presumed subsequent degeneration of its axons has no detectable effect on the development of the corresponding adult antenna.

Bate (1976b) has found exactly two pairs of pioneer neurons in the developing antennae of locust embryos. Their cell bodies lie against the inner wall near the tip of the antennal rudiment, and they are demonstrably the very first neurons to establish a connection between the antenna and the brain.

3.3.5 *Development of antennal lobes*

The postembryonic development of the antennal lobes in hemimetabolous insects parallels that of the antennae. The lobes are distinct and glomeruli are present at the time of hatching, and they grow in volume during subsequent development (Panov, 1961; Edwards, 1969). It is not clear whether the full number of second-order neurons is present at hatching (as is the case, for example, in abdominal ganglia as described below), or whether new neurons are added to accommodate the increasing number of receptor fibres.

In holometabolous insects, the antennal lobes of the adult arise from a small set of neuroblasts (Panov, 1961; Edwards, 1969) – initially only three in the butterfly *Danaus* (Nordlander and Edwards, 1970). The larval lobes may be entirely lacking or may be highly developed complete with glomeruli, according to the degree of development of the antennae of the larva. But among species which have been studied, those with prominent larval antennae and antennal lobes show degeneration of the larval structures, and the adult lobes are largely or entirely composed of cells derived from the neuroblasts (Panov, 1961). It should be pointed out that while the evidence for addition of new second- and higher-order cells is clear, the evidence that all of the larval cells degenerate is not, and would be exceedingly difficult to obtain.

The structural relationships between developing sensory neurons and interneurons have not been analysed.

3.3.6 *Supportive relationships between antennae and antennal lobes*

In view of the correlation just described between the degree of development of antennae and of antennal lobes, it is reasonable to ask whether the differentiation of antennal neuropile into glomeruli is dependent on the arrival of antennal fibres. Panov (1961) removed the antennae from pupal stages of *Tenebrio* and observed that the adults lacked discrete glomeruli. The entire antennal neuropile had the staining characteristics of a single glomerulus. In a more recent and thorough study, Sanes *et al.* (1977 and personal communication) found that in *Manduca* removal of the antennae prior to or during the arrival of their axons in the brain resulted in a marked reduction of the antennal neuropile and a loss of clear boundaries between the glomeruli, but retention of at least many of the postsynaptic neurons and of the lobulated appearance of the neuropile. Furthermore, the postsynaptic neurons still developed acetylcholine receptors as identified by α-bungarotoxin binding.

While the precise nature of the influence of receptor cells on their target neurons remains to be explored more thoroughly, there is no doubt that the receptors can differentiate fully in the absence of their targets. In addition to studying the morphogenesis of the *Manduca* antenna, Sanes and Hildebrand (1976c) studied the time course of development of acetylcholine and its metabolic enzymes, and Schweitzer *et al.* (1976) studied the development of electrical activity of the receptors by the electroantennogram technique. They then compared the development of structure, transmitter metabolism and electrical responsiveness of receptors in normal animals with those of animals whose brains had been removed before the time of arrival of the antennal sensory fibres (excluding, of course, the early or pioneer fibres). No differences of any kind were detected between the two groups (Sanes *et al.*, 1976).

3.3.7 *Homeotic transformation of antennae into legs*

There is a large literature on the anatomy, genetics and development of homeotic antenna-leg transformations (see Postlethwait and Schneiderman, 1971, and Gehring and Nöthiger, 1973, for extensive bibliographies). However, it was only in 1976 that the first papers dealing with the neurobiology of this, or any other, homeotic transformation appeared.

Deak (1976) asked whether the transformed appendage forms sensory neurons, and whether these form functional central connections, by examining the proboscis eversion response to chemical stimulation of the leg. When the tarsi of a normal thoracic leg are touched by a droplet of water containing sugar, the fly everts its proboscis in preparation for feeding. Similar stimulation of the antenna elicits no response. Proboscis eversion in response to stimulating the transformed antenna would mean both that its receptors were responding to chemical stimulation and that at least some of their connections within the c.n.s. could set in motion a reflex normally initiated by leg receptors. This was, in fact, the result of the experiment in the mutant *spineless-aristapedia* (ss^a). Some sugars are more effective than others in eliciting this reflex via normal legs. The rank order of their effectiveness in stimulating the homeotic appendage was the same as for the normal thoracic appendages of the same flies. Finally, NaCl added to sugar water inhibits the proboscis eversion response in wild type flies. It fails to block the reflex in the particular mutants used by Deak, both in the normal legs and in the homeotically transformed antennae. Thus, the chemical specificity of the reflex system is maintained in the homeotic appendages.

Concurrently, Stocker *et al.* (1976) studied the central projections of sensory fibres from the transformed antenna of a somewhat different mutant, *Antennapedia*^{73b}, by the anatomical method of experimental degeneration. This method revealed that sensory fibres from the transformed portions of the antenna terminate in the usual antennal projection areas, the antennal lobes and the posterior antennal centres (Fig. 32). They do not wander indiscriminately through the brain, or invade any area not normally supplied by antennal fibres. Two differences between normal and transformed projection patterns do occur, however, and can be seen in Fig. 32: 1. The normal distribution to the antennal lobes is bilateral, but in the mutants it was entirely ipsilateral (as is true for normal thoracic legs). 2. The neuropile of the antennal lobes of *Drosophila* does not form very discrete glomeruli and rather looks like a single, continuous glomerulus. Accordingly, antennal terminations are distributed throughout its substance. In the mutants, however, the endings were concentrated around the periphery.

These anatomical results do not immediately provide an explanation for the behavioural results of Deak (1976), which have since been confirmed and extended by Stocker (1977). The anatomy of the pathways by which the proboscis eversion reflex is accomplished is not known. Some primary sensory

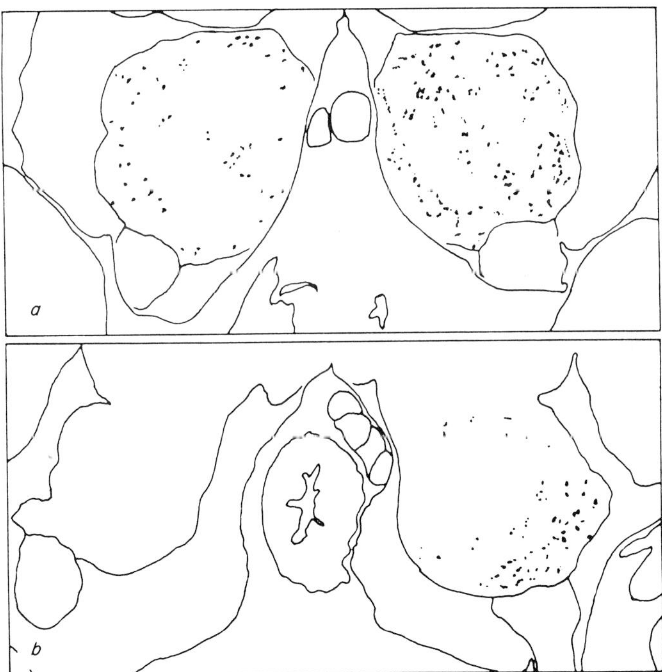

Fig. 32. Distribution of axonal terminals in the antennal lobes of wild type *Drosophila* and *Antennapedia*. The axons were revealed by cutting a single appendage (at the dashed line shown in Figs 4(*a*) and (*b*) and inducing degeneration. The distribution of degeneration spots was redrawn for clarity. The normal antenna (*a*) projects to both sides and the spots are distributed fairly uniformly. The transformed antenna (*b*) projects to the ipsilateral side only, and the spots are concentrated at the periphery of the lobe; other brain regions are not invaded (Stocker *et al.*, 1976).

fibres from the legs appear to reach the antennal lobes and the posterior antennal centre (Strausfeld, 1976). Conceivably, these might normally mediate the reflex and their postsynaptic targets would also be available to the mutant fibres. In addition, as in the visual system, some antennal fibres pass uninterrupted through the antennal glomeruli and terminate in the corpora pedunculata, which also receive second-order fibres from leg receptors. The possible role of the subesophageal ganglion in the proboscis eversion reflex has also not been explored in detail, and convergence between normal and homeotic leg fibres might occur here as well.

3.3.8 *Summary*

Several points of similarity between the antennal and visual systems of insects are apparent. The initial ingrowth of sensory fibres is along pre-existing fibre tracts. The receptors do not depend upon the central nervous system for their growth and differentiation. The first synaptic region is strongly affected by the absence of incoming sensory fibres (though, again as in the case of the visual system, the precise effect of deafferentiation is not known and may be variable).

As pointed out in Part I, particular parts of the *Antennapedia* homeotic appendage transform into particular parts of the mesothoracic leg, and this has been taken as a strong indication of the operation of the principle of *positional information*. The function or lack of function of a gene decides whether head appendage tissue will be antennal or leg in quality, but position on the appendage decides which particular structures should be formed. There is no evidence on the manner in which positional information might be encoded. The growth of axons along preformed neural paths suggests *differential adhesiveness* and is certainly compatible with a *gradient of adhesiveness*. I am not aware of any indications of *induction* within the antenna itself.

There is recent evidence (Lawrence, personal communication) that antennae are subdivided into *compartments*, and all homeotic transformations suggest the operation of controlling or selector genes. The neurobiological studies on homeotic appendages indicate that fibres from the transformed appendage enter and remain confined to the general areas of the c.n.s. appropriate to the normal appendage which has been replaced. However, their detailed distribution within the appropriate projection areas reflects the transformed nature of their cells of origin. Also, they form at least some functional connections which are appropriate to the transformed tissue. The distribution of sensory fibres from the *Antennapedia* appendage thus suggests that two different recognition mechanisms have been functionally separated: one which keeps the fibres within the appropriate territories and is not altered by the lack of function of the presumed selector gene (which lack results in the formation of the archetypal appendage, the mesothoracic leg); and another which affects the course of fibres within this territory and which is altered in the presence of the homeotic mutation. If this interpretation should be borne out, it would represent a first partitioning of the mechanisms which generate pathways in the central nervous system.

3.4 BITHORAX MUTANTS

Since the *bithorax* group of mutants has played such an important role in the elaboration of the theory of compartments, an analysis of the central projections of the wing and haltere sensilla in these mutants would seem to be worthwhile. The first results are currently in the course of publication (Palka, 1977; Ghysen, 1978; Palka *et al.*, 1979).

3.4.1 *Basic anatomy*

Two classes of sensilla are present on the surface of the wing, innervated bristles and campaniform sensilla. Their distribution has been redescribed in detail by Bryant (1975). The anterior margin bears about 400 bristles, whereas the 75 campaniform sensilla are located on the veins, the greatest concentration being on the radius, the principal proximal vein. The wing fibres reach the c.n.s. via the anterior dorsal mesothoracic nerve, the major mixed nerve of the dorsal mesothorax.

Both bristles and campaniform sensilla are also present on the surface of the haltere. The head of the haltere (capitellum) bears around 20 bristles, and the stalk (pedicellum and scabellum) around 200 campaniform sensilla. The haltere nerve is a purely sensory nerve which travels directly to the c.n.s. Both haltere and wing have internal chordotonal organs.

There is a single thoracico-abdominal nerve mass, but within it rather discrete divisions (neuromeres) corresponding to the pro-, meso-, and metathoracic ganglia can be recognized. A detailed anatomical account has been given by Power (1948). The abdominal ganglia are fused into a single lobe within this mass, but when abdominal nerves are selectively stained the composite nature of the neuropile can be detected (Schubiger, unpublished).

3.4.2 *Projections in wild type flies*

The projection of the wing sensory fibres, analysed with the cobalt (Palka, 1977; Palka *et al.*, 1979) and horse-radish peroxidase (Ghysen, 1978) backfilling methods, consists of several components which we group into two categories for our present purposes. Dorsally the main component is a bifurcating tract whose anterior branch travels through the cervical connective

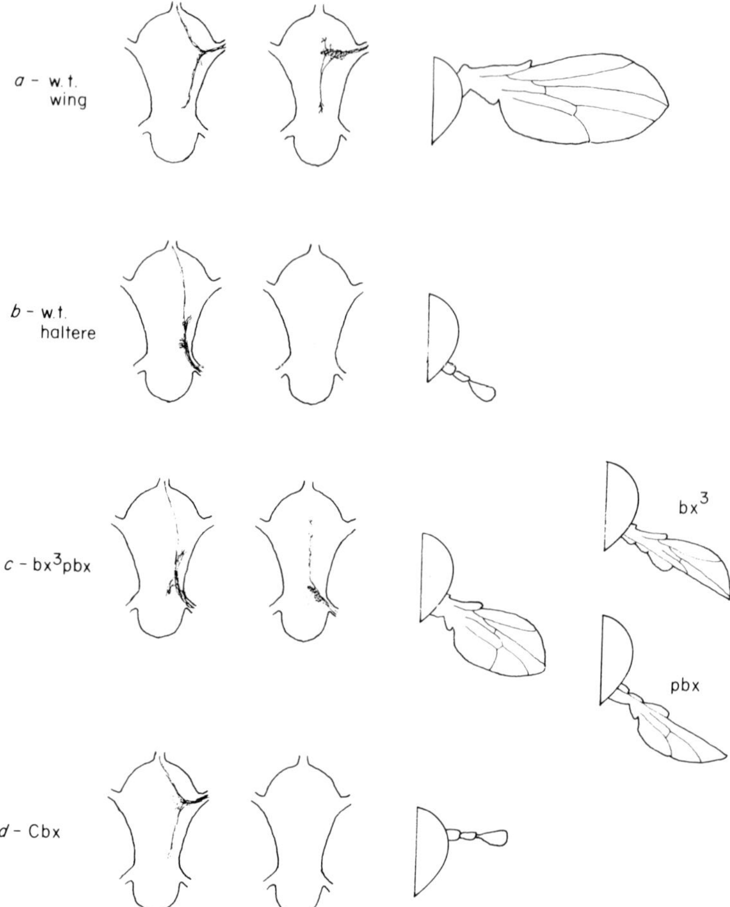

Fig. 33. Projection patterns of meso- and metathoracic appendages in wild type and mutant *Drosophila*. The dorsal components of the projections as determined by cobalt backfilling are shown on the left, the ventral components on the right. (*a*) and (*b*) wild type wing and haltere, (*c*) the double mutant *bithorax postbithorax* (bx^3pbx) in which both compartments of the haltere are replaced by wing tissue. (*d*) *Contrabithorax* (*Cbx*), wing replaced by haltere. The sketches on the right indicate the structure of the appendages.

to the head and whose posterior branch reaches the metathoracic neuromere (Fig. 33a, left). Ventrally there is an ovoid mass of fine fibres far anterior in the accessory mesothoracic neuromere of Power (1948), and a small group of large, intensely staining and widely ramifying axons (Fig. 33a, right). In addition, there are some smaller anterior and posterior tracts.

The projection of the haltere fibres is almost exclusively dorsal. Its major features are shown in Fig. 33b, left: a coarse tuft of fibres is given off towards the midline of the metathoracic neuromere; further anterior, a much smaller group of fibres is given off laterally in the main part of the mesothoracic neuromere; and the remainder of the fibres travel through the cervical connective in close association with the wing fibres. There is at best a minute projection into the ventral part of the ganglion, whose extent is in no way comparable to that of the wing projection in the mesothorax and is not indicated in Fig. 33b, right.

3.4.3 *Projections in homeotic mutants*

All of the sensilla on the wing and all but a few on the haltere lie in the anterior compartment. In the mutant *bithorax*, the anterior compartment of the haltere is replaced by anterior wing tissue (Fig. 33c); correlated with the presence of the wing sensilla, a ventral projection appears in the metathoracic neuromere and also sends fibres forward into the mesothorax. In *postbithorax*, posterior haltere is replaced by posterior wing, but this does not involve any of the sensilla; the projection of the normal haltere sensilla of the anterior compartment is much like that in wild type flies. When the two mutants are combined, yielding a fly with four wings and no halteres, the projection resembles that of *bithorax* but the ventral component appears to be better organized. In many cases it looks very much like the ventral projection which the wing forms in its own mesothoracic neuromere (Fig. 33c). In *Contrabithorax* the wing is converted into haltere to a variable extent. In individuals showing an extreme transformation, the ventral projection disappears entirely (Fig. 33d); in less extreme cases, in which some of the wing sensilla are retained, some of the ventral component is likewise retained.

It thus seems clear that some of the wing receptors are programmed to seek synaptic locations deep in the nervous system, and that they do so even in a foreign neuromere. Not all fibres of the ventral component show this behaviour, however. The ventral large fibres which enter from a mutant metathoracic appendage form a branching pattern in all three neuromeres (Fig. 33c, right) which very much resemble the ones they would have formed had they entered from the wild-type wing of the mesothorax (Fig. 33a, right; see especially Ghysen, 1978). We may guess that their programme includes instructions to grow ventrad, but not to limit their branching to the segment of entry.

Finally, as may be seen in Figure 33c left, a dorsal component with a course very much like that of the normal haltere projection is retained in the mutants *bithorax* and *bithorax postbithorax*, in which virtually no trace of haltere sensilla remains. The growth of fibres which enter this pathway seems to be governed more strongly by factors intrinsic to the ganglion or to particular tract regions. However, it must also be recognized that in these mutants some of the structures of the wing hinge are not quite perfectly formed, and their associated sensilla are not arranged quite as on a normal wing. The possibility of incomplete transformation of the receptor cells, while remote, has not been rigorously excluded.

To a first approximation it seems fair to describe these three sets of fibres as exhibiting three different behaviours: 1. the ventral fine fibres reproduce their ovoid pattern of organization but do so in a different neuromere so that a mutant fly has two ovoids, one in the meso- and the other in the metathorax; 2. the ventral large fibres reproduce their pattern of organization in all three neuromeres – perhaps they successfully seek out destination markers distributed throughout the composite ganglion; and 3. the dorsal fibres adopt the path which would have been followed by the axons of the untransformed appendage of the metathorax, the haltere. All available evidence indicates that these three classes of fibres originate in three different groups of sensilla: 1. from bristles, 2. from large campaniform sensilla and 3. from small campaniform sensilla.

3.4.4 *Projections from clones of bithorax tissue*

The above descriptions apply to flies in which every cell carries the mutant genotype, so we must suppose that the central nervous system might also show some segmental transformation. In fact, there are structural indications that this is so (King, personal communication). In order to eliminate this source of ambiguity from the results, we have produced clones of cells homozygous for *bithorax* in flies which are otherwise genotypically heterozygous for this recessive mutation, and are therefore phenotypically wild type. We have seen ventrally projecting fibres in almost all haltere-wing mosaics which include bristles and/or large campaniform sensilla, and have failed to see them in cases in which the clones do not include these receptor types. Thus, the formation of a ventral component by the appropriate sensilla does not require a homeotic transformation of the ganglion. However, the detailed distribution of the ventral fibres differs somewhat from that in mutant flies and suggests that some degree of transformation, important for pathfinding decisions other than the one to turn ventrad, probably does occur.

3.4.5 *Summary*

The wing has been a favourite experimental system for insect developmental biologists interested in pattern formation, and many of the theories reviewed in Part I refer explicitly to it. For example, the wing and haltere of *Drosophila* have contributed much of the evidence on which the theory of *compartments* is based. The relationship between certain proximal wing bristles and their associated bracts is generally regarded as an example of *induction*. The regular replacement of particular parts of the wing by particular haltere structures in *Contrabithorax* is conveniently described in terms of *positional information*. The wing of *Manduca* is the material on which the theory of a *gradient of adhesiveness* was developed.

For unknown reasons, in *Drosophila* all of the receptors on the wing and most on the haltere are in the anterior compartments, whereas they are present in both compartments in the corresponding legs. Otherwise, there is at present no evidence that the compartmentalization of the wing has any influence on its neuroanatomy or on the central destinations of its sensory fibres.

The initial results on projection patterns from mutant and mosaic appendages suggest different developmental programmes for different classes of fibres: fibres go ventral or dorsal, remain within one segment or ramify in several, depending on which peripheral receptors they originate in. At least some features of these programmes can be executed by homeotic fibres in wild-type ganglia, but the projection patterns in mosaic and in mutant animals are not identical and some homeotic transformation of the ganglia is indicated.

3.5 CERCI AND ASSOCIATED INTERNEURONS

The abdominal cerci of orthopteroid insects are purely sensory appendages arising from the 10th abdominal segment; an extensive comparative account has been given by Sihler (1924). In species in which they are particularly highly developed, their length approaches the length of the body. They are covered with a variety of sensilla; in *Acheta domesticus*, the house cricket, the number of sensilla on each cercus is about 3400 and the number of receptor cells and of axons in the cercal nerve between 10 000 and 11 000 (Edwards and Palka, 1974). The cerci have received attention partly because they are major sense organs, but especially because they provide input to the largest interneurons in the c.n.s., the abdominal giant fibres, which were among the very first and easiest single neurons to be studied in insects (e.g. Pumphrey and Rawdon-Smith, 1936, 1937). The pioneering analysis by Roeder of the role of cerci and the giant fibres in escape responses of cockroaches was summarized in his book of 1967. Access to the more recent literature may be had through the papers of Parnas and Dagan (1971) and Camhi (1978). The most detailed physiological study of the cercal mechanoreceptors is still that of Nicklaus (1965).

The present summary is based on studies on crickets, because virtually all of the experiments dealing with development and regeneration of the cerci have been done with these animals. The principal background papers describing the receptors (including the interesting phenomena associated with their moulting) are Gnatzy and Schmidt (1971, 1972a, b) and Schmidt and Gnatzy (1971, 1972); analysing the relationship of cercal receptors and giant fibres, Edwards and Palka (1974), Palka, Levine and Schubiger (1977), Murphey *et al.* (1977), Palka and Olberg (1977) and Matsumoto and Murphey (1977b); and describing the morphology of a number of the interneurons, Mendenhall and Murphey (1974).

3.5.1 *Basic anatomy*

The majority of sensilla covering the cercus of a cricket are multiply innervated short bristles set stiffly in their sockets and having a pore at the tip; they are presumed to be combined chemo- and mechanoreceptors. Nothing is known of their physiology or of the central connections of their axons except that they terminate in the ipsilateral hemiganglion. The best known sensilla are the filiform hairs, with long, slender, singly innervated shafts set delicately in their sockets in such a way that they oscillate only in a single plane when exposed to a sound or a wind stimulus. Associated with the sockets of the filiform hairs are one or several campaniform sensilla. On the proximal, medial face of each cercus is a cluster of clavate hairs with bulbous shafts believed to be gravity detectors (Bischof, 1974, 1975). The distribution of the filiform hairs on the surface of the cercus is ordered with respect to their plane of vibration: hairs vibrating close to the transverse plane of the cercus (T-hairs) are located dorsally and ventrally, and hairs vibrating longitudinally (L-hairs) are located medially and laterally (Fig. 34a).

The last abdominal ganglion of crickets, which receives almost all of the cercal sensory fibres, arises from the fusion of several of the most posterior ganglia during embryogenesis (Panov, 1966). It is organized in the fashion typical of insect ganglia – cell bodies around the outside, neuropile in the centre, and not much order evident unless selective staining techniques are used. The axons of the cercal nerve terminate almost entirely ipsilaterally,

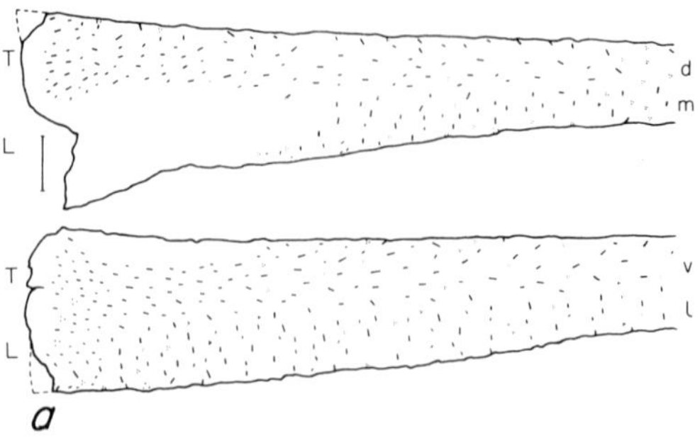

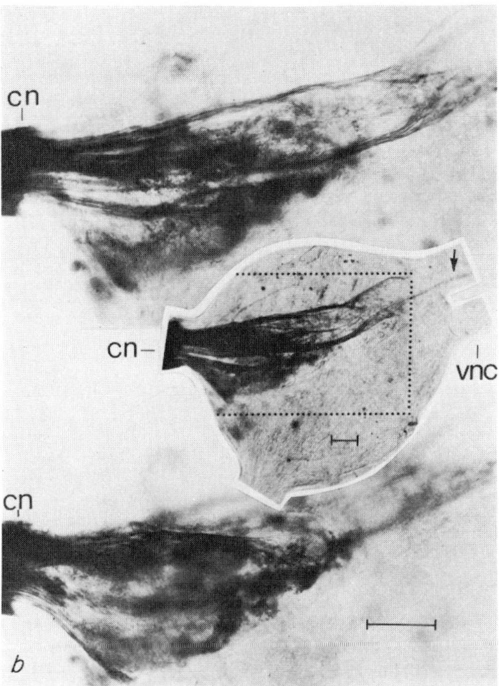

Fig. 34. Anatomy of cerci (the house cricket *Acheta*). (*a*) Map of the distribution of the T- and L-filiform hairs on the proximal 1/3 of the right cercus of an adult male. The short lines indicate the location, orientation and approximate relative size of each socket; the plane of vibration of the associated hair shaft is at right angles to the socket orientation. Dorsal (d) and ventral (v) quadrants contain hairs with predominantly transverse vibration planes, medial and lateral quadrants predominantly L-hairs. The empty area is occupied by clavate hairs (bar is 0.3 mm). (*b*) The projection of the cercal nerve in the terminal abdominal ganglion, dorsal view above and ventral view below. The projection is almost entirely ipsilateral and mainly posterior, and clusters of endings are apparent (bars are 100 μm) (Palka *et al.*, 1977).

most of them posteriorly but a significant number in a more anterior projection area. Their endings are conspicuously clustered (Fig. 34b).

Within the ganglion are found a number of bilaterally matched pairs of neurons recognizable by their dendritic configurations. By far the best known ones are the Medial and Lateral Giant Interneurons (MGI and LGI) shown in Fig. 35a.

3.5.2 *Physiological assessment of neural connections*

The electrophysiological responses of the MGI and LGI to sensory stimulation can be recorded both extra- and intracellularly (e.g. Murphey *et al.*,

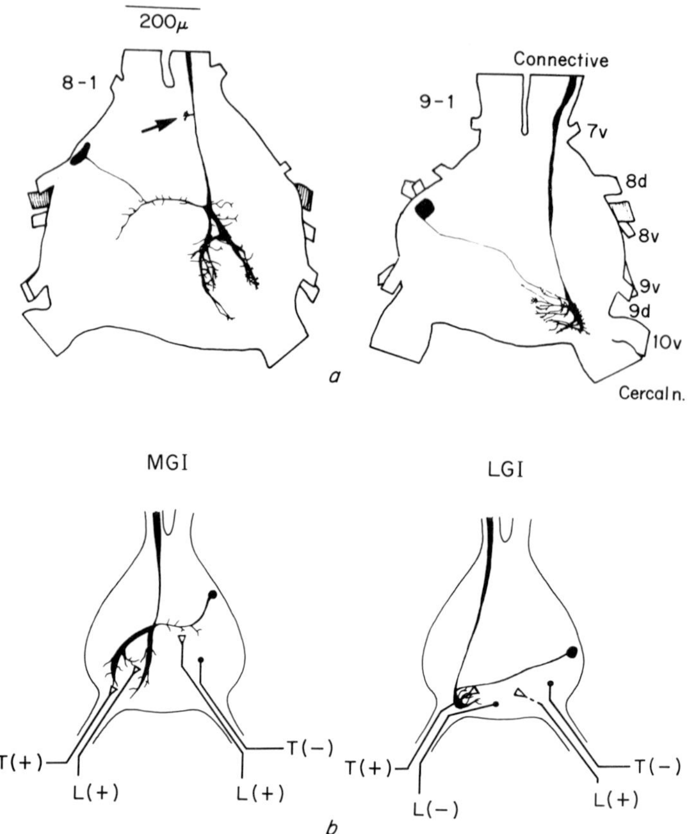

Fig. 35. Abdominal giant interneurons of crickets. (*a*) The anatomy of the MGI and LGI drawn from cobalt back-fills. The dorsal and ventral segmental nerves of the compound ganglion are indicated (bar is 200 μm) (Mendenhall and Murphey, 1974). (*b*) Model of the connections between the T- and L-hairs of the cerci and the MGI and LGI; open triangles indicate excitatory connections, closed circles inhibitory ones. Details in text (Palka and Olberg, 1977).

1977). To the extent that receptor classes or single receptors can be stimulated selectively, the responses of the interneurons can give evidence for the pattern of connections between the receptors and the interneurons (e.g. Palka and Olberg, 1977; Matsumoto and Murphey, 1977b). Furthermore, the same analytical methods can be used to evaluate the effects of surgical and other manipulations carried out during postembryonic development upon the receptor-to-interneuron connectivity patterns (see below).

The response of the MGI and LGI to sound stimulation is markedly directional, sound sources approximately transverse to the ipsilateral cercus being highly effective and sound sources aligned with the cercus very much less effective. This immediately suggests that the ipsilateral T-hairs provide the dominant excitatory input to these cells, and indeed a variety of tests based on selective stimulation or elimination of receptors show that ipsilateral T-hairs are strongly excitatory to the MGI, the ipsilateral L-hairs are weakly excitatory, the contralateral L-hairs even less excitatory, and the contralateral T-hairs inhibitory (Fig. 35a). The pattern for the LGI is somewhat different and also indicated in Fig. 35b. In addition, an unknown group of receptors not on the cerci and sensitive to substrate vibration also excites the giant interneurons (Edwards and Palka, 1974).

3.5.3 *Development of the ganglion and cerci*

Gymer and Edwards (1967) conducted a quantitative study of the postembryonic development of the terminal ganglion whose most important result was strong evidence that most or all of the nerve cells intrinsic to the ganglion are present at the time of hatching. The 40-fold increase in volume of the ganglion is attributed primarily to an increase in cell volume. The number of glial cells also increases greatly, from about 1000 to 17 000.

Edwards and Chen (1979 and unpublished) have examined the embryonic development of the cerci and of the terminal ganglion. They find that the cerci, like other innervated appendages, develop pioneer fibres long before the differentiation of ordinary sensory cells. The neuropile of the ganglion is very loosely textured with large extracellular spaces and dendritic processes arranged at right angles to each other prior to the arrival of the functional sensory axons, which is one of the last events of embryonic development. The cellular complement of the ganglion appears to form independently of cercal innervation; however, as is described below, the dimensions of interneurons are markedly affected if sensory input is removed.

During postembryonic development the cerci increase from 1 mm at hatching to about 10 mm in adult length and the number of sensilla increases about 50-fold. The mitoses which produce the new cells occur all over the cercus, though they may be concentrated near the base (Daniels, 1968; Chen, unpublished).

3.5.4 *Effects of rotation and exchange of cerci*

If a cercus is transected at its base, rotated clockwise or counter-clockwise, and grafted back into the same socket, it derotates back towards its original orientation in succeeding moults by as much as 90° per moult (Palka and Schubiger, 1975). If the right cercus is grafted to the left cercal socket or vice versa, lateral and medial supernumerary cerci are produced; if the cercus is inverted before grafting, dorsal and ventral supernumeraries develop as

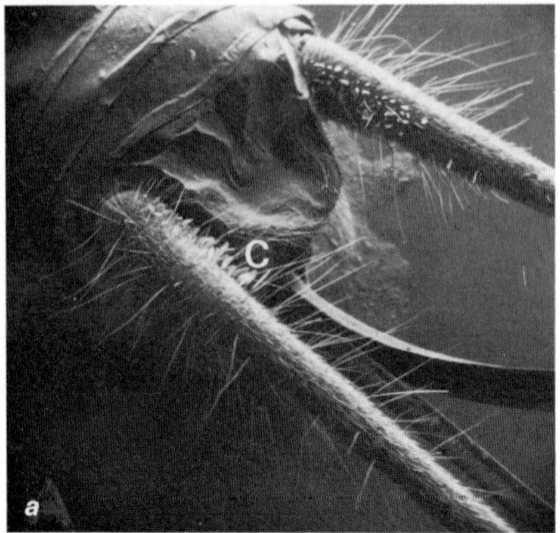

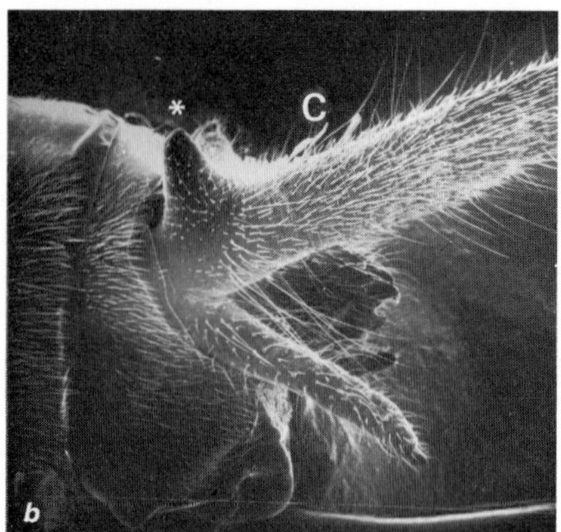

Fig. 36. (*a*) Proximal regions of the cerci of a normal cricket (*Acheta*). (*b*) Supernumerary cerci (asterisk) in a cricket produced by grafting a right cercus to the left socket following dorsoventral inversion. The clavate hairs on the medial face of the cercus base are marked with C (Palka and Schubiger, 1975).

shown in Fig. 36 (Palka and Schubiger, 1975). Long cercal fragments grafted to long stumps produce intercalary regeneration with reversed polarity (Schubiger, unpublished), as in leg grafts. Thus, the cerci respond to these operations just like legs do, and any theory to explain the responses of leg tissue is equally likely to apply to the cerci.

Cerci can be grafted to any location on the body surface, such as leg stumps (Edwards and Sahota, 1967). When they are grafted with careful regard to orientation, back-rotation of cerci upon leg stumps can be shown. Similarly, right cerci grafted to left leg stumps, etc., can produce supernumeraries as expected (Schubiger, unpublished). It seems reasonable to conclude that the tissues of cercus and leg can "talk to each other", and economical to describe this conversation as being about positional information. In contrast, it has not been possible to show predictable morphogenetic responses to the grafting of cerci to various locations directly upon the abdomen (Schubiger, unpublished).

Palka and Schubiger (1975) presented evidence, based on testing the directional characteristics of the MGI and LGI, that T-hairs always re-establish their dominance over L-hairs following rotation and exchange operations. The operations, of course, require cutting the cercal nerve, so the T-ness or L-ness of its parent sensillum is reflected in the behaviour of the growing axon. Since T- and L-hairs are grouped on the surface of the cercus, it was supposed that the same factors (positional information) which determine the orientational characteristic of the sensillum also determine the destination of its axon.

In exchange experiments, cercal axons were found to project to their own new hemiganglion and not to cross the midline to the side of the ganglion into which they had originally grown. Apparently, the sensory fibres discriminate among potential postsynaptic cells and between ipsilateral and contralateral sides of the ganglion, but not between left and right.

3.5.5 *Effects of deafferentation*

If a cricket is raised throughout its postembryonic life with one (or both) of its cerci removed, both anatomical and physiological effects can be detected (Palka and Edwards, 1974). The deafferented hemiganglion is much reduced in volume; its giant interneurons, deprived of their usual powerful ipsilateral cercal input, respond up to 20 times as well to the contralateral cercus as they would in a normal animal; and the normally rather weak input from the unknown vibration sensitive receptors is much enhanced.

The giant interneurons themselves are affected by deafferentation in a dendrite-specific way (Murphey *et al.*, 1975). The dendrites of the MGI and other identified cells on the deafferented side grow to much less than their normal length, the reduction sometimes amounting to as much as 50%. Dendrites of the very same neurons on the intact side reach their normal dimensions. In all cases, the basic morphology of the postsynaptic cells is preserved and the various dendrites can easily be recognized. Even if crickets are deprived of both their cerci for six or seven of the nine larval instars and regeneration is permitted only during the last two, the sensory fibres are still able to form functional connections with the giant interneurons (Edwards and Palka, 1971). In a more detailed anatomical-physiological study, Murphey *et al.* (1976) have shown that the deafferentation effects are reversible, that the period of greatest sensitivity to deafferentation is during the first two instars, and that anatomical and physiological effects respond differently during a recovery period.

3.5.6 *Effects of sensory deprivation*

Matsumoto and Murphey (1977a) have given very strong evidence that actual deafferentation, removal of the receptor cells accompanied by degeneration of their axons, is not necessary for the previously described physiological effects – simple silencing of undamaged hairs is sufficient. Silencing was accomplished by covering the cercus with a facial cream which hardened within the sockets even though the animal groomed off the excess on the surface. The treatment had to be repeated following each larval moult, of course; the hairs of the adult were left free to move. The axons of the cercal nerve looked normal, and recordings from receptor cells showed no differences between treated and control cerci. However, the intracellularly recorded responses of the MGIs on treated and control sides were drastically different (Fig. 37). Increased inhibition from the control side appeared to account for a large part of the unresponsiveness of treated MGI.

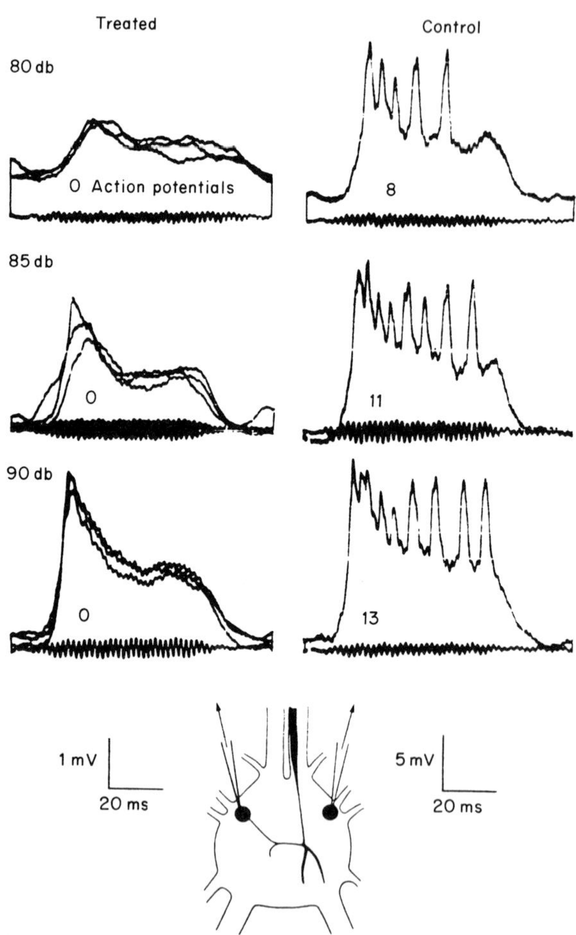

Fig. 37. Effect of sensory deprivation on the intracellularly recorded responses of the MGI in *Acheta*. On the treated side the cercal hairs were immobilized during larval stages but free to move in the adult; sound stimulation produced only a weak depolarization and no impulses. The cercal hairs on the control side were free to move throughout development, and sound stimulation produced numerous impulses. Note the different voltage scales (Matsumoto and Murphey, 1977).

A quite different method for producing functional deafferentation was employed by Bentley (1973; 1977). He isolated two mutants whose cercal sensilla lacked hair shafts. In the better studied of the two, *filiformless* (*fl*), filiform hairs were lacking from hatching. The stiff, appressed hairs also degenerated starting several weeks after the adult moult, so that the period for experimentation was confined to larval stages and early adulthood. Bentley found that the receptor cells of the affected hairs looked normal and sent axons into the cercal nerve, and that the MGI responded normally both to electrical stimulation of the cercal nerve and to intracellularly injected current. Within the limits of the analysis, therefore, the filiform receptors appeared to be quite normal but silent. Cobalt staining showed that the MGI dendrites were easily recognizable but were thinner than normal and looked stunted. When filiformless cerci were grafted to replace one cercus of normal crickets in the first instar and adults were tested physiologically, a much increased amount of contralateral excitation by the normal cercus was seen. The morphological effect is similar to that recently seen by Murphey (personal communication) in experimentally deprived animals, and the physiological effect is similar to that found by Palka and Edwards (1974) in surgically deafferented animals. Thus, sensory deprivation by genetic means yields results generally parallel to what is seen with experimental manipulation, reinforcing the conclusion that even the physiological activity of afferent axons exerts both physiological and anatomical effects on neural circuitry.

3.5.7 *Summary*

The morphogenesis of the sense organ itself, the cercus, follows the same rules as have been described for other parts of the body which are not specialized for neural functions. Every grafting experiment done on cerci has yielded the same result as the corresponding experiment on legs. We may, therefore, conclude that the theories of *positional information*, *longitudinal gradients* and *polar coordinate* based models apply with the same strength as they do to other regions of epidermis. Indeed, cercal grafts interact with leg stumps almost as vigorously as they do with the tissue of their own sockets, in accord with the view that the mechanisms actually encoding positional information should be widespread and perhaps universal.

As we have seen with the other systems reviewed thus far, primary sensory fibres reach the c.n.s. along the surface of a bundle of pioneer fibres. The development of particular dendrites of the cells with which they synapse is retarded if the sensory axons are removed, even postembryonically. The effect would appear to be quite specific, since one dendrite of a given postsynaptic cell may be retarded while another one grows normally. Because the target cells have been identified individually and studied physiologically, it has been possible to show effects of experimental removal of sensory input even in the presence of structurally normal receptors whose physiological function is restored in apparently normal form following the adult moult. Genetically produced sensory deprivation gives similar results.

The incoming sensory fibres respect the midline of the ganglion even if the cerci have been exchanged left-for-right. They seek out their appropriate synaptic partners according to the region of the cercus in which their somata are lodged, instar after instar and even in the face of re-orientation of the cercus relative to the body. When a postsynaptic cell is deprived of one of its usual strong inputs, the remaining inputs are strengthened implying some sort of *competition*.

The data on this system, more than on any other, emphasize the *plasticity* of neural anatomy and functional connections. One wonders whether a similar degree of plasticity will be shown in other systems when indentified postsynaptic neurons are studied.

3.6 ABDOMINAL SEGMENTS

We now come full circle in this essay, for many of the experiments described in Part I which led to the explicit formulation of the concept of gradients of positional information in insect development were done on the epidermis of abdominal segments, as were most of those which explored the particular hypothesis of gradients of diffusible materials. The literature on overall morphogenesis in abdominal segments is rich. The neurobiological literature is largely devoted to describing the distribution of sensilla on the surface of the segments and the pattern of cell divisions which leads to the production of the several cells comprising a sensillum. Very little is known of the specific central connections of the sensory fibres. However, a direct attack on the development of identified neurons within the ganglia is being made.

3.6.1 *Anatomy and development of sensilla*

Convenient reviews of the structure and development of the two types of sensilla, bristles and campaniform sensilla, which are most abundant on the surface of the abdomen are given by Dethier (1963) and Wigglesworth (1972), among others. The review of Schwartzkopff (1974) emphasizes sensory physiology. The material of the present section is based primarily on the descriptions of Wigglesworth (1953) and Lawrence (1966a) of *Rhodnius* and *Oncopeltus* respectively.

As described above for antennae, each sensillum contains a trichogen cell which forms the hair shaft (or the dome of a campaniform sensillum), a tormogen cell which forms the socket or stiffened rim, one or several nerve cells

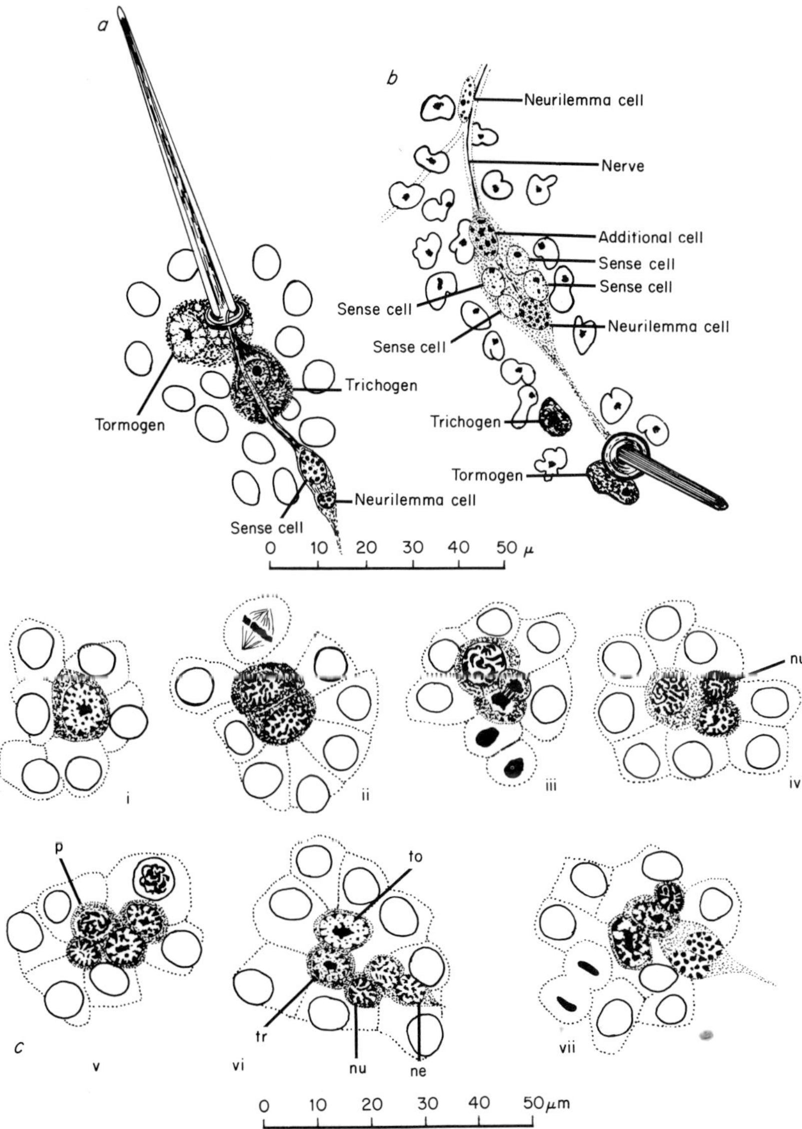

Fig. 38. Anatomy and development of abdominal bristles in *Oncopeltus*. (*a*) A singly innervated bristle showing the trichogen, tormogen, nerve and sheath (neurilemma) cells. (*b*) A multiply innervated bristle. (*c*) The sequence of cell divisions producing the four cells of a singly innervated bristle. The cells in dotted outline and with open nuclei are ordinary epidermal cells. Drawn from fixed and stained whole mounts (Lawrence, 1966a).

and a sheath or neurilemma cell which envelops them (Figs 38a, b). All of these cells are believed to arise from a single mother cell which is first recognizable among the ordinary cells of the epidermis when it swells and slightly changes its staining characteristics (Fig. 38c). When it divides, its spindle is oriented in a species-specific way, in many cases normal to the cuticle so that division produces peripheral and central daughter cells. The plane of the second division is also highly specific and often different for the two daughter cells. These two successive divisions yield the four cells typical of mechanoreceptors. Chemosensory sensilla generally have several nerve cells (Fig. 38b) which are believed to result from additional divisions of the primordial nerve cell. See Lawrence (1966a) for further details.

3.6.2 *Anatomy and development of the ganglia*

The free abdominal ganglia are among the simplest central nervous structures in insects, consisting of a rind of cell bodies and a core of neuropile not obviously organized into distinctively different regions. However, the branching patterns of neurons within the ganglia are as complex as anywhere in the nervous system. Figure 39, from the exquisite studies of Zawarzin (1924) on

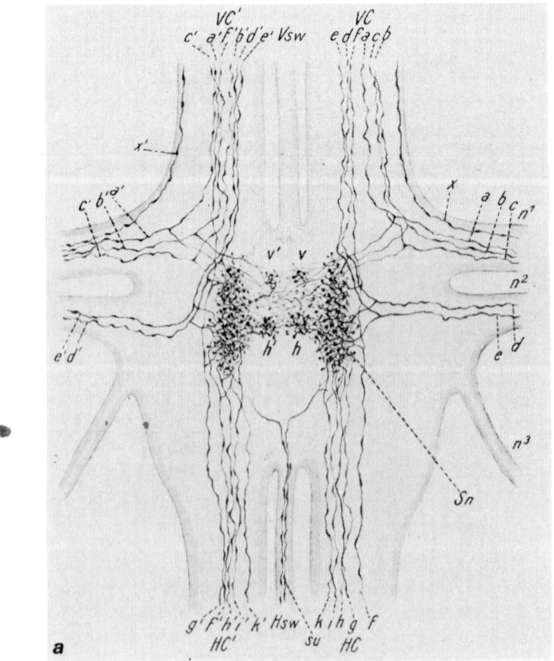

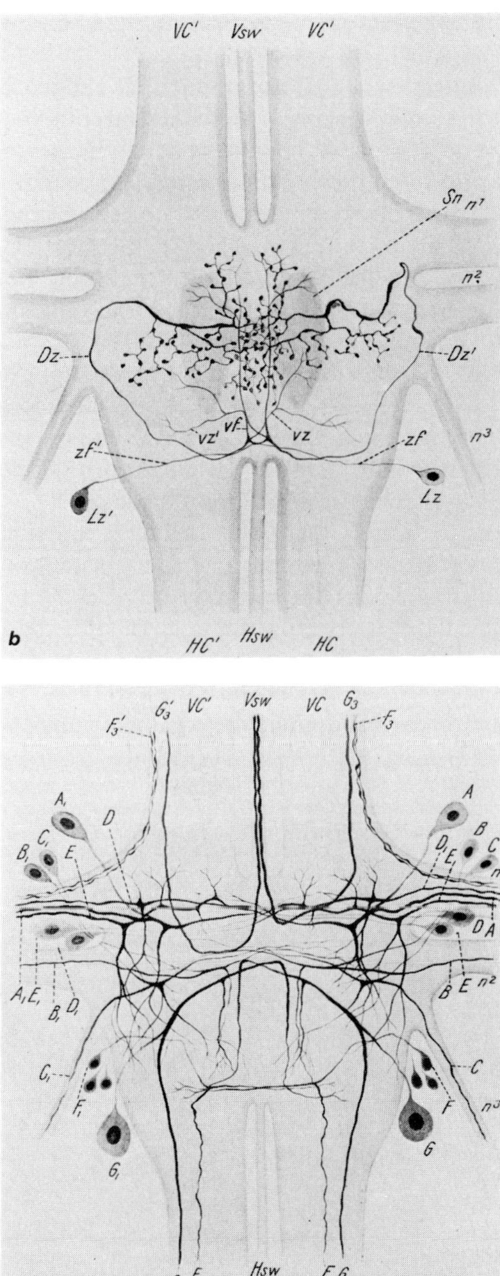

Fig. 39. Classes of neurons found in an abdominal ganglion of dragonfly larvae (*Aeshna*). (*a*) Sensory neurons; note that most of them bifurcate, one branch ramifying locally and the other ascending in the ipsilateral connective. (*b*) Intraganglionic interneurons. (*c*) Motor neurons. Drawn from methylene blue preparations (Zawarzin, 1924).

dragonfly larvae, illustrates types of sensory, inter- and motorneurons he saw in methylene blue stained preparations.

As is the case in other ganglia, the abdominal ganglia contain neurons with unique identities, recognizable in individual after individual of the same species. Not only do they have consistent adult branching patterns and physiological functions, but they undergo consistent changes of pattern during development. Taylor and Truman (1974) described how the population of motor neurons in an abdominal ganglion of the moth *Manduca* changes during metamorphosis. Of the 74 larval motor neurons, 70 are still present immmediately after metamorphosis. 40 of these die very soon after, and 12 new ones appear. Thus, the great majority of adult motor neurons were also present in the larva, and since the larva has both different musculature and different behaviour than does the adult, Truman and Reiss (1976) asked whether the morphology of the persisting neurons did not also change. They showed that at least for some neurons it changes profoundly (Fig. 40), and in a sequence which characterizes each particular cell.

Clearly, neurons with individual identities have individual life histories (see also Bentley, 1973). The beginning of the life history of a neuron is the neuroblast, and several efforts are currently being made to trace the entire cell lineage of particular neurons from the small set of embryonic neuroblasts through to the much larger set of neurons present in the larva or adult. The first step in this analysis has been published by Bate (1976). Embryonic locusts have 28 neuroblasts in each abdominal hemiganglion and 30 in each thoracic hemiganglion, arranged in a pattern which is repeated from segment to segment

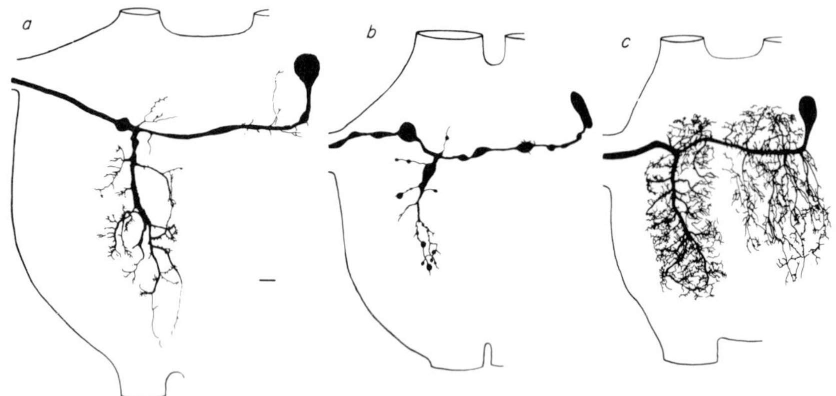

Fig. 40. Changes in the morphology of an identified abdominal motor neuron during metamorphosis in the moth *Manduca*. (*a*) The larval configuration, with a single major dendrite. (*b*) During pupation the dendrite is largely retracted. (*c*) In the adult the original dendrite has regrown and become much more elaborate, and a new dendritic arborization on the opposite side of the midline has formed (bar is 20 μm) (Truman and Reiss, 1976).

and animal to animal (Fig. 41c). The neuroblasts start to divide, cease dividing and ultimately die in a fixed sequence among the ganglia (Fig. 41a), one of whose consequences is the production of many more cells in the thoracic ganglia than in the abdominal, even though the number of neuroblasts is virtually the same. Cell lineage analyses are proceeding both by lesioning single, identified neuroblasts with an ultraviolet microbeam and looking for

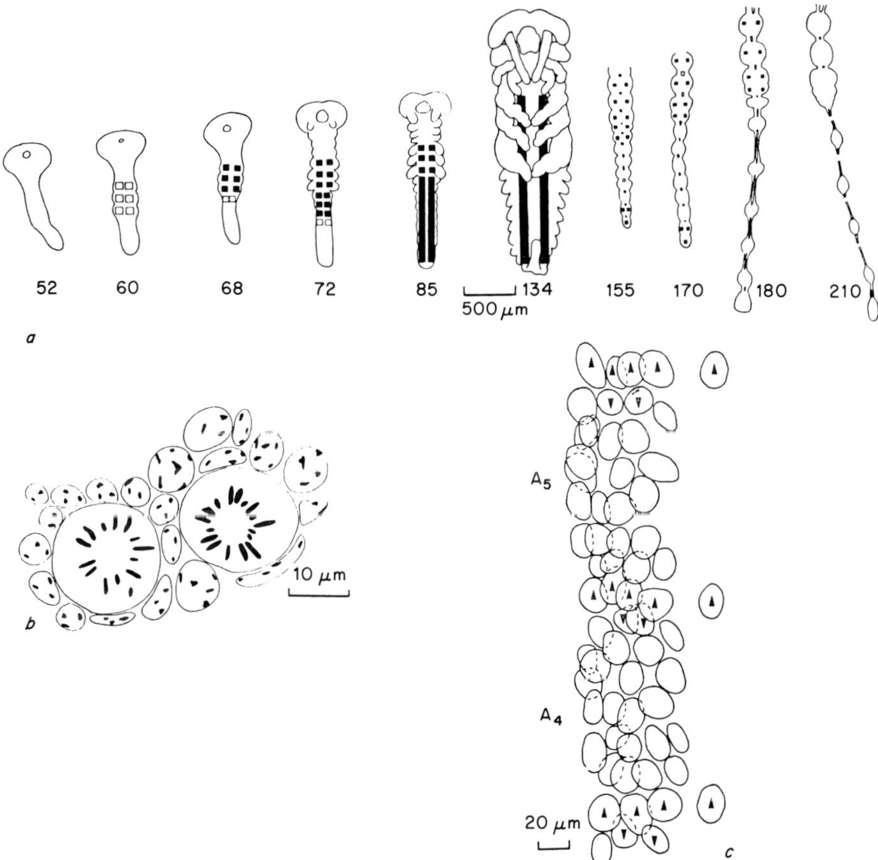

Fig. 41. Neuroblasts in the locust embryo (*Locusta*). (*a*) The distribution of neuroblast divisions at different stages of development. The squares represent ganglia, and ganglia containing dividing neuroblasts are shown black. Divisions in abdominal ganglia start later and cease earlier than in thoracic ganglia. (*b*) The appearance of two dividing neuroblast nuclei surrounded by ordinary epidermal cell nuclei. (*c*) A map of the neuroblasts of one side in two adjacent abdominal ganglia at 96 h of embryonic life. The cells marked with arrows are at the margins of the forming ganglia. The arrangement is complex but regular; for example, the anterior row contains four cells, the posterior row two, and there is a single unpaired median neuroblast aligned with the anterior row (Bate, 1976a).

cell deficiencies later in development (Bate, personal communication), and by injecting stable intracellular marker substances into single neuroblasts and studying their progeny (Bentley, personal communication).

3.6.3 *Growth of sensory axons*

Other than a change in staining characteristics, the first sign of differentiation among the four cells of the developing sensillum is the extrusion of an axon by the nerve cell (Fig. 38c, vi and vii). The developing axon grows at the base of the epidermis but without crossing the basal lamina until it encounters some other axon, whereupon it follows the surface of this pre-existing axon or its sheath cells. Single axons, most obviously ones added during postembryonic development, thus collect into bundles which ultimately pierce the basement lamina and reach the c.n.s.; an example of a network formed by

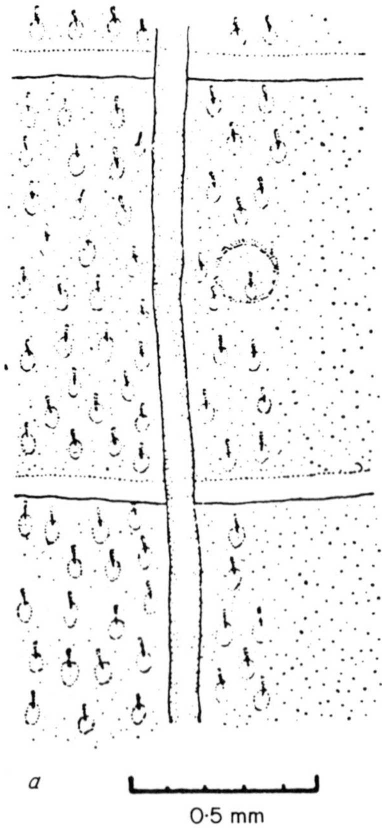

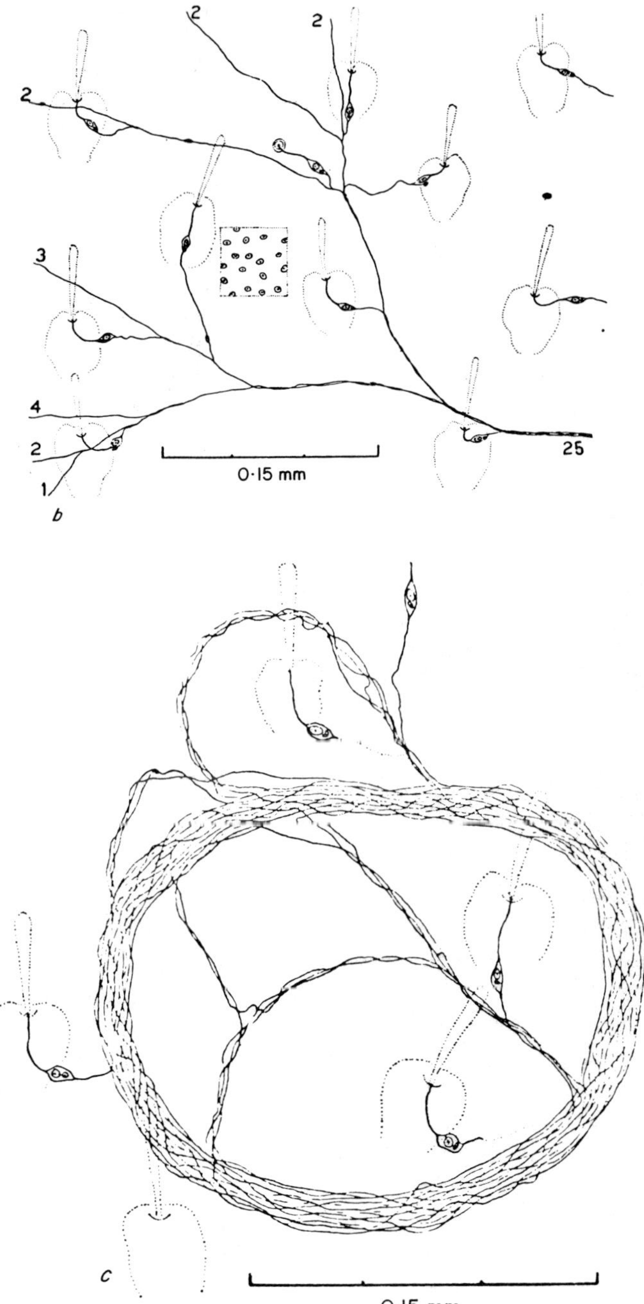

Fig. 42. Receptor cell axons in the integument of *Rhodnius*. (*a*) A sketch of the midline region of the abdomen showing the distribution of bristles, their lack in a region burned in the previous instar (stippled), and a circular nerve at the edge of the damaged area. (*b*) The pattern in which sensory axons collect to form larger bundles. The inset shows the nuclei of ordinary epidermal cells and the numbers indicate how many axons could be distinguished in various nerve strands. (*c*) Details of a circular nerve whose axons failed to penetrate the basal lamina. Drawn from silver impregnated whole mounts of epidermis. The dotted outlines indicate plaques of specialized cuticle surrounding individual bristles (Wigglesworth, 1953).

this process is shown in Fig. 42b. However, the question of how the very first axons reached their destination has not been studied.

If the cuticle of *Rhodnius* is damaged and regenerates, it initially forms no bristles. The axons from sensilla forming at the edge of such a bare, damaged area have a tendency to avoid it, and sometimes as they turn back the first nerve they encounter is an earlier segment of the same cell, which they follow as they would any other nerve. This, of course, results in a circular nerve trapped on the outside of the basement lamina (Fig. 42c) and is one of the clearest examples of the general tendency of axons to follow along the surface of other axons.

Wigglesworth (1953) originally described the initial growth of the axon as being randomly oriented. However, Lawrence (1975) (following Hasenfuss, 1973) found that axons very rarely and perhaps never cross from one segment to an adjacent one, whereas they cross the midline of the body freely (Fig. 43a). In *Oncopeltus* the intersegmental boundary is marked only by a subtle change in the shapes of the epidermal cells and by a sudden change in colour. There is no obvious barrier to the passage of axons, and yet they do not cross. Lawrence has given evidence (1973) that marked clones in the abdominal segments of *Oncopeltus* behave as if the intersegmental border were a compartment boundary in the sense defined in Part I; the observed distribution of axons would then be compartment-limited.

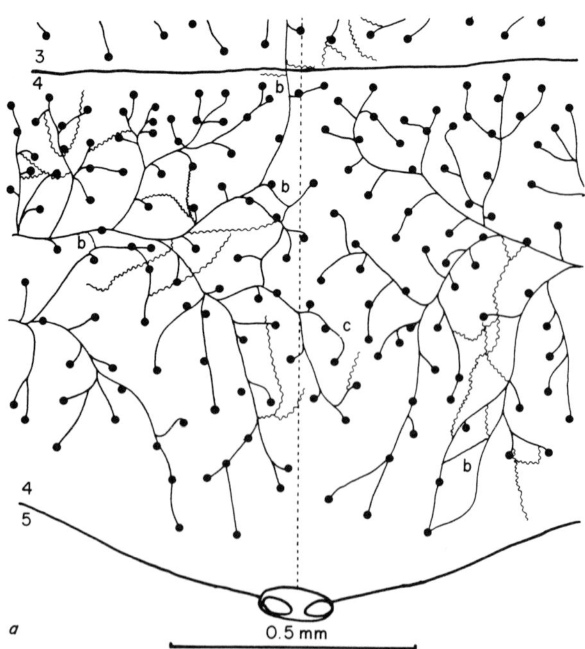

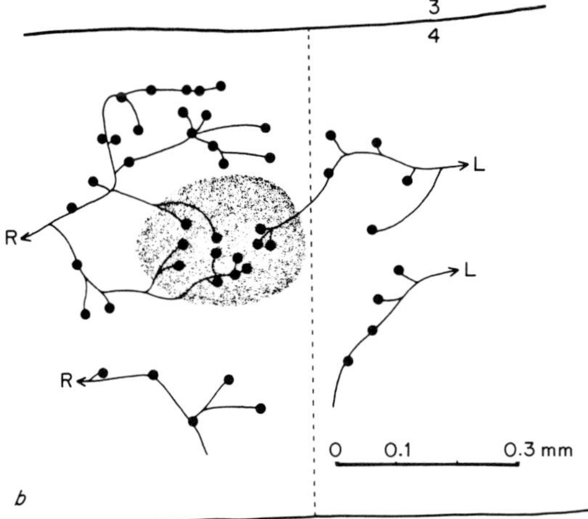

Fig. 43. Receptor cell axons in the integument of *Oncopeltus*. (*a*) A large field showing that axons cross the midline freely as in (c), but not the intersegmental boundary. The fibres marked (b) may not be axons. The wiggly lines indicate fibres interpreted as freely branching dendrites. (*b*) The course of axons from a colour-marked graft – there is no restriction to their passing out of the graft tissue (stippled) and joining a host nerve (Lawrence, 1975).

The mechanism by which sensory fibres become restricted to a single segment or compartment is not known. Figure 43a suggests that axons generally grow towards the middle of the proper segment, so that a sensory cell near the border must know which mid-segment is its correct one. Only one specific hypothesis can be ruled out. When colour-marked pieces of epidermis from the second abdominal segment were grafted into the middle of the third segment, axons from the graft crossed freely into the epidermis of the host (Fig. 43b). Thus, there is no hindrance to axonal growth associated with the particular identity of adjacent segments.

3.6.4 *Can a peripheral gradient influence central connections?*

On either side near the anterior end of each abdominal segment of a moth pupa is found a hair-lined pit, the gin trap. Touching the hairs within it elicits a sudden lateral bending of the abdomen towards the stimulated side. As one segment telescopes into the next, the horny posterior margin of the next anterior segment slides over the gin trap and crushes the offending object. A close analysis of the physiology of this system by Bate (1973a, b, c; Bate and Lawrence, 1973) yielded physiological indications that the central connections of the sensory bristles which trigger this response may be interpretable in terms of the segmental gradient which is believed to exist in the epidermis.

Sensory bristles are found not only within the gin trap but also near it on the general body surface. In young pupae, stimulation of a single bristle elicits the closure reflex. Bate found that not only the hairs within the trap elicited closure, but also hairs outside the pit but at the same antero-posterior level. More anterior and more posterior bristles were ineffective even though their physiological characteristics were the same. Thus, the critical spatial variable appeared to be not location within the trap, but rather antero-posterior level.

This observation certainly suggests the hypothesis that the position of a sensory cell body in the segmental gradient determines whether or not its axon forms connections with particular inter- or motor neurons. However, no further exploration of this system has been conducted, and the hypothesis has not been tested by any experimental manipulation of the system.

3.6.5 *Summary*

In this section, as in the previous one, we see clearly the difference between the types of analysis which have been applied to large populations of apparently interchangeable sensory cells, and small populations of central cells with unique identities. The abdominal sensilla have often served as markers for workers studying pattern formation in abdominal segments, and therefore all hypotheses concerning pattern formation would apply to their distribution on the body surface. Specifically, we have seen suggestions that *compartment* boundaries may influence the course of axons from the periphery to the c.n.s. In addition, the proposal has been made that the *segmental gradient* influences not only the location in which sensilla will develop, but also the central connections which their axons will make. The experiments of this section make no contribution to the difficult question of the nature of the segmental gradient. Finally, we have seen some of the strongest available evidence of *selective adhesiveness*, though only in the rather primitive form of nerves adhering to nerves rather than to other types of cells.

Within the ganglion, we have seen again the presence of uniquely identified neurons, in this case motor neurons. The attack on the question of their detailed origin, at least as far back as the stage at which identifiable neuroblasts are present, is currently one of the most exciting in insect developmental neurobiology. If particular neurons can be shown to arise from particular neuroblasts, we will have to ask the question, how did the neuroblasts acquire their identity? Was cell lineage the dominant mechanism here, or was positional information in the array of neuroblasts the immediate cause for differentiation among them?

4 Synthesis

At the end of this long excursion I want to offer some thoughts on the major features we have seen, to consider whether we have threaded our way through a random jumble of isolated peaks, of fascinating studies, or whether the peaks are grouped into ranges and the ranges give some hint of the forces that brought them into being. It does, in fact, seem to me that there are some clear patterns in the foregoing material, and that while no mechanisms are clearly established, some powerful hints about them do emerge.

4.1 CONSISTENT FINDINGS

(a) *Similar morphogenetic rules* To the extent that comparable experiments have been done, all of the sensory surfaces we have considered (retinae, antennae, wings, halteres, cerci and the general abdominal surface) show similar responses to such surgical manipulations as rotation and translocation of grafts. The abdominal segments, as well as the wings and halteres, have actually been among the main sources of experimental data for Part I. Particular theories of pattern formation are certainly open to challenge: one may argue about the strength of the evidence for a gradient of adhesiveness, for example, or whether gradient and polar coordinate models are fundamentally very different or very similar. But I do not know of any reason to suppose that the legs and the antennae or cerci develop according to basically different rules, even though the former are primarily organs of locomotion and the latter are organs of sensation. To the contrary, we know from direct experiment that cercal and leg tissues interact with each other according to the same rules as leg tissue and cercal tissue do when confronted with their own kind. The various theories we have discussed will stand or fall on their own merits, but there is no reason to suppose that different theories will apply to the general body surface than to the surface areas specialized for neural function.
(b) *Independence of the periphery* The body surface and its sensory elements develop independently of the underlying neural tissue. With very few exceptions, the reports in the literature agree that sensory surfaces can be isolated by any method at any postembryonic stage and they will nevertheless differentiate quite normally. It is more difficult to be sure about embryonic stages, but some evidence is available. Mutant optic lobes will not convert a retina to mutant phenotype if the retina itself is caused to be genotypically wild type; conversely, a wild type optic lobe cannot rescue a genotypically mutant retina from showing the mutant phenotype (Meyerowitz and Kankel, 1978). The procedures which create the difference in genotype between retina and optic lobe in such experiments are effective at very early times, at the very first nuclear division in the case of gynandromorphs.

(c) *Dependence of the centre* In contrast, the first synaptic region is profoundly dependent on the sensory periphery. In every system which has been examined on this point, there is clear evidence of powerful disruption in the first synaptic region if the sensory cells are removed or altered. In fact, the evidence in the cricket cercus-to-giant interneuron system indicates that actual removal of the sensory cells is not required – mere silencing of their physiological activity is sufficient. The details of the effects of deafferentation or sensory deprivation remain to be worked out and may well be different in different systems, but it may be that in extreme cases we will come to think of the second-order cells as competent to respond to an inductive signal from the first order cells in the same sense as we saw in the case of induction of bracts by bristle organs. I would expect one important variable to be simply the extent to which a second-order neuron's input is dominated by the particular sensory cells which are being manipulated in a given experiment.

(d) *Pioneer fibres* There are repeated indications that pioneer fibres are employed when long distance connections need to be formed. They have been identified in antennae, eyes, legs and cerci, and the wing disc in *Drosophila* has a neural stalk which could well serve the same function. The pioneer fibres always originate in the periphery, and a more detailed description of the manner of their growth would be very interesting. After all they, unlike all of their followers, do not grow along already formed neural tissue. Do they follow particular cells of a different type towards the c.n.s.? Bate (1976b) suggested pre-existing sheath cells as possible stepping stones for the pioneer fibres. One also wonders why sensory fibres formed later in development grow in the correct direction along the existing nerves. In the networks of Figs 42 and 43 there are many instances of single axons or small nerve bundles joining larger ones nearly at right angles. Do they perhaps know whether to grow towards the left or the right because there is a gradient of adhesiveness in the larger bundle?

4.2 DIFFERENCES AMONG SYSTEMS

There are many similarities among the systems we have discussed, and the foregoing list is only a sampling. But there are also important differences, of which I would like to emphasize three.

(a) *Timing* The retina and the underlying lamina of both hemi- and holometabolous insects grow in synchronous fashion, but this is far from a universal situation. For example, the hatchling cricket is born with one MGI and one LGI in each hemiganglion and 52 filiform hairs (Bentley, 1977). Hairs and their axons are added during each instar and the number in adults exceeds 700 (Palka *et al.*, 1977), but there is still just one interneuron of each kind, though of course there are many kinds in addition to the MGI and LGI. It seems reasonable to suppose, even though there is no experimental proof, that synchronous timing contributes to the formation of ordered connections between retina and lamina, but the relative role of timing among various simultaneously operating mechanisms must surely be different in the many regions of the nervous system in which increasing numbers of sensory cells synapse with a fixed number of postsynaptic neurons.

(b) *Mode of growth* The eye grows by induction at a growing margin. The antenna of hemimetabolous insects adds new segments at the base but also adds new receptors to existing segments, and the cercus grows more vigorously at the base than elsewhere. Abdominal segments apparently grow rather uniformly by cell division all over their surfaces. Again, the peripheral visual system is rather different from other sensory systems.

(c) *Degree of convergence* In the projection of the retina upon the lamina, every six retinula cells converge upon two giant monopolar cells which are apparently all alike except for their location. In the cercal system, each known postsynaptic neuron has an individual identity, and the same is probably true for neurons in other abdominal and thoracic segments. The antennal system is *terra incognita* in this respect. Again, the retina-to-lamina relationship appears to be distinctive. It might be fair to describe the lamina as a special layer of many similar cells interposed between the sensory surface and the main array of sensory interneurons which starts in the medulla; in fact, Cajal called the lamina the "intermediate retina". In the medulla we start to find many unique interneurons, as in other systems (e.g. Strausfeld, 1976).

The retina-to-lamina system is the most carefully studied of all, but in contemplating generalizations from its properties we must keep in mind its various rather special characteristics.

4.3 THE IMPORTANCE OF CONTACT

Over and over we have seen that cell-to-cell contact is an intimate part of pattern formation, both in general epidermal tissue and in nervous tissue. Nerve cells grow along nerve cells; epidermal cells from different locations, when artificially brought together, slide past each other until they reach some equilibrium configuration or are stabilized by intercalary regeneration; living cells make a barrier to the growth of the retina but dead cuticle becomes overgrown by living cells and cannot form a barrier; the growth cones of retinula cells appear to test each other as well as their postsynaptic cells. Morata and Lawrence (1975; also Lawrence and Morata, 1976) suppose that a difference in cellular adhesiveness normally separates the cells of adjacent compartments and that this difference is eliminated by certain mutations. The list could be long extended. A variety of types of contacts between cells has been described at the ultrastructural level and these are believed to have different roles in cell adhesiveness and communication. The finding of Eley and Shelton (1976) that the various types succeed each other in a regular sequence in the developing retina of locusts may open the way to a more exact description of the kinds of interactions that go on when cells are in contact.

4.4 SOME SHARED CHARACTERISTICS OF SEGMENTS

The existence of homeotic mutations suggests that the difference between one segment and another, one compartment and another, is under the (presumably indirect) control of a single gene. One might then suppose that the segments have many things in common, and we have seen some indications of such commonality in this essay. Legs of one segment interact as well with stumps of another segment as with their own. Cerci grafted to leg stumps interact with them as they would with cercal stumps. Homeotic leg fibres growing into the deutocerebrum distribute to the same regions as normal antennal fibres do. It is hard to imagine that a wing fibre would know how to grow ventrad in the strange territory of the metathoracic neuromere if it could not recognize and interact there with the same sort of spatial information as was available in its own proper segment. The number and arrangement of neuroblasts are similar throughout the thorax and abdomen. Some of the most persuasive indications of common attributes of segments are found in the literature on motor systems which has been so slighted here. For example, a metathoracic leg grafted to a mesothoracic stump becomes correctly innervated by mesothoracic motor neurons (Young, 1972).

A curious feature common to many ganglia is that their right and left sides seem to be interchangeable. A right cercus grafted to a left cercal socket will project as usual but now to the left hemiganglion. Motor neurons of a metathoracic ganglion which has been turned upside down (with connectives intact but peripheral nerve roots cut) will reinnervate the muscles of their new closest legs (Bate, 1976c).

We are still but groping for factors which steer the growth of axons and determine their synaptic connections, but whatever these are, it would seem very likely that many are segmentally repeated, both in peripheral organs and within the c.n.s.

4.5 POSITIONAL INFORMATION

(a) *The effect of position information in the periphery upon central connections*

In view of the orderly distribution of many receptors on the body surface, and the fact that information about stimulus location is generally conserved in the nervous system so that behaviour can be appropriately directed, it seems certain that information about receptor position is also conserved. But whether this means that positional information in the sense of Part I, as a causal influence in the differentiation of receptors in particular places in some larger region, has a direct effect on the establishment of synaptic connections by those receptors has not been tested directly in any case.

Among the systems discussed in Part II, the ones which have been studied with the deliberate intention of exploring this question include the trigger hairs in the gin trap system, the T- and L- hairs of the cerci, and the retina-to-lamina projection of dipterans. In attempting to evaluate the sparse relevant data, we encounter some difficulty with the basic notion of positional information. Wolpert (1971) states explicitly that according to his model there is no communication of PI from each cell to its neighbours when the system is in operation: each cell responds independently to the PI generally available in its location. The sort of system envisioned by Ready *et al.* (1976) for the retina, in which each additional cell recruited to an ommatidium is informed of its position, and thus of the connections which it should make to the lamina, by contact with the cells already present in that ommatidium would then not qualify as an example of the effect of peripheral positional information on central synapse formation. For a vigorous examination of the basic concept of positional information, the reader is referred to the discussion following Wolpert's paper of 1975 (pp. 119–130 of the Ciba Symposium on *Cell Patterning*).

(b) *Positional information in central ganglia* There has been much discussion of what might be the nature, or at least the spatial organization, of positional information (PI) on the body surface. We have examined the contrasting notions of PI organized in Cartesian form in gradients, or in polar coordinate form; as gradients of different nature; etc. Such discussions have been singularly lacking for central nervous tissue in insects.

We have seen at least one hint that spatially distributed systems of PI may not be the only ones of importance within the c.n.s., namely the differential behaviour of dorsally and ventrally projecting fibres in the homeotic metathoracic wings of *bithorax* mutants. One could easily imagine that the behaviour of the ventral fibres represents a classic case of the action of positional information in the Wolpertian sense: certain wing fibres turn ventrad in any ganglion they enter, because they are programmed to respond to PI in any ganglion in that fashion and this PI is provided by the same mechanism in all ganglia; similarly, haltere fibres always stay dorsal because they are programmed to seek dorsal levels in the very same PI system. But the dorsal fibres of the metathoracic wing go to precisely the same places, in two different neuromeres, that haltere fibres would go to if there were any of them present. One way of interpreting this observation is that the ventral fibres are responding to widely distributed, possibly orthogonally arranged PI, but that the dorsal fibres are responding to local factors which carry them along tracts normally followed by haltere fibres, and their wing nature is of secondary importance.

The ventral large fibres of the homeotic wing in *Drosophila* have been interpreted as seeking specific destinations, the same ones that are sought out by the co-existing large fibres of the normal wing. According to the hypothesis of Ghysen (1978), these destinations are reached because the growing axons are able to follow local substrate factors either in the normal or in the reverse direction to attain the correct destination.

Two quite different indications have recently appeared in the literature which also suggest the operation of local factors in the guidance of fibre growth. Altman and Tyrer (1977a, b) have shown that a particular large and well studied afferent cell, the mesothoracic wing stretch receptor of locusts, reaches particular regions of branching and termination by one of two different routes in different individuals. The finer branches of this cell compensate in their length and perhaps in the direction of their growth for the details of the route taken by the main axon, as if their ultimate destination rather than their precise route were under the most stringent control.

Similar examples have been provided by Goodman in a series of studies (1974, 1976, 1977, 1978) on the anatomy of ocellar interneurons in isogenic populations of locusts, some of which show (among other changes) altered branching patterns within the brain. In this material too, there are cases of identified cells reaching particular areas of branching by alternate routes. Furthermore, a given cell will branch in not just one but several alternate locations – if it fails to branch in its own correct region, it will branch instead not indiscriminately but in one or another of several well-defined alternate sites. A number of examples of this type of behaviour have led Goodman to propose the existence of multiple "epicentres", regions which stimulate branching by particular cells programmed to respond to those epicentres.

Altogether, these examples suggest that a ganglion contains a system of PI to which cells respond according to their previously determined nature, but a system with several different components: positional information is available not just in the form of map coordinates but as a map already showing at least major topographic features. Incoming cells programmed to search for peaks on whose summits to ramify might show a quite different behaviour, pay attention to different components, than those programmed to form tracts streaming along the river valleys. But one measure of our ignorance is the common occurrence of receptors, described already by Zawarzin (1924), forming one branch which ramifies locally in the ganglion of entry and another which enters a tract leading to more distant ganglia.

4.6 CELL LINEAGE

The information we have reviewed about the development of sensory systems indicates that cell lineage is not a major factor in determining the central distribution of sensory axons. Receptor cells arise by the division of mother cells which initially are indistinguishable from ordinary epidermal cells. There is no evidence that I am aware of to suggest that the mother cells are predetermined for their special function; on the contrary, evidence not presented here (Wigglesworth, 1940) suggests that the number of receptors formed on abdominal segments is adjusted to their dimensions, so that if the abdomen of a given animal is caused to grow larger than usual, more receptors will form out of the general epidermis and vice versa. The major possible exception to this generalization is the suggestion that the precursor cells of different classes of retinula cells in the eye may already be different from each other, possibly as a result of particular sequences of cell division.

Within the c.n.s., however, the story is incomplete. Many parts of the c.n.s. are dominated by the presence of neurons with unique identities and characteristic life histories. These neurons develop long before the sensory fibres arrive, and quite possibly make, or are themselves, the system of positional information to which incoming axons respond. Whether these cells arise via fixed cell lineages is an intriguing question which may soon be answered, and the answer may also shed light on the role of cell death in neural development (Edwards, 1969).

4.7 CONCLUSION

We are only beginning to learn how the nervous systems of insects develop. The task is difficult, and new information is much needed on every front. But I find it encouraging that there are so many similarities among the various subsystems we have explored, and especially that there seems to be no major distinction between the development of sense organs and of other parts of the body surface. The theories and methods of Part I, therefore, and any others which are devised for experimentally favourable systems, should become part of the intellectual and experimental armamentarium of neurobiologists interested in neural development. And conversely, since nervous systems, in spite of their complexity, share so much with other tissues, I look forward to finding ever increasing numbers of developmental biologists joining neurobiologists in the study of this most challenging of all components of the animal organism.

Acknowledgements

The initial impulse to prepare this essay came during a sabbatical year spent in Cambridge, England, in the stimulating laboratories of Peter Lawrence and Malcolm Burrows. Many people, starting with John Edwards, have helped me acquire some education in the field of insect development, but I want especially to express my thanks to Margrit Schubiger who has not only inspired and challenged my thinking, but who patiently and critically read every section of this essay as it was written, offered her comments, and discussed my revisions. The comments of Tom Abrams were also much appreciated. My stay in Cambridge was supported by the John Simon Guggenheim Foundation, and my research and writing has been supported for a number of years by Grant No. NS–07778 from the U.S. Public Health Service.

References

Altman, J. S. and Tyrer, N. M. (1977a). The locust wing hinge stretch receptors. I. Primary sensory neurones with enormous central arborizations. *J. comp. Neurol.* **172,** 409–430.

Altman, J. S. and Tyrer, N. M. (1977b). The locust wing hinge stretch receptors. II. Variation, alternative pathways and "mistakes" in the central arborizations. *J. comp. Neurol.* **172,** 431–440.

Armett-Kibel, C., Meinertzhagen, I. and Dowling, J. E. (1977). Cellular and synaptic organization in the lamina of the dragon-fly *Sympetrum rubicundulum*. *Proc. Roy. Soc. Lond.* **B196,** 385–413.

Bate, C. M. (1973a). The mechanism of the pupal gin trap. I. Segmental gradients and the connexions of the triggering sensilla. *J. exp. Biol.* **59,** 95–107.

Bate, C. M. (1973b). The mechanism of the pupal gin trap. II. The closure movement. *J. exp. Biol.* **59,** 109–119.

Bate, C. M. (1973c). The mechanism of the pupal gin trap. III. Interneurones and the origin of the closure mechanism. *J. exp. Biol.* **59,** 121–135.

Bate, C. M. (1976a). Embryogenesis of an insect nervous system. I. A map of the thoracic and abdominal neuroblasts in *Locusta migratoria*. *J. Embryol. exp. Morphol.* **35,** 107–123.

Bate, C. M. (1976b). Pioneer neurones in an insect embryo. *Nature* **260,** 54–56.

Bate, C. M. (1976c). Nerve growth in cockroaches (*Periplaneta americana*) with rotated ganglia. *Experientia*, **32,** 451–452.

Bate, C. M. and Lawrence, P. A. (1973). Gradients and the developing nervous system. *In* "Developmental Neurobiology of Arthropods" (Ed. D. Young), pp. 37–49. Cambridge University Press, Cambridge.

Bentley, D. (1973). Postembryonic development of insect motor systems. *In* "Developmental Neurobiology of Arthropods" (Ed. D. Young), pp. 147–177. Cambridge University Press, Cambridge.

Bentley, D. (1975). Single-gene cricket mutants: Effects on behaviour, sensilla, sensory neurons and identified interneurons. *Science*, **187,** 760–764.

Bentley, D. (1977). Development of insect nervous systems. *In* "Identified Neurons and Behaviour of Arthropods" (Ed. G. Hoyle), pp. 461–481. Plenum, New York.

Bischof, H.-J. (1974). Verteilung und Bewegungsweise der keulenförmigen Sensillen von *Gryllus bimaculatus* Deg. *Biol. Zbl.* **93,** 449–457.

Bischof, H.-J. (1975). Die keulenförmigen Sensillen auf den Cerci der Grille *Gryllus bimaculatus* als Schwererezeptoren. *J. comp. Physiol.* **98,** 277–288.

Bodenstein, D. (1953). Regeneration. *In* "Insect Physiology" (Ed. K. D. Roeder), pp. 866–878. Wiley, New York.

Boeckh, J., Sandri, C. and Akert, K. (1970). Sensorische Eingänge und synaptische Verbindungen im Zentralnervensystem von Insekten. *Z. Zellforsch.* **103,** 429–446.

Bohn, H. (1965). Analyse der Regenerationsfähigkeit der Insektenextremität durch Amputations- und Transplantationsversuche an Larven der Afrikanischen Schabe (*Leucophaea maderae* Fabr.) II. Achsendetermination. *Wilhelm Roux' Arch.* **156,** 449–503.

Bohn, H. (1970a). Interkalare Regeneration und segmentale Gradienten bei den Extremitäten von *Leucophaea*-Larven (Blattaria). I. Femur und Tibia. *Wilhelm Roux' Arch.* **165,** 303–341.

Bohn, H. (1970b). Interkalare Regeneration und segmentale Gradienten bei den Extremitäten von *Leucophaea*-Larven (Blattaria). II. Coxa und Tarsus. *Develop. Biol.* **23,** 355–379.

Bohn, H. (1971). Interkalare Regeneration und segmentale Gradienten bei den Extremitäten von *Leucophaea*-Larven (Blattaria). III. Die Herkunft des interkalaren Regenerates. *Wilhelm Roux' Arch.* **167,** 209–221.

Bohn, H. (1974). Pattern reconstitution in abdominal segment of *Leucophaea maderae* (Blattaria). *Nature,* **248,** 608–609.

Braitenberg, V. (1966). Unsymmetrische Projektion der Retinulazellen auf die Lamina ganglionaris bei der Fliege *Musca domestica. Z. vergl. Physiol.* **52,** 212–214.

Braitenberg, V. and Strausfeld, N. J. (1973). Principles of the mosaic organisation in the visual system's neuropil of *Musca domestica* L. *In* "Handbook of Sensory Physiology", **VII** pt. 3, *Central Visual Information A* (Ed. R. Jung), pp. 631–659. Springer-Verlag, Berlin, Heidelberg, New York.

Bryant, P. J. (1975). Pattern formation in the imaginal wing disc of *Drosophila melanogaster:* Fate map, regeneration and duplication. *J. exp. Zool.* **193,** 49–78.

Bryant, P. J., Bryant, S. V. and French, V. (1977). Biological regeneration and pattern formation. *Sci. Am.* **237,** 66–81.

Campos-Ortega, J. A. and Gateff, E. A. (1976). The development of ommatidal patterning in metamorphosed eye imaginal disc implants of *Drosophila melanogaster. Wilhelm Roux' Arch.* **179,** 373–392.

Campos-Ortega, J. A. and Hofbauer, A. (1977). Cell clones and pattern formation: On the lineage of photoreceptor cells in the compound eye of *Drosophila. Wilhelm Roux' Arch.* **181,** 227–245.

Counce, S. J. (1973). The causal analysis of insect embryogenesis. *In* "Developmental Systems: Insects" (Eds. S. J. Counce and C. H. Waddington), **2,** pp. 1–156. Academic Press, New York and London.

Crick, F. H. C. and Lawrence, P. A. (1975). Compartments and polyclones in insect development. *Science,* **189,** 340–347.

Daniels, C. J. (1968). Regeneration of cerci in the house cricket, *Acheta domesticus.* MS Thesis, Department of Biology, Case Western Reserve University.

Deak, I. I. (1976). Demonstration of sensory neurones in the ectopic cuticle of spineless-aristapedia, a homeotic mutant of *Drosophila. Nature,* **260,** 252–254.

Dethier, V. G. (1963). "The Physiology of Insect Senses". Methuen, London; Wiley, New York.

Edwards, J. S. (1969). Postembryonic development and regeneration of the insect nervous system. *Adv. Insect. Physiol.* **6,** 97–137.

Edwards, J. S. and Chen, S.-W. (1979). Embryonic development of an insect sensory system, the abdominal cerci of *Acheta domesticus. Wilhelm Roux' Arch.*, (in press).

Edwards, J. S. and Palka, J. (1971). Neural regeneration: Delayed formation of central contacts by insect sensory cells. *Science*, **172,** 591–594.

Edwards, J. S. and Palka, J. (1974). The cerci and abdominal giant fibres of the house cricket, *Acheta domesticus.* I. Anatomy and physiology of normal adults. *Proc. Roy. Soc. Lond.* **B185,** 83–103.

Edwards, J. S. and Palka, J. (1976). Neural generation and regeneration in insects. *In* "Simpler Networks and Behavior" (Ed. J. C. Fentress), pp. 167–185. Sinauer, Sunderland, Mass.

Edwards, J. S. and Sahota, (1967). Regeneration of a sensory system: The formation of central connections by normal and transplanted cerci of the house cricket *Acheta domesticus. J. exp. Zool.* **166,** 387–396.

Eley, S. and Shelton, P. M. J. (1976). Cell junctions in the developing compound eye of the desert locust *Schistocerca gregaria. J. Embryol. exp. Morph.* **36,** 409–423.

Ernst, K.-D., Boeckh, J. and Boeckh, V. (1977). A neuroanatomical study on the organization of the central antennal pathways in insects. II. Deutocerebral connections in *Locusta migratoria* and *Periplaneta americana. Cell Tiss. Res.* **176,** 285–308.

French, V., Bryant, P. J. and Bryant, S. V. (1976). Pattern regulation in epimorphic fields. *Science*, **193,** 969–981.

French, V. and Bullière, D. (1975a). Nouvelles données sur la détermination de la position des cellules épidermiques sur un appendice de Blatte. *Compt. Rend. Acad. Sci. Paris*, **280,** 53–56.

French, V. and Bullière, D. (1975b). Etude de la détermination de la position des cellules: ordonnance des cellules autour d'un appendice de Blatte; démonstration du concept de généatrice. *Compt. Rend. Acad. Sci. Paris*, **280,** 295–298.

García-Bellido, A. (1966). Pattern reconstruction by dissociated imaginal disk cells of *Drosophila melanogaster. Develop. Biol.* **14,** 278–306.

García-Bellido, A. (1975). Genetic control of wing disc development in *Drosophila. In* "Cell Patterning", Ciba Foundation Symposium 29 (new series), pp. 161–178. Elsevier-Excertpa Medica-North Holland, Amsterdam.

García-Bellido, A., Ripoll, P. and Morata, G. (1973). Developmental compartmentalisation of the wing disc of *Drosophila. Nature New Biol.* **245,** 251–253.

García-Bellido, A., Ripoll, P. and Morata, G. (1976). Developmental compartmentalization in the dorsal mesothoracic disc of *Drosophila. Develop. Biol.* **48,** 132–147.

Gehring, W. J. and Nöthinger, R. (1973). The imaginal discs of *Drosophila. In* "Developmental Systems: Insects" (Ed. S. J. Counce and C. H. Waddington), **2,** pp. 211–290. Academic Press, New York and London.

Ghysen, A. (1978). Sensory neurons recognise defined pathways in *Drosophila* central nervous system: *Nature*, **274,** 869–872.

Gnatzy, W. and Schmidt, K. (1971). Die Feinstruktur der Sinneshaare auf den Cerci von *Gryllus bimaculatus* Deg. (Saltatoria, Gryllidae). I. Faden- und Keulenhaare. *Z. Zellforsch.* **122,** 190–209.

Gnatzy, W. and Schmidt, K. (1972a). Die Feinstruktur der Sinneshaare auf den Cerci von *Gryllus bimaculatus* Deg. (Saltatoria, Gryllidae). IV. Die Häutung der kurzen Borstenhaare. *Z. Zellforsch.* **126,** 223–239.

Gnatzy, W. and Schmidt, K. (1972b). Die Feinstruktur der Sinneshaare auf den Cerci von *Gryllus bimaculatus* Deg. (Saltatoria, Gryllidae). V. Die Häutung der langen Borstenhaare an der Cercusbasis. *J. Microscopie*, **14,** 75–84.

Goodmann, C. S. (1974). Anatomy of locust ocellar interneurons: Constancy and variability. *J. comp. Physiol.* **95,** 185–201.

Goodman, C. S. (1976). Constancy and uniqueness in a large population of small interneurons. *Science*, **193,** 502–504.

Goodman, C. S. (1977). Neuron duplications and deletions in locust clones and clutches. *Science*, **197,** 1384–1386.

Goodman, C. S. (1978). Isogenic locusts: Genetic variability in the morphology of identified neurons. *J. comp. Neurol.* **182,** 681–705.

Green, S. M. and Lawrence, P. A. (1975). Recruitment of epidermal cells by the developing eye of *Oncopeltus* (Hemiptera). *Wilhelm Roux' Arch.* **177,** 61–65.

Gymer, A. and Edwards, J. S. (1967). The development of the insect nervous system. I. An analysis of postembryonic growth in the terminal ganglion of *Acheta domesticus*. *J. Morph.* **123,** 191–198.

Hasenfuss, I. (1973). Über die Beziehung zwischen sensorischer Innervierung und primären Segmentgrenzen bei Arthropoden. *Verh. Dtsch. Zool. Ges.* **66,** 71–75.

Hofbauer, A. and Campos-Ortega, J. A. (1976). Cell clones and pattern formation: Genetic eye mosaics in *Drosophila melanogaster*. *Wilhelm Roux' Arch.* **179,** 275–289.

Horridge, G. A. and Meinertzhagen, I. A. (1970). The accuracy of the patterns of connexions of the first- and second-order neurons of the visual system of *Calliphora*. *Proc. Roy. Soc. Lond.* **B175,** 69–82.

Hoyle, G. (1977). "Identified Neurons and Behavior of Arthropods". Plenum, New York and London.

Hyde, C. A. T. (1972). Regeneration, postembryonic induction and cellular interaction in the eye of *Periplaneta americana*. *J. Embryol. exp. Morph.* **27,** 367–379.

Kaissling, K.-E. (1971). Insect olfaction. *In* "Handbook of Sensory Physiology" **IV,** Pt. 1, *Olfaction* (Ed. L. M. Beidler,). pp. 351–431. Springer-Verlag, Berlin, Heidelberg, New York.

Kauffman, S. A. (1973). Control circuits for determination and transdetermination. *Science*, **181,** 310–318.

Kauffman, S. A. (1975). Control circuits for determination and transdetermination: Interpreting positional information in a binary epigenetic code. *In* "Cell Patterning", Ciba Foundation Symposium 29 (new series), pp. 201–221. Elsevier-Excerpta Medica-North Holland, Amsterdam.

Kauffman, S. A. (1977). Chemical patterns, compartments and a binary epigenetic code in *Drosophila*. *Amer. Zool.* **17,** 631–648.

Kauffman, S. A., Shymko, R. M. and Trabert, K. (1978). Control of sequential compartment formation in *Drosophila*. *Science*, **199,** 259–270.

Kühn, A. (1965). "Lectures on Developmental Physiology", 2nd Edn. (Milkman, R., transl). Springer-Verlag, New York, Heidelberg, Berlin.

Kopeć, S. (1922). Mutual relationship in the development of the brain and eyes of Lepidoptera. *J. exp. Zool.* **36,** 459–467.

Krafka, J. (1924). Development of the compound eye of *Drosophila melanogaster* and its bar-eyed mutant. *Biol. Bull.* **47,** 143–147.

Lawrence, P. A. (1966a). Gradients in the insect segment: The orientation of hairs in the milkweed bug *Oncopeltus fasciatus*. *J. exp. Biol.* **44,** 607–620.

Lawrence, P. A. (1966b). Development and determination of hairs and bristles in the milkweed bug, *Oncopeltus fasciatus* (Lygaeidae, Hemiptera). *J. Cell Sci.* **1,** 475–498.

Lawrence, P. A. (1973a). A clonal analysis of segment development in *Oncopeltus* (Hemiptera). *J. Embryol. exp. Morph.* **30,** 681–699.
Lawrence, P. A. (1973b). The development of spatial patterns in the integument of insects. *In* "Developmental Systems: Insects" (Ed. S. J. Counce and C. H. Waddington), **2,** pp. 157–209. Academic Press, New York and London.
Lawrence, P. A. (1974). Cell movement during pattern regulation in *Oncopeltus. Nature,* **248,** 609–610.
Lawrence, P. A. (1975). The structure and properties of a compartment border: The intersegmental boundary in *Oncopeltus. In* "Cell Patterning", Ciba Foundation Symposium 29 (new series), pp. 3–16. Elsevier-Excerpta Medica-North Holland, Amsterdam.
Lawrence, P. A., Crick, F. H. C. and Munro, M. (1972). A gradient of positional information in an insect, *Rhodnius. J. Cell Sci.* **11,** 815–853.
Lawrence, P. A. and Green, S. M. (1975). The anatomy of a compartment border: The intersegmental boundary in *Oncopeltus. J. Cell Biol.* **65,** 373–382.
Lawrence, P. A. and Morata, G. (1976a). Compartments in the wing of *Drosophila:* A study of the *engrailed* gene. *Develop. Biol.* **50,** 321–337.
Lawrence, P. A. and Morata, G. (1976b). The compartment hypothesis. *In* "Insect Development" (Ed. P. A. Lawrence), Symp. Roy. Ent. Soc. Lond. **8,** pp. 132–149. Blackwell Sci. Publ., Oxford.
Lawrence, P. A. and Shelton, P. M. J. (1975). The determination of polarity in the developing insect retina. *J. Embryol. exp. Morph.* **33,** 471–486.
Lewis, E. B. (1963). Genes and developmental pathways. *Am. Zool.* **3,** 33–56.
Locke, M. (1959). The cuticular pattern in an insect, *Rhodnius prolixus* Stål. *J. exp. Biol.* **36,** 459–478.
Locke, M. (1960). The cuticular pattern in an insect – The intersegmental membranes. *J. exp. Biol.* **37,** 398–407.
Locke, M. (1966a). The gradient concept in the development of the integument of insects. *Naturwiss.* **53,** 510.
Locke, M. (1966b). Hypotheses for gradient mechanisms in insect epidermis. *Naturwiss.* **53,** 510.
Locke, M. (1967). The development of patterns in the integument of insects. *Adv. Morphogen.* **6,** 33–88.
Matsumoto, S. G. and Murphey, R. K. (1977a). Sensory deprivation during development decreases the responsiveness of cricket giant interneurons. *J. Physiol.* **268,** 533–548.
Matsumoto, S. G. and Murphey, R. K. (1977b). The cercus-to-giant interneuron system of crickets. IV. Patterns of connectivity between receptors and the Medial Giant Interneuron. *J. comp. Physiol.* **119,** 319–330.
Meinertzhagen, I. A. (1972). Erroneous projection of retinula axons beneath a dislocation in the retinal equator of *Calliphora. Brain Res.* **41,** 39–49.
Meinertzhagen, I. A. (1973). Development of the compound eye and optic lobes of insects. *In* "Developmental Neurobiology of Arthropods" (Ed. D. Young), pp. 51–104. Cambridge University Press, Cambridge.
Meinertzhagen, I. A. (1975). The development of neuronal connection patterns in the visual systems of insects. *In* "Cell Patterning" Ciba Foundation Symposium 29 (new series), pp. 265–288. Elsevier-Excerpta Medica-North Holland, Amsterdam.
Meinertzhagen, I. A. (1976). The organization of perpendicular fibre pathways in the insect optic lobe. *Phil. Trans. Roy. Soc. Lond.* **B274,** 555–596.
Melamed, J. and Trujillo-Cenóz, O. (1975). The fine structure of the eye imaginal disks in muscoid flies. *J. Ultrastruct. Res.* **51,** 79–93.

Mendenhall, B. and Murphey, R. K. (1974). The morphology of cricket giant interneurons. *J. Neurobiol.* **5,** 565–580.
Meyerowitz, E. M. and Kankel, D. R. (1978). A genetic analysis of visual system development in *Drosophila melanogaster. Develop. Biol.* **62,** 63–93.
Morata, G. and Lawrence, P. A. (1975). Control of compartment development by the *engrailed* gene of *Drosophila. Nature,* **255,** 614–617.
Morata, G. and Lawrence, P. A. (1977). Homeotic genes, compartments and cell determination in *Drosophila. Nature,* **265,** 211–216.
Mouze, M. (1972a). Croissance et metamorphose de l'appareil visuel des Aeshnidae (Odonata). *Int. J. Insect. Morph. Embryol.* **1,** 181–200.
Mouze, M. (1972b). Étude éxperimentale des facteurs morphogénétiques et hormonaux réglant la croissance oculaire des Aeshnidae (Odonates, Anisoptères). *Odonatologica,* **1,** 221–232.
Mouze, M. (1974). Interactions de l'oeil et du lobe optique au cours de la croissance post-embryonnaire des Insectes odonates. *J. Embryol. exp. Morph.* **31,** 377–407.
Mouze, M. (1975). Croissance et régénération de l'oeil de la larve d'*Aeshna cyanea* Mull. (Odonate, Anisoptère). *Wilhelm Roux' Arch.* **176,** 267–283.
Murphey, R. K., Matsumoto, S. G. and Mendenhall, B. (1976). Recovery from deafferentation by cricket interneurons after reinnervation by their peripheral field. *J. comp. Neurol.* **169,** 335–346.
Murphey, R. K., Mendenhall, B., Palka, J. and Edwards, J. S. (1975). Deafferentation slows the growth of specific dendrites on identified giant interneurons. *J. comp. Neurol.* **159,** 407–418.
Murphey, R. K., Palka, J. and Hustert, R. (1977). The cercus-to-giant interneuron system of crickets. II. Response characteristics of two giant interneurons. *J. comp. Physiol.* **119,** 285–300.
Nardi, J. B. (1977). The construction of the insect compound eye: the involvement of cell displacement and cell surface properties in the positioning of cells. *Develop. Biol.* **61,** 287–298.
Nardi, J. B. and Kafatos, F. C. (1976a). Polarity and gradients in lepidopteran wing epidermis. I. Changes in graft polarity, form, and cell density accompanying transpositions and reorientations. *J. Embryol. exp. Morph.* **36,** 469–487.
Nardi, J. B. and Kafatos, F. C. (1976b). Polarity and gradients in lepidopteran wing epidermis. II. The differential adhesiveness model: gradient of a non-diffusible cell surface parameter. *J. Embryol. exp. Morph.* **36,** 489–512.
Nicklaus, R. (1965). Die Erregung einzelner Fadenhaare von *Periplaneta americana* in Abhängigkeit von der Grösse und Richtung der Auslenkung. *Z. vergl. Physiol.* **50,** 331–362.
Nordlander, R. H. and Edwards, J. S. (1969a). Postembryonic brain development in the monarch butterfly, *Danaus plexippus plexippus,* L. I. Cellular events during brain morphogenesis. *Wilhelm Roux' Arch.* **162,** 197–217.
Nordlander, R. and Edwards, J. S. (1969b). Postembryonic brain development in the monarch butterfly, *Danaus plexippus plexippus* L. II. The optic lobes. *Wilhelm Roux' Arch.* **163,** 197–220.
Nordlander, R. H. and Edwards, J. S. (1969b). Postembryonic brain development in the monarch butterfly, *Danaus plexippus plexippus* L. III. Morphogenesis of centers other than the optic lobes. *Wilhelm Roux' Arch.* **164,** 247–260.
Nübler-Jung, K. (1974). Cell migration during pattern reconstitution in the insect segment (*Dysdercus intermedius* Dist., Heteroptera). *Nature,* **248,** 610–611.
Nübler-Jung, K. (1977). Pattern stability in the insect segment. I. Pattern reconstitution

by intercalary regeneration and cell sorting in *Dysdercus intermedius* Dist. *Wilhelm Roux' Arch.* **183,** 17–40.

Palka, J. (1977a). Abnormal neural development in invertebrates. *In* "Function and Formation of Neural Systems" (Ed. G. Stent), pp. 139–159. Dahlem Konferenzen, Berlin.

Palka, J. (1977b). Neurobiology of homeotic mutants in *Drosophila. Soc. Neurosci. Abst.* **3,** 187.

Palka, J. and Edwards, J. S. (1974). The cerci and abdominal giant fibres of the house cricket, *Acheta domesticus*. II. Regeneration and effects of chronic deprivation. *Proc. Roy. Soc. Lond.* **B185,** 105–121.

Palka, J., Lawrence, P. A. and Hart, H. S. (1979). Neural projection patterns from hemeotic tissue of *Drosophila* studied in *bithorax* mutants and mosaics *Develop. Biol.* (in press).

Palka, J., Levine, R. and Schubiger, M. (1977). The cercus-to-giant interneuron system of crickets. I. Some attributes of the sensory cells. *J. comp. Physiol.* **119,** 267–283.

Palka, J. and Olberg, R. (1977). The cercus-to-giant interneuron system of crickets. III. Receptive field organization. *J. comp. Physiol.* **119,** 301–317.

Palka, J. and Schubiger, M. (1975). Central connections of receptors on rotated and exchanged cerci of crickets. *Proc. Nat. Acad. Sci. USA.* **72,** 966–969.

Panov, A. A. (1961). The structure of the insect brain at successive stages in postembryonic development. 4. The olfactory center. *Ent. Rev.* **40,** 140–145 (Eng. trans.).

Panov, A. A. (1963). The origin and fate of neuroblasts, neurons and neuroglial cells in the central nervous system of the China oak silkworm *Antheraea pernyi* Guér (Lepidoptera, Attacidae). *Ent. Rev.* **42,** 186–191 (Eng. trans.).

Panov, A. A. (1966). Correlations in the ontogenetic development of the central nervous system in the house cricket *Gryllus domesticus* L. and the mole cricket *Gryllotalpa gryllotalpa* L. (Orthoptera, Grylloidea). *Ent. Rev.* **45,** 179–185 (Eng. trans.).

Pareto, A. (1972). Die zentrale Verteilung der Fühlerafferenz bei Arbeiterinnen der Honigbiene, *Apis mellifera* L. *Z. Zellforsch.* **131,** 109–140.

Parnas, I., and Dagan, D. (1971). Functional organization of giant axons in the central nervous systems of insects: New Aspects. *Adv. Insect Physiol.* **8,** 95–114.

Pflugfelder, O. (1947). Die Entwicklung embryonaler Teile von *Carausius* (*Dixippus*) *morosus* in der Kopfkapsel von Larven und Imagines. *Biol. Zbl.* **66,** 372–387.

Pflugfelder, O. (1958). "Entwicklungsphysiologie der Insekten". 2nd Edn. Akademische Verlagsgesellschaft, Leipzig.

Piepho, H. (1955). Über die polare Orientierung der Bälge and Schuppen auf dem Schmetterlingsrumpf. *Biol. Zbl.* **74,** 467–474.

Postlethwait, J. and Schneiderman, H. A. (1971a). A clonal analysis of development in *Drosophila melanogaster*: Morphogenesis, determination and growth in the wild-type antenna. *Develop. Biol.* **24,** 477–519.

Postlethwait, J. and Schneiderman, H. A. (1971b). Pattern formation and determination in the antenna of the homeotic mutant *Antennapedia* of *Drosophila melanogaster. Develop. Biol.* **25,** 606–640.

Power, M. E. (1943). The effect of reduction in numbers of ommatidia upon the brain of *Drosophila melanogaster. J. exp. Zool.* **94,** 33–71.

Power, M. E. (1946). The antennal centers and their connections with the brain of *Drosophila melanogaster. J. comp. Neurol.* **85,** 485–517.

Power, M. E. (1948). The thoraco-abdominal nervous system of an adult insect, *Drosophila melanogaster. J. comp. Neurol.* **88,** 347–409.

Pumphrey, R. J. and Rawdon-Smith, A. F. (1936). Hearing in insects: The nature of the response of certain receptors to auditory stimuli. *Proc. Roy. Soc. Lond.* **B121,** 18–27.

Pumphrey, R. J. and Rawdon-Smith, A. F. (1937). Synaptic transmission of nervous impulses through the last abdominal ganglion of the cockroach. *Proc. Roy. Soc. Lond.* **B122,** 106–118.

Ready, D. F., Hanson, T. E. and Benzer, S. (1976). Development of the *Drosophila* retina, a neurocrystalline lattice. *Develop. Biol.* **53,** 217–240.

Richards, M. H. and Furrow, E. Y. (1925). The eye and optic tract in normal and "eyeless" *Drosophila*. *Biol. Bull.* **48,** 243–257.

Roeder, K. D. (1967). "Nerve Cells and Insect Behavior", Rev. Edn. Harvard University Press, Cambridge.

Sanes, J. R. and Hildebrand, J. G. (1975). Nerves in the antennae of pupal *Manduca sexta* Johanssen (Lepidoptera: Sphingidae). *Wilhelm Roux' Arch.* **178,** 71–78.

Sanes, J. R. and Hildebrand, J. G. (1976a). Structure and development of antennae in a moth, *Manduca sexta*. *Develop. Biol.* **51,** 282–299.

Sanes, J. R. and Hildebrand, J. G. (1976b). Origin and morphogenesis of sensory neurons in an insect antenna. *Develop. Biol.* **51,** 300–319.

Sanes, J. R. and Hildebrand, J. G. (1976c). Acetylcholine and its metabolic enzymes in developing antennae of the moth, *Manduca sexta*. *Develop. Biol.* **52,** 105–120.

Sanes, J. R., Hildebrand, J. G. and Prescott, D. J. (1976). Differentiation of insect sensory neurons in the absence of their normal synaptic targets. *Develop. Biol.* **52,** 121–127.

Sanes, J. R., Prescott, D. J. and Hildebrand, J. G. (1977). Cholinergic neurochemical development of normal and deafferented antennal lobes during metamorphosis of the moth, *Manduca sexta*. *Brain Res.* **119,** 389–402.

Schafer, R. (1971). Antennal sense organs of the cockroach, *Leucophaea maderae*. *J. Morph.* **134,** 91–104.

Schafer, R. (1973). Postembryonic development in the antenna of the cockroach, *Leucophaea maderae*: Growth, regeneration, and the development of the adult pattern of sense organs. *J. exp. Zool.* **183,** 353–364.

Schafer, R. and Sanchez, T. V. (1973). Antennal sensory system of the cockroach, *Periplaneta americana*: Postembryonic development and morphology of the sense organs. *J. comp. Neurol.* **149,** 335–354.

Schmidt, K. and Gnatzy, W. (1971). Die Feinstruktur der Sinneshaare auf den Cerci von *Gryllus bimaculatus* Deg. (Saltatoria, Gryllidae). II. Die Häutung der Faden- und Keulenhaare. *Z. Zellforsch.* **122,** 210–226.

Schmidt, K. and Gnatzy, W. 1972. Die Feinstruktur der Sinneshaare auf den Cerci von *Gryllus bimaculatus* Deg. (Saltatoria, Gryllidae). III. Die kurzen Borstenhaare. *Z. Zellforsch.* **126,** 206–222.

Schneider, D. (1964). Insect antennae. *Ann. Rev. Entomol.* **9,** 103–122.

Schoeller, J. (1964). Recherches descriptives et expérimentales sur la céphalogenèse de *Calliphora erythrocephala* (Meigen) au cours des développements embryonnaire et postembryonnaire. *Archs. Zool. exp. gen.* **103,** 1–216.

Schwartzkopff, J. (1974). Mechanoreception. *In* "The Physiology of Insecta" (Ed. M. Rockstein), 2nd Edn., **II,** pp. 273–352. Academic Press, New York and London.

Schweitzer, E. S., Sanes, J. R. and Hildebrand, J. C. (1976). Ontogeny of electroantennogram responses in the moth, *Manduca sexta*. *J. Insect Physiol.* **22,** 955–960.

Shelton, P. M. J. (1976). The development of the insect compound eye. *In* "Insect Development" (Ed. P. A. Lawrence), Symp. Roy. Ent. Soc. Lond. **8,** pp. 152–169. Blackwell Sci. Publ., Oxford.

Shelton, P. M. J., Anderson, H. J. and Eley, S. (1977). Cell lineage and cell determination in the developing compound eye of the cockroach, *Periplaneta americana*. *J. Embryol. exp. Morph.* **39,** 235–252.

Shelton, P. M. J. and Lawrence, P. A. (1974). Structure and development of ommatidia in *Oncopeltus fasciatus*. *J. Embryol. exp. Morph.* **32,** 337–353.

Sihler, H. (1924). Die Sinnesorgane an den Cerci der Insekten. *Zool. Jahrb. Anat.* **45,** 519–580.

Steiner, E. (1976). Establishment of compartments in the developing leg imaginal discs of *Drosophila melanogaster*. *Wilhelm Roux' Arch.* **180,** 9–30.

Stocker, R. F. (1977). Gustatory stimulation of a homeotic mutant appendage, *Antennapedia*, in *Drosophila melanogaster*. *J. comp. Physiol.* **115,** 351–361.

Stocker, R. F., Edwards, J. S., Palka, J. and Schubiger, G. (1976). Projection of sensory neurons from the homeotic mutant appendage, *Antennapedia*, in *Drosophila melanogaster*. *Develop. Biol.* **52,** 210–220.

Stumpf, H. (1965a). Die Deutung der Riefenmuster bei *Rhodnius prolixus* auf Grund eines Konzentrationsgefälles. *Naturwiss.* **52,** 500–501.

Stumpf, H. (1965b). Deutung der Richtungsmuster der Schuppen von *Galleria mellonella* auf Grund eines Konzentrationsgefälles. *Naturwiss.* **52,** 522.

Stumpf, H. (1966a). Über gefälleabhängige Bildungen des Insektensegmentes. *J. Insect Physiol.* **12,** 601–617.

Stumpf, H. (1966b). Mechanism by which cells estimate their location within the body. *Nature*, **212,** 430–431.

Stumpf, H. (1967). About the model of a concentration gradient in the insect segment. Replication to Locke. *Naturwiss.* **54,** 50–51.

Strausfeld, N. J. (1976). "Atlas of an Insect Brain". Springer-Verlag, Berlin, Heidelberg, New York.

Taylor, H. M. and Truman, J. W. (1974). Metamorphosis of the abdominal ganglia of the tobacco hornworm, *Manduca sexta*. *J. comp. Physiol.* **90,** 367–388.

Tobler, H. (1966). Zellspezifische Determination und Beziehung zwischen Proliferation und Transdetermination in Bein- und Flügelprimordien von *Drosophila melanogaster*. *J. Embryol. exp. Morph.* **16,** 609–633.

Tobler, H. and Maier, V. (1970). Zur Wirkung von Senfgaslösung auf die Differenzierung des Borstenoganes und auf die Transdeterminationsfrequenz bei *Drosophila melanogaster*. *Wilhelm Roux' Arch.* **164,** 303–312.

Tobler, H. and Pfluger, M. (1970). Untersuchungen zur Wirkung von Mitomycin C auf die Entwicklung der männlichen Vorderbeinscheibe und die Differenzierung des Borstenorgans von *Drosophila melanogaster* nach Transplantation in larvale Wirte. *Wilhelm Roux' Arch.* **164,** 293–302.

Tobler, H., Rothenbühler, V. and Nöthinger, R. (1973). A study of the differentiation of bracts in *Drosophila melanogaster* using two mutations, H^2 and sv^{de}. *Experientia*, **29,** 370–371.

Trujillo-Cenóz, O. (1972). The structural organization of the compound eye in insects. *In* "Handbook of Sensory Physiology" **VII,** Pt. 2, *Physiology of Receptor Organs* (Ed. M. G. F. Fuortes), pp. 5–62. Springer-Verlag, Berlin, Heidelberg, New York.

Trujillo-Cenóz, O. and Melamed, J. (1966). Compound eye of dipterans: Anatomical basis for integration – An electron microscope study. *J. Ultrastruct. Res.* **16,** 395–398.

Trujillo-Cenóz, O. and Melamed, J. (1973). The development of the retina-lamina complex in muscoid flies. *J. Ultrastruct. Res.* **42,** 554–581.

Truman, J. W. and Reiss, S. E. (1976). Dendritic reorganization of an identified motoneuron during metamorphosis of the tobacco hornworm moth. *Science*, **192**, 477–479.

Weiss, M. J. (1974). Neuronal connections and the function of the corpora pedunculata in the brain of the American cockroach, *Periplaneta americana*. *J. Morph.* **142**, 21–70.

White, R. H. (1961). Analysis of the development of the compound eye in the mosquito, *Aedes aegypti*. *J. exp. Zool.* **148**, 223–239.

White, R. H. (1963). Evidence for the existence of a differentiation center in the developing eye of the mosquito. *J. exp. Zool.* **152**, 139–147.

Wigglesworth, V. B. (1940). Local and general factors in the development of "pattern" in *Rhodnius prolixus* (Hemiptera). *J. exp. Biol.* **17**, 180–200.

Wigglesworth, V. B. (1953). The origin of sensory neurones in an insect, *Rhodnius prolixus* (Hemiptera). *Q. J. micr. Sci.* **94**, 93–112.

Wigglesworth, V. B. (1959). The histology of the nervous system of an insect, *Rhodnius prolixus* (Hemiptera). I. The peripheral nervous system. *Q. J. micr. Sci.* **100**, 285–298.

Wigglesworth, V. B. (1972). "The Principles of Insect Physiology". Chapman and Hall, London.

Wolpert, L. (1969). Positional information and the spatial pattern of cellular differentiation. *J. theoret. Biol.* **25**, 1–47.

Wolpert, L. (1971). Positional information and pattern formation. *Curr. Top. Develop. Biol.* **6**, 183–224.

Wolpert, L., Lewis, J. and Summerbell, D. (1975). Morphogenesis of the vertebrate limb. *In* "Cell Patterning" Ciba Foundation Symposium 29 (new series), pp. 95–119. Elsevier-Excerpts Medica-North Holland, Amsterdam.

Young, D. (1972). Specific re-innervation of limbs transplanted between segments in the cockroach, *Periplaneta americana*. *J. exp. Biol.* **57**, 305–316.

Zawarzin, A. (1924). Histologische Studien über Insekten. VI. Das Bauchmark der Insekten. *Z. wiss. Zool.* **122**, 323–424.

Zettler, F. and Weiler, R. (1976). "Neural Principles in Vision". Springer-Verlag, Berlin, Heidelberg, New York.

The literature survey on which this essay is based was completed in January, 1978.

The Scent Glands of Heteroptera

Brian W. Staddon

Zoology Department,
University College, Cardiff, Wales

1 Introduction

It is the main physiological role of the scent glands in Heteroptera to manufacture and store for subsequent release volatile, usually highly odoriferous, substances (the scent substances). It would appear to be the main but not the only biological role of the scent substances to provide scent producing Heteroptera with a chemical defence against predators.

The scent glands are epidermal glands and as invaginations formed from epidermis are lined with a cuticle continuous with that of body wall. The names given to the glands are from the parts of the body (abdomen, metathorax) in which they occur. The abdominal glands are dorsal and there may occur up to a maximum of four in a mid-dorsal metameric series above the heart and below abdominal tergites III to VI. The metamerically single metathoracic gland occupies a ventral position in the hind part of the metathorax; it is concealed from above by the abdominal nerves and over these nerves by the gut. The abdominal glands are essentially larval glands and only occasionally continue to function in the adults. The metathoracic gland is an exclusively adult structure and frequently shows a high degree of anatomical and physiological specialization of its own.

The abdominal and the metathoracic scent glands together form a system

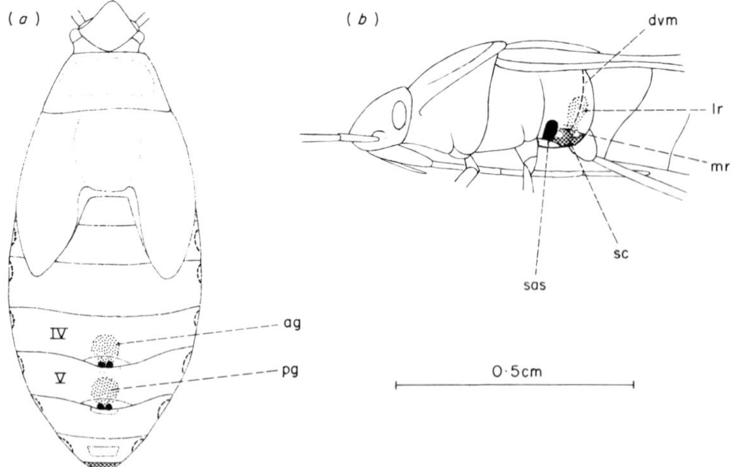

Fig. 1. The scent gland system of *Oncopeltus fasciatus* (Lygaeidae). (*a*) Dorsal view of fifth stage larva showing the position of the dorsal abdominal scent glands. The anterior gland (ag) opens at the posterior margin of abdominal tergite IV, the posterior gland (pg) at the posterior margin of abdominal tergite V. (*b*) Side view of head and thorax of adult showing the position of components of the metathoracic scent apparatus. dvm, dorso-ventral opener muscle; lr, lateral scent reservoir; mr, median scent reservoir; sas, scent accumulation surface; sc, scent canal.

of glands (the scent gland system) which constitutes one of the defining characteristics of the order Heteroptera (Carayon, 1962, 1971). Figure 1 shows the three glands (the two abdominal glands and the metathoracic gland) which together compose the scent gland system of the milkweed bug *Oncopeltus fasciatus*. There are other scent glands but these are restricted in their distribution to particular species or groups; Brindley's glands in Reduviidae for one example (Carayon, 1962).

There is wide variation in the details of scent gland structure and function amongst Heteroptera. The nature of these details and their significance for physiological and ecological studies on Heteroptera form the subject of this review.

2 Basic considerations of scent gland structure and function

2.1 STRUCTURE OF THE SCENT GLAND EPITHELIA

2.1.1 *General structure*

The scent glands consist of integument similar in basic structure to that forming the body wall. There are the same three basic layers; (1) a relatively thin, extracellular, basal layer called basement membrane; (2) a middle layer consisting of epithelial cells arranged as in a monolayer; (3) a relatively thick, extracellular, apical layer – the gland lining cuticle or gland cuticular intima.

The scent gland epithelia are made up of cells of several different types. Certain cells (secretory cells) manufacture scent substances from other substances acquired presumably by absorption from the haemolymph. Other cells (duct cells, intima cells) are devoted to the formation of structures (ducts, reservoirs) serving to convey and store the scent substances prior to release. Support (interstitial) cells are a notable feature of secretory epithelia in the abdominal scent glands. The secretory cells are in volume equal to around 50 duct cells or intima cells and the largest cells in scent gland epithelia. The secretory cells are of two main types. Type 1 secretory cells are associated with duct cells to form multicellular secretory units. Type 2 secretory cells are not associated with duct cells and form the unicellular secretory units. Multicellular but not unicellular secretory units are always present in scent gland epithelia.

2.1.2 *The secretory units*

A diagram giving the structure of the different types of secretory unit is shown in Fig. 2. The multicellular secretory units consist of one secretory cell

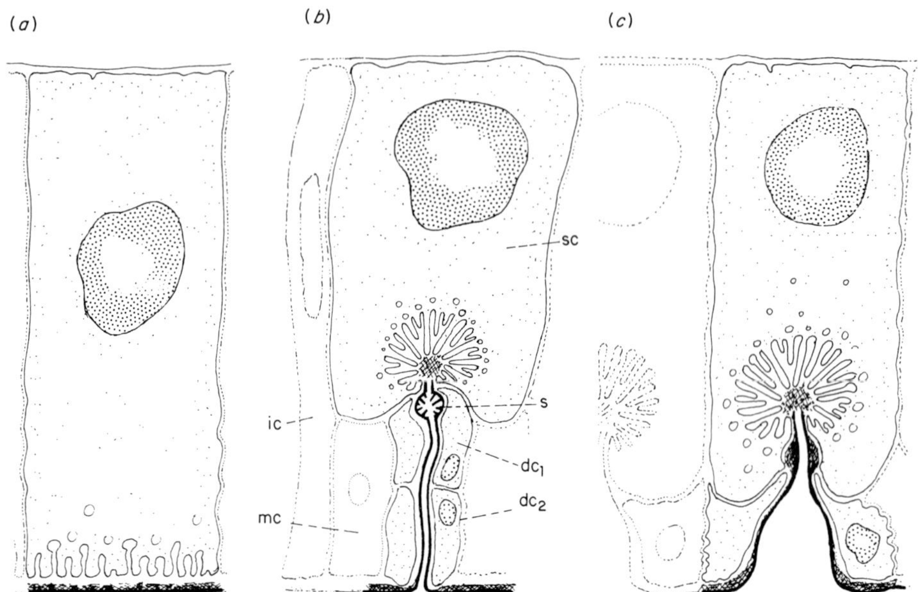

Fig. 2. Different types of secretory unit. (*a*) Unicellular secretory unit; (*b*) Three-cell secretory unit from abdominal scent gland; (*c*) Two-cell secretory unit from metathoracic secretory tubule. dc_1, basal duct cell; dc_2, apical duct cell; ic, interstitial (support) cell; mc, matrix cell; sc, secretory cell.

together with one or two duct cells; one duct cell as a rule in the case of multicellular secretory units from the metathoracic scent gland, two duct cells as a rule in the case of multicellular secretory units from the abdominal scent glands. However, three-cell as well as two-cell secretory units have occasionally been recorded from the metathoracic scent gland (Henrici, 1940). In the metathoracic gland, the two-cell secretory units are formed into unbranching or branched tubules (secretory tubules) around a central collecting duct. The secretory tubules may have a reservoir of their own (lateral reservoir); frequently they open into a separate "median" reservoir. Henrici (1940) observes that two-cell secretory units and the manner of their organization into tubules is an exclusively adult feature. The high degree of anatomical differentiation which only the metathoracic scent gland shows may be correlated with the absence of undifferentiated intima cells ("matrix" cells) and support cells.

2.1.3 *The cuticular intima*

Gluud (1968) supplies a description of the fine structure of the cuticular intima from the scent gland reservoir in pentatomids. In horizontal structure, the reservoir intima is similar in all essential respects to body wall cuticle. There is a difference in the vertical structure, however; pore canals which are a conspicuous feature of body wall cuticle are absent from the scent gland intima. It has been shown that the cuticular ductules in the multicellular secretory units of scent gland epithelia consist of epicuticle and little more (Filshie and Waterhouse, 1968).

2.2 METABOLISM OF THE SCENT GLAND EPITHELIAL CELLS

2.2.1 *Introduction*

The function of the abundant tracheal supply to the scent glands is twofold; 1, to provide the glands with a means of support; 2, to supply the gland epithelial cells with oxygen. The ducted secretory cells in particular have an abundant tracheal supply. It seems evident that the gland epithelial cells acquire substrates and other metabolites by absorption from the haemolymph.

Oenocytes have not been recorded from scent gland epithelia although a direct role for such cells in connexion with the biosynthesis of cuticular alkanes has been described in the locust *Schistocerca* (Diehl, 1973). Haemocytes may be found adhering to scent gland epithelia (Henrici, 1938, 1940). However, the formation of basement membrane excepted (Wigglesworth, 1973), there is no reason at present to suppose that blood cells are important in connexion with physiological processes in scent gland epithelial cells.

Secondary metabolism in scent gland epithelial cells presents three conspicuous aspects; 1, manufacture of cuticular materials; 2, pigment accumulation; 3, manufacture of the scent substances. All types of scent gland epithelial cell, the secretory cells not excepted, function at least for a short while during gland development in the manufacture of cuticular materials. Pigment accumulation is frequently observed in the cytoplasm of scent gland intima cells and ductless secretory units. Accumulations of pigment have not been recorded from the ducted secretory cells and above all other epithelial cells it is the ducted secretory cells that play a central role in the manufacture of the scent substances.

2.2.2 *The scent substances*

The scent substances are for the most part short to medium unbranched carbon-chain aliphatic substances; acids, aldehydes, ketoaldehydes, ketones, alcohols, and esters (Table 1). They have been fully listed by Dazzini and Finzi (1974) and several papers give species sources (Gilby and Waterhouse, 1965; Baggini *et al.*, 1966; Calam and Youdeowei, 1968).

TABLE 1 Unbranched aliphatic substances from the scent glands of Heteroptera and the number of different families from which each substance has been recorded

Classes of compound	C_n*													
	2	3	4	5	6	7	8	9	10	11	12	13	14	15
alkanoic acids	2		2		2									
alkanols			1		3									
alkenols					2		1		1					
alkanals	3	2	3		4		2							
alkenals		2	2	1	10	3	8		3					
alkadienals					1		1							
alkanones			2		1			1						
alkenones					1									
4-oxo-alkenals					8		4							
alkyl acetates					2		2							
alkyl butanoates			2		2							1	1	
alkyl hexanoates			2		2									
alkenyl acetates					5		5		2					
alkenyl butanoates					1									
alkadienyl acetates					1		1							
alkanes										1	3	4	1	2
alk-1-enes												1		

* Number of carbon atoms. In the case of the esters C_n denotes the number of carbon atoms in the alcohol chain

This table is based on data extracted from the following papers: Aldrich *et al.* (1972); Aldrich and Yonke (1975); Baggini *et al.* (1966); Baker *et al.* (1972); Baker and Jones (1969); Baker and Kemball (1967); Blum (1961); Blum and Traynham (1960); Blum *et al.* (1960); Blum *et al.* (1961); Butenandt (1955); Butenandt and Tâm (1957); Calam and Scott (1969); Calam and Youdeowei (1968); Cmelik (1969); Collins (1968); Collins and Drake (1965); Devakul and Maarse (1964); Everton *et al.* (1974); Everton *et al.* (1979); Games and Staddon (1973a, b); Games *et al.* (1973); Games *et al.* (1974); Gilby and Waterhouse (1964, 1965, 1967); Gilchrist *et al.* (1966); Ishiwatari (1974, 1976); Kathuria *et al.* (1974); Levinson *et al.* (1974); Macleod *et al.* (1975); Macleod *et al.* (1977); Maschwitz (1971); McCullough (1966, 1967, 1968, 1969, 1971, 1973, 1974a, b); Mukerji and Sharma (1966); Park and Sutherland (1962); Pattenden and Staddon (1968, 1970, 1972); Pinder and Staddon (1965a, b); Prestwich (1976); Roth (1961); Schildknecht *et al.* (1962); Smith (1974); Staddon (1973); Staddon and Weatherston (1967); Staddon *et al.* (1979); Tsuyuki *et al.* (1965); Waterhouse *et al.* (1961); Waterhouse and Gilby (1964); Youdeowei and Calam (1969).

Table 1 shows that carbon chains are mostly even and that the commonest carbon chain is the C_6 closely followed by the C_8 and the C_4. With the exception of tridecane (C_{13}), odd numbered constituents are rarely important. Tridecane has been recorded as a major constituent in secretions from the metathoracic gland of many pentatomid bugs (Blum *et al.*, 1960; Blum and Traynham, 1960; Waterhouse *et al.*, 1961; Park and Sutherland, 1962; Baggini *et al.*, 1966; Smith, 1974; Everton *et al.*, 1974; Kathuria *et al.*, 1974; Macleod *et al.*, 1975; Prestwich, 1976), from the abdominal glands of a larval pentatomid (Ishiwatari, 1974), and elsewhere recorded from the third abdominal gland of the pyrrhocorid *Dysdercus intermedius* (Calam and Youdeowei, 1968). Only one other odd numbered compound, *trans*-hept-2-enal, has been recorded as a major scent substance and as yet only from the metathoracic scent from the pentatomid *Oebalus pugnax* (Blum *et al.*, 1960; Blum and Traynham, 1960).

Aliphatic scent substances frequently possess one double bond and occasionally possess two (the 2,4-dienals and dienyl acetates). The arrangement of the double bond is *trans* as a rule. Interestingly, Gilby and Waterhouse (1965) found a small quantity of the less stable *cis* isomer together with *trans*-dec-2-enal in metathoracic scent from the pentatomid *Nezara viridula*. From the work of Macleod *et al.* (1977) on the C_8 compound it seems unlikely that the *cis*-isomers of the 4-oxo-alk-2-enals will be found as major constituents of the secretions from the scent glands. The *cis* isomers are thermodynamically very unstable and readily undergo *cis* to *trans* isomerization.

Aldrich and Yonke (1975) have recorded 2-methyl branched aliphatic scent substances from certain coreid bugs. Isobutyric acid has been identified as the only volatile substance in secretion from Brindley's glands in certain Reduviidae-Triatominae (Pattenden and Staddon, 1973; Games *et al.*, 1974; Kälin and Barrett, 1975).

There is evidence that Heteroptera manufacture their own scent substances. Gordon *et al.* (1963) injected ^{14}C labelled acetate into the haemolymph of *Nezara viridula* and later recorded significant radioactivity from the scent substances (the C_6 and the C_8 alk-2-enal, tridecane, 4-oxohex-2-enal). Significant radioactivity was also found in whole metathoracic scent isolated nine days after the injection of other 1-^{14}C labelled fatty acids (the C_3, C_6, and C_{10} *n*-alkanoic acids). The question of the form in which the gland secretory cells acquired the radioactive label was not answered.

It is possible that some Heteroptera acquire scent substances preformed from the diet. Baker and Kemball (1967) examined food plants of the coreid *Pternistria bispina* but found only one compound (hexanol) in both the *Pternistria* scent (hexanol is a minor constituent of the metathoracic scent) and the plant extracts. The plant extracts contained acetic acid. Acetic acid was not found free in the metathoracic scent from *Pternistria* but occurs widely as a major constituent in the metathoracic secretions from coreid bugs (Waterhouse and Gilby, 1964; McCullough, 1969, 1971, 1973, 1974a, 1974b; Aldrich and Yonke, 1975; Prestwich, 1976). In *Oncopeltus fasciatus*, cardiac glycosides from the diet (milkweed plants of the family Asclepiadaceae) have been recorded in the secretions from the scent glands (Duffey and Scudder, 1972). It is interesting that linalool production is common to the cotton plant and the cotton stainer *Dysdercus intermedius* (Everton *et al.*, 1979).

The C_6 and the C_8 alk-2-enals and 4-oxo-alk-2-enals are especially characteristic of the secretions from the scent glands of Heteroptera (Table 1). They are shown together with the acetates of the C_6 and C_8 alk-2-enols in Fig. 3.

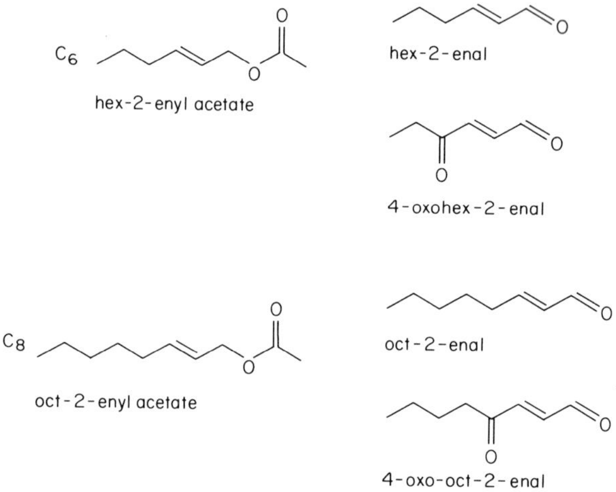

C_6

hex-2-enyl acetate

hex-2-enal

4-oxohex-2-enal

C_8

oct-2-enyl acetate

oct-2-enal

4-oxo-oct-2-enal

Fig. 3. Commonly found aliphatic scent substances in Heteroptera.

It seems probable that these similar compounds are formed by variations from the same biosynthetic pathway (section 5.1.2). The 4-oxo-alk-2-enals are not known elsewhere in nature and together with the corresponding alk-2-enals must have been acquired early in the evolution of the scent gland system in Heteroptera. All four compounds probably co-occur in the metathoracic scent from *Gelastocoris oculatus* (Staddon, 1973).

The evidence of structure points to the conclusion that the great majority of scent substances are formed from acetate and hence able to be classed as polyketides (Bu'Lock, 1965). It is unusual to find representatives from other biogenetic groups in the secretions.

The aromatic compounds shown in Fig. 4 have been recorded in the metathoracic secretions from the aquatic bugs *Ilyocoris cimicoides* and *Notonecta glauca* (Staddon and Weatherston, 1967; Pattenden and Staddon, 1968). The pattern of substitution (*p*-OH) is indicative that these substances

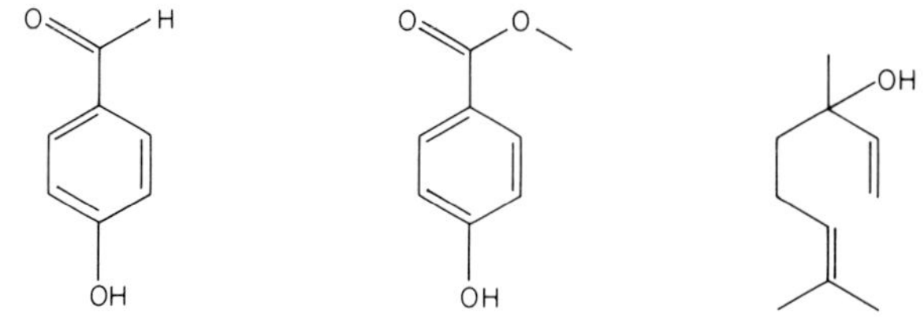

Fig. 4. Unusual compounds from the metathoracic scent gland in Heteroptera.

(*p*-hydroxybenzaldehyde and methyl *p*-hydroxybenzoate) are formed *via* the shikimic pathway. The shikimic pathway and the primary pathways of sugar metabolism are closely linked.

Calam and Scott (1969) recorded an isoprenoid ($C_{10}H_{16}$) from the metathoracic scent gland of *Dysdercus intermedius*. This compound has been identified as the monoterpene alcohol linalool (Everton *et al.*, 1979). Linalool (Fig. 4) accumulates in the lateral reservoir of the metathoracic gland. The concentration of linalool, at first negligible, rises to reach a maximum (*ca.* 92% of the total in the lateral reservoir) around 14 days after ecdysis. The secretions from *Dysdercus* also contain the C_6 and the C_8 alk-2-enals and alk-2-enyl acetates. It is assumed that *Dysdercus* manufactures its own linalool *via* a conventional isoprenoid pathway. Isoprenoid, aliphatic, and aromatic substances form the scent from the abdominal first gland of adult male *Podisus* (Asopinae) (Aldrich *et al.*, 1978a).

A miscellany of other substances have been found in the secretions; water (Aldrich and Yonke, 1975; Aldrich *et al.*, 1978b; Everton and Staddon, 1979), hydrogen peroxidae (Maschwitz, 1971), mucosubstances (Carayon, 1971; Everton and Staddon, 1979), enzyme proteins (Aldrich *et al.*, 1978b).

2.2.3 *Extracellular biochemistry*

Extracellular biochemical reactions are frequent in connexion with the manufacture of toxic substances by defence glands in arthropods (Schildknecht and Holoubek, 1961; Eisner *et al.*, 1963; Eisner and Meinwald, 1966; Happ, 1968; Schildknecht *et al.*, 1968; Eisner, 1970). Advantageously for the gland epithelial cells, the reaction steps which yield the toxic end substances are caused to take place at a distance from the cells in a cuticle-lined duct or some other cuticle-lined compartment of the gland. Similar extracellular systems are important in connexion with the tanning of cuticle (Brunet, 1967; Neville, 1975).

There is little to doubt that the final steps in the biosynthesis of the reactive scent carbonyls in Heteroptera occur outside the cells, but the evidence in support of this assumption is at present circumstantial and far from satisfactory. Betten (1943) detected differences in osmium staining in the ductules of the secretory cells of the abdominal glands of *Corixa*; the basal part of the ductule which includes the saccule, remained unstained unlike the longer apical part of the ductule which was blackened by osmium. Electron micrographs of multicellular secretory units often show a secretory cell lumen which is electron-light, in contrast to the ductule which contains a markedly electron-dense material (Stein and Schumacher, 1969; Walker, 1972). In *Dysdercus*, cytological appearances point to the conclusion that the duct cells of the multicellular secretory units of the metathoracic scent gland have a secretory function (Schumacher and Stein, 1971). Perhaps it is the role of the duct cells to supply enzymes capable of driving extracellular reactions in the lumen of the ductule. From studies on the metathoracic scent gland from diverse species of Heteroptera (pentatomid, lygaeid, pyrrhocorid and corixid bugs) it has been found that the scent carbonyls are confined to the lumen of the median reservoir (Gilby and Waterhouse, 1965; Games and Staddon, 1973b; Staddon *et al.*, 1979; Everton *et al.*, 1979).

Everton and Staddon (1979) used a silver impregnation technique at the electron microscope level of structure to look for reducing groups in the metathoracic accessory gland of *Oncopeltus fasciatus*. Reduced silver formed specifically in thin layers extracellularly along the margins of the thin cuticular intima of the accessory gland. Further deposition of silver after chromic acid oxidation indicated that the histologically determinate secretion from the accessory gland contains mucosubstances.

Baker *et al.* (1972) recorded the spontaneous formation of 2-*n*-butyloct-2-enal from hexanal in scent isolated from the metathoracic median reservoir from the coreid *Amblypelta nitida*. It is indicative that this aldol condensation reaction is promoted by a catalyst since no 2-*n*-butyloct-2-enal formed in an artificial scent consisting of hexanal and a number of other compounds (hexyl acetate, hexyl hexanoate, acetic acid) which also occur naturally in the scent from *Amblypelta*.

Extracellular reactions in connexion with the biosynthesis of the scent substances may occur in accumulations of secretory products even prior to their exocytosis from the secretory cells. In electron micrographs of unicellular secretory units from the third abdominal gland of *Dysdercus intermedius*, Stein (1969) recorded striking changes in the appearance of the secretory droplets prior to their exocytosis from the cells. It was noted that secretory droplets in the middle region of the cell were closely surrounded by mitochondria (section 4.2.1).

Enzyme proteins (two esterases, a dehydrogenase) have been isolated from the metathoracic secretion from the coreid *Leptoglossus* (Aldrich *et al.*, 1978b).

2.2.4 *Pigment metabolism*

The scent glands are frequently coloured yellow, orange, or red. However, multicellular secretory units are never pigmented. Because they consist of multicellular secretory units it is often difficult to distinguish the secretory tubules in dissections of the metathoracic scent gland. As in the epidermis so in scent gland epithelia the pigments are concentrated in granules in the cytoplasm. Electron micrographs of thin sections show that pigment granules have a membranous inner structure (Stein, 1969).

Pteridines mostly but carotenoids and other pigments also have been isolated from Heteroptera (Ziegler and Harmsen, 1969; Vuillaume, 1975). The red erythropterin in particular is widely present in Heteroptera (Merlini and Nasini, 1966). Erythropterin and several carotenoids (α-carotene, licopene, xanthophyl) have been isolated from the firebug *Pyrrhocoris apterus* (Merlini and Mondelli, 1962). The normally bright orange colour of the epidermis in *Oncopeltus fasciatus* is due to a mixture of two pteridines; erythropterin and the yellow YP2 (Lawrence, 1970). There are several other pteridines and these probably include xanthopterin, isoxanthopterin, leucopterin, and pterin-7 carbonic acid. The biological significance of pigments in scent gland epithelia awaits elucidation.

2.3 FUNCTIONS OF THE SCENT GLANDS

Postulated functions (biological) include; 1, defence against predators; 2, defence against microorganisms; 3, release of specific patterns of behaviour (alarm, aggregation, or mating). These suggestions are considered more fully in section 6. Specific co-adaptations include directional mechanisms for scent spray aiming in larvae (section 3.1.3) and the cuticular microsculpture of the metathoracic efferent system (section 3.4.1). The cuticle in scent producing Heteroptera may contain mechanisms specifically serving to resist the penetration of the scent substances (section 7).

There is wide variation in physiological function (chemical constitution of the secretions) between similar glands from different species and even between different scent glands within individuals from the same species. Organ specificity (qualitative and quantitative differences in scent constitution in different glands from the same individual) is extremely interesting in connexion with the problem of interrelating scent constitution and biological function and for that reason the known instances of it are fully catalogued below.

1. *Dysdercus intermedius* (Pyrrhocoridae). The C_6 and the C_8 4-oxo-alk-2-enals and tridecane from the third dorso-abdominal gland of the larva;

an unknown volatile substance (? tetradecane) from the first and second abdominal glands of the larva; a monoterpene alcohol (linalool) from the metathoracic scent gland. Only the C_6 and the C_8 alk-2-enals have been recorded from different glands (the metathoracic gland and the third larval abdominal gland) (Calam and Youdeowei, 1968; Calam and Scott, 1969; Youdeowei, 1969; Everton *et al.*, 1979).

2. *Oncopeltus fasciatus* (Lygaeidae). The C_6 and the C_8 4-oxo-alk-2-enals from the abdominal scent glands; the C_6 and the C_8 alka-2,4-dienals and -2,4-dienyl acetates from the metathoracic scent gland. Only the C_6 and the C_8 alk-2-enals (*cf. Dysdercus*) have been recorded from the metathoracic and abdominal scent glands (Games and Staddon, 1973a, 1973b; Games *et al.*, 1973). In *Oncopeltus*, there exist quantitative differences in scent constitution in the two abdominal glands; 4-oxo-oct-2-enal is the major constituent in the scent from the posterior gland, 4-octenal the major constituent in the scent from the anterior gland.

3. Coreid bugs. In many different species, the C_6 and the C_8 alk-2-enals and 4-oxohex-2-enal from the abdominal glands in the larvae; hexanal, hexyl acetate, and other saturated aliphatic compounds from the metathoracic scent gland (Baker and Kemball, 1967; Baker and Jones, 1969; Aldrich and Yonke, 1975; Prestwich, 1976).

It should be noted that no such differences have been recorded from *Pternistria bispina* (Baker and Kemball, 1967).

4. *Corixa dentipes* (Corixidae). 4-oxohex-2-enal from the metathoracic scent gland; 4-oxo-oct-2-enal from the abdominal scent glands (Pinder and Staddon, 1965a, 1965b; Staddon *et al.*, 1979).

5. *Tessaratoma aethiops* (Pentatomidae). Oct-2-enyl acetate from the metathoracic scent gland only (Baggini *et al.*, 1966). However, this qualitative difference between otherwise chemically similar metathoracic and abdominal scent gland secretions can be explained in terms of the special way the metathoracic scent gland is organized (section 5.2.1).

No organ specific differences were found in the pentatomid *Apodiphus amygdali* (Everton *et al.*, 1974).

Some of these differences may be associated with differences in mode of life between the larvae and their adults. Interestingly, the 4-oxo-alk-2-enals so far as is now known are more widespread in secretions from the abdominal scent glands; in the metathoracic scent gland they have been recorded only from Corixidae and Pentatomidae.

2.4 PHYSIOLOGICAL CONTROL MECHANISMS

This is a neglected aspect of the subject at all functional levels; cell, organ, and organism.

The secretory cells are not innervated; a mechanism of physiological control which uses hormones has not been demonstrated. A possibility to be examined is that secretion synthesis may be checked by stimuli or forces arising from the filling of the scent reservoir.

In all types of scent gland (possible exception Brindley's glands; Schofield and Upton, 1978) the gland orifice is surrounded by a muscle controlled valve (always the valve is caused to be closed by the elasticity of its cuticular components; the valve muscles are opener muscles). However, with the curious exception of the abdominal scent glands in Corixidae (Betten, 1943), the scent reservoir in all types of scent gland lacks an enveloping coat of muscles for compression. There are muscles which insert on the abdominal scent glands (stretch and other muscles) but no muscles (the opener muscles excepted) insert on the metathoracic scent gland. A rise in haemolymph pressure (such as could be caused by contractions in the abdominal muscles) perhaps assisted by the elasticity of the reservoir intima is presumably the source of the force required to expel scent from the metathoracic reservoir (Betten, 1943). Movements of the metacoxal muscles may be important in connexion with scent emission from the metathoracic gland in coreid bugs (Blum *et al.*, 1961). Scent ejection is usually triggered by handling; the scent vapour or visual disturbances are effective triggering stimuli for some species.

Spray aiming towards an attacker is shown by many land bugs (Remold, 1962). Presumably the overall pattern of physiological control in spray aiming mechanisms is "open loop". Visual stimuli perhaps assisted at close range by contact presumably enable the scent emitter to map the direction of the attacker.

In some species scent emission may be triggered by an internal stimulus. An apparently "voluntary" control of scent release is shown by the lesser backswimmer *Plea leachi*. From time to time, *Plea* climbs out of the water and causes secretion to be ejected from the metathoracic gland; the ejected secretion is caused to be rubbed over the body surface by movements of the limbs (Maschwitz, 1972). In *Plea*, it may be that internal stretch receptors sensitive to the filling of the metathoracic scent reservoir trigger scent emission; in turn grooming may be triggered by the ejected scent substances.

2.5 POSTEMBRYONIC DEVELOPMENT AND DEVELOPMENTAL FATE OF THE SCENT GLANDS

Descriptions of the postembryonic development and developmental fate of scent glands in different species of Heteroptera have been given by Henrici (1938, 1940).

2.5.1 *The abdominal scent glands*

The abdominal scent glands are complete by the end of embryonic development and commence to function in the newly emerged larva (*Pyrrhocoris*). At each moult there is an increase in gland size and the multicellular secretory units increase in number. The abdominal glands usually cease to function towards the end of the last larval instar, prior to apolysis. However, in some systems one gland in the abdominal system may cease to function at an earlier stage in larval development; and in some systems one or more glands continue to function in the adults (section 3.1.5).

Multicellular secretory units formed at one moult are caused to be destroyed at the next. New multicellular secretory units are formed from undifferentiated cells (the "matrix" cells) in the secretory epithelium. There are two cell divisions; 1, the matrix cell divides to produce the presumptive secretory cell and a second cell; 2, the second cell divides to form the two duct cells. It is possible that the presumptive secretory cell itself divides again, by unequal division, but if such a division occurs it was not recorded by Henrici.

The cells of the unicellular secretory units apparently remain intact during moults; they proliferate like the ordinary intima cells of the gland epithelium.

2.5.2 *The metathoracic scent gland*

Henrici (1940) followed the development of the metathoracic scent gland in the pentatomid *Carpocoris*. The rudiment of the gland makes it appearance at the last moult as an epithelial fold extending transversely between the metasternal apophyses. The lateral parts of the fold develop and differentiate relatively rapidly and soon show a small reservoir basally and several branching tubules distally. The median part of the fold supplies the median reservoir.

Betten (1943) described a similar pattern of development in the metathoracic scent gland of *Corixa*; but in *Corixa*, neither the median part of the fold nor the scent canals become permanently sealed off by a suture as happens in *Carpocoris*.

The idea of Henrici (1940) that the acquisition of two-cell secretory units by the metathoracic gland is a specialization specifically associated with the organization of multicellular secretory units into tubules is supported by the distribution of two-cell and three-cell secretory units in the metathoracic gland of *Oncopeltus fasciatus* (section 3.2.1); in *Oncopeltus*, the two-cell secretory units are restricted to the secretory tubules, the three-cell units to the wall of the lateral reservoir (Aldrich *et al.*, 1978a, find tubules in the abdominal first gland from the adult of the asopine *Alcaeorrhynchus*).

2.5.3 *Activation of the scent glands*

The scent glands (abdominal and metathoracic) are completed prior to ecdysis and the gland secretory cells begin to discharge secretion shortly after ecdysis (Henrici, 1938, 1940). The mechanism of activation of the gland secretory cells has not been investigated; perhaps the hormone "bursicon" is involved. However, it is not always the case that the secretory cells enter the secretion phase synchronously and in *Pyrrhocoris apterus* the time of onset of secretion synthesis differs in the different glands of the abdominal system (Henrici, 1938).

2.6 AUTONOMY OF FUNCTION AND THE NATURAL SELECTION OF OBSOLESCENCE

The subject is frankly speculative; and yet obsolescence in the scent gland system is widespread in Heteroptera. There are two ways of viewing autonomy; (a) physiological and (b) developmental. By physiological autonomy it is meant that the gland of interest is not a part of some physiological control system (homeostasis); by developmental autonomy it is meant that glands in different segments are free to evolve in different ways or to be deleted from the system (see Lawrence and Green, 1975). Since major evolutionary losses are irreversible (Dollo's law; statistical basis thereof in Monod, 1972) a renewed need for the advantages provided by scent glands stimulates the evolution of quite new structures (? Brindley's glands in Reduviidae).

Heteroptera constitute a large, ecologically diverse, and reasonably well studied group of insects (Poisson, 1951; Southwood and Leston, 1959; Miller, 1971; Woodward *et al.*, 1970) but virtually nothing is known of the circumstances that favour obsolescence in the scent gland system. It is at first glance a remarkable conclusion from the work of Schaeffer (1972) that a metathoracic scent gland has little or no value for phytophagous Heteroptera which lead conspicuous lives high on their food plants.

It seems that obsolescence may be acquired gradually, at first simply by means of an orderly decrease in size without loss of physiological function. A reduction in size in the case of the metathoracic scent gland is associated with the division of the gland (land bugs) into two symmetrical half-glands. Only when the gland is virtually vestigial is the original physiological function impaired. The abnormalities seen in greatly reduced scent glands probably signify a loss of function rather than the acquisition of some new function. For possible examples, the cytoplasmic peculiarities seen by Schumacher, 1971b, in the metathoracic tubules of *Pyrrhocoris apterus*; the secretion of water by the accessory gland in the metathoracic scent apparatus of *Oncopeltus fasciatus* (Everton and Staddon, 1979).

The natural selection of obsolescence may involve polygenes within a polygenic system or alleles at polymorphic loci; the latter may be but need not be polygenes. In the absence of pre-existing genetic variation, suitable mutants as they crop up will be favoured by natural selection.

3 Morphology of the scent gland system

3.1 THE ABDOMINAL GLANDS

3.1.1 *Basic structure*

The second abdominal scent gland from *Pyrrhocoris apterus* is shown in Fig. 5. The dorso-ventrally flattened gland sac opens at the hind margin of

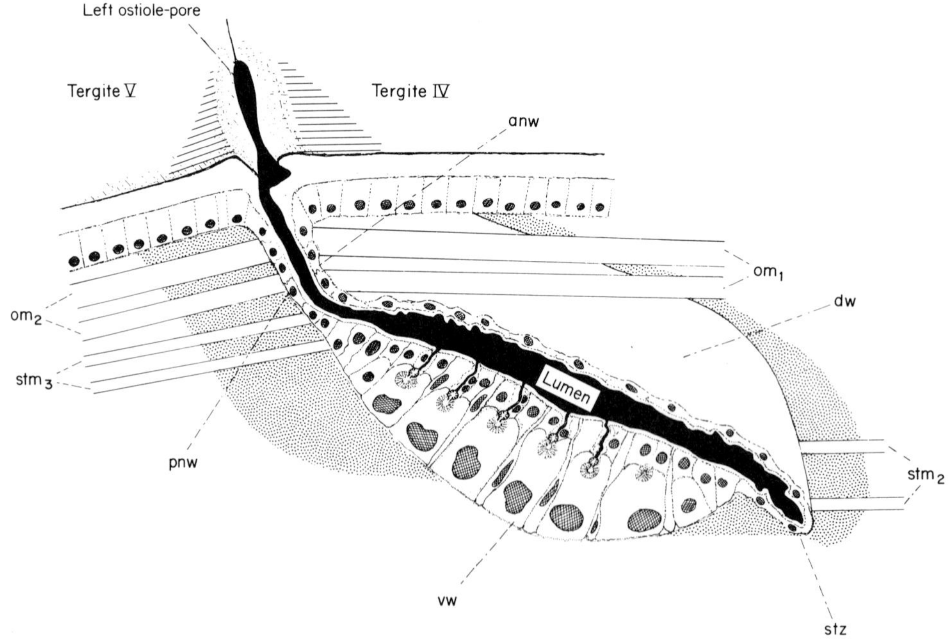

Fig. 5. Diagram indicating the structure of the second abdominal scent gland from *Pyrrhocoris apterus* (Pyrrhocoridae). anw and pnw, anterior gland neck wall and posterior gland neck wall; dw and vw, dorsal wall and ventral wall of gland sac; om_1 and om_2, left upper opener muscle and left lower opener muscle of first and second abdominal scent glands; stm_2 and stm_3, left stretch muscle and right stretch muscle of second and third abdominal scent gland; stz, stretch zone of gland sac wall. (Adapted from Henrici, 1938 and Stein, 1971.)

overlying tergite IV and ends blindly in the direction of the head. The upper and lower walls of the sac differ in histological structure. The multicellular secretory units are confined to the lower wall where they are distributed in a cellular matrix consisting of support cells and unmodified intima cells (matrix cells). The non-secretory epithelium of the upper wall consists of unmodified intima cells.

In Pyrrhocoridae there are interspecies differences in the unicellular secretory units; absent in all three abdominal glands from *Pyrrhocoris* (Henrici, 1938), present in the third abdominal scent gland from *Dysdercus intermedius* (Stein, 1969). In *Carpocoris pudicus* (Pentatomidae) and *Lygaeus saxatilis* (Lygaeidae) the whole of the dorsal wall of the gland sac is constituted from unicellular secretory units (Henrici, 1938). In *Coreus (Syromastes) marginatus*, cytological appearances indicate that the support cells from the ventral wall are secretory (Henrici, 1938).

The occlusion mechanism is below the visible orifice (ostiole) and is formed from the membranous walls (anterior, posterior) of the gland sac neck.

3.1.2 *The abdominal ostioles*

Remold (1962) correlated variation in the manner of scent emission from the abdominal glands with differences in the structure of the gland ostioles. Remold's three types of abdominal ostiole are shown in Fig. 6. Scent issues

Fig. 6. Three types of abdominal ostiole. (*a*) Type 1 (undivided) ostiole. (*b*) Type 2 (divided) ostiole. (*c*) Type 3 (divided) ostiole. (Simplified from Remold, 1962.)

relatively slowly through the undivided type 1 and divided type 2 ostioles to either accumulate in a droplet over the ostiole (e.g. scent from the second abdominal gland in *Oncopeltus fasciatus*) or spread out over the cuticle in a thin film (e.g. scent from the first abdominal gland in *Oncopeltus*). Only from the relatively minute lateral openings of divided type 3 ostioles can scent be squirted in such force as to form a spray.

Type 2 ostioles are not permanently sealed by actual fusion in the midline as this would preclude shedding of the gland lining intima at ecdysis. A permanent suture is seen in type 3 ostioles when the gland sac is completely divided into symmetrical half-glands (Henrici, 1938).

3.1.3 *The scent ejection mechanism*

The muscles of the abdominal scent glands have been described by Gulde (1902), Henrici (1938), and Gupta (1964). From position and mode of action the longitudinal gland muscles can be divided into two groups; (a) opener muscles and (b) stretch muscles. The gland openers are caused to be inserted on the gland sac neck (the anterior wall) while the insertion of the stretch muscles is on the basal extremity of the gland sac. The expulsion of scent is caused by the action of the stretch muscles, which flattens the gland sac, assisted by the elasticity of the cuticular intima (Henrici, 1938). A rise in haemolymph pressure may be the cause of the force required to eject scent in a spray through type 3 ostioles (Remold, 1962). In the aquatic Corixidae, the larval abdominal scent glands possess a unique "ring" muscle for compression (Betten, 1943). There are other gland muscles. Oblique muscles (not shown in Fig. 6) to the gland neck have been described. The significance of these muscles for scent emission is not understood; perhaps it is connected with the need to hold the gland orifice firm against the pull of the opener muscles.

The control of scent jet aiming in lygaeid, alydid, and pentatomid larvae is accomplished with the aid of a special occlusion plate. The occlusion plate is located below the lateral pore in the ostiole and is formed by sclerotization from cuticle in the back of the gland sac neck. The semicircularly shaped plate receives the upper longitudinal opener muscle and can be turned to occupy different positions during scent emission. By its ability to use the occlusion plate as a deflector, a jet producing larva can eject scent in any direction horizontally within an angle of about 95° from the side of the body.

Cuticular specializations in the tergites surrounding the abdominal ostioles have been described in different species (Remold, 1962; Carayon, 1971; Aldrich *et al.*, 1972; Cobben, 1978).

3.1.4 *Differences in number and structure*

The abdominal glands differ in number and metameric distribution in different species of Heteroptera (Dupuis, 1947a, 1947b; Cobben, 1978). They also differ in size, shape and degree of pigmentation. A complete metameric series of four abdominal scent glands is rarely seen. The series is complete in *Tessaratoma papillosa* (Pentatomidae) but the first and last glands may be functionless, at least in mature larvae (Kershaw and Muir, 1907). Organ specific differences are well displayed by pyrrhocorids (*Pyrrhocoris apterus*, *Dysdercus intermedius*); the third abdominal scent gland is the largest and the only one of the three glands in the system to possess a conspicuously pigmented apithelium (Henrici, 1938; Youdeowei and Calam, 1969). In many Pentatomidae the first gland in the system shows a complete division into half-glands (Henrici, 1938; Dupuis, 1947a, 1947b). It is an unusual feature in Corixidae that the multicellular secretory units and support cells are organized into compact bundles; the gland consists of a pair of pigmentless secretory bundles and a pigmented median non-secretory storage sac (Betten, 1943). This arrangement in Corixidae is associated with a "ring" muscle surrounding the gland reservoir (section 3.1.3).

3.1.5 *Developmental fate*

Differences in developmental fate have been documented by Henrici (1938).

1. *Lygaeus saxatilis* (Lygaeidae). The gland sacs and gland muscles persist through to the adult and new multicellular secretory units are formed, but the secretory cells fail to develop and the glands cease to function in the adults.
2. *Syromastes marginatus* (Coreidae). As in *Lygaeus*, the gland sacs and gland muscles persist through to the adult, but the multicellular units are not renewed and the gland sacs become permanently sealed by fusion of the neck walls.
3. *Cimex lectularius* (Cimicidae). The abdominal glands are lost at metamorphosis to the adult; there are no vestiges of these glands in the adults in *Cimex*.
4. *Pyrrhocoris apterus* (Pyrrhocoridae). The reduced first and second abdominal glands but not the third continue to function for a while in the adults. A similar fate has been recorded from *Dysdercus intermedius* (Calam and Scott, 1969).
5. *Carpocoris pudicus* (Pentatomidae). The divided first gland continues to function in the adults.

A loss of function at an early stage has been observed in corixids (*Corixa* and *Sigara*, but not *Micronecta*); the first gland in the abdominal series of three ceases to function towards the end of the second larval instar (Poisson, 1924; Betten, 1943). In mature larvae of *Scaptocoris divergens* (Cydnidae) the first gland lacks scent (Roth, 1961) but perhaps functions in younger larvae. A similar occurrence has been recorded from *Apodiphus amygdali* (Pentatomidae) (Everton *et al.*, 1974).

A remarkable sexual dimorphism is shown by the first abdominal scent gland in the adults of at least some Pentatomidae-Asopinae; in the males, at metamorphosis to the adult, the divided first gland becomes transformed into a pair of potentially voluminous sacs and soon fills up with secretion after ecdysis (Dupuis, 1949, 1952, 1959; Aldrich *et al.*, 1978). The first gland persists only as a relatively rudimentary structure in the female adults.

3.2 THE METATHORACIC GLAND

3.2.1 *Basic structure*

Johansson (1957) was the first to give a description of the metathoracic scent gland of the milkweed bug *Oncopeltus fasciatus*. Figure 7 depicts the structure

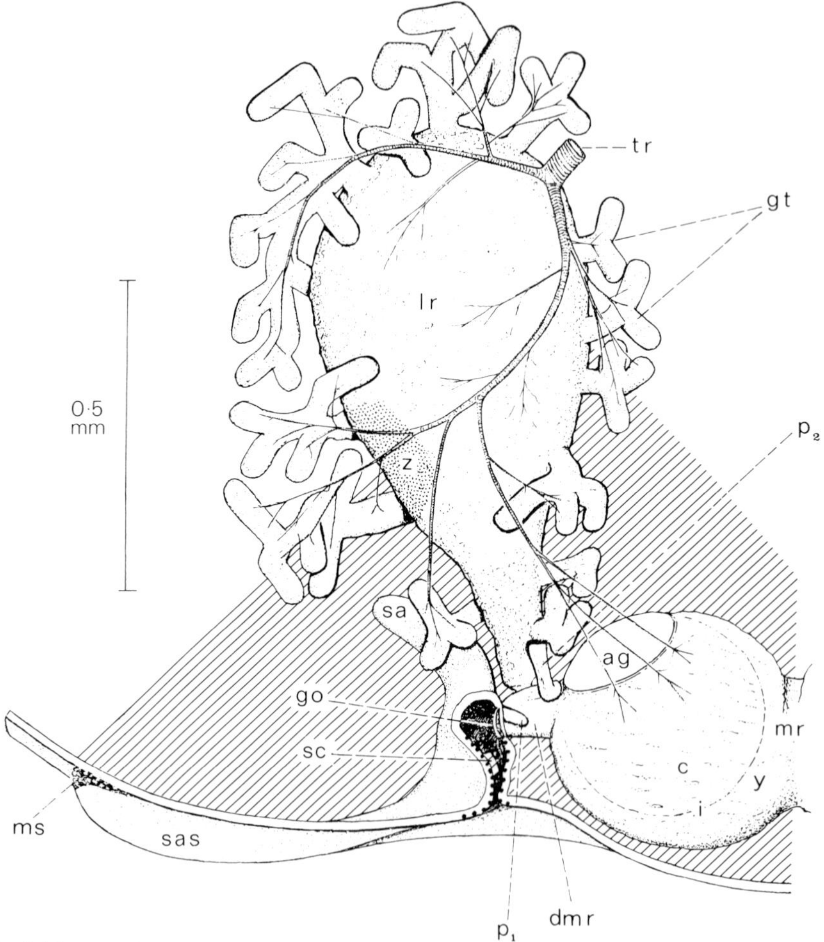

Fig. 7. The metathoracic scent gland system of *Oncopeltus fasciatus* (Lygaeidae). Left half-gland from a mature male adult, viewed from behind. ag, accessory gland; dmr, duct from median reservoir; go, gland external orifice; gt, secretory tubule; lr, lateral reservoir; mr, median reservoir; ms, cuticular microsculpture border; p_1, valve opener apodeme; p_2, apodeme from orifice from the lateral reservoir; sa, metasternal apophysis; sas, scent accumulation surface; sc, scent canal; tr, tracheal supply to gland; z, area in wall of lateral reservoir occupied by 3-cell secretory units. Secretion in median reservoir; c, clear aqueous phase; i, interface between the two liquid phases of the secretion; y, yellow lipid (scent) phase.

of the left half-gland from a mature male adult of *Oncopeltus*. The lateral part consists of branching secretory tubules and a reservoir (lateral reservoir); the median part consists of a second reservoir (median reservoir) and an "accessory" gland. The secretory tubules consist of two-cell secretory units; small areas in the wall of the lateral reservoir consist of three-cell secretory units and support cells (hatched areas in Fig. 7). The median reservoir is pigmented an orange colour but the gland elsewhere is pigmentless. In some individuals the accessory gland is pigmentless and of a glass-like transparency. Three-cell secretory units are unusual in the metathoracic scent gland; they have been recorded previously only from the metathoracic gland in *Reduvius personatus* (Henrici, 1940). In *Oncopeltus*, the metathoracic three-cell secretory units are similar in structure to those from the larval abdominal scent glands.

How the lateral and the median parts of the metathoracic scent gland functionally interrelate in *Oncopeltus* will be considered further in section 5.2. How the different parts are structurally interconnected is shown more clearly than is possible in a natural representation of the gland in the box diagram given in Fig. 8.

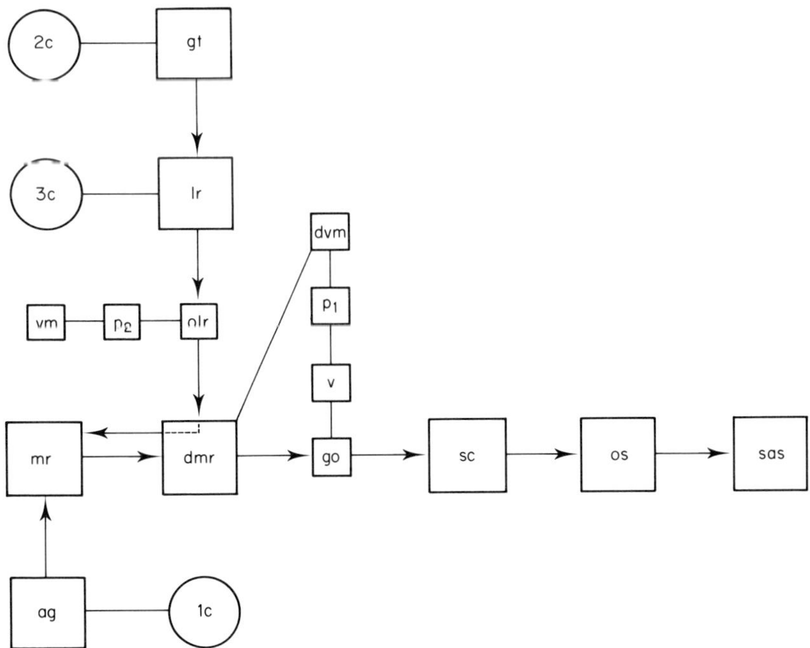

Fig. 8. Box diagram showing the interrelationships of the component parts of the metathoracic scent gland system of *Oncopeltus fasciatus* (Lygaeidae). dvm, dorso-ventral opener muscle, olr, orifice from lateral reservoir; os, metathoracic ostiole; v, valve plate; vm, gland ventral muscle. lc, 2c, 3c; unicellular, two-cell, and three-cell secretory units. For other terms see key in Fig. 7.

3.2.2 *Interspecies variation in the morphology of the metathoracic gland*

Comparative aspects of metathoracic scent gland morphology form the subject of Figs 9*a–f*. The variation shown is remarkable. But while differences in gross morphology are usually clear-cut and easy to determine it is not so easy to discern differences in the secretory units.

a *Differences in gross morphology* Most variants have a paired component in the secretory tubules but the metathoracic scent gland in Enicocephalidae (Fig. 9*d*) is a median and unpaired tube. One gland showing an unpaired component (the reservoir) and a pair of branching secretory tubules is that of *Ilyocoris cimicoides* (Fig. 9*b*). A pair of relatively simple unbranching tubules alone form the metathoracic scent gland in Belostomatidae-Lethocerinae (Fig. 9*a*). The tubules shown in Fig. 9*a* are those from a female adult; male adults have much larger tubules. The metathoracic gland in *Omania* also (Fig. 9*c*) is completely divided. In *Omania*, each half-gland consists of two reservoirs and an interconnecting duct; the wall of the lateral reservoir contains the secretory units.

b *Differences from the unicellular secretory units* The metathoracic accessory gland consists of unicellular secretory units; it is restricted to the median reservoir. The form of the accessory gland differs in different species; compact in *Oncopeltus* (Fig. 7), it is a strip-like organ in Pentatomidae (Fig. 9, f_1), and altogether absent in *Ilyocoris* (Fig. 9*b*). Differences in the metathoracic gland in coreids are shown in Fig. 9e_1 to e_3; e_1, one pair of accessory glands; e_2, accessory gland absent; e_3, two pairs of accessory glands.

Flask-like indentations are a feature of the intima of the accessory gland in Pentatomidae (Fig. 9, f_2); similar indentations are found in Nabidae (Henrici, 1940) and Cimicidae (Henrici, 1940; Carayon, 1966). The indentations contain a histologically determinate secretory product. In some glands the median reservoir shows two different types of unicellular secretory unit. In Pentatomidae (*Carpocoris*) almost the whole of the ventral epithelium consists of unicellular secretory units of a second type (Fig. 9, f_2). Unlike the unicellular units of the accessory gland, which are pigmentless, the type 2 unicellular units of the ventral epithelium contain pigment; and the intima formed by the type 2 units shows no flask-like indentations. In Corixidae, the cells of the accessory gland can be divided from differences in form and location into two types; (*a*) central cells and (*b*) encapsulating cells (Betten, 1943; Walker, 1972). A differentiated accessory gland is absent in Saldidae but the ventral epithelium of the median reservoir probably contains secretory cells (Brindley, 1930; Cobben, 1978).

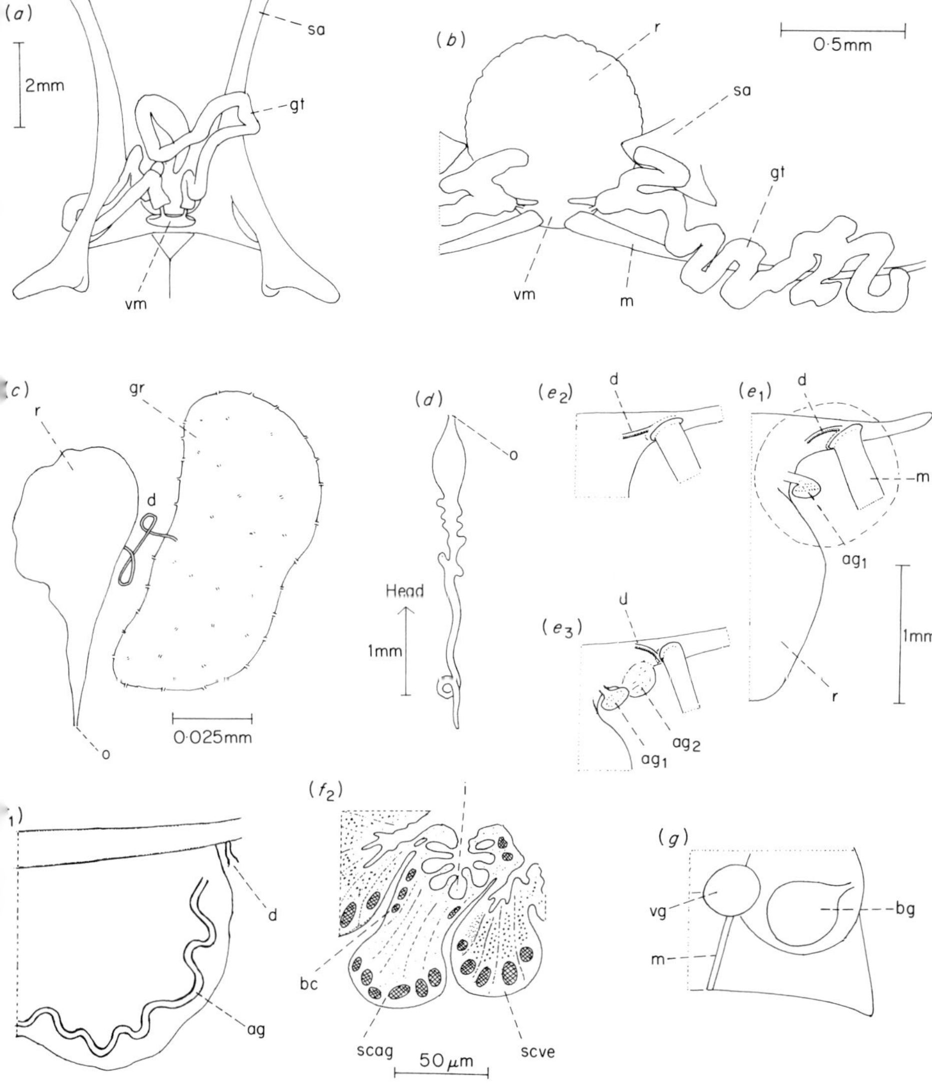

Fig. 9. (*a*) to (*f*) comparative morphology of the metathoracic gland. (*g*) the scent gland system of *Themonocoris*.

(*a*) *Lethocerus indicus* (Belostomatidae-Lethocerinae) Female gland *in situ* observed after treatment with hot caustic potash. gt, gland tubule; sa, sternal apophysis; vm, valve-opener membrane. (Redrawn from Pattenden and Staddon, 1970.)

(*b*) *Ilyocoris cimicoides* (Naucoridae) Metathoracic gland *in situ* left-hand side incompletely shown. gt, gland secretory tubule of right-hand side; m. opener muscle; r, median reservoir; sa, sternal apophysis: vm, valve opener membrane. (Redrawn from Staddon and Thorne, 1973.)

(*Contd. p. 374*)

3.2.3 *Carayon's classification*

Carayon (1971) divides the many variations of metathoracic gland into two groups; (*a*) omphalian and (*b*) diastomian. The form of the gland orifice is defining (the true valved orifice, not the ostiole); either median and unpaired (omphalian) or consisting of a pair of relatively widely spaced openings (diastomian). Paired but closely associated openings such as occur in Belostomatidae-Lethocerinae (Fig. 9*a*) and Omaniidae (Fig. 9*c*) are in the omphalian class.

The two types of gland are distributed as follows; 1, *omphalian*, the aquatic bugs, surface aquatic bugs, two littoral families (Saldidae, Omaniidae), and two minor terrestrial families (Enicocephalidae, Dipsocoridae); 2, *diastomian*, with the exception of Enicocephalidae and Dipsocoridae, the terrestrial families. In diastomian glands the openings usually occur in the wall of the scent canal at a point near the base of the metasternal apophyses (see Fig. 7).

There is no doubt of the descriptive utility of Carayon's terms, omphalian and diastomian, but it may be questioned whether the distinction between the two types is so clear-cut as Carayon implies. From the work of Staddon and Thorne (1974), it is evident that the morphologically omphalian gland of *Notonecta glauca* is functionally diastomian; from the work of Henrici

---

(*Caption contd. from p. 373*)

(*c*) *Omania coleoptrata* (Omaniidae). "Half" gland consisting of a "glandular reservoir" (gr); reservoir (r) and inter-connecting duct (d). o, orifice. (Redrawn from Cobben, 1970.)

(*d*) *Dydimocephalus curculio* (Enicocephalidae). Male gland. o, gland orifice. (Redrawn from Carayon, 1948a.)

(e_1)–(e_3) Details of the metathoracic scent gland of some coreoid bugs.

(e_1) *Pachycolpura mance* (Coreidae). Right-hand side of gland shown after removal of the gland secretory tubules (from the area within the dashed line). ag_1, accessory gland; d, common duct from gland secretory tubules; m, opener muscle; r, median reservoir.

(e_2) *Aulacosternum nigrorubrum* (Coreidae). Detail for comparison with (e_1). Note the absence of an accessory gland.

(e_3) *Hyocephalus*, spp. (Hyocephalidae). Detail for comparison with (e_1) and (e_2). Note the presence of an additional accessory gland, ag_2. (Redrawn from Gilby and Waterhouse, 1964.)

(f_1) and (f_2) Details of the metathoracic scent gland reservoir of pentatomid bugs.

(f_1) *Eurydema ventralis* (Pentatomidae). Ventral view right-hand side of median reservoir showing the position of the strip-like accessory gland (ag). d, common duct from gland secretory tubules. (Redrawn from Bonnemaison, 1952.)

(f_2) *Carpocoris pudicus* (Pentatomidae) Section through ventral side of median reservoir showing the unpigmented type 2 secretory cells of the accessory gland (scag), the small cells (bc) which border the accessory gland, and the pigmented type 2 secretory cells (scve) which compose most of the ventral epithelium. Note the flask-like indentations (i) of the intima of the accessory gland secretory cells. (Redrawn from Henrici, 1940.)

(*g*) View of the right inner side of metathorax of *Themonocoris kinkulanus* (Reduviidae-Phymatinae) showing the position of the ventral gland (vg), the opener muscle (m) of the ventral gland, and Brindley's gland (bg). (Redrawn from Carayon *et al.*, 1958.)

(1940) it is clear that the diastomian gland in many Pentatomidae is ontologically omphalian (section 2.4.2).

3.2.4 *Evolution of obsolescence*

A widely seen trend in the terrestrial families is the complete division, reduction in size, and ultimately total loss of the metathoracic scent gland. The metathoracic scent gland of *Oncopeltus fasciatus* is of the reduced, but incompletely divided, diastomian type. A completely divided metathoracic gland is frequently seen in Pyrrhocoridae, Lygaeidae-Lygaeinae, Aradidae, Miridae, Cimicidae, and Rhopalidae; it is also seen very occasionally in Pentatomidae (Carayon, 1971). Complete division is the rule in Reduviidae, Joppeicidae, Tingidae, Vianidae, and Piesmatidae (Carayon, 1971).

3.2.5 *Sexually dimorphic glands*

A conspicuous sexual dimorphism in the metathoracic scent gland has been recorded from Enicocephalidae (the metathoracic gland is confined to the male adults), Belostomatidae-Lethocerinae (the tubules of the metathoracic gland are much larger in the male adults), many Lygaeidae (well-developed lateral reservoirs are present only in the male adults), and a few Nabidae (*Alloeorhynchus* shows a sexual dimorphism similar to that found in Lygaeidae). Carayon (1971) gives a comprehensive account of the phenomenon; Staddon (1971) has shown that the sexual dimorphism in Lethocerinae is a characteristic of this subfamily. Some observations of physiological interest follow.

a *Lethocerinae* Pattenden and Staddon (1970 show that the metathoracic tubules (Fig. 9*a* shows the tubules from a female adult; the male tubules are much larger) of both sexes in *Lethocerus cordofanus* produce chemically similar, virtually identical secretions (*trans*-hex-2-enyl acetate is the main secretory product).

b *Lygaeidae* The dimorphism is restricted to the lateral part of the metathoracic gland. It is not obvious in newly ecdysed adults because the potentially voluminous lateral reservoirs (Fig. 7) are empty at the start of adult life. Only some few days after ecdysis and coinciding with the attainment of sexual maturity does the full extent of this dimorphism become clearly apparent. Although the gland in the female adults has no obvious lateral reservoir, a small reservoir is sometimes to be discerned in the basal part of the lateral complex. The dimorphism has a physiological component. The metathoracic secretory tubules of the female adults but not those of the male undergo regressive changes during the first few days after adult emergence. A shrinkage of the secretory cells leads in the females to a decrease in the diameter of the metathoracic secretory tubules (Carayon, 1948a; Johansson, 1957).

Carayon (1971) records a sexual dimorphism of this type in four subfamilies from the Lygaedae; Lygaeinae, Blissinae, Heterogastrinae, and Rhyparochrominae. In certain subfamilies the dimorphism occurs in some but not in all genera; for an example from Heterogastrinae, the metathoracic dimorphism is present in *Platypax* but not in *Heterogaster*. It is perhaps the case that a sexual dimorphism in the metathoracic gland is universal in Lygaeinae.

3.3 THE SCENT GLANDS OF REDUVIIDAE

The scent gland system of Reduviidae is interesting because it includes other glands (Brindley's glands, ventral glands) in addition to metathoracic and abdominal scent glands. Brindley's glands and the ventral glands are exclusively adult glands and frequently co-occur with the metathoracic scent gland.

3.3.1 *The metathoracic gland*

The structure of this gland in *Rhodnius neglectus* is shown in Fig. 10*a*. Each half-gland (the metathoracic gland in Reduviidae is of the reduced completely divided diastomian type) consists of a small pear-shaped reservoir

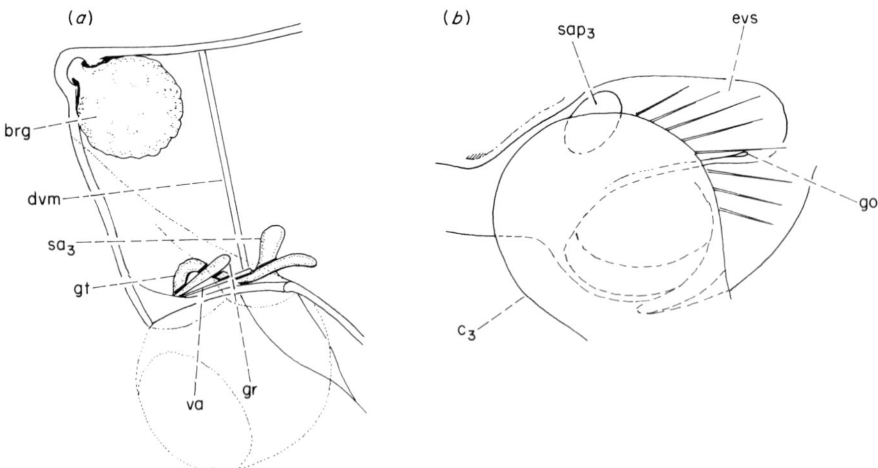

Fig. 10. Diagrams of the structure of the scent gland system of *Rhodnius neglectus*. (*a*) Metathoracic cavity, left-hand side, viewed from behind, showing Brindley's gland and internal components of the metathoracic scent apparatus; (*b*) Metathorax, left-hand side, viewed from below, showing external components of the metathoracic scent apparatus. brg, Brindley's gland. Components of the metathoracic scent apparatus; dvm, dorso-ventral opener muscle; evs, scent evaporation (apophyseal) surface; go, gland external orifice; gr, scent reservoir; gt, secretory tubule; va, valve opener apodeme. Other structures; c_3, metacoxa; sa_3, metasternal apophysis; sap_3, metasternal apophyseal pit. Note the hair fan on c_3.

and an unbranched secretory tubule; it is entirely pigmentless and there is no accessory gland. The orifice is minute and unusually situated, appreciably lateral to the sternal apophyseal pit. The dorso-ventral opener muscle exerts its effect through a valve-opener apodeme which in length greatly exceeds that found in the metathoracic apparatus in other groups of Heteroptera. The biological role of this greatly reduced gland in Reduviidae is not clear; that it is probably slight is emphasized by frequent absences (e.g. in *Themonocoris*; Fig. 9*g*).

3.3.2 *Brindley's glands*

These adult structures are the most conspicuous of the scent glands in a majority of Reduviidae; they are simple sac-like structures (Figs 9*b* and 10*a*). The glands occur in the haemocoele below the first visible abdominal tergite, towards the lateral margins, but in origin they are metathoracic, and open onto the metathoracic episternum. Descriptions of these glands have been given by Brindley (1930), Carayon *et al.* (1958), Kälin and Barrett (1975), Schofield and Upton (1978). Glands similar in structure and position to Brindley's glands have also been recorded from Pachynomidae (Carayon, 1962), Tingidae (Carayon, 1962), and Thaumastellidae (Cobben, 1968; p. 369). In Reduviidae, Brindley's glands have been looked for but not found in Tribocephalinae, certain Emesinae, and certain Saicinae (Carayon *et al.*, 1958).

3.3.3 *The ventral glands*

The ventral glands, which I propose should be called Carayon's glands in honour of J. Carayon, are sac-like structures. Although paired and occupying a similar position to the metathoracic gland they are readily distinguished from the metathoracic gland in three essential respects; 1, position of the external openings in the ventral articular membrane between the thorax and abdomen; 2, possession of ventral opener muscles (the origin of these openers is near the posterior boundary of the first visible abdominal sternite, the insertion on the posterior lip of the gland orifice); 3, absence of secretory tubules separate from a reservoir (in the ventral glands, the secretory units are distributed in the wall of the gland sacs).

The paired ventral gland has been recorded from Phymatinae, Elasmodemiinae, and Holoptilinae (Carayon *et al.*, 1958). The position of the ventral gland in *Themonocoris* is shown in Fig. 9*g*; Brindley's but not the metathoracic half-glands are also present in the adults of *Themonocoris*.

3.3.4 *The abdominal glands*

There may occur up to three unpaired abdominal scent glands in larvae in the Reduviidae; the unique dorsal scent gland of *Themonocoris* continues to function in the adults (Carayon *et al.*, 1958).

3.4 THE METATHORACIC EFFERENT SYSTEM

The metathoracic efferent system can be considered in three parts; 1, the valve mechanism surrounding the gland external orifice; 2, the scent canals formed by folding from integument; 3, (in the land bugs) the pleural or so-called scent "evaporation" surfaces. A cuticular microsculpture of a highly distinctive type is common to the surfaces of all three components and for that reason will be considered first. Studies of the metathoracic efferent system are specifically important in connexion with the manner of scent emission in Heteroptera adults.

3.4.1 *Cuticular microsculpture of the metathoracic efferent system*

A scanning electron micrograph of a small part of the pleural scent area from *Eurydema ventrale* (Pentatomidae) is shown in Fig. 11. The mushroom-shaped pillars are about 5 nm high, 9 nm across, and 10 to 15 nm apart. The flat tops, which lie in the same level as cuticular surfaces adjoining the pleural scent areas, dip down slightly towards the centre; the surfaces lining the depressions between pillars are folded in an intricate manner. Filshie and Waterhouse (1969) describe the fine structure and morphogenesis of this pattern in *Nezara viridula* (Pentatomidae). There are two different types of epidermal cell in the pleural scent areas. Relatively unmodified type 1 epidermal cells pro-produce the mushroom-shaped pillars; specialized type 2 epidermal cells produce the depressions. The main features of the pattern are determined by folding in the surface of the pleural epidermis prior to the commencement of epicuticle deposition. The definitive details are determined only later during the initial phases of epicuticle deposition; at this time the flat tops of the pillars are caused to become expanded laterally and the surfaces lining the depressions to become folded. The process of pattern formation is evidently assisted by a deposition of numerous microtubules in the apical cytoplasm specifically in the type 1 pillar forming cells and by the secretion of a gel-like "moulting" fluid specifically by the depression-forming cells.

Differences in cuticular microsculpture have been recorded from different parts of the same system and in the same part of the system in different species. A polygonal pattern is widespread; specifically in this pattern the

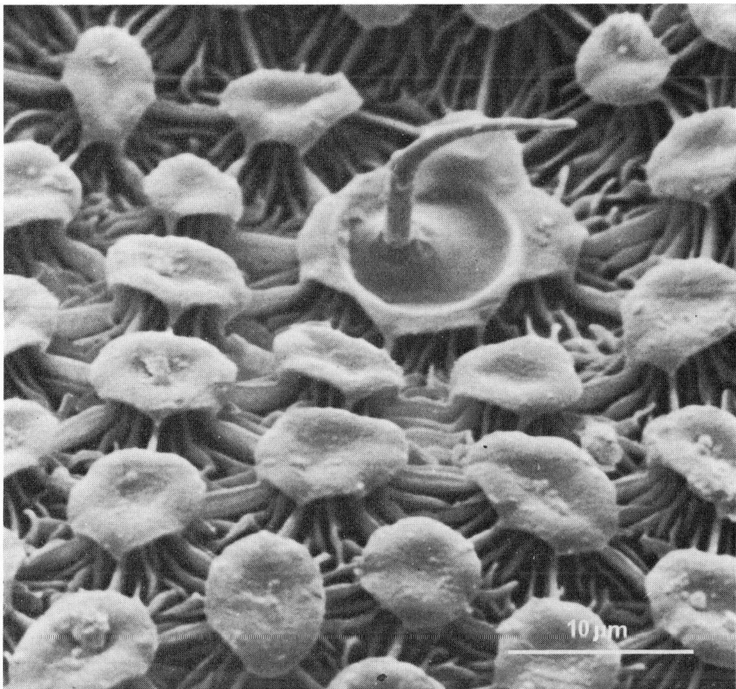

Fig. 11. Scanning electron micrograph of part of pleural scent area from *Eurydema ventrale* (Pentatomidae).

mushroom-shaped pillars are interconnected by ridges such as divide the depressed surface up into polygonal depressions. Carayon (1971) supplies numerous details from different species with the aid of scanning electron micrographs of the efferent system and several other authors supply similar details (Filshie and Waterhouse, 1969; Johansson and Bråten, 1970; Staddon and Thorne, 1974; Dethier, 1974).

3.4.2 *The metathoracic scent valves*

A diagram indicating the structure and mode of action of the valve mechanism of the metathoracic external orifice in *Oncopeltus fasciatus* is shown in Fig. 12. The dorsal origin of the dorso-ventral opener is on the metathoracic phragma, the ventral insertion below, one branch of the opener on the valve opener apodeme, the other branch on the upper wall of the exit-duct from the metathoracic median reservoir. The pull of the valve opener branch is transmitted *via* the valve plate to the valve lip (the upper moveable lip of the gland orifice) and the orifice is caused to gape, like a mouth. The

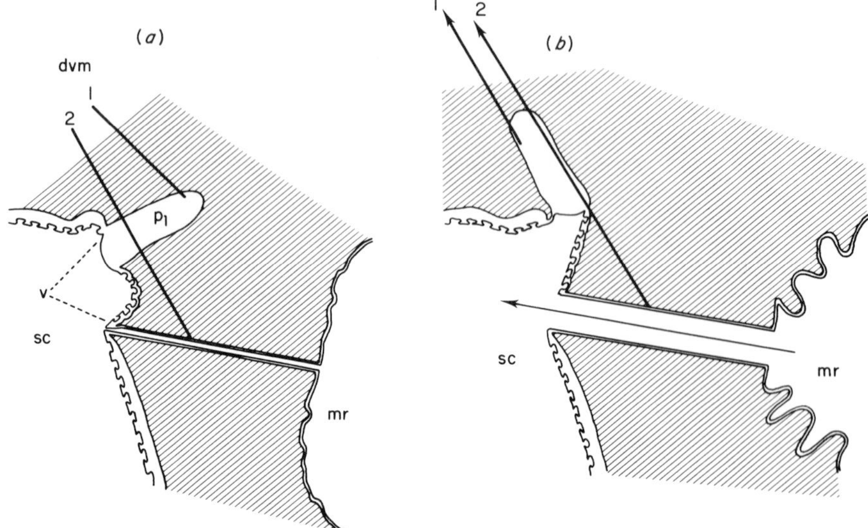

Fig. 12. Diagram showing the mode of action of the opener muscle on the metathoracic valve from *Oncopeltus fasciatus* (Lygaeidae). (*a*) Gland external orifice closed. (*b*) Gland external orifice open. dvm_1 and dvm_2, the two branches of the opener muscle; mr, median reservoir; p_1, valve opener apodeme; sc, scent canal; v, valve plate.

pull of the second branch (the duct opener) lifts the dorsal wall of the exit-duct away from the ventral wall. There are no closer muscles; closure is automatic by virtue of elastic forces in the valve plate. The elasticity of the valve hinge in the upper corner of the orifice especially may be important as a component of the occlusion mechanism (see Fi in Fig. 2 from Henrici, 1940).

Figure 13 shows a scanning electron micrograph of the metathoracic valve from *Oncopeltus fasciatus*. The pattern of cuticular microsculpture is of the polygonal type; it is similar to that occurring elsewhere in the metathoracic efferent system. There is comparatively little cuticle deposition specifically in the valve plate; hence the valve plate does not lose the flexibility essential to its function.

The valve mechanism in coreid and pentatomid bugs is similar in basic plan of structure to that of *Oncopeltus* and a similar division of the insertion of the dorso-ventral opener muscle is shown (Moody, 1930; Henrici, 1940; Akbar, 1957; Remold, 1962; Gupta, 1964; Hepburn and Yonke, 1972). However, the gland orifice extends further along the margin of the valve plate and so becomes U-shaped; moreover, it seems that the valve plate is relatively stiff so that when acted on by the valve lifter it is raised much as the gate in a sluice-valve is raised (the valve plate in the "sluice valve" in Remold's Fig. 8 is, in my view, open at an improbable angle).

As is the case for other components of the metathoracic efferent system,

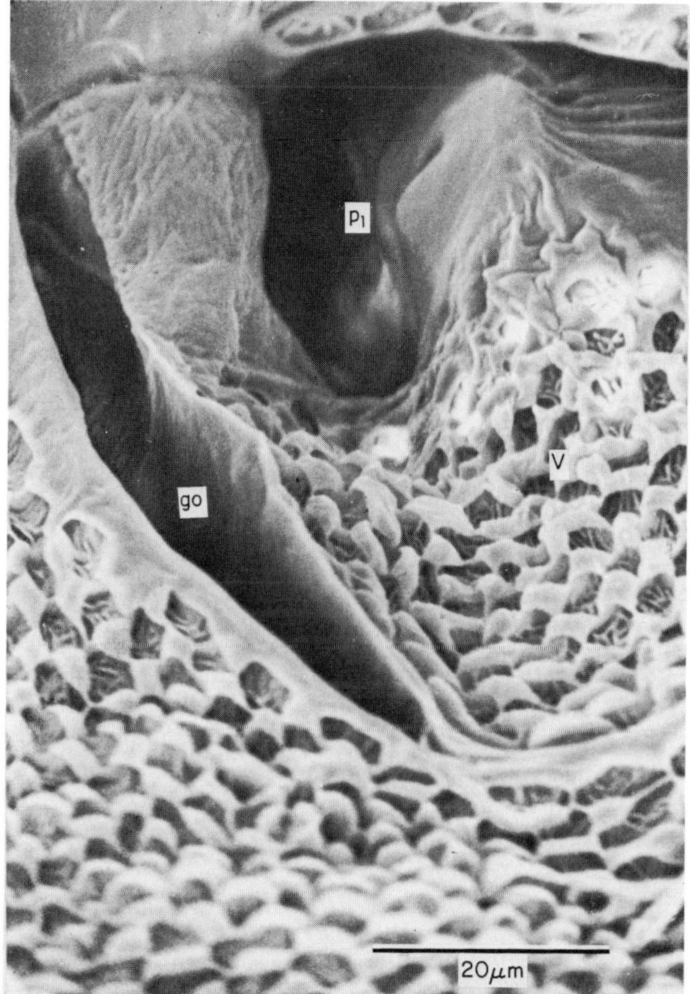

Fig. 13. Scanning electron micrograph of the metathoracic gland external orifice and valve plate from *Oncopeltus fasciatus* (Lygaeidae). go, metathoracic orifice; p_1, insertion of valve opener apodeme; v, valve plate.

the valve mechanism differs in different species of Heteroptera. It takes several different forms in the aquatic bugs; a median "flap-valve" in Corixidae (Betten, 1943); a median "lip-valve" in Naucoridae (Staddon and Thorne, 1973); paired "stop-valves" in Notonectidae (Staddon and Thorne, 1974). A surprising feature of the metathoracic valve in Reduviidae is the great length of the valve-opener apodeme (Fig. 10*a*).

In some species there is a second controller. This second controller is associated with the duct from the lateral reservoir and is acted on by a ventral muscle (Johansson, 1957; Remold, 1962). It has been recorded from *Oncopeltus fasciatus* (Fig. 7) and other lygaeids and from berytids.

3.4.3 *Scent canals and pleural scent surfaces*

For a specific example, it is convenient to describe the metathoracic efferent system of *Oncopeltus*. A cross-section through the system is shown in Fig. 7; scanning electron micrographs of the system are shown in Fig. 14. The scent canal commences near the base of the metasternal apophysis, proceeds forwardly, and turns laterally to open anterior to the third coxa onto the episternum (Fig. 1). The centre strip ("apophyseal" surface) originates in the metasternal apophyseal pit and lateral to the ostiole forms the raised drop-shaped surface (scent accumulation surface). It is quite smooth within the scent canal but roughened by spines where it is transformed into the scent accumulation surface. The two side-walls of the scent canal show the usual pattern of microsculpture (the polygonal type) which continues out through the ostiole to form a border (pleural scent surface) surrounding the scent accumulation surface.

The efferent system varies widely between species (Henrici, 1940; Remold, 1962; Carayon, 1971). In Reduviidae-Triatominae, it is greatly reduced. There is no closed canal and only the apophyseal component remains distinct to supply a differentiated border along the anterior margin of the third coxa (Fig. 10*b*). It seems to be more frequently the case that the apophyseal component is concealed throughout by enclosure within the scent canal. The pleural scent surfaces are frequently extensive in the pentatomoid families; in Plataspidae the pleural scent areas in fact extend beyond the pleura to cover the sterna as well as the pleura in all three thoracic segments (Remold, 1962; Carayon, 1971; Dethier, 1974).

It is a feature of diastomian systems that the scent canals are in no way connected across the mid-ventral line; the canals in diastomian systems are morphologically and hence functionally separate. In many omphalian systems (the aquatic bugs) the scent canals are united between the metasternal apophyseal pits by a transverse canal (vestibule) (Betten, 1943; Staddon and Thorne, 1973, 1974, 1979). In the acquatic bugs, the scent canals typically open on the boundary between the meso-epimeron and meta-episternum, towards the lateral extremity (Brindley, 1929; Betten, 1943; Parsons, 1960; Staddon and Thorne, 1973, 1974).

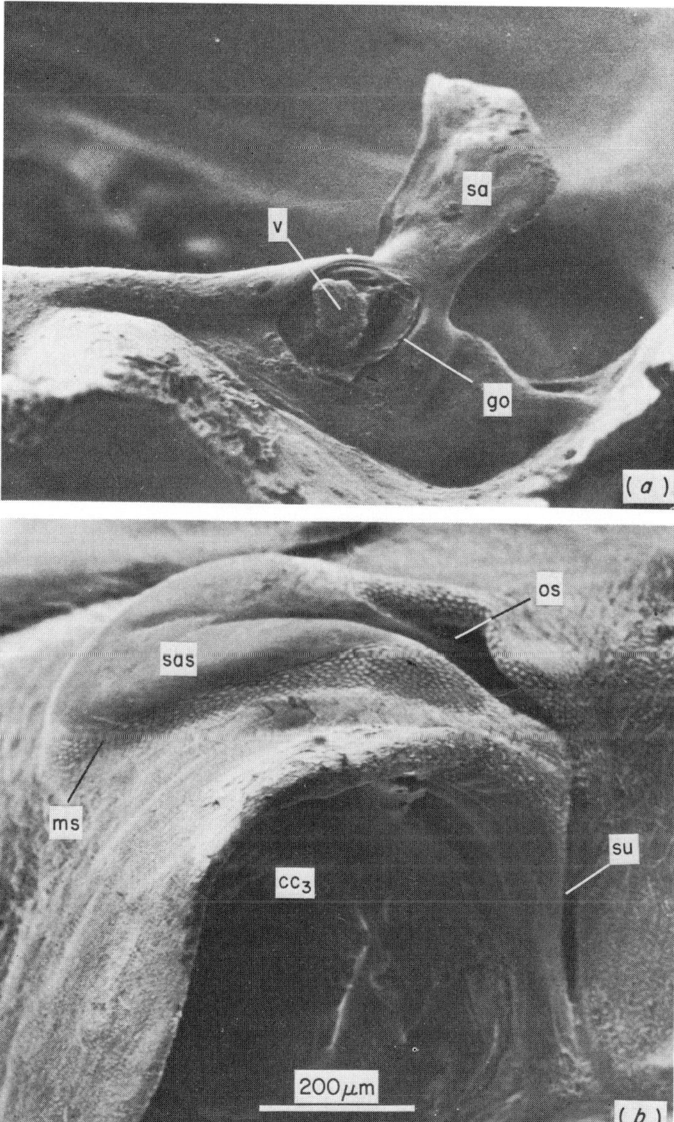

Fig. 14. Scanning electron micrographs of the metathoracic scent efferent system in *Oncopeltus fasciatus* (Lygaeidae). (*a*) View from inside of valve plate and scent canal after removal of the gland (→, anterior). (*b*) View from outside of ostiole and scent accumulation surface (↑, anterior). sas, scent accumulation surface; cc_3, third coxal articular cavity; go, gland external orifice; ms, cuticular microsculpture; os, metathoracic ostiole; sa, sternal apophysis; su, sulcus; v, valve plate.

3.4.4 *Functional considerations*

It is possible to suppose that the external scent surface (pleural, apophyseal) have different functions in different species. There are broadly three main ways in which they might be useful; 1, to promote evaporation of the scent substances (evaporation surface); 2, to cause secretion to be retained on the surface in a droplet (scent accumulation surface); 3, to facilitate shedding from the surface (scent shedding surface). A critical appraisal on the subject of function has been given by Carayon (1971).

Scent evaporation surfaces. In *Triatoma phyllosoma* (Reduviidae) the metathoracic secretion flows out over the apophyseal surface in a thin film as is required if evaporation is to be promoted. The hair fan on the hind coxa perhaps has a function in connexion with the evaporation of the scent (isobutyric acid); if the hairs are caused by rotatory movements of the hind coxa to brush over the apophyseal surface the rate of evaporation could be greatly increased (whether this actually happens has not been observed).

Scent accumulation surfaces. In *Oncopeltus fasciatus*, the ejected secretion accumulates on the apophyseal surface in a droplet; it is prevented from spreading to neighbouring pleural surfaces by the band of cuticular microsculpture which borders the apophyseal surface. The apophyseal surface itself is wetted by the secretion; the secretion extends over it and up to the microsculpture border.

Scent shedding surfaces. In *Ilyocoris cimicoides*, the ejected secretion accumulates temporarily in a spherical droplet on the hair-pile surrounding the metathoracic ostiole (Staddon and Thorne, 1973). Hence the hair-pile, although not specially differentiated from that of the body venter, functions as a scent shedding surface. In *Corixa dentipes* also, the scent oil is ejected to accumulate in a spherical droplet over the ostiole; the ostiole notch and setal tuft presumably have some as yet undefined role in connexion with scent emission.

In *Ilyocoris* and *Corixa* there is no cuticular microsculpture of the usual type in the metathoracic efferent system; the scent canal walls are lined with hydrofuge hairs similar to those which hold the "plastron" of the body venter.

Staddon and Thorne (1974) suppose that the cuticular microsculpture of the efferent system in *Notonecta glauca* increases the resistance to wetting by surface forces; it enables the scent fluid to be evacuated from the scent canals finally and quantitatively by capillarity. It is possible that pleural scent surfaces in land bugs differ in function according to the behaviour of the emitting secretion in different species. Filshie and Waterhouse (1969) suggest that the depressions in the cuticular microsculpture in pentatomids increase the protection time of the scent by trapping and delaying the evaporation of the scent substances; but such surfaces could equally well function as scent shedding surfaces or as scent evaporation surfaces depending on the behaviour of the ejected scent.

The results of descriptive studies of the metathoracic valve mechanism in *Oncopeltus* indicate that a selective release of scent from the metathoracic lateral reservoir is not possible; but the possibility that the controlling element in the orifice from the lateral reservoir can be used to prevent scent loss from the lateral reservoir is not excluded. Johannson (1957) showed that the effect of cutting the dorso-ventral opener muscle was to prevent scent emission from the affected side. Cutting the ventral muscle, on the other hand, had no discernible effect; and the median reservoir filled up with secretion as in the normal animal. This result agrees with the evidence of structure that the lateral reservoir is normally open for the passage of secretion from the gland secretory tubules to the median reservoir. It is possible that the essential effector function of the control element in the orifice of the lateral reservoir has to do with adjusting tension in the exit system from the metathoracic gland.

Remold (1962) has shown that the adults of many pentatomids, like their larvae, can eject scent in a jet in any direction within a wide angle horizontally from the side of the body. In the adult, however, it seems that only by a change in body stance could an emission be aimed accurately towards an attacker. The occlusion apparatus of the metathoracic gland, unlike that of the abdominal glands, is precluded, by virtue of its location at the base of a relatively long scent canal, from being manipulated to influence the direction of the emitted secretion; and it is unlikely that the metathoracic pleuron, which bears the ostiole, can be turned through a dorso-ventral axis independently of the rest of the body.

Remold (1962) described an interesting specialization in *Rhopalus*. A special groove conveys the emitted scent to a small trough high on the pleuron in which it accumulates in a droplet. Thence *Rhopalus* is able to transfer the scent droplet on the metatarsus to the body surface of an attacker.

4 Morphology and fine structure of the secretory units

4.1 THE MULTICELLULAR SECRETORY UNITS

Figure 15 shows a section through a part of the wall of a secretory tubule from the metathoracic scent gland of *Ilyocoris cimicoides* (Naucoridae). The tubule is composed almost exclusively of similar two-cell secretory units. The prominent lumen of the secretory cell is lined with microvilli. The wall of

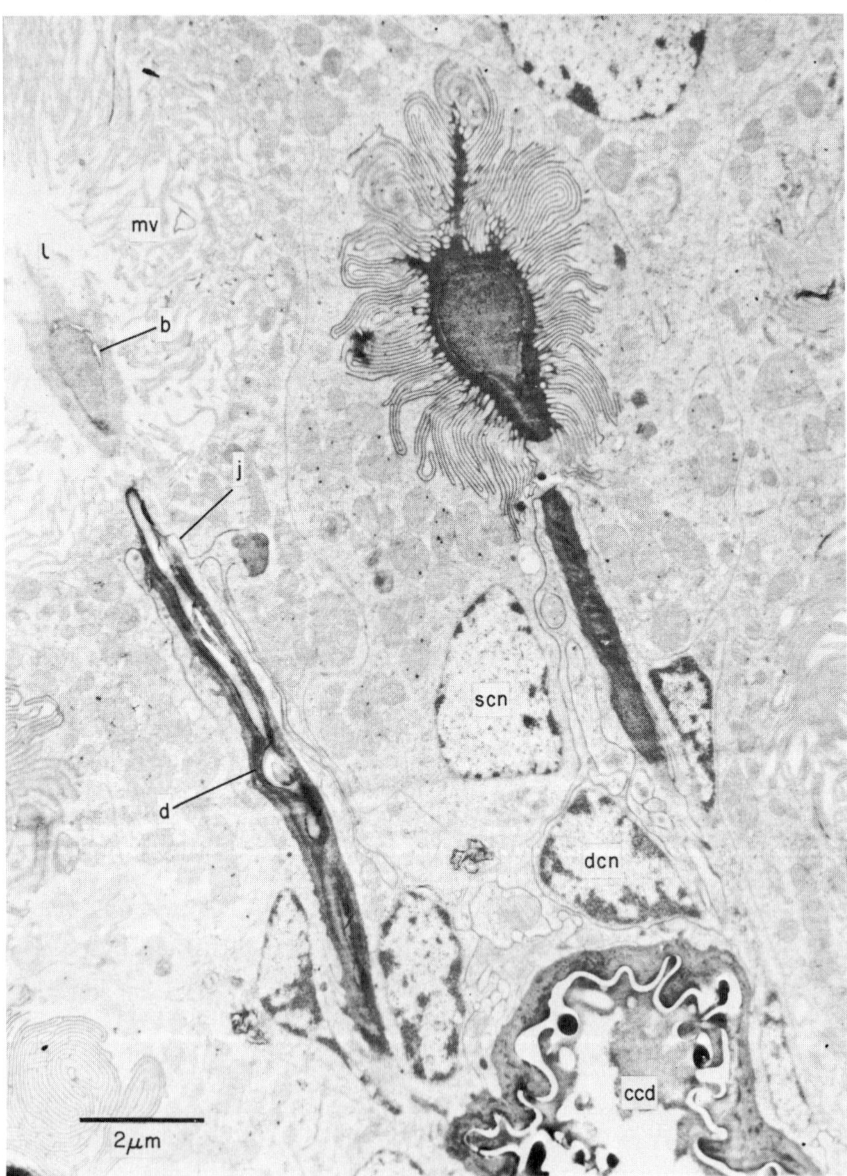

Fig. 15. Electron micrograph of cross section through the wall of a secretory tubule from the metathoracic scent gland of *Ilyocoris cimicoides* (Naucoridae). b, cuticular bulb; ccd, central collecting duct; d, cuticular ductule; dcn, duct cell nucleus; j, junction between duct cell and secretory cell; l, lumen of secretory cell; mv, microvilli; scn, secretory cell nucleus. (Micrograph by Dr. M. J. Thorne.)

the thin-walled cuticular bulb (end-sac) which occupies the lumen of the secretory cell is perforated by small pores (Staddon and Thorne, 1973). From the evidence of position it is concluded that the end-sac and the epicuticular funnel ("neck") which lines the opening from the lumen of the secretory cell is formed during development from materials secreted specifically by the presumptive secretory cell. The ductules and central collecting duct are mostly epicuticle formed by the duct cells.

From Fig. 15 it can be seen that even adjacent secretory cells are in different phases in the secretory cycle. A swollen lumen, wide spacings between microvilli, and generally lighter contents distinguish the cell occupying the top left part of the section.

The sources of the secretory products (the definitive secretory products include *p*-hydroxybenzaldehyde and methyl *p*-hydroxybenzoate) have not been elucidated. The difference in contrast across the membrane of the microvilli (lighter on the cell side in the cell in the top left of the section) is an indication that the final steps in the biosynthesis of the secretory products may take place in the membranes of the microvilli. In high resolution micrographs these membranes appear as a regularly and closely spaced row of granules. The metathoracic gland of *Notonecta glauca* produces a secretion which is chemically very similar to that from *Ilyocoris* but the mechanism of secretion of the scent phenols may be different. In *Notonecta*, the cytoplasm of the secretory cell contains numerous intracytoplasmic vesicles (Stein and Walker, 1970; unpublished observations). A second conspicuous difference from *Ilyocoris* consists in the presence of a basal zone of infolding which penetrates for about a third of the way from the basal to the apical surface of the secretory cell. In *Ilyocoris* the basal surface shows relatively little infolding. There are other differences; in *Notonecta* the microvilli show a typical "unit membrane" and the apical wall of the ductule is thickened to form a bulb-like structure. Numerous microvilli indicate that the duct cells of *Notonecta* are secretory (unpublished data).

Most fine structure studies have been of secretory cells from glands engaged in the manufacture of aliphatic scent substances. As will be shown in the following section (under 5.1), aliphatic esters in diverse pentatomid and lygaeid bugs are formed by the secretory tubules. A section of a part of the gland secretory tubule from the metathoracic gland of *Oncopeltus fasciatus* is shown in Fig. 16. The cytoplasm of the secretory cell is distinguished from that of the duct cell by the presence of numerous small membrane-bounded vesicles. An abundance of vesicles is the rule in the cytoplasm of secretory cells engaged in the manufacture of aliphatic scent substances. But there is no clear agreement on the origin and fate of these secretory vesicles. As to origin, the following suggestions have been made; 1. dictyosomes of the Golgi apparatus (Filshie and Waterhouse, 1968; Stein, 1969;

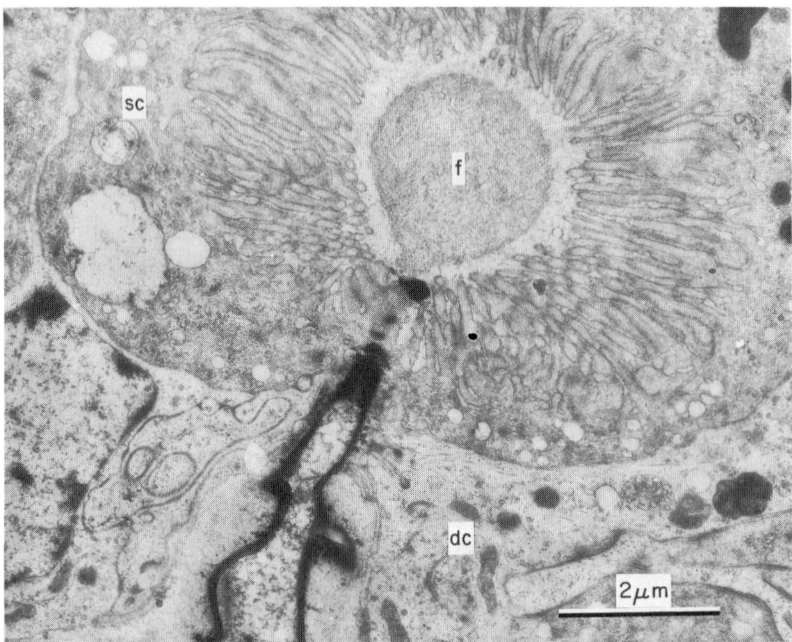

Fig. 16. Electron micrograph of a cross-section through the wall of a secretory tubule from the metathoracic gland of *Oncopeltus fasciatus* (Lygaeidae). The secretory cell (sc), lumen contains a mass of fibrillae (f); numerous membrane-bounded vesicles are present in the cytoplasm. Few vesicles occur in the cytoplasm of the duct cells (dc).

Stein and Schumacher, 1969; Stein and Walker, 1970; Schumacher, 1970, 1971b; Schumacher and Stein, 1971); 2. endoplasmic reticulum (Stein, 1966a; Walker, 1972); 3. Golgi dictyosomes *and* endoplasmic reticulum (Stein, 1969). Yet a third and additional origin for secretory vesicles in *Dysdercus* – pinocytosis in the zone of basal infolding – has been described by Schumacher (1970). The secretory vesicles are electron-light or at most contain only traces of stainable material, which is an indication that they lose their contents by extraction during processing for electron microscopy (Filshie and Waterhouse, 1968).

The secretory vesicles frequently show considerable size differences within the cytoplasm of a given cell. Sometimes the largest vesicles are to be seen near the base of the cell, sometimes they are concentrated in the apical area of the cell. In all cases, however, numerous small vesicles are to be seen crowded around the bases of the microvilli. Whether these small vesicles are simply engaged in the processes of membrane expansion and retrieval or whether they are in the nature of exocytotic vesicles is not at present clear.

It has been suggested that secretory products are exocytosed through canals opening out between the bases of adjacent microvilli (Schumacher, 1970; Schumacher and Stein, 1971). A second route through the microvilli themselves may be present in *Corixa* (Walker, 1972).

Deposits of glycogen and lipid droplets have been seen in the basal area of the secretory cell in the secretory tubules from the metathoracic gland of *Nezara* (Filshie and Waterhouse, 1968). In cells from tubules removed soon after adult emergence, the glycogen droplets and lipid droplets were seen to be surrounded by concentric fenestrated shells of endoplasmic reticulum. The lipid droplets may contain the immediate precursors of the scent substances. Perhaps the latter are formed in endoplasmic reticulum and subsequently sequestered by the Golgi vesicles (Filshie and Waterhouse, 1968).

Age related changes in the cytoplasmic organization of the secretory cells have been observed. In *Dysdercus*, the acquisition of an extensive system of basal infolding coincides with the onset of secretion synthesis (Schumacher, 1970; Schumacher and Stein, 1971). In *Nezara*, there occur changes in the distribution of the cytoplasmic organelles and in the distribution of the deposits of reserve materials (Filshie and Waterhouse 1968). Walker (1972) has recorded similar cytoplasmic changes from *Corixa*. In *Pyrrhocoris*, structures indicative of cell dissolution make their appearance soon after the cells have reached their peak in secretion synthesis (Stein, 1966a; Schumacher, 1971b).

The cuticular efferent system differs in different species; it also differs in the metathoracic and abdominal glands from the same species. In the metathoracic secretory tubules, an end-sac is frequently (*Notonecta*, *Lethocerus*, *Gelastocoris*, *Ilyocoris*) but not always (*Oncopeltus*, *Nezara*, *Dysdercus*, *Pyrrhocoris*) present (Filshie and Waterhouse, 1968; Schumacher, 1971c; Stein and Schumacher, 1969; Walker, 1972; Staddon and Thorne, 1973; unpublished observations). If an end-sac is absent, the secretory cell lumen is seen to contain a mass of fibrillae (Fig. 16). This material is evidently cuticular; it is laid down prior to secretion synthesis (Filshie and Waterhouse, 1969) and shed with the old cuticle at ecdysis (Henrici, 1938). Differences in the morphology of the ductules are frequent; a relatively straight and simple tube in *Nezara* (Filshie and Waterhouse 1968); with a prominent flask-shaped exit-chamber in *Corixa* (Betten, 1943; Walker, 1972); with a cone-shaped exit-chamber in *Ilyocoris* (Staddon and Thorne, 1973). In *Dysdercus*, septa are present in the lumen of the apical part of the ductule (Schumacher, 1971c; Stein and Schumacher, 1969). The presence of a globe-shaped dilatation (saccule) in the basal region of the ductule is a constant feature of the ductules in 3-cell secretory units (Henrici, 1938, 1940). The lumen of the saccule is frequently broken up by the presence of solid finger-like processes of cuticle (Henrici, 1938; Stein, 1966a, b, 1967, 1969); however, finger-like structures are absent in *Rinocoris* (Henrici, 1938). The evidence of occurrence (restriction to the 3-cell secretory units) may mean that the saccule has a purely supporting rather than a special physiological role in connexion with the biosynthesis of the scent substances.

Cross-sections through the duct cells (Walker, 1972; Staddon and Thorne, 1973) show that they ensheath the ductules; a well-defined junction ("mesduct"; Lai-Fook, 1970) occurs where the opposing surfaces of the duct cell meet. Even the duct cell nucleus may be pulled into lobes to surround the ductule (Staddon and Thorne, 1973).

It is possible to conceive that the definitive scent substances are not finally formed until the "primary" secretion has left the secretory cell lumen; but the evidence for the most part consists in the presence of osmiophilic deposits specifically in the lumen in the ductules and as such it is clearly far from convincing (section 2.2.3). It is perhaps indicative of a secretory function for the duct cells in the metathoracic secretory tubules of *Dysdercus* that the surface ensheathing the basal third of the ductule bears prominent microvilli (Stein and Schumacher, 1969). Prominent microvilli have been seen in the surface ensheathing the prominent apical swelling in the ductules in the metathoracic secretory tubules of *Notonecta* (unpublished observations).

4.2 THE UNICELLULAR SECRETORY UNITS

4.2.1 *The abdominal glands*

Stein (1969) gives a good description of the fine structure of the unicellular secretory units of the third abdominal gland from *Dysdercus intermedius*. The cytoplasm contains numerous secretory droplets. The smallest and presumably the most recently formed droplets occur in the basal area of the cell; they have strongly osmiophilic contents. The droplets increase in size from the basal to the apical part of the cell. They also undergo a striking change in appearance. Lighter areas appear within each droplet and finally the whole droplet becomes electron-light. In the apical part of the cell, the droplets fragment into smaller droplets. It is suggested that the secretion is exocytosed through the apical cell surface between the bases of the microvilli; although no obvious routes have been detected, it is clear that the secretion must penetrate the intima to reach the gland sac lumen.

The secretory droplets, unlike some other droplets associated with endoplasmic reticulum, perhaps form spontaneously from loci within the cytoplasm. Mitochondria may be involved in the changes which occur in the droplets in the middle region of the cell. Here they are closely associated with the droplets. The identity of the secretory product is not known but tridecane is the obvious possibility (section 5.1.3).

Diagrams comparing secretory processes in ducted and ductless secretory cells from *Dysdercus* are shown in Fig. 17.

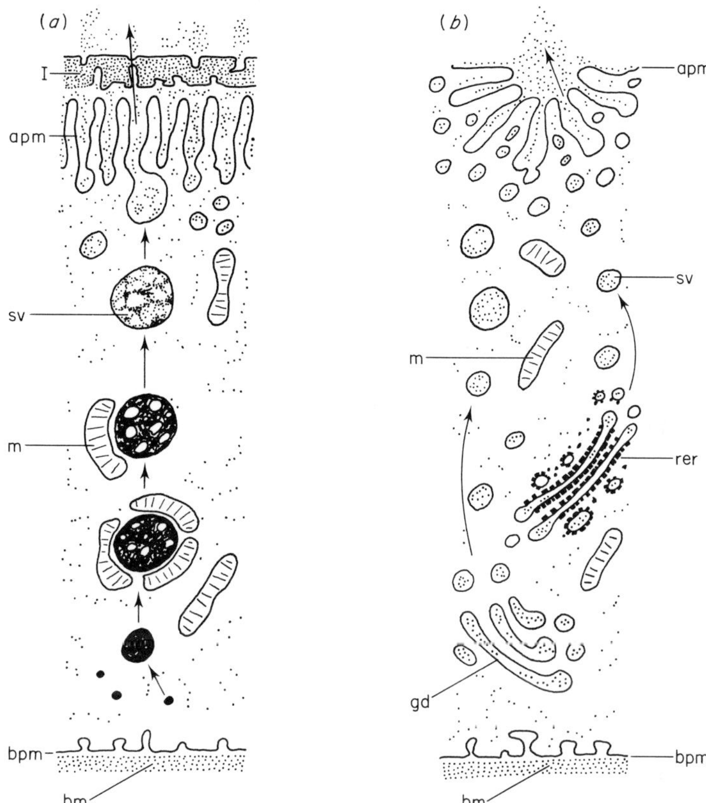

Fig. 17. Diagrams indicating possible secretory processes in ducted and ductless secretory cells from the third abdominal scent gland of *Dysdercus intermedius* (Pyrrhocoridae). (*a*) Unicellular secretory unit. The osmiophilic droplets which originate in the basal region of the cell change to become electron-light in the apical region. Mitochondria may be involved in the mechanism of transformation. The secretion is exocytosed into the space between the apical plasma membrane and the cuticular intima and thence is extruded through the intima into the gland sac lumen. (*b*) Secretory cell from three-cell secretory unit. A dual origin for the secretory vesicles in the Golgi dictyosomes and rough endoplasmic reticulum is indicated. apm, apical plasma membrane; bm, basement membrane; bpm, basal plasma membrane; gd, Golgi dictyosomes; I, cuticular intima; m, mitochondria; rer, rough endoplasmic reticulum; sv, secretory vesicle. Adapted from Stein (1969).

4.2.2 *The metathoracic accessory gland*

Atypical material may have been chosen for electron microscope studies of the unicellular secretory units of the metathoracic accessory gland. Schumacher (1971a) finds a multilayered structure in *Dysdercus* although previous light microscope studies of this organ from other species reveal it as consisting of only a single layer of secretory cells (Henrici, 1940; Johansson, 1957; Carayon, 1966). Microtubules are numerous; they are oriented perpendicularly to the surface of the gland. Mitochondria are large and characterized by a tubular or a laminate inner structure. Small endoplasmic reticular vesicles are indicative of a secretory function. The intima, however, is not equipped with obvious pores for the passage of secretion. The accessory gland of *Corixa*, as has been noted, is unusual in that it consists of two types of cells; (*a*) central cells and (*b*) encapsulating cells. Walker (1972) saw little evidence of secretion synthesis in the encapsulated central cells. The apical surfaces showed poorly developed microvilli; intracytoplasmic vesicles were few and small. A secretory function for the encapsulating cells is possible but the evidence of structure alone is far from convincing. It has been suggested that the accessory gland in *Corixa* has a function prior to ecdysis, in the pharate adult (Betten, 1943).

5 Physiology of the scent glands

Two questions will be considered; 1, the cytological sources of the scent substances; 2, the physiological function of the metathoracic accessory gland. The approach to these questions has been mostly comparative; 1, comparisons of scent constitution in different parts (lateral, median) of the metathoracic system in the same individual; 2, comparisons of scent constitution in homologous glands from different species.

5.1 CYTOLOGICAL SOURCES OF THE SCENT SUBSTANCES

There is evidence that the metathoracic scent esters are synthesized by the ducted secretory cells of the secretory tubules: the scent carbonyls are restricted to the median reservoir. The facts now known are detailed below; quantitative findings from *Nezara viridula* and *Oncopeltus fasciatus* are given in Tables 2 and 3.

1. *Leptoglossus phyllopus* (Coreidae). Hexyl acetate from the secretory tubules; hexyl acetate, hexanal, and related compounds from the median reservoir (Aldrich *et al.*, 1978b).

2. *Nezara viridula* (Pentatomidae). Dec-2-enyl acetate from the secretory tubules (Table 2); dec-2-enal from the reservoir. *Commius elegans* is like *Nezara*; in *Musgraveia sulciventris*, the tubules produce oct-2-enyl acetate and oct-2-enal accumulates in the reservoir (Gilby and Waterhouse, 1967).

3. *Oncopeltus fasciatus* (Lygaeidae). The C_6 and the C_8 alk-2-enyl acetates

TABLE 2 Constitution (%) of scents from the metathoracic scent gland of *Nezara viridula* (Pentatomidae) (from Gilby and Waterhouse, 1967)

Source of scent	Alkanes		Alk-2-enals			Alk-2-enyl acetates			4-Oxo-hex-2-enal
	C_{12}	C_{13}	C_6	C_8	C_{10}	C_6	C_8	C_{10}	
Secretory tubules	2.4	18.1	–	–	3.3	–	–	76.2	–
Median reservoir	2.3	54.1	1	0.5	25.8	2.9	0.5	3.9	7.1
Ratio $\frac{\text{Tubule scent}}{\text{Median reservoir scent}}$		0.33			0.13			19.6	

TABLE 3 Constitution (%) of scents from the metathoracic scent gland of mature male *Oncopeltus fasciatus* (Lygaeidae) (Games and Staddon, 1973a; Games *et al.*, 1973)

Source of scent	Alk-2-enals		Alka-2,4-dienals		Alk-2-enyl acetates		Alka-2,4-dienyl acetates	
	C_6	C_8	C_6	C_8	C_6	C_8	C_6	C_8
Lateral reservoir	–	–	–	–	9	24	10	55
Median reservoir	23	25	5	46	tr	tr	tr	tr

tr, trace amounts

and alka-2, 4-dienyl acetates from the lateral reservoir (Table 3); in specifically similar relative proportions, the C_6 and the C_8 alk-2-enals and alka-2, 4-dienals from the median reservoir (Games and Staddon, 1973b).

4. *Corixa dentipes* (Corixidae). An as yet unidentified volatile substance, but not a carbonyl, has been detected in the metathoracic secretory tubules; the scent carbonyls (in addition to 4-oxohex-2-enal there is a small quantity of ? hex-2-enal) are confined to the median reservoir (Staddon *et al.*, 1978). It is postulated that the tubule scent consists of hex-2-enyl acetate but mass spectral confirmation is awaited.

5. *Dysdercus intermedius* (Pyrrhocoridae). The C_6 and the C_8 alk-2-enyl

acetate from the secretory tubules; the C_6 and the C_8 alk-2-enals from the median reservoir (Everton *et al.*, 1979).

In addition to the scent carbonyls, the metathoracic median reservoir usually contains a small quantity of the tubule esters.

Thus there is evidence (circumstantial) from several different species of Heteroptera indicating that the scent aldehydes and other scent carbonyls are formed extracellularly in the metathoracic median reservoir from esters synthesized in the secretory tubules. This hypothesis was first formulated by Gilby and Waterhouse (1967) from their work on *Nezara*. Recently, age related changes in scent constitution have been recorded from the two main parts (lateral, median) of the metathoracic scent gland in *Dysdercus intermedius* (see Fig. 18). The scent esters appear first, in the lateral part of the gland; later, in the median reservoir, the fall in concentration coincides with a rise in concentration in the scent aldehydes. (Curiously, linalool remains restricted mostly to the lateral reservoir.) All species showing a relationship between the production of specific scent carbonyls in the median reservoir and the production of specific esters in the secretory tubules possess an accessory gland in the wall of the median reservoir.

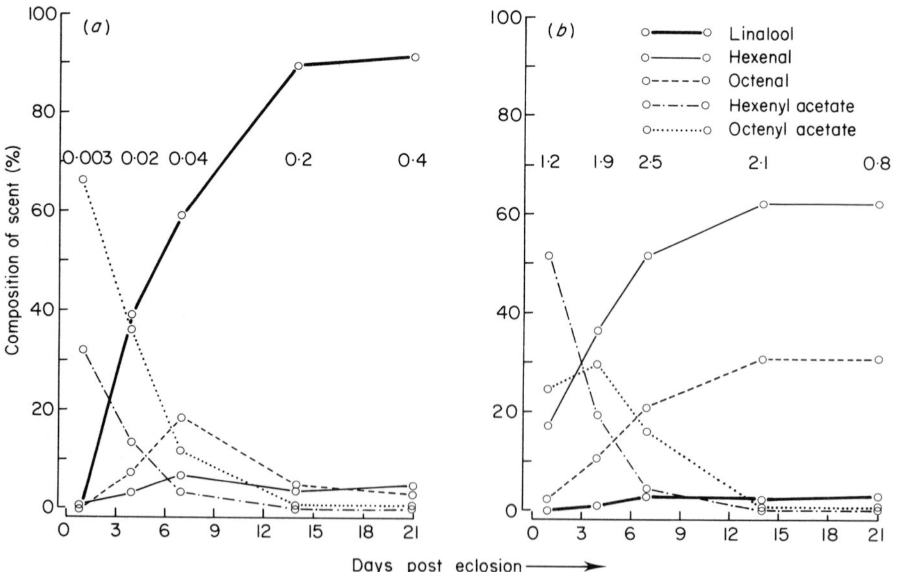

Fig. 18. Graphs showing age-dependent changes in constitution and weight of scent in the scents from the lateral and median parts of the metathoracic scent gland of *Dysdercus intermedius*. (*a*) Lateral reservoir and secretory tubules; (*b*) median reservoir and accessory gland. (From Everton *et al.*, 1979.) The small quantity of aldehyde recorded from the lateral part of the gland is almost certainly an accident of sampling.

There is no accessory gland in the metathoracic scent gland in the water bugs *Ilyocoris cimicoides* and *Notonecta glauca* and tests on these glands with 1, Schiff's reagent or 2, an acidic solution of 2,4-dinitrophenylhydrazine show that the scent carbonyls (*p*-hydroxybenzaldehyde, and perhaps other compounds, in *Ilyocoris*) are formed in the lumen of the secretory tubules or within the ductules from the secretory cells. A similar system is to be looked for in the abdominal scent glands; scent esters have not so far been found in the abdominal secretions.

It is suggested that unicellular secretory units in the metathoracic median reservoir in Pentatomidae are the source of tridecane. From Table 2 it may be seen that the concentration of tridecane differs in different parts of the metathoracic system in *Nezara viridula*; the concentration is high in the median reservoir (*ca* 54% tridecane), the concentration low in the secretory tubules. In many Pentatomidae, the metathoracic median reservoir contains secretory units of two structurally distinct types (Fig. 9f_2). Since the accessory gland is not engaged in the production of tridecane in other Heteroptera (*Dysdercus intermedius*, *Oncopeltus fasciatus*) it is suggested that the source of tridecane in Pentatomidae resides in the unicellular secretory units composing the ventral secretory epithelium of the metathoracic median reservoir. Unicellular secretory units are present in the third abdominal scent gland in *Dysdercus intermedius* (Stein, 1969) and these units rather than the ducted secretory cells are the probable source of tridecane.

A sex difference in scent constitution has been recorded from *Lygus lineolaris* (Gueldner and Parrott, 1978). Hex-2-enyl butanoate is more abundant than hexyl butanoate in the male scent. Perhaps this dimorphism is based in the presence and unequal distribution of two functionally distinct types of secretory cell. The greater abundance of tridecane in scent from the first abdominal gland in adult male *Nezara viridula* (Aldrich *et al.*, 1978a) perhaps arises from a greater relative abundance of unicellular secretory units in the male.

5.2 PHYSIOLOGICAL FUNCTION OF THE METATHORACIC ACCESSORY GLAND

The secretion from the metathoracic median reservoir, as has just been observed, frequently differs qualitatively from that formed in the secretory tubules. In general it is distinguished by the presence of the scent carbonyls; hence from this the hypothesis that the scent carbonyls are formed extracellularly in the metathoracic median reservoir from precursor materials (esters) synthesized in the secretory tubules (Gilby and Waterhouse, 1967). Recent observations indicate that reducing groups are present specifically along the margins of the epicuticle in the accessory gland (Everton and Staddon, 1979); this evidence is in line with the suggestion that the role of the accessory gland is to supply the mechanism for producing the scent aldehydes from the scent esters (Games and Staddon, 1973a). However, these observations do not rule out the possibility that the accessory gland itself manufactures the scent aldehydes from substances absorbed directly from the haemolymph.

Aldrich *et al.* (1978b) isolated enzyme proteins from the aqueous phase of the secretion from the metathoracic gland of the coreid *Leptoglossus phyllopus*. Two proteins isolated by gel electrophoresis showed strong esterase activity; a third showed strong dehydrogenase activity when hexanol was used as substrate. The results indicate that hexanal is formed from hexyl acetate in two steps; 1, hydrolysis of hexyl acetate to hexanol and acetic acid; 2, oxidation of hexanol to hexanal. The fate of the acid was not studied. It is interesting that the alcohol and acid components of the scent esters have rarely (the acids never) been detected free in the metathoracic scents from non-coreoid bugs. Macleod *et al.* (1975) find 1% dec-2-enol in metathoracic scent from the pentatomid *Biprorulus bibax*. Using naphthyl acetate as substrate, positive staining for esterase was obtained in the accessory gland of *Leptoglossus* (Aldrich *et al.*, 1978b). It is probable that the scent reservoir enzyme proteins are or closely associated with mucoproteins or glycoproteins (Carayon, 1971; Everton and Staddon, 1979).

An apparent anomaly concerning the generalization that the scent aldehydes and scent esters from a species share the same carbon chain (the alcohol chain in the case of the esters) is contained in work by Prestwich (1976). The secretions from a number of pentatomids were found to contain the C_6 and C_8 alk-2-enals and 4-oxoalk-2-enals; but the alk-2-enyl acetates present were the C_6 and the C_{10}.

The physiological and biological significance of the sometimes observed aqueous phase in scent gland secretions is not at present clear. Possible functions for water in the metathoracic secretion (probable source of water, the accessory gland; Everton and Staddon, 1979) include; 1, vehicle for the secretion of enzyme proteins (Aldrich *et al.*, 1978b); 2, volumetric compensation for a diminished production of scent repellent (Everton and Staddon, 1979). It should be noted that the thorough work of Gilby and Waterhouse (1965) on the metathoracic secretion from *Nezara viridula* contains no mention of an aqueous phase and that Everton *et al.* (1979) found no aqueous phase in the metathoracic secretion from *Dysdercus intermedius*. Interestingly, whereas the metathoracic secretion from *Corixa dentipes* is a homogeneous yellow oil the larval secretions probably consist of a mixture of aqueous and scent oil phases (Staddon *et al.*, 1979).

6 Biological functions of the scent glands

6.1 DEFENCE AGAINST PREDATORS

A protective role for the scent glands against small vertebrate and invertebrate predators has been established for coreid and pentatomid bugs (Conradi, 1904; Blum *et al.*, 1960, 1961; Roth, 1961; Waterhouse *et al.*, 1961; Remold, 1962). Corixid larvae similarly may be protected against the small predator *Plea* (Hagemann, 1910). There are three different types of defence behaviour in land bugs; 1, emission of scent onto the cuticle surrounding the scent ostiole; 2, transference of the emitted scent from the scent ostiole *via* the tarsus to the body surface of the attacker; 3, ejection of scent in a jet, unilaterally or bilaterally, and often accurately towards the attacker (Remold, 1962). The tarsal defence mechanism may have evolved by modification from normal self-grooming behaviour. The same emitter may be capable of variations in response; a long or a short range scent jet from *Coptosoma scutellatum* (Plataspidae); tarsal and spray mechanisms of scent transference in *Troilus luridus* (Pentatomidae) (Remold, 1962).

Not a great deal of effort has been expended on the subject of possible relationships between scent constitution and function. Known instances of organ specificity within species were listed in section 2.5. Table 4 gives a list of major scent substances and their species sources. Multicomponent scents are widespread (examples in Tables 2 and 3). In many pentatomids the scent carbonyls exceed their solubility in tridecane so that a two-phase scent forms (Blum and Traynham, 1960; Gilby and Waterhouse, 1965). Two phases, one aqueous, also form if water is abundantly present in the secretion (Aldrich and Yonke, 1979; Everton and Staddon, 1979). The "ester type" scents of coreoid bugs form a special class of their own (Waterhouse and Gilby, 1964; Tsuyuke *et al.*, 1965; Baker and Kemball, 1967; Baker *et al.*, 1972; Aldrich and Yonke, 1975; Prestwich, 1976).

Two special functions have been claimed for tridecane; 1, to promote the penetration of the toxic scent carbonyls through cuticle in arthropod predators (Remold, 1962); 2, acting as a fixative, to delay the evaporation of the scent carbonyls from the body surface of the scent emitter (Blum and Traynham, 1960). It is also to be considered as possible that tridecane has a defensive function in its own right.

The abdominal system of scent glands may supply more than one chemical line of defence. In *Oncopeltus fasciatus*, it is often the second gland that is caused to discharge its contents first; the ejected scent mixes with a toxic fluid ejected simultaneously from the rectum and a large fluid droplet forms around the tip of the abdomen (Games and Staddon, 1973a; Graham and

TABLE 4 Catalogue of "monocomponent" secretions

1. BUTANOIC ACID
Merocoris distinctus; Coreidae, metathoracic gland (almost pure butanoic acid) (Aldrich and Yonke, 1975).

2. ISOBUTYRIC ACID
Rhodnius prolixus, *Triatoma phyllosoma*, and *Panstrongylus megistus*; Reduviidae, Brindley's gland (isobutyric acid the only volatile detected) (Pattenden and Staddon, 1972; Games *et al.*, 1974).

3. HEXANAL
Hyocephalus spp.; Hyocephalidae, metathoracic gland (98% hexanal) (Waterhouse and Gilby, 1964). *Agriopocoris froggatti*; Coreidae, metathoracic gland (hexanal the only volatile in a sample from an old individual) (Waterhouse and Gilby, 1964). *Anoplocnemis montandoni*; Coreidae, metathoracic gland (92% hexanal) (Prestwich, 1975).

4. HEX-2-ENAL
Brochynema quadripustulata (Blum, 1961) and probably *Piezodorous teretipes* (Gilchrist *et al.*, 1966); Pentatomidae, metathoracic gland. *Acanthocephala femorata* (Blum *et al.*, 1961); Coreidae, metathoracic gland. *Holopterna allata*; Coreidae, dorsal abdominal glands (90% hexenal) (Prestwich, 1976).

5. OCT-2-ENAL
Oncopeltus fasciatus; Lygaeidae, anterior abdominal gland (87% octenal) (Games and Staddon, 1973b).

6. 4-OXOHEX-2-ENAL
Sigara falleni and *Corixa dentipes*; Corixidae, metathoracic gland (95% 4-oxohex-2-enal) (Pinder and Staddon, 1965a, b).

7. 4-OXO-OCT-2-ENAL
Oncopeltus fasciatus; Lygaeidae, posterior abdominal gland (81% 4-oxo-oct-2-enal) (Games and Staddon, 1973b). *Corixa dentipes*; Corixidae, the second and third functional abdominal glands (*ca* 95% 4-oxo-oct-2-enal) (Staddon *et al.*, 1979).

8. HEXYL ACETATE
Aulacosternum nigrorubrum; Coreidae, metathoracic gland (90% hexyl acetate) (Waterhouse and Gilby, 1964).

9. HEX-2-ENYL ACETATE
Lethocerus indicus and *L. cordofanus*; Belostomatidae, metathoracic gland (almost pure hexenyl acetate) (Butenandt and Tâm, 1957; Pattenden and Staddon, 1970).

10. TRIDECANE
Ceratocoris cephalicus; Plataspidae, ether washings (87% tridecane) (Baggini *et al.*, 1966). *Nezara viridula*; Pentatomidae, abdominal glands (the only volatile detected) (Ishiwatari, 1974).

Staddon, 1974). The first gland, which usually responds only in response to further stimulation, produces a secretion which spreads in a thin film over the cuticle surrounding the ostiole and so is caused to evaporate relatively rapidly. It is interesting that 4-oxo-oct-2-enal, the more polar of the two major scent substances, is predominant in the secretion from the second gland; oct-2-enal, is predominant in the secretion from the first gland (Games and Staddon, 1973a).

Intraspecies responses to the scent substances have been described and categorized as alarm behaviour; aggregated larvae of *Dysdercus intermedius* disperse (Calam and Youdeowei, 1969); larvae and adults of *Cimex lectularius* seek a place of concealment (Levinson and Bar Ilan, 1971); larvae of *Eurydema rugosa* (Pentatomidae) release their hold and drop off the food plant (Ishiwatari, 1974); other individuals in field populations of *Musgraveia sulciventris* release scent from their own scent glands (Macleod *et al.*, 1975). Figure 19 shows aggregated individuals of *Oncopeltus fasciatus* dispersing in response to the vapour stimulus from hex-2-enal.

Levinson *et al.* (1974) have made a special study of the alarm pheromone system of *Cimex lectularius*. Alarm pheromone receptors are present in the antennae in sensilla types E_1 and E_2. The physiological response threshold of the alarm pheromone receptors to hex-2-enal was determined as 2×10^{10} molecules of hexenal per ml of air. From observations on the physiological responses of the alarm pheromone receptors to different chemical substances (the C_4 and C_5 *n*-alkanals; the C_6 and C_8 *trans*-alk-2-enes; the C_6 and C_8 *trans*-alk-2-enals) it was concluded that a minimum carbon chain length of C_6 and a terminal carbonyl group are essential functional characteristics of the natural alarm substances (hex-2-enal, oct-2-enal) but that a *trans*-double bond is not. Behavioural thresholds to hex-2-enal and oct-2-enal were determined. For hex-2-enal a value of 6×10^{15} molecules per ml of air was obtained for this threshold; the threshold for oct-2-enal was 9×10^{14} molecules per ml of air.

It is possible that for some Heteroptera the scent substances exert a protective action by releasing alarm behaviour in ant predators. Releasers of alarm behaviour in the african weaver ant *Oecophylla longinoda* include hexanol, hexanal, and 2-*n*-butyloct-2-enal; these substances from the mandibular gland of *Oecophylla* have been recorded in Heteroptera from coreid bugs. Hex-2-enal, which is widespread in Heteroptera, has been isolated from the African black cocktail ant *Crematogaster africana* (Bevan *et al.*, 1961). Linalool from *Dysdercus intermedius* (Everton *et al.*, 1979) similarly may have an alarm releaser function against ants.

Schaeffer (1972) points out that the metathoracic scent gland is well developed in ant mimics and grass feeders (e.g. Blissinae). Perhaps the metathoracic secretion from ant mimicking bugs is attractive to the ants. From a

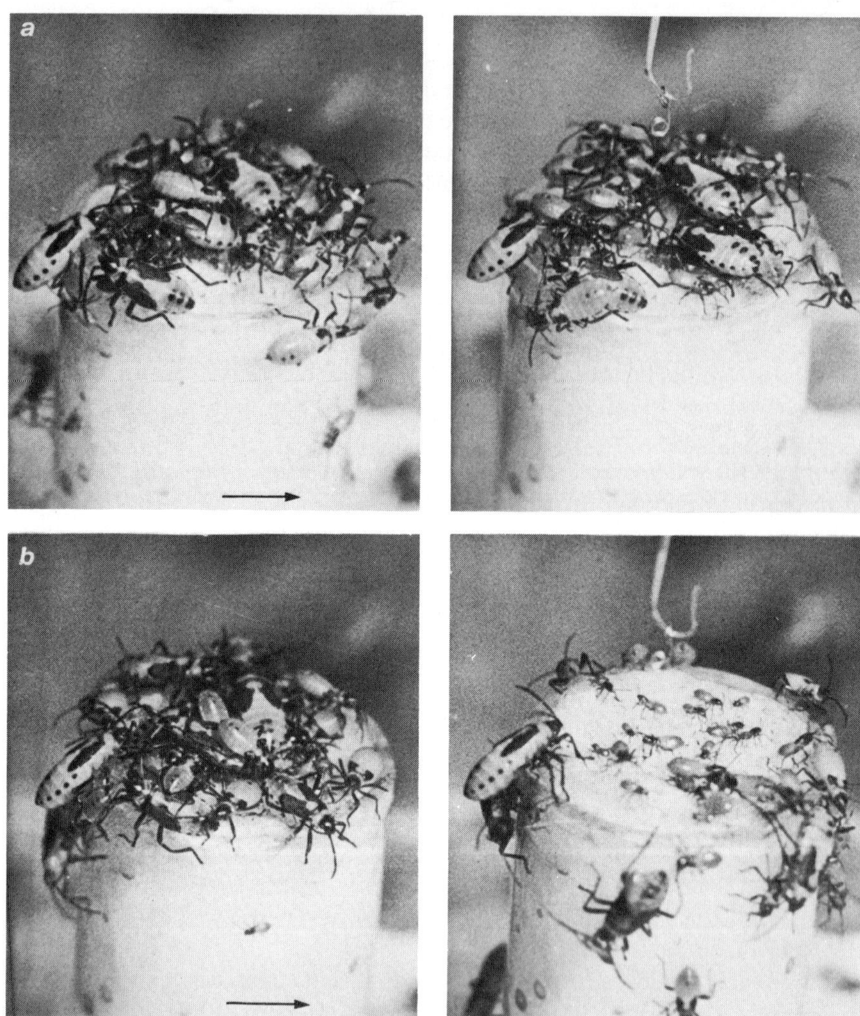

Fig. 19. Dispersion of *Oncopeltus fasciatus* aggregated for drinking in response to stimulation from hex-2-enal vapour. (*a*) No dispersion 3 min after suspending water droplet (control) over the aggregation (see (*a*) right). (*b*) Dispersion of aggregated bugs within 6 s after suspending droplet of hex-2-enal over the aggregation (photographs by Mr. I. J. Everton).

unique ventral abdominal gland the Javanese assassin bug *Ptilocerus venosus* (Reduviidae-Holoptilinae) produces a secretion which is both attractive and toxic to the ants (*Dolichoderus bituberculatus*) on which it feeds (Jacobson, 1911). The metathoracic secretions ejected from populations may function in the defence of grass-feeding bugs as a gustatory repellent against large grazing animals. Miller (1971) noted that the rice bug *Leptocorisa* frequently reveals its presence by the emission of an odour; whether this release of scent is an alarm reaction has not been established.

The biological value of scent glands as a protection is clearly far from complete. Heteroptera populations are effectively controlled by predators and parasites of many different kinds (Remold, 1962; Carayon, 1971; Miller, 1971). The complete absence of scent glands in many different species is an indication perhaps that the advantages frequently only marginally exceed the disadvantages. Possible disadvantages include drain on metabolic pool, source of olfactory host recognition signals for parasites and predators (see Mitchell and Mau, 1971). Mites of a special kind occupy the metathoracic scent canals in some coreid bugs (Fain, 1970).

It is a fact of great interest that the metathoracic scent apparatus is frequently greatly reduced in aposematic bugs which live high off the ground on their food plants (Schaeffer, 1972). The metathoracic scent apparatus in the brightly coloured milkweed bugs is greatly reduced compared to that in other members of their family (Lygaeidae). A probable component of any biological explanation of this great reduction of the metathoracic scent apparatus in aposematic bugs is that insectivorous birds in general are not repelled by the secretions (Heikertinger, 1922; Miller, 1971). Some aposematic species have acquired other mechanisms of protection against avian and other vertebrate predators. The milkweed bug *Oncopeltus fasciatus* sequesters cardiac glycosides from the milkweed plants (Asclepiadaceae) on which it feeds (Duffey and Scudder, 1972); it also contains other pharmacologically active compounds of its own (Graham and Staddon, 1974). Lygaeids from oleander (*Nerium oleander*) similarly sequester cardiac glycosides (Euw *et al*., 1971). In contrast, the conspicuously coloured *Pyrrhocoris apterus* appears to have no special chemical mechanism of defence against predators (Heikertinger, 1922; Schumacher, 1971c). Aposematism and mimicry systems in Heteroptera should be given much greater attention than hitherto (see Stride, 1956; Rettenmayer, 1970; Aldrich and Blum, 1978).

6.2 DEFENCE AGAINST MICROORGANISMS

The evidence is circumstantial and derives from studies on soil-dwelling bugs and water bugs.

6.2.1 *Soil dwellers*

The metathoracic scent from the cydnid *Scaptocoris divergens* has antimicrobial properties (inhibition of certain soil microorganisms including the spores and mycelium of *Fusarium*, a pathogen of bananas causing banana wilt) (Roth, 1961) but there is no evidence that antimicrobial properties are of value to *Scaptocoris* in soil.

6.2.2 *Aquatic bugs*

Defence against microorganisms has been suggested as the function of the metathoracic secretion for aquatic bugs from four different families; Corixidae (Pinder and Staddon, 1965a), Naucoridae (Staddon and Weatherston, 1967), Notonectidae (Pattenden and Staddon, 1968), and Pleidae (Maschwitz, 1971). It is interesting that the production of antiseptic phenols (*p*-hydroxybenzaldehyde and methyl *p*-hydroxybenzoate) is common to water bugs of the families Naucoridae and Notonectidae (the metathoracic secretion) and water beetles of the family Dytiscidae (the pygidial secretion) (see Table 5). The water beetle secretions differ only in the presence of additional materials; benzoic acid, sometimes together with *p*-hydroxybenzoic acid, and proteinaceous materials (Schildknecht, 1970). The secretion from *Plea* contains hydrogen peroxide and is rubbed over the body when grooming occurs,

TABLE 5 Aromatic substances from some aquatic bugs and beetles. The relative proportions (%) are from gas chromatographic analyses (5% SE-30 at 110 °C) of samples from individual adults. The identities of the substances are from Staddon and Weatherston (1967), Pattenden and Staddon (1968), and Schildknecht (1970)

Group and species (glandular source)	Benzoic acid	*p*-Hydroxybenzaldehyde	Methyl *p*-hydroxybenzoate
WATER BUGS (metathoracic gland)			
Notonectidae			
Notonecta glauca	ND	70	30
Naucoridae			
Ilyocoris cimicoides	ND	55	45
WATER BEETLES (pygidial gland)			
Dytiscidae			
Colymbetes fuscus	24*	38	38
Acilius sulcatus	S_1 35	65	ND
	S_2 21	71	7
Dytiscus marginalis	40	40	220

ND, not detected

* Probably a mixture. *p*-Hydroxybenzoic acid has been recorded from *Colymbetes fuscus* (Schildknecht, 1970) but the column used in this work did not separate this compound from benzoic acid

S_1 and S_2, samples from two different adult individuals

out of water, which is a strong indication that the metathoracic secretion is employed specifically as a defence against microorganisms (Maschwitz, 1971).

6.3 COMMAND FUNCTIONS

Suggestions that the scent substances in some species have command or releaser functions rest on circumstantial evidence except those made in connexion with alarm behaviour. Here the possibility that the scent substances are utilized as command signals in aggregation behaviour, sexual and non-sexual, will be considered.

6.3.1 *Sexual behaviour*

A sexual function for the scent substances has not been demonstrated although sexual dimorphisms in gland structure (sections 3.1.5 and 3.2.5) and function (Games and Staddon, 1973a; Aldrich *et al.*, 1978; Gueldner and Parrott, 1978) have been recorded. Shorey (1973) lists possible behavioural effects.

Mature male adults of *Lethocerus indicus* are strongly scented with their own scent during that period of the year when mating occurs (Caillot and Boisson, 1954); mature male adults of *Oncopeltus fasciatus* become strongly scented with the esters from the lateral reservoir of the metathoracic gland but only when reared separately from the females (Lener, 1969; Games and Staddon, 1973a). Perhaps the natural sexual function of the scent is to repel other males. Macleod *et al.* (1975) employed an olfactometer technique to test the possibility that oct-2-enyl acetate is a sex attractant of *Musgraveia sulciventris*. The results indicated rather that this compound might be repellent to *Musgraveia*. As noted in an earlier section (5.1.1) octenyl acetate in *Musgraveia* is produced in the metathoracic secretory tubules.

6.3.2 *Aggregation*

Levinson and Bar-Ilan (1971) demonstrate convincingly that a volatile pheromone is a cause of aggregation in *Cimex lectularius*. The anatomical source of the aggregation pheromone was not established. Neither hex-2-enal nor oct-2-enal (the major scent substances of *Cimex*) whether separately tested or mixed promoted aggregation formation under test conditions; but the possibility that the natural scent itself contained as a minor constituent the aggregation pheromone was not tested. Calam and Youdeowei (1969) claim that the abdominal secretions from *Dysdercus intermedius* are aggregation promoting. In the absence of full experimental data it is difficult to evaluate this suggestion. Ishiwatari (1976) produced experimental evidence that hex-2-enal from larvae of *Eurydema rugosa* (Pentatomidae) functions as an aggregation pheromone at low concentrations. In my view the experimental findings are inconclusive on this point; the results perhaps demonstrate the importance of hex-2-enal as a feeding stimulus for *Eurydema* larvae (hex-2-enal had been well named leaf-aldehyde). Eggerman and Bongers (1972) suggest that glycosides sequestered from the diet function as aggregation pheromones for *Oncopeltus fasciatus*; but milkweed bugs as readily aggregate when reared on sunflower seeds (glycosides absent). Aldrich and Blum (1978) present field evidence for an aggregation pheromone in the coreid *Thasus acutangulus*.

6.4 OTHER FUNCTIONS

The suggestion that the secretion from the metathoracic gland is employed as a waterproofing agent in surface aquatic bugs has been tested on *Aquarius* (*Gerris*) *najas* (Staddon, 1972). Sealing the external median opening from the gland had no detectable effect on the ability of the hair-pile of the body venter to resist wetting by surface forces.

7 Mechanisms of self-protection against the scent substances

The main ideas on this subject can be grouped under three headings; 1, structural; 2, biochemical; 3, behavioural.

7.1 STRUCTURAL ADAPTATIONS

The pore canal–wax canal system provides the most important potential route for the penetration of ejected scent substances through cuticle. Epicuticular canals in dermal glands and chemosensilla supply other potential routes. Gluud (1968) showed that the superficial cement and wax layers over epicuticle are destroyed (*Pyrrhocoris apterus*) by an artificial scent consisting of hex-2-enal and tridecane and thus of negligible value as a permeability barrier (a role for the cement layer as permeability barrier was postulated by Remold, 1962, from meagre evidence). The structural layers of epicuticle are resistant to the scent substances but only after tanning; cuticle from newly ecdysed individuals (*Carpocoris*, *Pyrrhocoris*) is unresistant (Remold, 1962; Gluud, 1968).

Variations in the arrangement of wax canals in epicuticle have been observed (Gluud, 1968). Two contrasting arrangements ("open" and "closed") are shown in Fig. 20. Gluud estimates that the "closed" arrange-

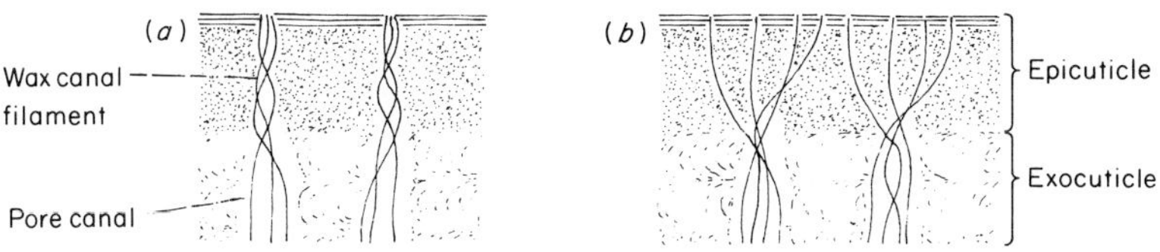

Fig. 20. Diagram of sections through epicuticle showing two arrangements of wax canals; (*a*) "closed" arrangement and (*b*) "open" arrangement. (From Gluud, 1968.)

ment reduces the area of cuticle occupied by pore canals (*ca* 1% of the total area) to about $\frac{1}{6}$ of that in cuticle showing an "open" arrangement. Although the data at present available are too meagre to safely generalize it is possible that wax canals specifically in scent producing Heteroptera are of the "closed" type.

The cuticular microsculpture of the metathoracic pleural scent areas may be important in resisting the spread of scent oil from the metathoracic ostioles to the metathoracic spiracles (Remold, 1962). The evidence appearing to support this suggestion has been criticized by Carayon (1971).

7.2 BIOCHEMICAL MECHANISMS

Two possibilities under this heading are; 1, removal by a reaction mechanism in the wax canals (*Pyrrhocoris*); 2, removal by solution in a solvent trap below the epicuticle (Pentatomids) (Gluud, 1968). The hypothetical reaction mechanism involves interactions between the penetrating scent substances and other substances in the pore canals; the reaction product forms a plug serving to resist the further penetration of scent substances down the wax canals. A similarly effective removal of penetrating scent substances could be achieved by deposition of suitable solvent materials below the epicuticle. The widespread occurrence of dicarbonyls particularly in the larval scents may be connected with the ease with which these compounds polymerize to resins on exposure to air.

7.3 BEHAVIOURAL RESPONSES

Grooming reactions with the limbs have been observed in larvae released after scent emission, in different species of Heteroptera (Calam and Youdeowei, 1969; Games and Staddon, 1973a; Aldrich and Yonke, 1975). Grooming reactions serve to remove residual scent materials from the body surface of the emitter. In *Oncopeltus fasciatus*, grooming reactions may be elicited by stimulation merely with the vapour of the scent substances (Games and Staddon, 1973a).

8 Comparisons between scent glands and other epidermal glands

The epidermal glands called dermal glands and the multicellular secretory units of the scent glands are similar in structure. The pygidial glands of female *Pyrrhocoris apterus* consist of groupings of multicellular secretory units and support cells and as such show a close resemblance in structure to the secretory bundles of the abdominal scent glands in Corixidae (Henrici, 1938; Betten, 1943). Epidermal glands consisting of unicellular secretory units have not been found in Heteroptera but have been recorded from other kinds of insects (Noirot and Quennedy, 1974; Percy and Weatherston, 1974; Grassé, 1975). There is a superficial resemblance between the unicellular secretory units of the scent glands and the "chloride cells" from aquatic Heteroptera (see Komnick and Wichard, 1975a, 1975b) and between the unicellular secretory units and the "socket glands" from the cotton stainer *Dysdercus fasciatus* (Lawrence and Staddon, 1975). However, the socket glands at least appear to be modified hair organs if not entirely new structures. There is a strikingly close resemblance between the multicellular secretory units of Brindley's glands and the type "B" dermal glands (Lai-Fook, 1970; Kälin and Barrett, 1975; Schofield and Upton, 1978). There is also a close resemblance between the three-cell secretory units of the abdominal scent glands and the multicellular secretory units of the subgenital glands (the "uradenie" of Thouvenin, 1965).

Delachambre (1973) has drawn attention to some basic similarities in structure between dermal glands and trichoid sensilla; like the gland cells, the sheath cells (trichogen cell and tormogen cell) of trichoid sensilla are secretory (Slifer, 1970). Exclusively secretory hairs have been recorded from diverse Heteroptera (for one example, from mirid bugs; Aryeety and Kumar, 1973). Such secretory hairs are presumably derived during evolution by modification (including loss of sense cells) from trichoid sensilla.

The dermal glands, like the multicellular secretory units of the abdominal scent glands, are to be classed as three-cell secretory units (see Wigglesworth, 1933; Lai-Fook, 1970; Lawrence and Staddon, 1975) but there is evidence that they pass through a four-cell stage during development. In *Rhodnius* type "B" dermal glands, the fourth cell persists for a while as a sheath surrounding the secretory cell (Wigglesworth, 1933); in *Dysdercus* the fourth cell degenerates prior to ecdysis (Lawrence and Staddon, 1975). Like the three-cell secretory units of the abdominal scent glands, the dermal glands formed at one moult are caused to be destroyed at the next; they are caused to be replaced by new dermal glands formed from unmodified epithelial cells.

Dermal glands may differ within a species. Two different types have been recorded from *Rhodnius* (Wigglesworth, 1933); two different types have been recorded from *Dysdercus fasciatus* (Lawrence and Staddon, 1975). There is a striking sexual dimorphism in the abundance of dermal glands in the abdominal sternites in *D. fasciatus*; dermal glands of one type (the "floral" glands) are extremely abundant specifically over the male sternites. This dimorphism parallels that seen in diverse Heteroptera in the metathoracic scent gland.

A close physical association between dermal glands and oenocytes has been described in the abdominal tergites, from *Graphosoma italicum* (Henrici, 1938). Three different cell layers can be distinguished in the tergites of *Graphosoma*; 1, a basal layer formed from the secretory cells of the dermal glands; 2, a middle layer formed from oenocytes; 3, an apical layer formed from shrunken epithelial cells together with the duct cells of the dermal glands.

The secretory products from the dermal glands are not as well known as those from the scent glands. Mucosubstances have been recorded from the type "B" dermal glands in *Rhodnius prolixus* (Baldwin and Salthouse, 1959; Lai-Fook, 1972). Reducing substances of an unknown nature have been recorded from the small type "A" glands of *Rhodnius* (Wigglesworth, 1948).

Dermal glands, like the secretory units from the scent glands, presumably enter the secretion phase around the time of ecdysis. From cytological evidence it has been concluded that the type "B" dermal glands of *Rhodnius* are maximally active at ecdysis, either shortly after ecdysis (Wigglesworth, 1933, 1947; Lai-Fook, 1972) or shortly before (Baldwin and Salthouse, 1959). However, dermal glands from other Heteroptera may continue to function for a long while after ecdysis (Henrici, 1938). In *Dysdercus fasciatus*, whereas the dermal glands of one type ("simple") are maximally active at ecdysis to the adult those of the other main type ("floral") become maximally active when the adults become sexually mature, a few days after ecdysis (Lawrence and Staddon, 1975).

Suggestions concerning possible functions of the dermal glands in Heteroptera are numerous; 1, lubrication of the cuticle (Henrici, 1938); 2, lubrication of the exuviae prior to ecdysis (Baldwin and Salthouse, 1959; 3, deposition of a protective "cement layer" over the waterproofing wax layer (Wigglesworth, 1947); 4, chemical defence against attack by microorganisms (Henrici, 1938); 5, production of sex pheromones (Lawrence and Staddon, 1975). Evidence for sex pheromones in Heteroptera is accumulating rapidly (Edwards, 1966; Osmani and Naidu, 1966; Antich, 1968; Scales, 1968; Strong *et al.*, 1970; Zdarek, 1970; Baldwin *et al.*, 1971; Mitchell and Mau, 1971; Brunt, 1971; King, 1973; Ubik *et al.*, 1975; Zdarek and Kontev, 1975; Brennan *et al.*, 1977). It is possible that the pygidial glands of female *Pyrrhocoris apterus* have a use specifically in connexion with oviposition (see Pendergrast, 1953).

Aldrich *et al.* (1976) investigated the chemical constitution of the secretion from the subgenital gland in male adults of the coreid *Leptoglossus phyllopus*. The presence in this secretion of vanillin (4-hydroxy-3-methoxybenzaldehyde) is interesting in view of evidence that this compound functions as a male produced sex attractant in *Eurygaster integriceps* (Ubik *et al.*, 1975). The list of other compounds included methyl *p*-hydroxybenzoate and this also is interesting in view of the fact that methyl *p*-hydroxybenzoate has been recorded as a major constituent in secretions from the metathoracic scent gland in certain aquatic Heteroptera (section 2.2.2.).

The work of Zdarek and Kontev (1975) on *Eurygaster integriceps* points towards a possible general explanation (see Aldrich *et al.*, 1976) of the fact that the frequent sex dimorphisms to be seen in the scent or other epidermal glands in Heteroptera frequently favour the male adults. In *E. integriceps*, the male adults become sexually active earlier than and they colonize new host plants before the females. Evidently the males have the ability to attract females from a distance by means of pheromone from some as yet unidentified anatomical source. There is considerable interest in the possibility that male specific epidermal glands supply pheromones for use in sex behaviour systems.

Descriptions of other epidermal glands from Heteroptera can be found in papers by Barth (1961), Carayon (1948b, 1950, 1954), Cobben (1961, 1978), Lansbury (1965), Leston (1953), and Pendergrast (1953).

Acknowledgements

The work was in part assisted by a research grant (ref. no. B/RG/0800.1) from the Science Research Council and in part by a Scientific Investigations Grant from the Royal Society. I am indebted to Dr. M. J. Thorne for supplying the photograph on which Fig. 15 is based. Mr. I. J. Everton kindly supplied the photographs for Fig. 19.

References

Akbar, S. S. (1957). The morphology and life history of *Leptocorisa varicornis* Fabr. (Coreidae: Hemiptera) – a pest of paddy crop in India. Part 1: Head and thorax. *Alig. Musl. Univ. Publs.* (*Zool. Ser.*) **5,** 1–53.

Aldrich, J. R. and Blum, M. S. (1978). Aposematic aggregation of a bug (Hemiptera: Coreidae): the defensive display and formation of aggregations. *Biotropica* **10,** 58–61.

Aldrich, J. R., Yonke, T. R. and Oetting, R. D. (1972). Histology and morphology of the abdominal scent apparatus in three alydids. *J. Kans. ent. Soc.* **45,** 162–171.

Aldrich, J. R. and Yonke, T. R. (1975). Natural products of abdominal and metathoracic scent glands of coreoid bugs. *Ann. ent. Soc. Am.* **68,** 955–960.

Aldrich, J. R., Blum, M. S., Duffey, S. S. and Fales, H. M. (1976). Male specific natural products in the bug *Leptoglossus phyllopus*: chemistry and possible function. *J. Insect Physiol.* **22,** 1201–1206.

Aldrich, J. R., Blum, M. S., Lloyd, H. and Fales, H. M. (1978a). Pentatomid natural products. Chemistry and morphology of the III and IV dorsal abdominal glands of the adults. *J. chem. Ecol.* **4,** 161–172.

Aldrich, J. R., Blum, M. S., Hefetz, A., Fales, H. M., Lloyd, H. A. and Roller, P. (1978b). Proteins in a non venomous defensive secretion: biosynthetic significance. *Science, N.Y.* **201,** 452–454.

Antich, A. V. (1968). Attraccion por olor en ninfas y adultos de *Rhodnius prolixus. Rev. Inst. Med. trop. São Paulo,* **10,** 242–246.

Aryeetey, E. A. and Kumar, R. (1973). Structure and function of the dorsal abdominal gland and defence mechanism in cocoa-capsids (Miridae: Heteroptera). *J. Ent.* **A47,** 181–189.

Baggini, A., Bernardi, R., Casnati, G., Pavan, M. and Ricca, A. (1966). Richerche sulle secrezioni difensive di insetti Emitteri Eterotteri (Hem. Heteroptera). *Eos, Madrid,* **42,** 7–26.

Baker, J. T., Blake, J. D., MacCleod, J. K., Ironside, D. A. and Johnson, I. C. (1972). The volatile constituents of the scent gland reservoir of the fruit-spotting bug, *Amblypelta nitida. Aust. J. Chem.* **25,** 393–400.

Baker, J. T. and Jones P. A. (1969). Volatile constituents of the scent gland reservoir of the nymph of the coreoid, *Pternistria bispina* Stal. *Aust. J. Chem.* **22,** 1793–1796.

Baker, J. T. and Kemball, P. A. (1967). Volatile constituents of the scent gland reservoir of the coreoid, *Pternistria bispina* Stal. *Aust. J. Chem.* **20,** 395–398.

Baldwin, W. F. and Salthouse, T. N. (1959). Dermal glands and mucin in the moulting cycle of *Rhodnius prolixus* Stål. *J. Insect Physiol.* **3,** 345–348.

Baldwin, W. F., Knight, A. G. and Lynn, K. R. (1971). A sex pheromone in the insect *Rhodnius prolixus* (Hemiptera: Reduviidae). *Can. Ent.* **103,** 18–22.

Barth, R. (1960). Sôbre o órgão abdominal glandular de *Arilus carinatus* (Forster, 1771) (Heteroptera, Reduviidae). *Mem. Inst. Osw. Cruz.* **59,** 37–43.

Betten, H. (1943). Die Stinkdrüsen der Corixiden. *Zool. Jb., anat.* **68,** 137–176.

Bevan, C. W. L., Birch, A. J. and Caswell, H. (1961). An insect repellent from black cocktail ants. *J. Chem. Soc.*, 488.

Blum, M. S. (1961). The presence of 2-hexenal in the scent gland of the pentatomid *Brochymena quadripustulata. Ann. ent. Soc. Am.* **54,** 410–412.

Blum, M. S. and Traynham, J. G. (1960). The chemistry of the pentatomid scent gland. *XI Int. Kongr. Ent. Vien., 1960, Verhandlungen,* **3,** 48–52.

Blum, M.. S., Crain, R. D. and Chidester, J. B. (1961). *Trans*-2-hexenal in the scent gland of the hemipteran *Acanthocephala femorata. Nature, Lond.* **189,** 245–246.

Blum, M. S., Traynham, J. G., Chidester, J. B. and Boggus, J. D. (1960). *n*-Tridecane and trans-2-heptenal in scent gland of the rice stink bug *Oebalus pugnax* (F). *Science, N.Y.* **132,** 1480–1481.

Bonnemaison, L. (1952). Morphologie et biologie de la punaise ornée du chou (*Eurydema ventralis* Kol.) *Annals Epiphyt.* **3,** 127–272.

Bradshaw, J. W. S., Baker, R. and Howse, P. E. (1975). Multicomponent alarm pheromones of the weaver ant. *Nature, Lond.* **258,** 230–231.

Brennan, B. M., Chang, F. and Mitchell, W. C. (1977). Physiological effects on sex pheromone, communication in the southern green stink bug, *Nezara viridula. Environ. Ent.* **6,** 169–173.

Brindley, M. D. H. (1929). On the repugnatorial glands of *Corixa. Trans. ent. Soc. Lond.* **77,** 7–15.

Brindley, M. D. H. (1930). On the metasternal scent-glands of certain Heteroptera. *Trans. ent. Soc. Lond.* **78,** 199–207.

Brunet, P. C. J. (1967). Sclerotins. *Endeavour,* **26,** 68–74.

Brunt, A. M. (1971). The reproductive behaviour of *Dysdercus fasciatus* Signoret (Hem., Pyrrhocoridae) in culture. *Entomologist's mon. Mag.* **107,** 18–23.

Bu'Lock, J. D. (1965). "The Biosynthesis of Natural Products". McGraw Hill.

Butenandt, A. (1955). Wirkstoffe des Insektenreiches. *Nova Acta Leopoldina*, **17**, 445–471.

Butenandt, A. and Tâm, N. (1957). Über einen geschlechtsspezifischen Duftstoff der Wasserwanze *Belostoma indica* Vitalis (*Lethocerus indicus* Lep.). *Hoppe-Seyler's Z. physiol. Chem.* **308,** 277–283.

Caillot, Y. and Boisson, C. (1954). Développement larvaire du Bélostome (*Lethocerus indicus* Lep.) Insecte Hémiptère, Hydrocoryse, Cryptocérate. *Annls Sci. nat.* **16,** 51–64.

Calam, D. H. and Scott, G. C. (1969). The scent gland complex of the adult cotton stainer bug, *Dysdercus intermedius. J. Insect Physiol.* **15,** 1695–1702.

Calam, D. H. and Youdeowei, A. (1968). Identification and functions of secretion from the posterior scent gland of fifth instar larva of the bug *Dysdercus intermedius. J. Insect Physiol.* **14,** 1147–1158.

Carayon, J. (1948a). Dimorphisme sexuel des glandes odorantes métathoraciques chez quelques Hémiptères. *C.r. hebd. Séanc. Acad. Sci. Paris*, **227,** 303–305.

Carayon, J. (1948b). Les organes parastigmatiques des Hémiptères Nabidae. *C.r. hebd. Séanc. Acad. Sci., Paris*, **227,** 864–866.

Carayon, J. (1950). Caractères anatomiques et position systématique des Hémiptères Nabidae (note préliminaire). *Bull. Mus. Hist. nat., Paris*, **22,** 95–101.

Carayon, J. (1954). Un type nouveau d'appareil glandulaire propre aux mâles de certains Hémiptères Anthocoridae. *Bull. Mus. Hist. nat., Paris*, **26,** 602–606.

Carayon, J. (1962). Observations sur l'appareil odorifique des Hétéroptères particulièrement celui des Tingidae, Vianaididae et Piesmatidae. *Cah. Nat.* **18,** 1–16.

Carayon, J. (1966). Metathoracic scent apparatus, *In* "Monograph of Cimicidae" (Ed. R. L. Usinger) **VII** *Ent. Soc. Am.* edit pp. 69–80. Thomas Say Foundation.

Carayon, J. (1971). Notes et documents sur l'appareil odorant metathoracique des Hémiptères. *Annls. Soc. ent. Fr.* (*N.S.*) **7,** 737–770.

Carayon, J., Usinger, R. L. and Wygodzinsky, P. (1958). Notes on the higher classification of the Reduviidae, with the description of a new tribe of the Phymatinae. *Rev. Zool. Bot. afr.* **57,** 256–281.

Cmelik, S. (1969). Volatile aldehydes in the odoriferous secretion of the stink bug *Libyaspis angolensis. Hoppe-Seyler's Z. Physiol. Chem.* **350,** 1076–1080.

Cobben, R. H. (1961). A new genus and four new species of Saldidae (Heteroptera). *Ent. Ber.* **21,** 96–107.

Cobben, R. H. (1968). Evolutionary trends in Heteroptera. Part 1. Eggs. Architecture of the shell, gross embryology and eclosion. Centre for agricultural publishing and documentation, Wageningen, The Netherlands.

Cobben, R. H. (1970). Morphology and taxonomy of intertidal dwarfbugs (Heteroptera: Omaniidae fam. nov.). *Tijdschr. Ent.* **113,** 61–90.

Cobben, R. H. (1978). Evolutionary trends in Heteroptera. Part II. Mouthpart-structures and feeding strategies. Meded. handb. No. 289, pp. 407. Hoogesch. Wageningen.

Collins, R. P. (1968). Carbonyl compounds produced by the bed bug, *Cimex lectularius. Ann. ent. Soc. Am.* **61,** 1338–1339.

Collins, R. P. and Drake, T. H. (1965). Carbonyl compounds produced by the meadow plant bug, *Leptopterna dolabrata* (Hemiptera: Miridae). *Ann. ent. Soc. Am.* **58,** 764–765.

Conradi, A. F. (1904). Variations in the protective value of odoriferous secretions of some Heteroptera. *Science*, **19,** 393–394.

Dazzini, M. V. and Finzi, P. V. (1974). Chemically known constituents of arthropod defensive secretions. *Atti Accad. naz. Lincei. Memorie.* **12,** 109–146.

Delachambre, J. (1973). L'ultrastructure des glandes dermiques de *Tenebrio molitor* L. (Insecta, Coleoptera). *Tissue & Cell*, **5,** 243–257.

Dethier, M. (1974). Les organes odoriférants métathoraciques des Cydnidae. *Bull. Soc. vaud. Sc. nat.* **72,** 127–140.

Devakul, V. and Maarse, H. (1964). A second compound in the odorous gland liquid of the giant water bug *Lethocerus indicus* (Lep. and Serv.). *Analyt. Biochem.* **7,** 269–274.

Diehl, P. A. (1973). Paraffin synthesis in the oenocytes of the desert locust. *Nature, Lond.* **243,** 468–470.

Duffey, S. S. and Scudder, G. C. E. (1972). Cardiac glycosides in north American Asclepiadaceae, a basis for unpalatability in brightly coloured Hemiptera and Coleoptera. *J. Insect Physiol.* **18,** 63–78.

Dupuis, C. (1947a). Données sur la morphologie des glandes dorso-abdominales des Hémiptères-Hétéroptères. *Feuille Nat.* **2,** 13–21.

Dupuis, C. (1947b). Nouvelles données sur les glandes dorso-abdominales des Hémiptères Hétéroptères; relations entre les urites et les glandes dorso-abdominales. *Feuille Nat.* **2,** 49–52.

Dupuis, C. (1949). Données nouvelles sur la morphologie abdominale des Hémiptères Hétéroptères et en particulier des Pentatomoidea. *XIIIe Congr. intern. Zool. Paris*, 1947, 471–472.

Dupuis, C. (1952). Notes, remarques et observations diverses sur les Hémiptères; première série: notes I–IV. III – Dimorphisme sexuel de la glande dorso-abdominale anterieure de certains Asopinae (Pentatomidae). *Feuille Nat. N.S.* **7,** 1–4.

Dupuis, C. (1959). Notes, remarques et observations diverse sur les Hemiptères; quatrième série: notes IX–XII. XII – Dimorphisme de la glande dorsale antérieure des Asopinae. *Cahiers Nat., Bull. N.P. N.S.* **15,** 45–52.

Edwards, J. S. (1966). Observations on the life history and predatory behaviour of *Zelus exsanquis* (Stål) (Heteroptera: Reduviidae). *Proc. R. ent. Soc. Lond.* **A41,** 21–24.

Eggerman, W. and Bongers, J. (1972). Die Wirtswal von *Oncopeltus fasciatus* Dall. (Heteroptera: Lygaeidae): Bindung an Asclepiadaceen durch wirtsspezifische Glykoside. *Oecologia, Berl.* **9,** 363–370.

Eisner, T. (1970). Chemical defense against predation in arthropods. *In* "Chemical Ecology" (Eds. E. Sondheimer and J. B. Simeone), pp. 157–217, Academic Press, New York and London.

Eisner, T. and Meinwald, J. (1966). Defensive secretions of arthropods. *Science*, **153,** 1341–1350.

Eisner, T., Eisner, H. E., Hurst, J. J., Kafatos, F. C. and Meinwald, J. (1963). Cyanogenic glandular apparatus of a millipede. *Science, N.Y.* **139,** 1218–1220.

Euw, J. Von, Reichstein, T. and Rothschild, M. (1971). Heart poisons (cardiac glycosides) in the lygaeid bugs *Caenocoris nerii* and *Spilostethus pandurus*. *Insect Biochem.* **1,** 373–384.

Everton, I. J. and Staddon, B. W. (1979). The accessory gland and metathoracic scent gland function in *Oncopeltus fasciatus*. *J. Insect Physiol.* (in press).

Everton, I. J., Games, D. E. and Staddon, B. W. (1974). Composition of scents from *Apodiphus amygdali*. *Ann. ent. Soc. Am.* **67,** 815–816.

Everton, I. J., Knight, D. W. and Staddon, B. W. (1979). Linalool from the metathoracic scent gland of the cotton stainer *Dysdercus intermedius* Distant (Heteroptera: Pyrrhocoridae). *Comp. Biochem. Physiol.* **63B,** 157–161.

Fain, A. (1970). *Coreitarsonemus*, un nouveau genre d'Acariens parasitant la glande odoriférante des Hémiptères Coreidae (Tarsonemidae: Trombidiformes). *Revue Zool. Bot. afr.* **82,** 315–334.

Filshie, B. K. and Waterhouse, D. F. (1968). The fine structure of the lateral scent glands of the green vegetable bug, *Nezara viridula* (Hemiptera, Pentatomidae). *J. Microscopie* **7,** 231–244.

Filshie, B. K. and Waterhouse, D. F. (1969). The structure and development of a surface pattern on the cuticle of the green vegetable bug *Nezara viridula*. *Tissue & Cell,* **1,** 367–385.

Games, D. E. and Staddon, B. W. (1973a). Chemical expression of a sexual dimorphism in the tubular scent glands of the milkweed bug *Oncopeltus fasciatus* (Dallas) (Heteroptera; Lygaeidae). *Experientia,* **29,** 532–533.

Games, D. E. and Staddon, B. W. (1973b). Composition of scents from the larva of the milkweed bug *Oncopeltus fasciatus*. *J. Insect Physiol.* **19,** 1527–1532.

Games, D. E., Schofield, C. J. and Staddon, B. W. (1974). The secretion from Brindley's scent glands in Triatominae. *Ann. ent. Soc. Am.* **67,** 820.

Games, D. E., Jackson, A. H., Millington, D. S. and Staddon, B. W. (1973). Mass spectral studies of insect secretions. *In* "Advances in Mass Spectrometry" (Ed. A. R. West) **6,** pp. 207–213. Applied Science, London.

Gilby, A. R. and Waterhouse, D. F. (1964). The identification of *trans*-4-ketohex-2-enal by its proton magnetic resonance spectrum. *Aust. J. Chem.* **17,** 1311–1313.

Gilby, A. R. and Waterhouse, D. F. (1965). The composition of the scent of the green vegetable bug, *Nezara viridula*. *Proc. R. Soc. Lond.* **B162,** 105–120.

Gilby, A. R. and Waterhouse, D. F. (1967). Secretions from the lateral scent glands of the green vegetable bug, *Nezara viridula*. *Nature, Lond.* **216,** 90–91.

Gilchrist, T. L., Stansfield, F. and Cloudsley-Thompson, J. L. (1966). The odoriferous principle of *Piezodorus teretipes* (Stål) (Hemiptera: Pentatomoidea). *Proc. R. ent. Soc. Lond.* **A41,** 55–56.

Gluud, A. (1968). Zur Feinstruktur der Insektencuticula – Ein Beitrag zur Frage des Eigengiftschutzes der Wanzencuticula. *Zool. Jb. Anat.* **85,** 191–227.

Gordon, H. T., Waterhouse, D. F. and Gilby, A. R. (1963). Incorporation of ^{14}C-acetate into scent constituents by the green vegetable bug. *Nature, Lond.* **197,** 818.

Graham, J. D. P. and Staddon, B. W. (1974). Pharmacological observations on body fluids from the milkweed bug *Oncopeltus fasciatus* (Dallas) (Heteroptera: Lygaeidae). *J. Ent.* **A48,** 177–183.

Grassé, P.-P. (1975). Les glandes tégumentaires des insectes. *In* "Traite de Zoologie" (Ed. Grassé, P.-P.) **8 (3),** pp. 199–320 and pp. 859–866. Masson et Cie, Paris.

Gueldner, R. C. and Parrott, W. L. (1978). Volatile constituents of the tarnished plant bug. *Insect Biochem.* **8,** 389–391.

Gulde, J. (1902). Die Dorsaldrüsen der larven der Hemiptera-Heteroptera. *Ber. senckenb. naturf. Ges.* 85–136.

Gupta, A. P. (1964). Musculature and mechanism of the nymphal scent apparatus of *Riptortus linearis* H.S. (Heteroptera: Alydidae) with comments on the number, variation, and homology of the abdominal scent glands in other Heteroptera. *Proc. ent. Soc. Wash.* **66,** 12–19.

Hageman, J. (1910). Beitrage zur Kenntnis von *Corixa*. *Zool. Jb.* **30,** 373–426.

Happ, G. M. (1968). Quinone and hydrocarbon production in the defensive glands of *Eleodes longicollis* and *Tribolium castaneum* (Coleoptera, Tenebrionidae). *J. Insect Physiol.* **14,** 1821–1837.

Heikertinger, F. (1922). Sind die Wanzen (Hemiptera Heteroptera) durch Ekelgeruch geschützt. *Biol. Zbl.* **42,** 441–464.

Henrici, H. (1938). Die Hautdrüsen der Landwanzen (Geocorisae) ihre mikroskopische Anatomie, ihre Histologie und Entwicklung. Teil 1. Die abdominalen Stinkdrüsen, die Drüsenpakete und die zerstreuten Hautdrüsen. *Zool. Jb., Anat.* **65,** 141–228.

Henrici, H. (1940). Die Hautdrüsen der Landwanzen (Geocorisae), ihre mikroskopische Anatomie, ihre Histologie und Entwicklung. Teil 2. Die thorakalen stinkdrüsen. *Zool. Jb., Anat.* **66,** 371–402.

Hepburn, H. R. and Yonke, T. R. (1971). The metathoracic scent glands of coreoid Heteroptera. *J. Kans. ent. Soc.* **44,** 187–210.

Hepburn, H. R., Berman, N. J., Jacobson, H. J. and Fatti, L. P. (1973). Trends in arthropod defensive secretions, an aquatic predator assay. *Oecologia Berl.* **12,** 373–382.

Ishiwatari, T. (1974). Studies on the scent of stink bugs (Hemiptera: Pentatomidae) I. Alarm pheromone activity. *Appl. Ent. Zool.* **9,** 153–158.

Ishiwatari, T. (1976). Studies on the scent of stink bugs (Hemiptera: Pentatomidae) II. Aggregation pheromone activity. *Appl. Ent. Zool.* **11,** 38–44.

Jacobson, E. (1911). Biological notes on the hemipteron *Ptilocerus ochraceus*. *Tijdschr. Ent.* **54,** 175–179.

Johansson, A. S. (1957). The functional anatomy of the metathoracic scent glands of the milkweed bug, *Oncopeltus fasciatus* (Dallas) (Heteroptera: Lygaeidae). *Norsk ent. Tidsskr.* **10,** 95–109.

Johansson, A. S. and Bråten, T. (1970). Cuticular morphology of the scent gland areas of some heteropterans. *Ent. Scand.* **1,** 158–162.

Kälin, M. and Barrett, F. M. (1975). Observations on the anatomy, histology, release site, and function of Brindley's glands in the blood-sucking bug, *Rhodnius prolixus* (Heteroptera: Reduviidae). *Ann. ent. Soc. Am.* **68,** 126–134.

Kathuria, O. P., Brown, W. V. and Gilby, A. R. (1974). The defensive secretion of *Aspongopus janus* (Fabricius) (Hemiptera: Pentatomidae). *Indian J. Ent.* **36,** 31–33.

Kemper, H. (1929). Beitrag zur Kenntnis des Stinkapparates von *Cimex lectularius* L. *Z. Morph. Ökol. Tiere,* **15,** 524–546.

Kershaw, J. C. and Muir, F. (1907). Life-history of *Tessaratoma papillosa*, Thunberg. With notes on the stridulating organ and stink-glands. *Trans. ent. Soc. Lond.* 253–258.

King, A. B. S. (1973). Studies of sex attraction in the cocoa capsid *Distantiella theobroma* (Heteroptera: Miridae). *Entomologia exp. appl.* **16,** 243–254.

Komnick, H. and Wichard, W. (1975a). Chloride cells of larval *Notonecta glauca* and *Naucoris cimicoides* (Hemiptera, Hydrocorisae). *Cell. Tiss. Res.* **156,** 539–549.

Komnick, H. and Wichard, W. (1975b). Histochemischer nachweis von Chloridzellen bei Wasserwanzen (Hemiptera: Hydrocorisae) und ihre Feinstruktur bei *Hesperocorixa sahlbergi* Fieb. (Hemiptera: Corixidae). *Int. J. Insect Morphol. & Embryol.* **4,** 89–105.

Lai-Fook, J. (1970). The fine structure of developing type 'B' dermal glands in *Rhodnius prolixus*. *Tissue & Cell,* **2,** 119–138.

Lai-Fook, J. (1972). A comparison between the dermal glands in two insects *Rhodnius prolixus* (Hemiptera) and *Calpodes ethlius* (Lepidoptera). *J. Morph.* **136,** 495–504.

Lansbury, I. (1965). New organ in Stenocephalidae (Hemiptera-Heteroptera). *Nature, Lond.* **205,** 106.

Lawrence, P. A. (1970). Some new mutants of the large milkweed bug *Oncopeltus fasciatus* Dall. *Genet. Res., Camb.* **15,** 347–350.

Lawrence, P. A. and Green, S. M. (1975). The anatomy of a compartment border. *J. Cell Biol.* **65,** 373–382.
Lawrence, P. A. and Staddon, B. W. (1975). Peculiarities of the epidermal gland system of the cotton stainer *Dysdercus fasciatus* Signoret (Heteroptera: Pyrrhocoridae). *J. ent.* **A49,** 121–136.
Lener, W. (1969). Pheromone secretion in the large milkweed bug, *Oncopeltus fasciatus. Am. Soc. Zool.* **9,** 1143.
Leston, D. (1953). Notes on the Ethiopian Pentatomoidea (Hem.): XIV, on a new, infra-rectal, organ in *Boerias* Kirkaldy (Pentatomidae). *The Entomologist*, **86,** 152–155.
Levinson, H. Z. and Bar Ilan, A. R. (1971). Assembling and alerting scents produced by the bedbug *Cimex lectularius* L. *Experientia*, **27,** 102–103.
Levinson, H. Z., Levinson, A. R., Müller, B. and Steinbrecht, R. A. (1974). Structure of sensilla, olfactory perception, and behaviour of the bedbug, *Cimex lectularius*, in response to its alarm pheromone. *J. Insect Physiol.* **20,** 1231–1248.
Macleod, J. K., Bott, G. and Cable, J. (1977). Synthesis of (Z)- and (E)-4-oxooct-2-enal. *Aust. J. Chem.* **30,** 2561–2564.
Macleod, J. K., Howe, I., Cable, J., Blake, J. D., Baker, J. T. and Smith, D. (1975). Volatile scent gland components of some tropical Hemiptera. *J. Insect Physiol.* **21,** 1219–1224.
Maschwitz, U. (1971). Wasserstoffperoxid als Antiseptikum bei einer Wasserwanze. *Naturwissenschaften*, **58,** 572.
McCullough, T. (1966). Carbonyl and acidic compounds produced by *Acanthocephala granulosa* (Hemiptera: Coreidae). *Ann. ent. Soc. Am.* **59,** 410.
McCullough, T. (1967). Quantitative determination of trans-2-hexenal in the defensive scent fluid of *Acanthocephala declivis* and *A. granulosa* (Hemiptera: Coreidae). *Ann. ent. Soc. Am.* **60,** 862.
McCullough, T. (1968). Acid and aldehyde compounds in the scent fluid of *Leptoglossus oppositus. Ann. ent. Soc. Am.* **61,** 1044.
McCullough, T. (1969). Chemical analysis of the scent fluid of *Leptoglossus clypeatus. Ann. ent. Soc. Am.* **62,** 673–674.
McCullough, T. (1971). Chemical analysis of the defensive scent fluid released by *Hypselonotus punctiventris* (Hemiptera: Coreidae). *Ann. ent. Soc. Am.* **64,** 749.
McCullough, T. (1973). Chemical analysis of the defensive scent fluid from the bug *Mozena obtusa* (Hemiptera: Coreidae). *Ann. ent. Soc. Am.* **66,** 231–232.
McCullough, T. (1974a). Chemical analysis of the defensive scent fluid produced by *Mozena lunata* (Hemiptera: Coreidae). *Ann. ent. Soc. Am.* **67,** 298.
McCullough, T. (1974b). Chemical analysis of the defensive scent fluid from the cactus bug *Chelinidea vittiger. Ann. ent. Soc. Am.* **67,** 300.
Merlini, L. and Mondelli, R. (1962). Sui pigmenti di *Pyrrhocoris apterus* L. *Gazz. chim. ital.* **92,** 1251–1261.
Merlini, L. and Nasini, G. (1966). Insect pigments – IV. Pteridines and colour in some Hemiptera. *J. Insect Physiol.* **12,** 123–127.
Miller, N. C. E. (1971). The biology of the Heteroptera. E. W. Classey Ltd.
Mitchell, W. C. and Mau, R. F. L. (1971). Response of the female southern green stink bug and its parasite, *Trichopoda pennipes*, to male stink bug pheromones. *J. econ. Ent.* **64,** 856–859.
Monod, J. (1972). "Chance and Necessity", Collins.
Moody, D. L. (1930). The morphology of the repugnatory glands of *Anasa tristis* De Geer. *Ann. ent. Soc. Am.* **23,** 81–104.

Mukerji, S. K. and Sharma, H. L. (1966). Investigations on the offensive odour of Hemiptera-bugs. *Tetrahedron Lett.* **22,** 2479–2481.

Neville, A. C. (1975). "Biology of the Arthropod cuticle". Springer-Verlag.

Noirot, C. and Quennedy, A. (1974). Fine structure of insect epidermal glands. *A. Rev. Ent.* **19,** 61–80.

Osmani, Z. and Naidu, M. B. (1966). Evidence of sex attractant in female *Dysdercus cingulatus* Fabr. *Ind. J. exp. Biol.* **5,** 51.

Park, R. J. and Sutherland, M. D. (1962). Volatile constituents of the bronze orange bug, *Rhoecocoris sulciventris*. *Aust. J. Chem.* **15,** 172–174.

Parsons, M. C. (1960). Skeleton and musculature of the thorax of *Gelastocoris oculatus* (Fabricius) (Hemiptera-Heteroptera). *Bull. Mus. comp. Zool. Harv.* **122,** 299–357.

Pattenden, G. and Staddon, B. W. (1968). Secretion of the metathoracic glands of the waterbug *Notonecta glauca* L. (Heteroptera: Notonectidae). *Experientia*, **24,** 1092.

Pattenden, G. and Staddon, B. W. (1970). Observations on the metasternal scent glands of *Lethocerus* spp. (Heteroptera: Belostomatidae). *Ann. ent. Soc. Am.* **63,** 900–901.

Pattenden, G. and Staddon, B. W. (1972). Identification of *iso*-butyric acid in secretion from Brindley's scent glands in *Rhodnius prolixus* (Heteroptera: Reduviidae). *Ann. ent. Soc. Am.* **65,** 1240–1241.

Pendergrast, J. G. (1953). Setose areas of the abdomen in females of some Acanthosominae (Heteroptera, Pentatomidae). *The Entomologist*, **86,** 135–138.

Percy, J. E. and Weatherston, J. (1974). Gland structure and pheromone production in insects. *In* "Pheromones" (Ed. M. C. Birch), pp. 11–34. North-Holland, Amsterdam and London.

Pinder, A. R. and Staddon, B. W. (1965a). Trans-4-oxohex-2-enal in the odoriferous secretion of *Sigara falleni* (Fieb.) (Hemiptera-Heteroptera). *Nature, Lond.* **205,** 106–107.

Pinder, A. R. and Staddon, B. W. (1965b). The odoriferous secretion of the water bug, *Sigara falleni* (Fieb.). *J. chem. Soc.* **530,** 2955–2958.

Poisson, R. (1924). Contributions à l'étude des Hémiptères Aquatiques. *Bull. biol. Fr. Belg.* **58,** 49–305.

Poisson, R. (1951). Hétéroptères. *In* "Traité de Zoologie" (Ed. P.-P. Grassé) **10,** 1657–1803. Masson et Cie, Paris.

Prestwich, G. D. (1976). Composition of the scent of eight east african hemipterans. Nymph-adult chemical polymorphism in coreids. *Ann. ent. Soc. Am.* **69,** 812–814.

Remold, H. (1962). Über die biologische Bedeutung der Duftdrüsen bei den Landwanzen (Geocorisae). *Z. vergl. Physiol.* **45,** 636–694.

Rettenmeyer, C. W. (1970). Insect mimicry. *A. Rev. Ent.* **15,** 43–74.

Roth, L. M. (1961). A study of the odoriferous glands of *Scaptocoris divergens* (Hemiptera: Cydnidae). *Ann. ent. Soc. Am.* **54,** 900–911.

Scales, A. L. (1968). Female tarnished plant bugs attract males. *J. econ. Ent.* **61,** 1466–1467.

Schaeffer, C. W. (1972). Degree of metathoracic scent-gland development in the trichophorous Heteroptera (Hemiptera). *Ann. ent. Soc. Am.* **65,** 810–821.

Schildknecht, H. (1970). The defensive chemistry of land and water beetles. *Angew. Chem.* **9,** 1–9.

Schildknecht, H. and Holoubek, K. (1961). Die Bombardierkäfer und ihre Explosionschemie. V. Mitteilung über Insekten-Abwehrstoffe. *Angew. Chem.* **73,** 1–7.

Schildknecht, H., Weis, K. H. and Vetter, H. (1962). XI. Mitteilung über Insektenabwehrstoffe. α, β-Ungesättigte Aldehyde als Inhaltsstoffe der Stinkblasen der Blattwanze *Dolycoris baccarum* L. *Z. Naturf.* **17b,** 350–351.

Schildknecht, H., Maschwitz, E. and Maschwitz, U. (1968). Die Explosionschemie der Bombardierkäfer (Coleoptera, Carabidae) III. Mitt.: Isolierung und Charakterisierung der Explosionskatalysatoren. *Z. Naturf.* **23b,** 1213–1218.

Schofield, C. J. and Upton, C. P. (1978). Brindley's scent-glands and the metasternal scent glands of *Panstrongylus megistus* (Hemiptera, Reduviidae, Triatominae). *Rev. Brasil. Biol.* **38,** 665–678.

Schumacher, R. (1970). Endocytosevorgänge während der erhöhten Sekretionsaktivät in den Drüsenzellen des imaginalen Duftdrüsenkomplexes der Baumwollwanze *Dysdercus intermedius* Dist. (Pyrrhocoridae). *Z. Naturf.* **25b,** 435.

Schumacher, R. (1971a). Zur funktionellen Morphologie der imaginalen Duftdrüsen zweier Landwanzen II. Mitteilung: Das Reservoir und das "Nierenförmige Organ" des imaginalen Duftdrüsenkomplexes der Baumwollwanze *Dysdercus intermedius.* Dist. *Z. wiss. Zool.* **182,** 411–426.

Schumacher, R. (1971b). Zur funktionellen Morphologie der imaginalen Duftdrüsen zweier Landwanzen III. Mitteilung: Die Drüsenzelle des imaginalen Duftdrüsenkomplexes der Feuerwanze *Pyrrhocoris apterus* L. (Geocorisae, Fam.: Pyrrhocoridae). *Z. wiss. Zool.* **183,** 71–82.

Schumacher, R. (1971c). Zur funktionellen Morphologie der imaginalen Duftdrüsen zweier Landwanzen IV. Mitteilung: Das ableitende Kanalsystem und das Reservoir des imaginalen Duftdrüsenkomplexes der Feuerwanze *Pyrrhocoris apterus* L. (Geocorisae, Fam.: Pyrrhocoridae). *Z. wiss. Zool.* **183,** 83–96.

Schumacher, R. and Stein, G. (1971). Zur funktionellen Morphologie der imaginalen Duftdrüsen zweier Landwanzen I. Mitteilung: Drüsenzellen und ableitendes Kanalsystem des imaginalen Duftdrüsenkomplexes der Baumwollwanze *Dysdercus intermedius* Dist. *Z. wiss. Zool.* **182,** 395–410.

Shorey, H. H. (1973). Behavioural responses to insect pheromones. *A. Rev. Ent.* **18,** 349–380.

Slifer, E. H. (1970). The structure of arthropod chemoreceptors. *A. Rev. Ent.* **15,** 131–142.

Smith, R. M. (1974). The defensive secretion of *Vitellus insularis* (Heteroptera: Pentatomidae). *N.Z. J. Zoo.* **1,** 375–376.

Southwood, T. R. E. and Leston, D. (1959). "Land and water bugs of the British Isles". F. Warne & Co. Ltd.

Staddon, B. W. (1971). Metasternal scent glands in Belostomatidae (Heteroptera). *J. Ent.* **A46,** 69–71.

Staddon, B. W. (1972). On the suggestion that the secretion from the metathoracic scent glands of a surface-dwelling aquatic insect, *Gerris najas* (De Geer) (Heteroptera; Gerridae) has a waterproofing function. *J. exp. Biol.* **57,** 765–769.

Staddon, B. W. (1973). A note on the composition of the scent from the metathoracic scent glands of *Gelastocoris oculatus* (Fabricius) (Heteroptera: Gelastocoridae). *The Entomologist,* **106,** 253–255.

Staddon, B. W., Everton, I. J. and Games, D. E. (1979). Organ specificity and scent constitution in Corixidae (Heteroptera: Hydrocorisae). *Comp. Biochem. Physiol. 62 B.* 259–262.

Staddon, B. W. and Thorne, M. J. (1973). The structure of the metathoracic scent gland system of the water bug *Ilyocoris cimicoides* (L.) (Heteroptera: Naucoridae). *Trans. R. ent. Soc. Lond.* **124,** 343–363.

Staddon, B. W. and Thorne, M. J. (1974). Observations on the metathoracic scent gland system of the backswimmer, *Notonecta glauca* L. (Heteroptera: Notonectidae). *J. Ent.* **A48,** 223–227.

Staddon, B. W. and Thorne, M. J. (1979) The metathoracic scent gland system of Hydrocorisae (Heteroptera: Nepomorpha). *Syst. Ent.* (in press).

Staddon, B. W. and Weatherston, J. (1967). Constituents of the stink gland of the water bug *Ilyocoris cimicoides* (L.) (Heteroptera: Naucoridae). *Tetrahedron Lett.* **46,** 4567–4571.

Stein, G. (1966a). Über den Feinbau der Duftdrüsen von Feuerwanzen (*Pyrrhocoris apterus* L., Geocorisae) I. Mitteilung: Zur funktionellen Morphologie der Drüsenzelle. *Z. Zellforsch. mikrosk. Anat.* **74,** 271–290.

Stein, G. (1966b). Über den Feinbau der Duftdrüsen von Feuerwanzen (*Pyrrhocoris apterus* L., Geocorisae) II. Mitteilung: Das ableitende Kanalsystem und die nichtdrüsigen Anteile. *Z. Zellforsch. mikrosk. Anat.* **75,** 501–516.

Stein, G. (1967). Über den Feinbau der Duftdrüsen von Feuerwanzen (*Pyrrhocoris apterus* L., Geocorisae). Die 2. Larvale Abdominaldrüse. *Z. Zellforsch. mikrosk. Anat.* **79,** 49–63.

Stein, G. (1969). Uber den Feinbau der Duftdrüsen von Heteropteren. Die hintere larvale Abdominaldrüse der Baumwollwanze *Dysdercus intermedius* Dist. (Insecta, Heteroptera). *Z. Morph. Tiere,* **65,** 374–391.

Stein, G. (1971). Giftige Duftstoffe bei Wanzen. *Umschau,* **71,** 52–53.

Stein, G. and Schumacher, R. (1969). Über den Feinbau der Duftdrüsen von Baumwollwanzen (*Dysdercus intermedius* Dist., Pyrrhocoridae). *Z. Naturf.* **24b,** 148–149.

Stein, G. and Walker, S. (1970). Über den Feinbau der Duftdrüsen des Rückenschwimmers *Notonecta glauca* L. (Notonectidae). *Z. Naturf.* **25b,** 562.

Stride, G. O. (1956). On the mimetic association between certain species of *Phonoctonus* (Hemiptera, Reduviidae) and the Pyrrhocoridae. *J. ent. Soc. S. Africa,* **19,** 12–28.

Strong, F. E., Sheldahl, J. A., Hughes, P. R. and Ussein, E. M. K. (1970). Reproductive biology of *Lygus hesperus* Knight. *Hilgardia,* **40,** 105–147.

Thouvenin, M. (1965). Étude préliminaire des "uradénies" chez certains Hétéroptères pentatomorphes. *Ann. Soc. ent. Fr.* (*N.S.*) **1,** 973–988.

Tsuyuki, T., Ogata, Y., Yamamoto, I. and Shimi, K. (1965). Stink bug aldehydes. *Agr. Biol. Chem.* **29,** 419–427.

Ubik, K., Vrkoč, J., Žďárek, J. and Kontev, C. (1975). Vanillin: Possible sex pheromone of an insect. *Naturwiss.* **62,** 348.

Vuillaume, M. (1975). Pigments des insectes. *In* "Traité de Zoologie" (Ed. P.-P. Grassé) **8 (3),** pp. 77–184. Masson et Cie, Paris.

Walker, S. (1972). Die Feinstruktur der imaginalen Duftdrúsen von *Corixa punctata* (Heteroptera, Hydrocorisae). *Z. wiss. Zool. Leipzig,* **183,** 190-205.

Waterhouse, D. F., Forss, D. A. and Hackman, R. H. (1961). Characteristic odour components of the scent of stink bugs. *J. Insect Physiol.* **6,** 113–121.

Waterhouse, D. F. and Gilby, A. R. (1964). The adult scent glands and scent of nine bugs of the superfamily Coreoidea. *J. Insect Physiol.* **10,** 977–987.

Wigglesworth, V. B. (1933). The physiology of the cuticle and of ecdysis in *Rhodnius* prolixus (Triatomidae, Hemiptera); with special reference to the function of the oenocytes and of the dermal glands. *Q. Jl microsc. Sci.* **76,** 269–318.

Wigglesworth, V. B. (1947). The epicuticle in an insect, *Rhodnius prolixus* (Hemiptera). *Proc. R. Soc. B.* **134,** 163–180.

Wigglesworth, V. B. (1973). Haemocytes and basement membrane formation in *Rhodnius. J. Insect Physiol.* **19,** 831–844.

Woodward, T. E., Evans, J. W. and Eastop, V. F. (1970). Hemiptera. *In* "The insects of Australia". Ch. 26, pp. 387–457. C.S.I.R.O.

Youdeowei, A. and Calam, D. H. (1969). The morphology of the scent glands of *Dysdercus intermedius* Distan (Hemiptera: Pyrrhocoridae) and a preliminary analysis of

the scent gland secretions of the fifth instar larvae. *Proc. R. ent. Soc. Lond.* **A44,** 38–44.
Žďárek, J. (1970). Mating behaviour in the bug, *Pyrrhocoris apterus* L. (Heteroptera): ontogeny and its environmental control. *Behaviour,* **37,** 253–268.
Žďárek, J. and Kontev, C. (1975). Some ethological aspects of reproduction in *Eurygaster integriceps* (Heteroptera, Scutelleridae). *Acta ent. Bohemoslov.* **72,** 239–248.
Ziegler, I. and Harmsen, R. (1969). The biology of pteridines in insects. *Adv. Insect Physiol.* **6,** 139–203.

NOTES ADDED IN PROOF (Franz Engelmann)

Recent progress in the research on insect vitellogenin and vitellin has been reviewed in two publications (Wyatt and Pan, 1978; Hagedorn and Kunkel, 1979). For *Locusta* it is now shown that the primary translation product of vitellogenin consists of two macromolecules of 265 000 and 250 000 daltons (Chen *et al.*, 1978) which are processed to smaller polypeptides before reaching the haemolymph. In contrast to this, no such primary products are found in *Drosophila melanogaster*, where a total of three polypeptides of 44 700, 45 700, 47 000 daltons could be identified in haemolymph and eggs by SDS-PAGE (Bownes and Hames, 1977; Hames and Bownes, 1978). These molecules are distinct polypeptides as shown by their PI, peptide maps, and immunoprecipitations (Warren and Mahowald, 1979). Also in *Leucophaea* were three subunits found in vitellogenin from the microsomes or haemolymph, as well as in vitellin from the eggs (Engelmann, 1978). Processing of one of the subunits occurs during uptake into the oocytes.

Hagedorn *et al.* (1978) discussed in detail and provided new evidence for the validity of their filter technique for identification of vitellogenin in *Aedes aegypti*.

Bownes, M. and Hames, B. D. (1977). Accumulation and degradation of three major yolk proteins in *Drosophila melanogaster. J. exp. Zool.* **200,** 149–156.
Chen, T. T., Strahlendorf, P. W. and Wyatt, G. R. (1978). Vitellin and vitellogenin from Locusts (*Locusta migratoria*). *J. biol. Chem.* **253,** 5325–5331.
Engelmann, F. (1978). Juvenile hormone regulated vitellogenin synthesis in an insect. *In* "Comparative Endocrinology" (Eds. P. J. Gaillard and H. H. Boer), pp. 17–20. Elsevier, North Holland, Amsterdam.
Hagedorn, H. H. and Kunkel, J. G. (1979). Vitellogenin and vitellin in insects. *Ann. Rev. Entomol.* **24,** 475–505.
Hagedorn, H. H., Kunkel, J. G. and Wheelock, G. (1978). The specificity of an antiserum against mosquito vitellogenin and its use in a radioimmunological precipitation assay for protein synthesis. *J. Insect Physiol.* **24,** 481–489.
Hames, B. D. and Bownes, M. (1978). Synthesis of yolk proteins in *Drosophila melanogaster. Insect Biochem.* **8,** 319–328.
Warren, T. G. and Mahowald, A. P. (1979). Isolation and chemical characterization of the three major yolk polypeptides from *Drosophila melanogaster. Devl. Biol.* **68,** 130–139.
Wyatt, G. R. and Pan, M. L. (1978). Insect plasma proteins. *Ann. Rev. Biochem.* **47,** 779–817.

Subject Index

Cumulative List of Authors

Numbers in bold face indicate the volume numbers of the series

Cumulative List of Chapter Titles

Numbers in bold face indicate the volume number of the series